Cell Biology and Genetics

Cell Biology and Genetics
Starr Taggart Evers Starr
Biology The Unity and Diversity of Life

Twelfth Edition

BROOKS/COLE
CENGAGE Learning

Australia • Brazil • Japan • Korea • Mexico • Singapore • Spain • United Kingdom • United States

BROOKS/COLE
CENGAGE Learning™

Cell Biology and Genetics
Biology: The Unity and Diversity of Life, Twelfth Edition
Cecie Starr, Ralph Taggart, Christine Evers, Lisa Starr

Publisher: Yolanda Cossio

Managing Development Editor: Peggy Williams

Assistant Editor: Elizabeth Momb

Editorial Assistant: Samantha Arvin

Technology Project Manager: Kristina Razmara

Marketing Manager: Amanda Jellerichs

Marketing Assistant: Katherine Malatesta

Marketing Communications Manager: Linda Yip

Project Manager, Editorial Production: Andy Marinkovich

Creative Director: Rob Hugel

Art Director: John Walker

Print Buyer: Karen Hunt

Permissions Editor: Bob Kauser

Production Service: Grace Davidson & Associates

Text and Cover Design: John Walker

Photo Researcher: Myrna Engler Photo Research Inc.

Copy Editor: Anita Wagner

Illustrators: Gary Head, ScEYEnce Studios, Lisa Starr

Compositor: Lachina Publishing Services

Cover Image: Biologist/photographer Tim Laman took these photos of mutualisms in Indonesia. *Top:* A wrinkled hornbill (*Aceros corrugatus*) eats fruits of a strangler fig (*Ficus stupenda*). The plant provides food for the bird, and the bird disperses its seeds. *Below:* Two species of sea anemone, each with its own species of anemone fish. Anemones provide a safe haven for anemonefish, who chase away other fish that would graze on the anemone's tentacles. www.timlaman.com

For product information and technology assistance, contact us at
Cengage Learning Customer & Sales Support, 1-800-354-9706.

For permission to use material from this text or product, submit all requests online at **cengage.com/permissions**. Further permissions questions can be emailed to **permissionrequest@cengage.com**.

Library of Congress Control Number: 2008930414

ISBN-13: 978-0-495-55798-2

ISBN-10: 0-495-55798-6

Brooks/Cole
10 Davis Drive
Belmont, CA 94002
USA

Cengage Learning is a leading provider of customized learning solutions with office locations around the globe, including Singapore, the United Kingdom, Australia, Mexico, Brazil, and Japan. Locate your local office at: **international.cengage.com/region**.

Cengage Learning products are represented in Canada by Nelson Education, Ltd.

For your course and learning solutions, visit **academic.cengage.com**.

Purchase any of our products at your local college store or at our preferred online store **www.ichapters.com**.

Printed in the United States of America
1 2 3 4 5 6 7 12 11 10 09 08

CONTENTS IN BRIEF

Highlighted chapters are not included in Cell Biology and Genetics

DETAILED CONTENTS

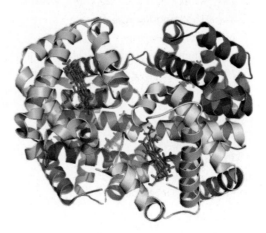

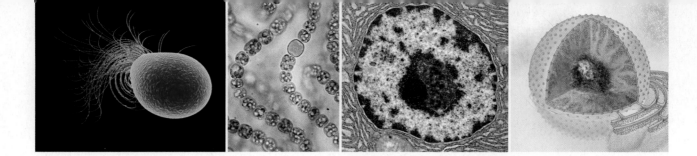

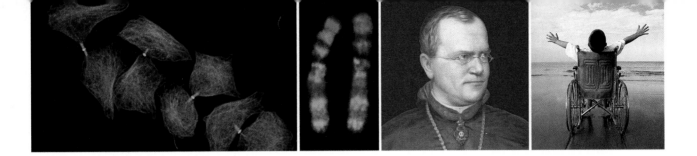

Preface

In preparation for this revision, we invited instructors who teach introductory biology for non-majors students to meet with with us and discuss the goals of their courses. The main goal of almost every instructor was something like this: "To provide students with the tools to make informed choices as consumers and as voters by familiarizing them with the way science works." Most students who use this book will not become biologists, and many will never take another science course. Yet for the rest of their lives they will have to make decisions that require a basic understanding of biology and the process of science.

Our book provides these future decision makers with an accessible introduction to biology. Current research, along with photos and videos of the scientists who do it, underscore the concept that science is an ongoing endeavor carried out by a diverse community of people. The research topics include not only what the researchers discovered, but also how the discoveries were made, how our understanding has changed over time, and what remains undiscovered. The role of evolution is a unifying theme, as it is in all aspects of biology.

As authors, we feel that understanding stems mainly from making connections, so we are constantly trying to achieve the perfect balance between accessibility and level of detail. A narrative with too much detail is inaccessible to the introductory student; one with too little detail comes across as a series of facts that beg to be memorized. Thus, we revised every page to make the text in this edition as clear and straightforward as possible, keeping in mind that English is a second language for many students. We also simplified many figures and added tables that summarize key points.

CHANGES IN THIS EDITION

Impacts, Issues To make the *Impacts, Issues* essays more appealing, we shortened and updated them, and improved their integration throughout the chapters. Many new essays were added to this edition.

Key Concepts Introductory summaries of the *Key Concepts* covered in the chapter are now enlivened with eye-catching graphics taken from relevant sections. The links to earlier concepts now include descriptions of the linked concepts in addition to the section numbers.

Take Home Message Each section now concludes with a *Take Home Message* box. Here we pose a question that reflects the critical content of the section, and we also provide answers to the question in bulleted list format.

Figure It Out *Figure It Out Questions* with answers allow students to check their understanding of a figure as they read through the chapter.

Data Analysis Exercise To further strengthen a student's analytical skills and provide insight into contemporary research, each chapter includes a *Data Analysis Exercise*. The exercise includes a short text passage—

usually about a published scientific experiment—and a table, chart, or other graphic that presents experimental data. The student must use information in the text and graphic to answer a series of questions.

Chapter-Specific Changes Every chapter was extensively revised for clarity; this edition has more than 250 new photos and over 300 new or updated figures. A page-by-page guide to content and figures is available upon request, but we summarize the highlights here.

• *Chapter 1, Invitation to Biology* New essay about the discovery of new species. Greatly expanded coverage of critical thinking and the process of science; new section on sampling error.

• *Chapter 2, Life's Chemical Basis* Sections on subatomic particles, bonding, and pH simplified; new pH art.

• *Chapter 3, Molecules of Life* New essay about *trans* fats. Structural representations simplified and standardized.

• *Chapter 4, Cell Structure and Function* New essay about foodborne *E. coli*; microscopy section updated; new section on cell theory and history of microscopy; two new focus essays on biofilms and lysosome malfunction.

• *Chapter 5, A Closer Look at Cell Membranes* Membrane art reorganized; new figure illustrating cotransport.

• *Chapter 6, Ground Rules of Metabolism* Energy and metabolism sections reorganized and rewritten; much new art, including molecular model of active site.

• *Chapter 7, Where It Starts—Photosynthesis* New essay about biofuels. Sections on light-dependent reactions and carbon fixing adaptations simplified; new focus essay on atmospheric CO_2 and global warming.

• *Chapter 8, How Cells Release Chemical Energy* All art showing metabolic pathways revised and simplified.

• *Chapter 9, How Cells Reproduce* Updated micrographs of mitosis in plant and animal cells.

• *Chapter 10, Meiosis and Sexual Reproduction* Crossing over, segregation, and life cycle art revised.

• *Chapter 11, Observing Patterns in Inherited Traits* New essay about inheritance of skin color; mono- and dihybrid cross figures revised; new Punnett square for coat color in dogs; environmental effects on *Daphnia* phenotype added.

• *Chapter 12, Chromosomes and Human Inheritance* Chapter reorganized; expanded discussion and new figure on the evolution of chromosome structure.

• *Chapter 13, DNA Structure and Function* New opener essay on pet cloning; adult cloning section updated.

• *Chapter 14, From DNA to Protein* New art comparing DNA and RNA, other art simplified throughout; new micrographs of transcription Christmas tree, polysomes.

• *Chapter 15, Controls Over Genes* Chapter reorganized; eukaryotic gene control section rewritten; updated X chromosome inactivation photos; new lac operon art.

• *Chapter 16, Studying and Manipulating Genomes* Text extensively rewritten and updated; new photos of *bt* corn, DNA fingerprinting; sequencing art revised.

• *Chapter 17, Evidence of Evolution* Extensively revised, reorganized. Revised essay on evidence/inference; new focus essay on whale evolution; updated geologic time scale correlated with grand canyon strata.

• *Chapter 18, Processes of Evolution*   Extensively revised, reorganized. New photos showing sexual selection in stalk-eyed flies, mechanical isolation in sage.

• *Chapter 19, Organizing Information About Species* Extensively revised, reorganized. New comparative embryology photo series; updated tree of life.

• *Chapter 20, Life's Origin and Early Evolution*   Information about origin of agents of metabolism updated. New discussion of ribozymes as evidence for RNA world.

• *Chapter 21, Viruses and Prokaryotes*   Opening essay about HIV moved here, along with discussion of HIV replication. New art of viral structure. New section describes the discovery of viroids and prions.

• *Chapter 22, Protists—The Simplest Eukaryotes*   New opening essay about malaria. New figures show protist traits, how protists relate to other groups.

• *Chapter 23, The Land Plants*   Evolutionary trends revised. More coverage of liverworts and hornworts.

• *Chapter 24, Fungi*   New opening essay about airborne spores. More information on fungal uses and pathogens.

• *Chapter 25, Animal Evolution—The Invertebrates* New summary table for animal traits. Coverage of relationships among invertebrates updated.

• *Chapter 26, Animal Evolution—The Chordates*   New section on lampreys. Human evolution updated.

• Material previously covered in the *Biodiversity in Prespective* chapter now integrated into other chapters.

• *Chapter 27, Plants and Animals—Common Challenges* New section about heat-related illness.

• *Chapter 28, Plant Tissues*   Secondary structure section simplified; new essay on dendroclimatology.

• *Chapter 29, Plant Nutrition and Transport*   Root function section rewritten and expanded; new translocation art.

• *Chapter 30, Plant Reproduction*   Extensively revised. New essay on colony collapse disorder; new table showing flower specializations for specific pollinators; new section on flower sex; many new photos added.

• *Chapter 31, Plant Development*   Sections on plant development and hormone mechanisms rewritten.

• *Chapter 32, Animal Tissues and Organ Systems*   Essay on stem cells updated. New section on lab-grown skin.

• *Chapter 33, Neural Control*   Reflexes integrated with coverage of spinal cord. Section on brain heavily revised.

• *Chapter 34, Sensory Perception*   New art of vestibular apparatus, image formation in eyes, and accommodation. Improved coverage of eye disorders and disease.

• *Chapter 35, Endocrine Control*   New section about pituitary disorders. Tables summarizing hormone sources now in appropriate sections, rather than at end.

• *Chapter 36, Structural Support and Movement*   Improved coverage of joints and joint disorders.

• *Chapter 37, Circulation*   Updated opening essay. New section about hemostasis. Blood cell diagram simplified. Blood typing section revised for clarity.

• *Chapter 38, Immunity*   New essay on HPV vaccine; new focus essays on periodontal-cardiovascular disease and allergies; vaccines and AIDS sections updated.

• *Chapter 39, Respiration*   Better coverage of invertebrate respiration and of Heimlich maneuver.

• *Chapter 40, Digestion and Human Nutrition*   Nutritional information and obesity research sections updated.

• *Chapter 41, Maintaining the Internal Environment*   New figure of fluid distribution in the human body. Improved coverage of kidney disorders and dialysis.

• *Chapter 42, Animal Reproductive Systems*   New essay on intersex conditions. Coverage of reproductive anatomy, gamete production, intercourse, and fertilization.

• *Chapter 43, Animal Development*   Information about principles of animal development streamlined.

• *Chapter 44, Animal Behavior*   More on types of learning.

• *Chapter 45, Population Ecology*   Exponential and logistic growth clarified. Human population material updated.

• *Chapter 46, Community Structure and Biodiversity* New table of species interactions. Competition section heavily revised.

• *Chapter 47, Ecosystems*   New figures for food chain and food webs. Updated greenhouse gas coverage.

• *Chapter 48, The Biosphere*   Improved coverage of lake turnover, ocean life, coral reefs, and threats to them.

• *Chapter 49, Human Impacts on the Biosphere*   Covers extinction crisis, conservation biology, ecosystem degradation, and sustainable use of biological wealth.

Appendix V, Molecular Models   New art and text explain why we use different types of molecular models.

Appendix VI, Closer Look at Some Major Metabolic Pathways   New art shows details of electron transport chains in thylakoid membranes.

ACKNOWLEDGMENTS

No list can convey our thanks to the team of dedicated people who made this book happen. The professionals who are listed on the following page helped shape our thinking. Marty Zahn and Wenda Ribeiro deserve special recognition for their incisive comments on every chapter, as does Michael Plotkin for voluminous and excellent feedback. Grace Davidson calmly and tirelessly organized our efforts, filled in our gaps, and put all of the pieces of this book together. Paul Forkner's tenacious photo research helped us achieve our creative vision. At Cengage Learning, Yolanda Cossio and Peggy Williams unwaveringly supported us and our ideals. Andy Marinkovich made sure we had what we needed, Amanda Jellerichs arranged for us to meet with hundreds of professors, Kristina Razmara continues to refine our amazing technology package, Samantha Arvin helped us stay organized, and Elizabeth Momb managed all of the print ancillaries.

CECIE STARR, CHRISTINE EVERS, AND LISA STARR
June 2008

MARC C. ALBRECHT
University of Nebraska at Kearney

ELLEN BAKER
Santa Monica College

SARAH FOLLIS BARLOW
Middle Tennessee State University

MICHAEL C. BELL
Richland College

LOIS BREWER BOREK
Georgia State University

ROBERT S. BOYD
Auburn University

URIEL ANGEL BUITRAGO-SUAREZ
Harper College

MATTHEW REX BURNHAM
Jones County Junior College

P.V. CHERIAN
Saginaw Valley State University

WARREN COFFEEN
Linn Benton

LUIGIA COLLO
Universita' Degli Studi Di Brescia

DAVID T. COREY
Midlands Technical College

DAVID F. COX
Lincoln Land Community College

KATHRYN STEPHENSON CRAVEN
Armstrong Atlantic State University

SONDRA DUBOWSKY
Allen County Community College

PETER EKECHUKWU
Horry-Georgetown Technical College

DANIEL J. FAIRBANKS
Brigham Young University

MITCHELL A. FREYMILLER
University of Wisconsin - Eau Claire

RAUL GALVAN
South Texas College

NABARUN GHOSH
West Texas A&M University

JULIAN GRANIRER
URS Corporation

STEPHANIE G. HARVEY
Georgia Southwestern State University

JAMES A. HEWLETT
Finger lakes community College

JAMES HOLDEN
Tidewater Community College - Portsmouth

HELEN JAMES
Smithsonian Institution

DAVID LEONARD
Hawaii Department of Land and Natural Resources

STEVE MACKIE
Pima West Campus

CINDY MALONE
California State University - Northridge

KATHLEEN A. MARRS
Indiana University - Purdue University Indianapolis

EMILIO MERLO-PICH
GlaxoSmithKline

MICHAEL PLOTKIN
Mt. San Jacinto College

MICHAEL D. QUILLEN
Maysville Community and Technical College

WENDA RIBEIRO
Thomas Nelson Community College

MARGARET G. RICHEY
Centre College

JENNIFER CURRAN ROBERTS
Lewis University

FRANK A. ROMANO, III
Jacksonville State University

CAMERON RUSSELL
Tidewater Community College - Portsmouth

ROBIN V. SEARLES-ADENEGAN
Morgan State University

BRUCE SHMAEFSKY
Kingwood College

BRUCE STALLSMITH
University of Alabama - Huntsville

LINDA SMITH STATON
Pollissippi State Technical Community College

PETER SVENSSON
West Valley College

LISA WEASEL
Portland State University

DIANA C. WHEAT
Linn-Benton Community College

CLAUDIA M. WILLIAMS
Campbell University

MARTIN ZAHN
Thomas Nelson Community College

Introduction

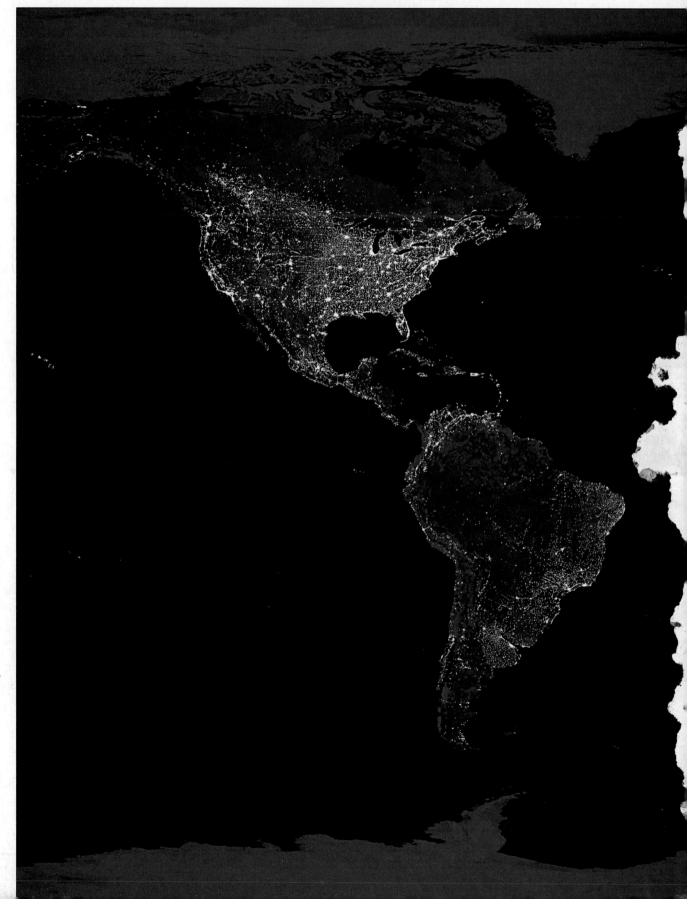

Current configurations of the Earth's oceans and land masses—the geologic stage upon which life's drama continues to unfold. This composite satellite image reveals global energy use at night by the human population. Just as biological science does, it invites you to think more deeply about the world of life—and about our impact upon it.

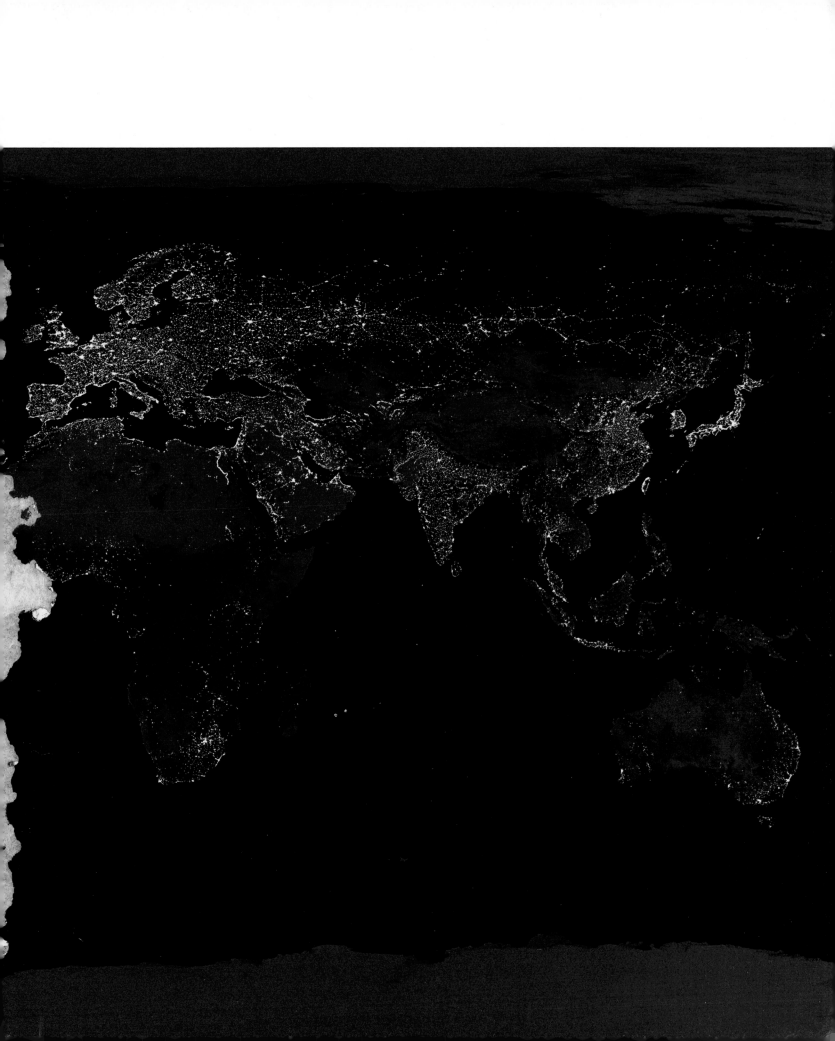

1 Invitation to Biology

Lost Worlds and Other Wonders

In this era of satellites, submarines, and global positioning systems, could there possibly be any more places on Earth that we have not explored? Well, yes. In 2005, for instance, helicopters dropped a team of biologists into a swamp in the middle of a vast and otherwise inaccessible tropical forest in New Guinea. Later, team member Bruce Beehler remarked, "Everywhere we looked, we saw amazing things we had never seen before. I was shouting. This trip was a once-in-a-lifetime series of shouting experiences."

The team discovered dozens of animals and plants that had been unknown to science, including a rhododendron with plate-sized flowers. They found animals that are on the brink of extinction in other parts of the world, and a bird that was supposedly extinct.

The expedition fired the imagination of people all over the world. It is not that finding new kinds of organisms is such a rare event. Almost every week, biologists discover many kinds of insects and other small organisms. However, the animals in this particular rain forest—mammals and birds especially—seem too big to have gone unnoticed. Had people just missed them? Perhaps not. No trails or other human disturbances cut through that part of the forest. The animals had never learned to be afraid of humans, so the team members could simply walk over and pick them up (Figure 1.1).

Many other animals have been discovered in the past few years, including lemurs in Madagascar, monkeys in India and Tanzania, cave-dwelling animals in two of California's national parks, carnivorous sponges near Antarctica, and whales and giant jellylike animals in the seas. Most came to light during survey trips similar to the New Guinea expedition—when biologists simply were attempting to find out what lives where.

Exploring and making sense of nature is nothing new. We humans and our immediate ancestors have been at it for thousands of years. We observe, come up with explanations about what the observations mean, and then test the explanations. Ironically, the more we learn about nature, the more we realize how much we have yet to learn.

You might choose to let others tell you what to think about the world around you. Or you might choose to develop your own understanding of it. Perhaps, like the New Guinea explorers, you are interested in animals and where they live. Maybe you are interested in aspects that affect your health, the food you eat, or your home and family. Whatever your focus may be, the scientific study of life—biology—can deepen your perspective on the world.

Throughout this book, you will find examples of how organisms are constructed, where they live, and what they do. These examples support concepts that, when taken together, convey what "life" is. This chapter gives you an overview of basic concepts. It sets the stage for upcoming descriptions of scientific observations and applications that can help you refine your understanding of life.

See the video! Figure 1.1 Biologist Kris Helgen and a rare golden-mantled tree kangaroo in a tropical rain forest in the Foja Mountains of New Guinea. There, in 2005, explorers discovered forty previously unknown species.

Key Concepts

Links to Earlier Concepts

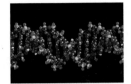

Levels of organization

We study the world of life at different levels of organization, which extend from atoms and molecules to the biosphere. The quality of "life" emerges at the level of cells. **Section 1.1**

Life's underlying unity

All organisms consist of one or more cells, which stay alive through ongoing inputs of energy and raw materials. All sense and respond to change; all inherited DNA, a type of molecule that encodes information necessary for growth, development, and reproduction. **Section 1.2**

Life's diversity

Many millions of kinds of organisms, or species, have appeared and disappeared over time. Each kind is unique in some aspects of its body form or behavior. **Section 1.3**

Explaining unity in diversity

Theories of evolution, especially a theory of evolution by natural selection, help explain why life shows both unity and diversity. Evolutionary theories guide research in all fields of biology. **Section 1.4**

How we know

Biologists make systematic observations, predictions, and tests in the laboratory and in the field. They report their results so others may repeat their work and check their reasoning. **Sections 1.5–1.8**

■ This book parallels nature's levels of organization, from atoms to the biosphere. Learning about the structure and function of atoms and molecules primes you to understand the structure of living cells. Learning about processes that keep a single cell alive can help you understand how multicelled organisms survive, because their many living cells all use the same processes. Knowing what it takes for organisms to survive can help you see why and how they interact with one another and with their environments.

At the start of each chapter, we will use this space to remind you of such connections. Within chapters, cross-references will link you to relevant sections in earlier chapters.

How would you vote? The discoverer of a new species usually is the one who gives it a scientific name. In 2005, a Canadian casino bought the right to name a monkey species. Should naming rights be sold? See CengageNOW for details, then vote online.

1.1 Life's Levels of Organization

■ We understand life by thinking about nature at different levels of organization.

■ Nature's organization begins at the level of atoms, and extends through the biosphere.

■ The quality of life emerges at the level of the cell.

Making Sense of the World

Most of us intuitively understand what nature means, but could you define it? **Nature** is everything in the universe except what humans have manufactured. It encompasses every substance, event, force, and energy —sunlight, flowers, animals, bacteria, rocks, thunder, humans, and so on. It excludes everything artificial.

Researchers, clerics, farmers, astronauts, children— anyone who is of the mind to do so attempts to make sense of nature. Interpretations differ, for no one can be expert in everything learned so far or have foreknowledge of all that remains hidden. If you are reading this book, you are starting to explore how a subset of scientists, the biologists, think about things, what they found out, and what they are up to now.

A Pattern in Life's Organization

Biologists look at all aspects of life, past and present. Their focus takes them all the way down to atoms, and all the way up to global relationships among organisms and the environment. Through their work, we glimpse a great pattern of organization in nature.

The pattern starts at the level of atoms. **Atoms** are fundamental building blocks of all substances, living and nonliving (Figure 1.2*a*).

At the next level of organization, atoms join with other atoms, forming **molecules** (Figure 1.2*b*). Among the molecules are complex carbohydrates and lipids, proteins, and nucleic acids. Today, only living cells make these "molecules of life" in nature.

The pattern crosses the threshold to life when many molecules are organized as cells (Figure 1.2*c*). A **cell** is the smallest unit of life that can survive and reproduce on its own, given information in DNA, energy inputs, raw materials, and suitable environmental conditions.

An **organism** is an individual that consists of one or more cells. In larger multicelled organisms, trillions

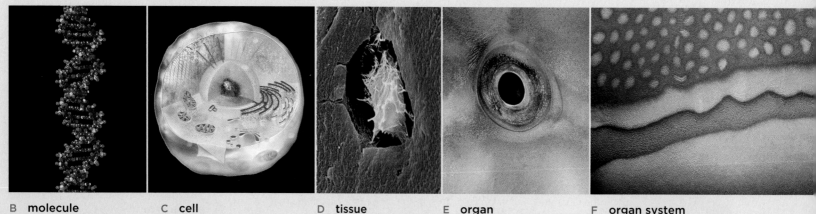

B molecule

Two or more atoms joined in chemical bonds. In nature, only living cells make the molecules of life: complex carbohydrates and lipids, proteins, and nucleic acids.

C cell

Smallest unit that can live and reproduce on its own or as part of a multicelled organism. A cell has DNA, an outermost membrane, and other components.

D tissue

Organized array of cells and substances that are interacting in some task. Bone tissue consists of secretions (*brown*) from cells such as this (*white*).

E organ

Structural unit of two or more tissues that interact in one or more tasks. This parrotfish eye is a sensory organ used in vision.

F organ system

Organs that interact in one or more tasks. The skin of this parrotfish is an organ system that consists of tissue layers, organs such as glands, and other parts.

single-celled organisms can form populations

A atom

Atoms are fundamental units of all substances. This image shows a model of a single hydrogen atom.

Figure 1.2 Animated Levels of organization in nature.

of cells organize into tissues, organs, and organ systems, all interacting in tasks that keep the whole body alive. Figure 1.2*d–g* defines these body parts.

Populations are at a greater level of organization. Each **population** is a group of individuals of the same kind of organism, or species, living in a specified area (Figure 1.2*h*). Examples are all heavybeak parrotfish living on Shark Reef in the Red Sea or all California poppies in California's Antelope Valley Poppy Reserve.

Communities are at the next level. A **community** consists of all populations of all species in a specified area. As an example, Figure 1.2*i* shows a sampling of the Shark Reef's species. This underwater community includes many kinds of seaweeds, fishes, corals, sea anemones, shrimps, and other living organisms that make their home in or on the reef. Communities may be large or small, depending on the area defined.

The next level of organization is the **ecosystem**: a community interacting with its physical and chemical environment. The most inclusive level, the **biosphere**, encompasses all regions of Earth's crust, waters, and atmosphere in which organisms live.

Bear in mind, life is more than the sum of its individual parts. In other words, some emergent property occurs at each successive level of life's organization. An **emergent property** is a characteristic of a system that does not appear in any of its component parts. For example, the molecules of life are themselves not alive. Considering them separately, no one would be able to predict that a particular quantity and arrangement of molecules will form a living cell. Life—an emergent property—appears first at the level of the cell but not at any lower level of organization in nature.

Take-Home Message

How does "life" differ from "nonlife"?

■ The building blocks—atoms—that make up all living things are the same ones that make up all nonliving things.

■ Atoms join as molecules. The unique properties of life emerge as certain kinds of molecules become organized into cells.

■ Higher levels of organization include multicelled organisms, populations, communities, ecosystems, and the biosphere.

G multicelled organism

Individual composed of different types of cells. Cells of most multicelled organisms, such as this parrotfish, form tissues, organs, and organ systems.

H population

Group of single-celled or multicelled individuals of a species in a given area. This is a population of one fish species in the Red Sea.

I community

All populations of all species in a specified area. These populations belong to a coral reef community in a gulf of the Red Sea.

J ecosystem

A community that is interacting with its physical environment through inputs and outputs of energy and materials. Reef ecosystems flourish in warm, clear seawater throughout the Middle East.

K biosphere

All regions of Earth's waters, crust, and atmosphere that hold organisms. Earth is a rare planet. Life as we know it would be impossible without Earth's abundance of free-flowing water.

- Continual inputs of energy and the cycling of materials maintain life's complex organization.
- Organisms sense and respond to change.
- DNA inherited from parents is the basis of growth and reproduction in all organisms.

Energy and Life's Organization

Eating supplies your body with energy and nutrients that keep it organized and functioning. **Energy** is the capacity to do work. A **nutrient** is a type of atom or molecule that has an essential role in growth and survival and that an organism cannot make for itself.

All organisms spend a lot of time acquiring energy and nutrients, although different kinds get such inputs from different sources. These differences allow us to classify organisms into one of two broad categories: producers or consumers.

Producers acquire energy and simple raw materials from environmental sources and make their own food. Plants are producers. By the process of **photosynthesis**, they use sunlight energy to make sugars from carbon dioxide and water. Those sugars function as packets of immediately available energy or as building blocks for larger molecules.

Consumers cannot make their own food; they get energy and nutrients indirectly—by eating producers and other organisms. Animals fall within the consumer category. So do decomposers, which feed on wastes or remains of organisms. We find leftovers of their meals in the environment. Producers take up the leftovers as sources of nutrients. Said another way, nutrients cycle between producers and consumers.

Energy, however, is not cycled. It flows through the world of life in one direction—from the environment, through producers, then through consumers. This flow maintains the organization of individual organisms, and it is the basis of life's organization within the biosphere (Figure 1.3). It is a one-way flow, because with each transfer, some energy escapes as heat. Cells do not use heat to do work. Thus, energy that enters the world of life ultimately leaves it—permanently.

Organisms Sense and Respond to Change

Organisms sense and respond to changes both inside and outside the body by way of receptors. A **receptor** is a molecule or cellular structure that responds to a specific form of stimulation, such as the energy of light or the mechanical energy of a bite (Figure 1.4).

Stimulated receptors trigger changes in activities of organisms. For example, after you eat, the sugars from

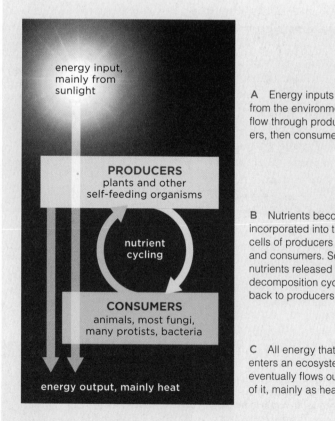

A Energy inputs from the environment flow through producers, then consumers.

B Nutrients become incorporated into the cells of producers and consumers. Some nutrients released by decomposition cycle back to producers.

C All energy that enters an ecosystem eventually flows out of it, mainly as heat.

Figure 1.3 Animated The one-way flow of energy and cycling of materials through an ecosystem.

Figure 1.4 A roaring response to signals from pain receptors, activated by a lion cub flirting with disaster.

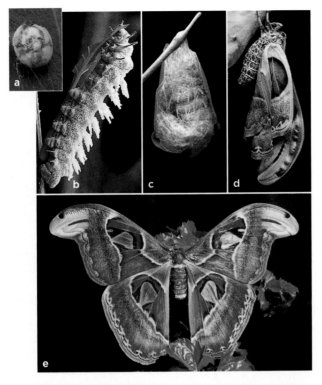

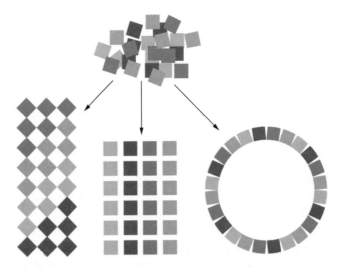

Figure 1.5 Development of the atlas moth. Instructions in DNA guide the development of this insect through a series of stages, from a fertilized egg (**a**), to a larval stage called a caterpillar (**b**), to a pupal stage (**c**), to the winged adult form (**d,e**).

Figure 1.6 Animated Three examples of objects assembled in different ways from the same materials.

your meal enter your bloodstream, and then your blood sugar level rises. The added sugars bind to receptors on cells of the pancreas (an organ). Binding sets in motion a series of events that causes cells throughout the body to take up sugar faster, so the sugar level in your blood returns to normal.

In multicelled organisms, the internal environment is all fluid inside of the body but outside of cells. Unless the composition of the internal environment is kept within certain ranges, body cells will die. By sensing and adjusting to change, organisms keep conditions in their internal environment within a range that favors cell survival. This process is called **homeostasis**, and it is a defining feature of life. All organisms, whether single-celled or multicelled, undergo homeostasis.

Organisms Grow and Reproduce

DNA, a nucleic acid, is the signature molecule of life. No chunk of rock has it. Why is DNA so important? It is the basis of growth, survival, and reproduction in all organisms. It is also the source of each individual's distinct features, or **traits**.

In nature, an organism inherits DNA—the basis of its traits—from parents. **Inheritance** is the transmission of DNA from parents to offspring. Moths look like moths and not like chickens because they inherited moth DNA, which differs from chicken DNA. **Reproduction** refers to actual mechanisms by which

parents transmit DNA to offspring. For all multicelled individuals, DNA has information that guides growth and **development**—the orderly transformation of the first cell of a new individual into an adult (Figure 1.5).

DNA contains instructions. Cells use some of those instructions to make proteins, which are long chains of amino acids. There are only 20 kinds of amino acids, but cells string them together in different sequences to make a tremendous variety of proteins. By analogy, just a few different kinds of tiles can be organized into many different patterns (Figure 1.6).

Different proteins have structural or functional roles. For instance, certain proteins are enzymes—functional molecules that make cell activities occur much faster than they would on their own. Without enzymes, such activities would not happen fast enough for a cell to survive. There would be no more cells—and no life.

Take-Home Message

How are all living things alike?

■ A one-way flow of energy and a cycling of nutrients through organisms and the environment sustain life, and life's organization.

■ Organisms maintain homeostasis by sensing and responding to change. They make adjustments that keep conditions in their internal environment within a range that favors cell survival.

■ Organisms grow, develop, and reproduce based on information encoded in their DNA, which they inherit from their parents.

1.3 | Overview of Life's Diversity

■ Of an estimated 100 billion kinds of organisms that have ever lived on Earth, as many as 100 million are with us today.

Each time we discover a new **species**, or kind of organism, we assign it a two-part name. The first part of the name specifies the **genus** (plural, genera), which is a group of species that share a unique set of features. When combined with the second part, the name designates one species. Individuals of a species share one or more heritable traits, and they can interbreed successfully if the species is a sexually reproducing one.

Genus and species names are always italicized. For example, *Scarus* is a genus of parrotfish. The heavy-beak parrotfish in Figure 1.2*g* is called *Scarus gibbus*. A different species in the same genus, the midnight parrotfish, is *S. coelestinus*. Note that the genus name may be abbreviated after it has been spelled out one time.

We use various classification systems to organize and retrieve information about species. Most systems group species together on the basis of their observable characteristics, or traits. Table 1.1 and Figure 1.7 show a common system in which more inclusive groupings above the level of genus are phylum (plural, phyla), kingdom, and domain. Here, all species are grouped into domains Bacteria, Archaea, and Eukarya. Protists, plants, fungi, and animals make up domain Eukarya.

All **bacteria** (singular, bacterium) and **archaeans** are single-celled organisms. All of them are prokaryotic, which means they do not have a nucleus. In other organisms, this membrane-enclosed sac holds and protects a cell's DNA. As a group, prokaryotes have the most diverse ways of procuring energy and nutrients. They are producers and consumers in nearly all of the biosphere, including extreme environments such as frozen desert rocks, boiling sulfur-clogged lakes, and nuclear reactor waste. The first cells on Earth may have faced similarly hostile challenges to survival.

Cells of **eukaryotes** start out life with a nucleus. Structurally, **protists** are the simplest kind of eukaryote. Different protist species are producers or consumers. Many are single cells that are larger and more complex than prokaryotes. Some of them are tree-sized, multi-

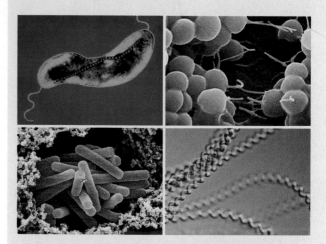

A **Bacteria** These prokaryotes tap more diverse sources of energy and nutrients than all other organisms. *Clockwise from upper left*, a magnetotactic bacterium has a row of iron crystals that acts like a tiny compass; bacteria that live on skin; spiral cyanobacteria; and *Lactobacillus* cells in yogurt.

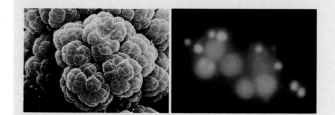

B **Archaea** Although they often appear similar to bacteria, these prokaryotes are evolutionarily closer to eukaryotes. *Left*, a colony of methane-producing cells. *Right*, two species from a hydrothermal vent on the seafloor.

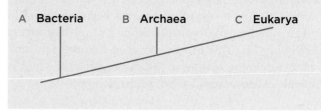

A **Bacteria**   B **Archaea**   C **Eukarya**

Figure 1.7 Animated Representatives of diversity from the three most inclusive branchings of the tree of life.

celled seaweeds. Protists are so diverse that they are now being reclassified into a number of separate major lineages based on emerging biochemical evidence.

Cells of fungi, plants, and animals are eukaryotic. Most **fungi**, such as the types that form mushrooms, are multicelled. Many are decomposers, and all secrete enzymes that digest food outside the body. Their cells then absorb the released nutrients.

Table 1.1	Comparison of Life's Three Domains
Bacteria	Single cells, prokaryotic (no nucleus). Most ancient lineage.
Archaea	Single cells, prokaryotic. Evolutionarily closer to eukaryotes.
Eukarya	Eukaryotic cells (with a nucleus). Single-celled and multicelled species categorized as protists, plants, fungi, and animals.

Protists are single-celled and multicelled eukaryotic species that range from the microscopic to giant seaweeds. Many biologists are now viewing the "protists" as many major lineages.

Plants are multicelled eukaryotes, most of which are photosynthetic. Nearly all have roots, stems, and leaves. Plants are the primary producers in land ecosystems. Redwood trees and flowering plants are examples.

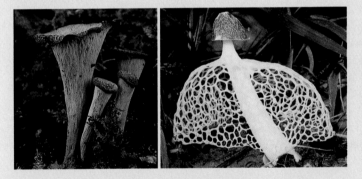

Fungi are eukaryotes. Most are multicelled. Different kinds are parasites, pathogens, or decomposers. Without decomposers such as fungi, communities would become buried in their own wastes.

Animals are multicelled eukaryotes that ingest tissues or juices of other organisms. Like this basilisk lizard, they actively move about during at least part of their life.

Plants are multicelled species. Most of them live on land or in freshwater environments. Nearly all plants are photosynthetic: They harness the energy of sunlight to drive the production of sugars from carbon dioxide and water. Besides feeding themselves, photosynthesizers also feed much of the biosphere.

The **animals** are multicelled consumers that ingest tissues or juices of other organisms. Herbivores graze, carnivores eat meat, scavengers eat remains of other organisms, and parasites withdraw nutrients from the tissues of a host. Animals grow and develop through a series of stages that lead to the adult form. Most kinds actively move about during at least part of their lives.

From this quick overview, can you get a sense of the tremendous range of life's variety—its diversity?

Take-Home Message

How do living things differ from one another?

■ Organisms differ in their details; they show tremendous variation in observable characteristics, or traits.

■ Various classification systems group species on the basis of shared traits.

■ A theory of evolution by natural selection is an explanation of life's diversity.

Individuals of a population are alike in certain aspects of their body form, function, and behavior, but the details of such traits differ from one individual to the next. For instance, humans (*Homo sapiens*) characteristically have two eyes, but those eyes come in a range of color among individuals.

Most traits are the outcome of information encoded in DNA, so they can be passed to offspring. Variations in traits arise through **mutations**, which are small-scale changes in DNA. Most mutations have neutral or negative effects, but some cause a trait to change in a way that makes an individual better suited to its environment. The bearer of such an **adaptive trait** has a better chance of surviving and passing its DNA to offspring than other individuals of the population. The naturalist Charles Darwin expressed the concept of "survival of the fittest" like this:

First, a natural population tends to increase in size. As it does, the individuals of the population compete more for food, shelter, and other limited resources.

Second, individuals of a population differ from one another in the details of shared traits. Such traits have a heritable basis.

Third, adaptive forms of traits make their bearers more competitive, so those forms tend to become more common over generations. The differential survival and reproduction of individuals in a population that differ in the details of their heritable traits is called **natural selection**.

Think of how pigeons differ in feather color and other traits (Figure 1.8a). Imagine that a pigeon breeder prefers black, curly-tipped feathers. She selects birds with the darkest, curliest-tipped feathers, and allows only those birds to mate. Over time, more and more pigeons in the breeder's captive population will have black, curly-tipped feathers.

Pigeon breeding is a case of artificial selection. One form of a trait is favored over others under contrived, manipulated conditions—in an artificial environment. Darwin saw that breeding practices could be an easily understood model for natural selection, a favoring of some forms of a given trait over others in nature.

Just as breeders are "selective agents" that promote reproduction of certain pigeons, agents of selection act on the range of variation in the wild. Among them are pigeon-eating peregrine falcons (Figure 1.8b). Swifter or better camouflaged pigeons are more likely to avoid falcons and live long enough to reproduce, compared with not-so-swift or too-flashy pigeons.

When different forms of a trait are becoming more or less common over successive generations, evolution is under way. In biology, **evolution** simply means change in a line of descent.

Take-Home Message

How did life become so diverse?

■ Individuals of a population show variation in their shared, heritable traits. Such variation arises through mutations in DNA.

■ Adaptive traits improve an individual's chances of surviving and reproducing, so they tend to become more common in a population over successive generations.

■ Natural selection is the differential survival and reproduction among individuals of a population that differ in the details of their shared, heritable traits. It and other evolutionary processes underlie the diversity of life.

wild rock pigeon

Figure 1.8 (**a**) Outcome of artificial selection: a few of the hundreds of varieties of domesticated pigeons descended from captive populations of wild rock pigeons (*Columba livia*). (**b**) A peregrine falcon (*left*) preying on a pigeon (*right*) is acting as an agent of natural selection in the wild.

1.5 Critical Thinking and Science

- Critical thinking means judging the quality of information.
- Science is limited to that which is observable.

Thinking About Thinking

Most of us assume that we do our own thinking—but do we, really? You might be surprised to find out just how often we let others think for us. For instance, a school's job, which is to impart as much information as possible to students, meshes with a student's job, which is to acquire as much knowledge as possible. In this rapid-fire exchange of information, it is easy to forget about the quality of what is being exchanged. If you accept information without question, you allow someone else to think for you.

Critical thinking means judging information before accepting it. "Critical" comes from the Greek *kriticos* (discerning judgment). When you think this way, you move beyond the content of information. You look for underlying assumptions, evaluate the supporting statements, and think of possible alternatives (Table 1.2).

How does the busy student manage this? Be aware of what you intend to learn from new information. Be conscious of bias or underlying agendas in books, lectures, or online. Consider your own biases—what you want to believe—and realize those biases influence your learning. Respectfully question authority figures. Decide whether ideas are based on opinion or evidence. Such practices will help you decide whether to accept or reject information.

The Scope and Limits of Science

Because each of us is unique, there are as many ways to think about the natural world as there are people. **Science**, the systematic study of nature, is one way. It helps us be objective about our observations of nature, in part because of its limitations. We limit science to a subset of the world—only that which is observable.

Science does not address some questions, such as "Why do I exist?" Most answers to such questions are subjective; they come from within as an integration of the personal experiences and mental connections that shape our consciousness. This is not to say subjective answers have no value: No human society functions for very long unless its individuals share standards for making judgments, even if they are subjective. Moral, aesthetic, and philosophical standards vary from one society to the next, but all help people decide what is important and good. All give meaning to what we do.

Also, science does not address the supernatural, or anything that is "beyond nature." Science does not

---

Table 1.2 A Guide to Critical Thinking

What message am I being asked to accept?

What evidence supports the message? Is the evidence valid?

Is there another way to interpret the evidence?

What other evidence would help me evaluate the alternatives?

Is the message the most reasonable one based on the evidence?

---

assume or deny that supernatural phenomena occur, but scientists may still cause controversy when they discover a natural explanation for something that was thought to be unexplainable. Such controversy often arises when a society's moral standards have become interwoven with traditional interpretations of nature.

For example, Nicolaus Copernicus studied the planets centuries ago in Europe, and concluded that Earth orbits the sun. Today this conclusion seems obvious, but at the time it was heresy. The prevailing belief was that the Creator made Earth—and, by extension, humans—as the center of the universe. Galileo Galilei, another scholar, found evidence for the Copernican model of the solar system and published his findings. He was forced to publicly recant his publication, and to put Earth back at the center of the universe.

Exploring a traditional view of the natural world from a scientific perspective might be misinterpreted as questioning morality even though the two are not the same. As a group, scientists are no less moral, less lawful, or less compassionate than anyone else. As you will see in the next section, however, their work follows a particular standard: Explanations must be testable in the natural world in ways that others can repeat.

Science helps us communicate experiences without bias; it may be as close as we can get to a universal language. We are fairly sure, for example, that laws of gravity apply everywhere in the universe. Intelligent beings on a distant planet would likely understand the concept of gravity. We might well use such concepts to communicate with them—or anyone—anywhere. The point of science, however, is not to communicate with aliens. It is to find common ground here on Earth.

Take-Home Message

What is science?

- Science is the study of the observable—those objects or events for which valid evidence can be gathered. It does not address the supernatural.

■ Scientists make and test potentially falsifiable predictions about how the natural world works.

Observations, Hypotheses, and Tests

To get a sense of how science works, consider Table 1.3 and this list of common research practices:

1. Observe some aspect of nature.

2. Frame a question that relates to your observation.

3. Read about what others have discovered concerning the subject, then propose a **hypothesis**, a testable answer to your question.

4. Using the hypothesis as a guide, make a **prediction**: a statement of some condition that should exist if the hypothesis is not wrong. Making predictions is called the if–then process: "if" is the hypothesis, and "then" is the prediction.

5. Devise ways to test the accuracy of the prediction by conducting experiments or gathering information. Experiments may be performed on a **model**, or analogous system, if experimenting directly with an object or event is not possible.

6. Assess the results of the tests. Results that confirm the prediction are evidence—data—in support of the hypothesis. Results that disprove the prediction are evidence that the hypothesis may be flawed.

7. Report all the steps of your work, along with any conclusions you drew, to the scientific community.

Table 1.3 Example of a Scientific Approach

1. Observation	People get cancer.	
2. Question	Why do people get cancer?	
3. Hypothesis	Smoking cigarettes may cause cancer.	
4. Prediction	If smoking causes cancer, then individuals who smoke will get cancer more often than those who do not.	
5. Gather information	Conduct a survey of individuals who smoke and individuals who do not smoke. Determine which group has the highest incidence of cancers.	
Laboratory experiment	Establish identical groups of laboratory rats (the model system). Expose one group to cigarette smoke. Compare the incidence of new cancers in each of the two groups.	
6. Assess results	Compile test results and draw conclusions from them.	
7. Report	Submit the results and the conclusions to the scientific community for review and publication.	

You might hear someone refer to these practices as "the scientific method," as if all scientists march to the drumbeat of a fixed procedure. They do not. There are different ways to do research, particularly in biology (Figure 1.9). Some biologists do surveys; they observe without making hypotheses. Others make hypotheses and leave tests to others. Some stumble onto valuable information they are not even looking for. Of course, it is not only a matter of luck. Chance favors a mind that is already prepared, by education and experience, to recognize what the new information might mean.

Regardless of the variation, one thing is constant: Scientists do not accept information simply because someone says it is true. They evaluate the supporting evidence and find alternative explanations. Does this sound familiar? It should—it is critical thinking.

About the Word "Theory"

Most scientists avoid the word "truth" when discussing science. Instead, they tend to talk about evidence that supports or does not support a hypothesis.

Suppose a hypothesis has not been disproven even after years of tests. It is consistent with all of the evidence gathered to date, and it has helped us to make successful predictions about other phenomena. When any hypothesis meets these criteria, it is considered to be a **scientific theory**.

To give an example, observations for all of recorded history have supported the hypothesis that gravity pulls objects toward Earth. Scientists no longer spend time testing the hypothesis for the simple reason that, after many thousands of years of observation, no one has seen otherwise. This hypothesis is now a scientific theory, but it is not an "absolute truth." Why not? An infinite number of tests would be necessary to confirm that it holds under every possible circumstance.

A single observation or result that is *not* consistent with a theory opens that theory to revision. For example, if gravity pulls objects toward Earth, it would be logical to predict that an apple will fall down when dropped. However, a scientist might well see such a test as an opportunity for the prediction to fail. Think about it. If even one apple falls up instead of down, the theory of gravity would come under scrutiny. Like every other theory, this one remains open to revision.

A well-tested theory is as close to the "truth" as scientists will venture. Table 1.4 lists a few scientific theories. One of them, the theory of natural selection, holds after more than a century of testing. Like all other scientific theories, we cannot be sure that it will hold under all possible conditions, but we can say it

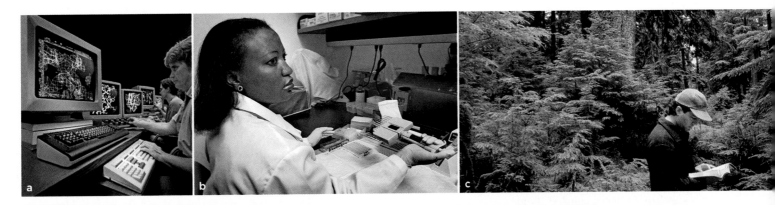

Figure 1.9 Scientists doing research in the laboratory and in the field. (**a**) Analyzing data with computers. (**b**) At the Centers for Disease Control and Prevention, Mary Ari testing a sample for the presence of dangerous bacteria. (**c**) Making field observations in an old-growth forest.

has a very high probability of not being wrong. If any evidence turns up that is inconsistent with the theory of natural selection, then biologists will revise it. Such a willingness to modify or discard even an entrenched theory is one of the strengths of science.

You may hear people apply the word "theory" to a speculative idea, as in the phrase "It's just a theory." Speculation is opinion or belief, a personal conviction that is not necessarily supported by evidence. A scientific theory is not an opinion: By definition, it must be supported by a large body of evidence.

Unlike theories, many beliefs and opinions cannot be tested. Without being able to test something, there is no way to disprove it. Even though personal conviction has tremendous value in our lives, it should not be confused with scientific theory.

Table 1.4 Examples of Scientific Theories

Atomic theory	All substances are composed of atoms.
Gravitation	Objects attract one another with a force that depends on their mass and how close together they are.
Cell theory	All organisms consist of one or more cells, the cell is the basic unit of life, and all cells arise from existing cells.
Germ theory	Microorganisms cause many diseases.
Plate tectonics	Earth's crust is cracked into pieces that move in relation to one another.
Evolution	Change occurs in lines of descent.
Natural selection	Variation in heritable traits influences differential survival and reproduction of individuals of a population.

Some Terms Used in Experiments

Careful observations are one way to test predictions that flow from a hypothesis. So are experiments. You will find examples of experiments in the next section. For now, just get acquainted with some of the important terms that researchers use:

1. **Experiments** are tests that can support or falsify a prediction.

2. Experiments are usually designed to test the effects of a single variable. A **variable** is a characteristic that differs among individuals or events.

3. Biological systems are an integration of so many interacting variables that it can be difficult to study one variable separately from the rest. Experimenters often test two groups of individuals, side by side. An **experimental group** is a set of individuals that have a certain characteristic or receive a certain treatment. This group is tested side by side with a **control group**, which is identical to the experimental group except for one variable—the characteristic or the treatment being tested. Ideally, the two groups have the same set of variables, except for the one being tested. Thus, any differences in experimental results between the two groups should be an effect of changing the variable.

Take-Home Message

How does science work?

■ Scientific inquiry involves asking questions about some aspect of nature, formulating hypotheses, making and testing predictions, and reporting the results.

■ Researchers design experiments to test the effects of one variable at a time.

■ A scientific theory is a long-standing, well-tested concept of cause and effect that is consistent with all evidence, and is used to make predictions about other phenomena.

■ Researchers unravel cause and effect in complex natural processes by changing one variable at a time.

Potato Chips and Stomach Aches

In 1996 the FDA approved Olestra®, a type of synthetic fat replacement made from sugar and vegetable oil, as a food additive. Potato chips were the first Olestra-laced food product on the market in the United States. Controversy soon raged. Some people complained of intestinal cramps after eating the chips and concluded that Olestra caused them.

Two years later, four researchers at Johns Hopkins University School of Medicine designed an experiment to test the hypothesis that this food additive causes cramps. They predicted that *if* Olestra causes cramps, *then* people who eat Olestra will be more likely to get cramps than people who do not.

A Hypothesis
Olestra® causes intestinal cramps.

B Prediction
People who eat potato chips made with Olestra will be more likely to get intestinal cramps than those who eat potato chips made without Olestra.

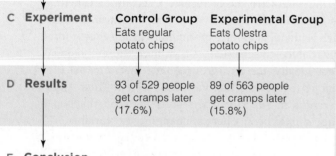

C Experiment

	Control Group Eats regular potato chips	Experimental Group Eats Olestra potato chips
D Results	93 of 529 people get cramps later (17.6%)	89 of 563 people get cramps later (15.8%)

E Conclusion
Percentages are about equal. People who eat potato chips made with Olestra are just as likely to get intestinal cramps as those who eat potato chips made without Olestra. These results do not support the hypothesis.

Figure 1.10 Animated The steps in a scientific experiment to determine if Olestra causes cramps. A report of this study was published in the *Journal of the American Medical Association* in January of 1998.

To test the prediction, they used a Chicago theater as the "laboratory." They asked more than 1,100 people between ages thirteen and thirty-eight to watch a movie and eat their fill of potato chips. Each person got an unmarked bag that contained 13 ounces of chips. The individuals who got a bag of Olestra-laced potato chips were the experimental group. Individuals who got a bag of regular chips were the control group.

Afterward, researchers contacted all of the people and tabulated the reports of gastrointestinal cramps. Of 563 people making up the experimental group, 89 (15.8 percent) complained about problems. However, so did 93 of the 529 people (17.6 percent) making up the control group—who had munched on the regular chips! This simple experiment disproved the prediction that eating Olestra-laced potato chips at a single sitting can cause gastrointestinal cramps (Figure 1.10).

Butterflies and Birds

Consider the peacock butterfly, a winged insect that was named for the large, colorful spots on its wings. In 2005, researchers published a report on their tests to identify factors that help peacock butterflies defend themselves against insect-eating birds. The researchers made two observations. First, when a peacock butterfly rests, it folds its ragged-edged wings, so only the dark underside shows (Figure 1.11a). Second, when a butterfly sees a predator approaching, it repeatedly flicks its paired forewings and hindwings open and closed. At the same time, each forewing slides over the hindwing, which produces a hissing sound and a series of clicks.

The researchers asked this question, "Why does the peacock butterfly flick its wings?" After they reviewed earlier studies, they formulated three hypotheses that might explain the wing-flicking behavior:

1. When folded, the butterfly wings resemble a dead leaf. They may camouflage the butterfly, or help it hide from predators in its forest habitat.

2. Although the wing-flicking probably attracts predatory birds, it also exposes brilliant spots that resemble owl eyes (Figure 1.11b). Anything that looks like owl eyes is known to startle small, butterfly-eating birds, so exposing the wing spots might scare off predators.

3. The hissing and clicking sounds produced when the peacock butterfly rubs the sections of its wings together may deter predatory birds.

The researchers decided to test hypotheses 2 and 3. They made the following predictions:

a   b   c

Table 1.5   Results of Peacock Butterfly Experiment*

Wing Spots	Wing Sound	Total Number of Butterflies	Number Eaten	Number Survived
Spots	Sound	9	0	9 (100%)
No spots	Sound	10	5	5 (50%)
Spots	No sound	8	0	8 (100%)
No spots	No sound	10	8	2 (20%)

* *Proceedings of the Royal Society of London, Series B* (2005) 272: 1203–1207.

Figure 1.11 Peacock butterfly defenses against predatory birds. (**a**) With wings folded, a resting peacock butterfly looks like a dead leaf. (**b**) When a bird approaches, the butterfly repeatedly flicks its wings open and closed. This defensive behavior exposes brilliant spots. It also produces hissing and clicking sounds.

Researchers tested whether the behavior deters blue tits (**c**). They painted over the spots of some butterflies, cut the sound-making part of the wings on other butterflies, and did both to a third group; then the biologists exposed each butterfly to a hungry bird.

The results are listed in Table 1.5. **Figure It Out:** Which defense, wing spots or sounds, more effectively deterred the tits?

Answer: wing spots

1. *If* brilliant wing spots of peacock butterflies deter predatory birds, *then* individuals with no wing spots will be more likely to get eaten by predatory birds than individuals with wing spots.

2. *If* the sounds that peacock butterflies produce deter predatory birds, *then* individuals that do not make the sounds will be more likely to be eaten by predatory birds than individuals that make the sounds.

The next step was the experiment. The researchers painted the wing spots of some butterflies black, cut off the sound-making part of the hindwings of others, and did both to a third group. They put each butterfly into a large cage with a hungry blue tit (Figure 1.11*c*) and then watched the pair for thirty minutes.

Table 1.5 lists the results of the experiment. All of the butterflies with unmodified wing spots survived, regardless of whether they made sounds. By contrast, only half of the butterflies that had spots painted out but could make sounds survived. Most of the butterflies with neither spots nor sound structures were eaten quickly.

The test results confirmed both predictions, so they support the hypotheses. Birds are deterred by peacock butterfly sounds, and even more so by wing spots.

Asking Useful Questions

Researchers try to design single-variable experiments that will yield quantitative results, which are counts or some other data that can be measured or gathered objectively. Even so, they risk designing experiments and interpreting results in terms of what they want to find out. Particularly when studying humans, isolating a single variable is not often possible. For example, by thinking critically we may realize that the people who participated in the Olestra experiment were chosen randomly. That means the study was not controlled for gender, age, weight, medications taken, and so on. Such variables may well have influenced the results.

Scientists expect one another to put aside bias. If one individual does not, others will, because science works best when it is both cooperative and competitive.

Take-Home Message

Why do biologists do experiments?

■ Natural processes are often influenced by many interacting variables.

■ Experiments help researchers unravel causes of such natural processes by focusing on the effects of changing a single variable.

1.8 | Sampling Error in Experiments

- Biology researchers experiment on subsets of a group.
- Results from such an experiment may differ from results of the same experiment performed on the whole group.

A Natalie, blindfolded, randomly plucks a jelly bean from a jar. There are 120 green and 280 black jelly beans in that jar, so 30 percent of the jelly beans in the jar are green, and 70 percent are black.

B The jar is hidden from Natalie's view before she removes her blindfold. She sees only one green jelly bean in her hand and assumes that the jar must hold only green jelly beans.

C Blindfolded again, Natalie picks out 50 jelly beans from the jar and ends up with 10 green and 40 black jelly beans.

D The larger sample leads Natalie to assume that one-fifth of the jar's jelly beans are green (20 percent) and four-fifths are black (80 percent). The sample more closely approximates the jar's actual green-to-black ratio of 30 percent to 70 percent. The more times Natalie repeats the sampling, the greater the chance she will come close to knowing the actual ratio.

Rarely can researchers observe all individuals of a group. For example, remember the explorers you read about in the chapter introduction? They did not survey the entire rain forest, which cloaks more than 2 million acres of New Guinea's Foja Mountains. Even if it were possible, doing so would take unrealistic amounts of time and effort. Besides, tromping about even in a small area can damage delicate forest ecosystems.

Given such constraints, researchers tend to experiment on subsets of a population, event, or some other aspect of nature that they select to represent the whole. They test the subsets, and then use the results to make generalizations about the whole population.

Suppose researchers design an experiment to identify variables that influence the population growth of golden-mantled tree kangaroos. They might focus only on the population living in one acre of the Foja Mountains. If they identify only 5 golden-mantled tree kangaroos in the specified acre, then they might extrapolate that there are 50 in every ten acres, 100 in every twenty acres, and so forth.

However, generalizing from a subset is risky because the subset may not be representative of the whole. If the only population of golden-mantled tree kangaroos in the forest just happens to be living in the surveyed acre, then the researchers' assumptions about the number of kangaroos in the rest of the forest will be wrong.

Sampling error is a difference between results from a subset and results from the whole. It happens most often when sample sizes are small. Starting with a large sample or repeating the experiment many times helps minimize sampling error (Figure 1.12). To understand why, imagine flipping a coin. There are two possible results: The coin lands heads up, or it lands tails up. You might predict that the coin will land heads up as often as it lands tails up. When you actually flip the coin, though, often it will land heads up, or tails up, several times in a row. If you flip the coin only a few times, the results may differ greatly from your prediction. Flip it many times, and you probably will come closer to having equal numbers of heads and tails.

Sampling error is an important consideration in the design of most experiments. The possibility that it occurred should be part of the critical thinking process as you read about experiments. Remember to ask: If the experimenters used a subset of the whole, did they select a large enough sample? Did they repeat the experiment many times? Thinking about these possibilities will help you evaluate the results and conclusions reached.

Figure 1.12 Animated Demonstration of sampling error.

Lost Worlds and Other Wonders

Almost every week, another new species is discovered and we are again reminded that we do not yet know all of the organisms on our own planet. We don't even know how many to look for. The vast information about the 1.8 million species we do know about changes so quickly that collating it has been impossible—until now. A new web site, titled the Encyclopedia of Life, is intended to be an online reference source and database of species information maintained by collaborative effort. See it at www.eol.org.

How would you vote? Discovered in Madagascar in 2005, this tiny mouse lemur was named *Microcebus lehilahytsara* in honor of primatologist Steve Goodman (lehilahytsara is a combination of the Malagasy words for "good" and "man"). Should naming rights be sold? See CengageNOW for details, then vote online.

Summary

Section 1.1 There are **emergent properties** at each level of organization in **nature**. All matter consists of **atoms**, which combine as **molecules**. **Organisms** are one or more **cells**, the smallest units of life. A **population** is a group of individuals of a species in a given area; a **community** is all populations of all species in a given area. An **ecosystem** is a community interacting with its environment. The **biosphere** includes all regions of Earth that hold life.

■ *Explore levels of biological organization with the interaction on CengageNOW.*

Section 1.2 All living things have similar characteristics (Table 1.6). All organisms require inputs of **energy** and **nutrients** to sustain themselves. **Producers** make their own food by processes such as **photosynthesis**; **consumers** eat producers or other consumers. By **homeostasis**, organisms use molecules and structures such as **receptors** to help keep the conditions in their internal environment within ranges that their cells tolerate. Organisms grow, **develop**, and **reproduce** using information in their **DNA**, a nucleic acid **inherited** from parents. Information encoded in DNA is the source of an individual's **traits**.

■ *Use instructions with the animation on CengageNOW to see how different objects are assembled from the same materials. Also view energy flow and materials cycling.*

Section 1.3 Each type of organism is given a name that includes **genus** and **species** names. Classification systems group species by their shared, heritable traits. All organisms can be classified as **bacteria**, **archaea**, or **eukaryotes**. **Plants**, **protists**, **fungi**, and **animals** are eukaryotes.

■ *Use the interaction on CengageNOW to explore characteristics of the three domains of life.*

Section 1.4 Information encoded in DNA is the basis of traits that an organism shares with others of its species. **Mutations** are the original source of variation in traits.

Some forms of traits are more adaptive than others, so their bearers are more likely to survive and reproduce. Over generations, such **adaptive traits** tend to become more common in a population; less adaptive forms of traits tend to become less common or are lost.

Thus, the traits that characterize a species can change over generations in evolving populations. **Evolution** is change in a line of descent. The differential survival and reproduction among individuals that vary in the details of their shared, heritable traits is an evolutionary process called **natural selection**.

Section 1.5 **Critical thinking** is judging the quality of information as one learns. **Science** is one way of looking at the natural world. It helps us minimize bias in our judgments by focusing on only testable ideas about observable aspects of nature.

Section 1.6 Researchers generally make observations, form **hypotheses** (testable assumptions) about it, then make **predictions** about what might occur if the hypothesis is correct. They test predictions with **experiments**, using **models**, **variables**, **experimental groups**, and **control groups**. A hypothesis that is not consistent with results of scientific tests (evidence) is modified or discarded. A **scientific theory** is a long-standing hypothesis that is used to make useful predictions.

Section 1.7 Scientific experiments simplify interpretations of complex biological systems by focusing on the effect of one variable at a time.

Section 1.8 Small sample size increases the likelihood of **sampling error** in experiments. In such cases, a subset may be tested that is not representative of the whole.

Table 1.6 Summary of Life's Characteristics

Shared characteristics that underlie life's unity

Organisms grow, develop, and reproduce based on information encoded in DNA, which is inherited from parents.

Ongoing inputs of energy and nutrients sustain all organisms, as well as nature's overall organization.

Organisms maintain homeostasis by sensing and responding to changes inside and outside of the body.

Basis of life's diversity

Mutations (heritable changes in DNA) give rise to variation in details of body form, the functioning of body parts, and behavior.

Diversity is the sum total of variations that have accumulated, since the time of life's origin, in different lines of descent. It is an outcome of natural selection and other processes of evolution.

Data Analysis Exercise

The photographs to the *right* represent the experimental and control groups used in the peacock butterfly experiment from Section 1.7.

See if you can identify each experimental group, and match it with the relevant control group(s). *Hint:* Identify which variable is being tested in each group (each variable has a control).

 a Wing spots painted out

 b Wing spots visible; wings silenced

 c Wing spots painted out; wings silenced

 d Wings painted but spots visible

 e Wings cut but not silenced

 f Wings painted but spots visible; wings cut but not silenced

Self-Quiz

Answers in Appendix III

1. _____ are fundamental building blocks of all matter.

2. The smallest unit of life is the _____ .

3. _____ move around for at least part of their life.

4. Organisms require _____ and _____ to maintain themselves, grow, and reproduce.

5. _____ is a process that maintains conditions in the internal environment within ranges that cells can tolerate.

6. Bacteria, Archaea, and Eukarya are three _____ .

7. DNA _____ .
   a. contains instructions for building proteins
   b. undergoes mutation
   c. is transmitted from parents to offspring
   d. all of the above

8. _____ is the transmission of DNA to offspring.
   a. Reproduction
   b. Development
   c. Homeostasis
   d. Inheritance

9. _____ is the process by which an organism produces offspring.

10. Science only addresses that which is _____ .

11. _____ are the original source of variation in traits.

12. A trait is _____ if it improves an organism's chances to survive and reproduce in its environment.

13. A control group is _____ .
   a. a set of individuals that have a certain characteristic or receive a certain treatment
   b. the standard against which experimental groups can be compared
   c. the experiment that gives conclusive results

14. Match the terms with the most suitable description.
   ___ emergent property
   ___ natural selection
   ___ scientific theory
   ___ hypothesis
   ___ prediction
   ___ species

   a. statement of what a hypothesis leads you to expect to see
   b. type of organism
   c. occurs at a higher organizational level in nature, not at levels below it
   d. time-tested hypothesis
   e. differential survival and reproduction among individuals of a population that vary in details of shared traits
   f. testable explanation

■ *Visit CengageNOW for additional questions.*

Critical Thinking

1. Why would you think twice about ordering from a cafe menu that lists only the second part of the species name (not the genus) of its offerings? *Hint:* Look up *Ursus americanus, Ceanothus americanus, Bufo americanus, Homarus americanus, Lepus americanus,* and *Nicrophorus americanus.*

2. How do prokaryotes and eukaryotes differ?

3. Explain the relationship between DNA and natural selection.

4. Procter & Gamble makes Olestra and financed the study described in Section 1.7. The main researcher was a consultant to Procter & Gamble during the study. What do you think about scientific information that comes from tests financed by companies with a vested interest in the outcome?

5. Once there was a highly intelligent turkey that had nothing to do but reflect on the world's regularities. Morning always started out with the sky turning light, followed by the master's footsteps, which was always followed by the appearance of food. Other things varied, but food always followed footsteps. The sequence of events was so predictable that it eventually became the basis of the turkey's theory about the goodness of the world. One morning, after more than 100 confirmations of the goodness theory, the turkey listened for the master's footsteps, heard them, and had its head chopped off.

   Any scientific theory is modified or discarded when contradictory evidence becomes available. The absence of absolute certainty has led some people to conclude that "facts are irrelevant—facts change." If that is so, should we stop doing scientific research? Why or why not?

6. In 2005 a South Korean scientist, Woo-suk Hwang, reported that he made immortal stem cells from eleven human patients. His research was hailed as a breakthrough for people affected by currently incurable degenerative diseases, because such stem cells might be used to repair a person's own damaged tissues. Hwang published his results in a respected scientific journal. In 2006, the journal retracted his paper after other scientists discovered that Hwang and his colleagues had faked their results. Does the incident show that the results of scientific studies cannot be trusted? Or does it confirm the usefulness of a scientific approach, because other scientists quickly discovered and exposed the fraud?

I PRINCIPLES OF CELLULAR LIFE

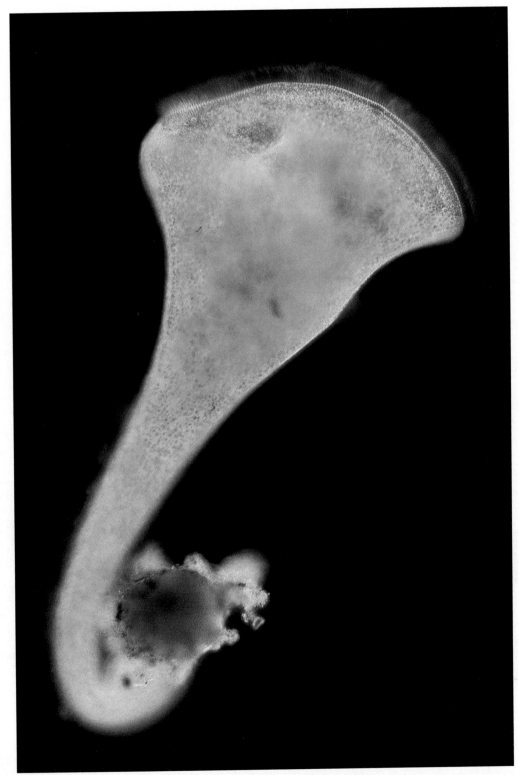

Staying alive means securing energy and raw materials from the environment. Shown here, a living cell of the genus *Stentor*. This protist has hairlike projections around an opening to a cavity in its body, which is about 2 millimeters long. Its "hairs" of fused-together cilia beat the surrounding water. They create a current that wafts food into the cavity.

2   Life's Chemical Basis

What Are You Worth?

Hollywood thinks actor Keanu Reaves is worth $30 million plus per movie, the Yankees think shortstop Alex Rodriguez is worth $252 million per decade, and the United States thinks the average public school teacher is worth $46,597 per year. How much is one human body really worth? You can buy the entire collection of ingredients that make up an average 70-kilogram (150-pound) body for about $118.63 (Figure 2.1). Of course, all you have to do is watch Keanu, Alex, or any teacher to know that a human body is far more than a collection of those ingredients. What makes us worth more than the sum of our parts?

The fifty-eight pure substances listed in Figure 2.1 are called elements. You will find the same elements that make up the human body in, say, dirt or seawater. However, the proportions of those elements differ between living and nonliving things. For example, a human body contains far more carbon. Seawater and most rocks have no more than a trace of it.

We are only starting to understand the processes by which a collection of elements becomes assembled as a living body. We do know that life's unique organization starts with the properties of atoms that make up certain elements. This is your chemistry. It makes you far more than the sum of your body's ingredients—a handful of lifeless chemicals.

Elements in a Human Body		
Element	Number of Atoms (x 10^{15})	Retail Cost
Hydrogen	41,808,044,129,611	$ 0.028315
Oxygen	16,179,356,725,877	0.021739
Carbon	8,019,515,931,628	6.400000
Nitrogen	773,627,553,592	9.706929
Phosphorus	151,599,284,310	68.198594
Calcium	150,207,096,162	15.500000
Sulfur	26,283,290,713	0.011623
Sodium	26,185,559,925	2.287748
Potassium	21,555,924,426	4.098737
Chlorine	16,301,156,188	1.409496
Magnesium	4,706,027,566	0.444909
Fluorine	823,858,713	7.917263
Iron	452,753,156	0.054600
Silicon	214,345,481	0.370000
Zinc	211,744,915	0.088090
Rubidium	47,896,401	1.087153
Strontium	21,985,848	0.177237
Bromine	19,588,506	0.012858
Boron	10,023,125	0.002172
Copper	6,820,886	0.012961
Lithium	6,071,171	0.024233
Lead	3,486,486	0.003960
Cadmium	2,677,674	0.010136
Titanium	2,515,303	0.010920
Cerium	1,718,576	0.043120
Chromium	1,620,894	0.003402
Nickel	1,538,503	0.031320
Manganese	1,314,936	0.001526
Selenium	1,143,617	0.037949
Tin	1,014,236	0.005387
Iodine	948,745	0.094184
Arsenic	562,455	0.023576
Germanium	414,543	0.130435
Molybdenum	313,738	0.001260
Cobalt	306,449	0.001509
Cesium	271,772	0.000016
Mercury	180,069	0.004718
Silver	111,618	0.013600
Antimony	98,883	0.000243
Niobium	97,195	0.000624
Barium	96,441	0.028776
Gallium	60,439	0.003367
Yttrium	40,627	0.005232
Lanthanum	34,671	0.000566
Tellurium	33,025	0.000722
Scandium	26,782	0.058160
Beryllium	24,047	0.000218
Indium	20,972	0.000600
Thallium	14,727	0.000894
Bismuth	14,403	0.000119
Vanadium	12,999	0.000322
Tantalum	6,654	0.001631
Zirconium	6,599	0.000830
Gold	6,113	0.001975
Samarium	2,002	0.000118
Tungsten	655	0.000007
Thorium	3	0.004948
Uranium	3	0.000103
Total	67,179,218,505,055 x 10^{15}	**$118.63**

See the video! Figure 2.1 Composition of an average-sized adult human body, by weight and retail cost. Manufacturers commonly add fluoride to toothpaste. Fluoride is a form of fluorine, one of several elements with vital functions—but only in trace amounts. Too much can be toxic.

Key Concepts

Atoms and elements

Atoms are particles that are the building blocks of all matter. They can differ in their numbers of component protons, electrons, and neutrons. Elements are pure substances, each consisting entirely of atoms that have the same number of protons. **Sections 2.1, 2.2**

Why electrons matter

Whether one atom will bond with others depends on the element, and the number and arrangement of its electrons. **Section 2.3**

Atoms bond

Atoms of many elements interact by acquiring, sharing, and giving up electrons. Ionic, covalent, and hydrogen bonds are the main interactions between atoms in biological molecules. **Section 2.4**

Water of life

Life originated in water and is adapted to its properties. Water has temperature-stabilizing effects, cohesion, and a capacity to act as a solvent for many other substances. These properties make life possible on Earth. **Section 2.5**

The power of hydrogen

Life is responsive to changes in the amounts of hydrogen ions and other substances dissolved in water. **Section 2.6**

Links to Earlier Concepts

- With this chapter, we turn to the first of life's levels of organization—atoms and energy—so take a moment to review Section 1.1.

- Life's organization requires continuous inputs of energy (1.2). Organisms store that energy in chemical bonds between atoms.

- You will come across a simple example of how the body's built-in mechanisms maintain homeostasis (1.2).

■ The behavior of elements, which make up all living things, starts with the structure of individual atoms.

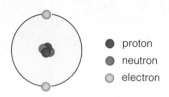

● proton
● neutron
○ electron

Characteristics of Atoms

Atoms are particles that are the building blocks of all substances. Even though they are about one billion times smaller than basketballs, atoms consist of even smaller subatomic particles called **protons** (p+), which carry a positive charge; **neutrons**, which carry no charge; and **electrons** (e^-), which carry a negative charge. **Charge** is an electrical property that attracts or repels other subatomic particles. Protons and neutrons cluster in an atom's central core, or **nucleus**. Electrons move around the nucleus (Figure 2.2).

Atoms differ in the number of subatomic particles. The number of protons, which is the **atomic number**, determines the element. **Elements** are pure substances, each consisting only of atoms with the same number of protons. For example, a chunk of carbon contains only carbon atoms, all of which have six protons in their nucleus. The atomic number of carbon is 6. All atoms with six protons in their nucleus are carbon atoms, no matter how many electrons or neutrons they have. Each element has a symbol that is an abbreviation of its Latin name. Carbon's symbol, C, is from *carbo*, the Latin word for coal—which is mostly carbon.

Figure 2.2 Atoms. Electrons move about a nucleus of protons and neutrons. Models such as this do not show what an atom really looks like. A more accurate rendering would show electrons occupying fuzzy, three-dimensional shapes about 10,000 times larger than the nucleus.

All elements occur in different forms called **isotopes**. Atoms of isotopes have the same number of protons, but different numbers of neutrons. We refer to isotopes by **mass number**, which is the total number of protons and neutrons in their nucleus. The mass number of an isotope is shown as a superscript to the left of an element's symbol. For instance, the most common isotope of carbon is ^{12}C (six protons, six neutrons). Another is ^{13}C (six protons, seven neutrons).

The Periodic Table

Today, we know that the numbers of electrons, protons, and neutrons determine how an element behaves, but scientists were classifying elements by chemical behavior long before they knew about subatomic particles. In 1869, the chemist Dmitry Mendeleev arranged all of the elements known at the time into a table based on their chemical properties. He had constructed the first **periodic table of the elements**.

Elements are ordered in the periodic table by their atomic number (Figure 2.3). Those in each vertical column behave in similar ways. For instance, all of the elements in the far right column of the table are inert gases; their atoms do not interact with other atoms. In nature, such elements occur only as solitary atoms.

We find the first ninety-four elements in nature. The others are so unstable that they are extremely rare. We know they exist because they can be made, one atom at a time, for a fraction of a second. It takes a nuclear physicist to do this, because an atom's nucleus cannot be altered by heat or other ordinary means.

Figure 2.3 Periodic table of the elements and its creator, Dmitry Mendeleev. Until he came up with the table, Mendeleev was known mainly for his extravagant hair; he cut it only once per year.

Atomic numbers are shown above the element symbols. Some of the symbols are abbreviations for their Latin names. For instance, Pb (lead) is short for plumbum; the word "plumbing" is related—ancient Romans made their water pipes with lead. Appendix IV has a more detailed table.

Take-Home Message

What are the basic building blocks of all matter?

■ Atoms are tiny particles, the building blocks of all substances.

■ Atoms consist of electrons moving around a nucleus of protons and (except for hydrogen) neutrons.

■ An element is a pure substance. Each kind consists only of atoms with the same number of protons.

2.2 Putting Radioisotopes to Use

■ Some radioactive isotopes—radioisotopes—are used in research and in medical applications.

In 1896, Henri Becquerel made a chance discovery. He left some crystals of a uranium salt in a desk drawer, on top of a metal screen. Under the screen was an exposed film wrapped tightly in black paper. Becquerel developed the film a few days later and was surprised to see a negative image of the screen. He realized that "invisible radiations" coming from the uranium salts had passed through the paper and exposed the film around the screen.

Becquerel's images were evidence that uranium has **radioisotopes**, or radioactive isotopes. So do many other elements. The atoms of radioisotopes spontaneously emit subatomic particles or energy when their nucleus breaks down. This process, **radioactive decay**, can transform one element into another. For example, ^{14}C is a radioisotope of carbon. It decays when one of its neutrons spontaneously splits into a proton and an electron. Its nucleus emits the electron, and so an atom of ^{14}C (with eight neutrons and six protons) becomes an atom of ^{14}N (nitrogen 14, with seven neutrons and seven protons).

Radioactive decay occurs independently of external factors such as temperature, pressure, or whether the atoms are part of molecules. A radioisotope decays at a constant rate into predictable products. For example, after 5,730 years, we can predict that about half of the atoms in any sample of ^{14}C will be ^{14}N atoms. This predictability can be used to estimate the age of rocks and fossils by their radioisotope content. We return to this topic in Section 17.6.

Researchers and clinicians also introduce radioisotopes into living organisms. Remember, isotopes are atoms of the same element. All isotopes of an element generally have the same chemical properties regardless of the number of neutrons in their atoms. This consistent chemical behavior means that organisms use atoms of one isotope (such as ^{14}C) the same way that they use atoms of another (such as ^{12}C). Thus, radioisotopes can be used in tracers.

A **tracer** is any molecule with a detectable substance attached. Typically, a radioactive tracer is a molecule in which radioisotopes have been swapped for one or more atoms. Researchers deliver radioactive tracers into a biological system such as a cell or a multicelled body. Instruments that can detect radioactivity let researchers follow the tracer as it moves through the system.

For example, Melvin Calvin and his colleagues used a radioactive tracer to identify specific reaction steps of photosynthesis. The researchers made carbon dioxide with ^{14}C, then let green algae (simple aquatic organisms) take up the radioactive gas. Using instruments that detected the radioactive decay of ^{14}C, they tracked carbon through steps by which the algae—and all plants—make sugars.

Radioisotopes have medical applications as well. PET (short for *Positron-Emission Tomography*) helps us "see" cell activity. By this procedure, a radioactive sugar or other tracer is injected into a patient, who is then moved into a PET scanner (Figure 2.4a). Inside the patient's body, cells with differing rates of activity take up the tracer at different rates. The scanner detects radioactive decay wherever the tracer is, then translates that data into an image. Such images can reveal abnormal cell activity (Figure 2.4b).

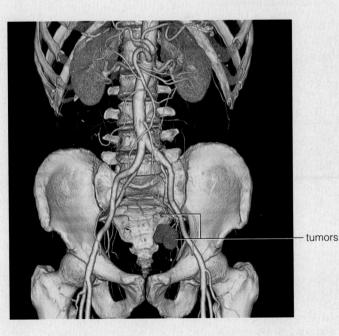

A A patient is injected with a radioactive tracer and moved into a scanner like this one. Detectors that intercept radioactive decay of the tracer surround the body part of interest.

B Radioactive decay detected by the scanner is converted into digital images of the body's interior. Two tumors (*blue*) in and near the bowel of a cancer patient are visible in this PET scan.

tumors

Figure 2.4 Animated PET scanning.

Why Electrons Matter

- Atoms acquire, share, and donate electrons.
- Whether an atom will interact with other atoms depends on how many electrons it has.

Electrons and Energy Levels

Electrons are really, really small: If they were as big as apples, you would be 3.5 times taller than our solar system is wide. Simple physics explains the motion of, say, an apple falling from a tree. Electrons are so tiny that everyday physics does not explain their behavior, but that behavior underlies interactions among atoms.

vacancy

no vacancy

A typical atom has about as many electrons as protons, so a lot of electrons may be zipping around one nucleus. Those electrons never collide, despite moving at nearly the speed of light (300,000 kilometers per second, or 670 million miles per hour). Why not? They travel in different orbitals, which are defined volumes of space around the nucleus.

Imagine that an atom is a multilevel apartment building, with rooms available for rent by electrons. The nucleus is in the basement, and each "room" is an orbital. No more than two electrons can share a room at the same time. An orbital with only one electron has a vacancy, and another electron can move in.

Each floor in the apartment building corresponds to one energy level. There is only one room on the first floor: one orbital at the lowest energy level, closest to the nucleus. It fills up first. In hydrogen, the simplest atom, a single electron occupies that room. Helium has two electrons, so it has no vacancies at the lowest energy level. In larger atoms, more electrons rent the second-floor rooms. When the second floor fills, more electrons rent third-floor rooms, and so on. Electrons fill orbitals at successively higher energy levels.

The farther an electron is from the basement (the nucleus), the greater its energy. An electron in a first-floor room cannot move to the second or third floor, let alone the penthouse, unless an input of energy gives it a boost. Suppose an electron absorbs enough energy from sunlight to get excited about moving up. Move it does. If nothing fills that lower room, though, the electron immediately moves back down, emitting its extra energy as it does. In later chapters, you will see how some types of cells harvest that released energy.

Why Atoms Interact

Shells and Electrons We use a **shell model** to help us check an atom for vacancies (Figure 2.5). With this model, nested "shells" correspond to successive energy levels. Each shell includes all rooms on one floor of the

○ electron

C **Third shell** This shell corresponds to the third energy level. It has four orbitals with room for eight electrons. Sodium has one electron in the third shell; chlorine has seven. Both have vacancies, so both form chemical bonds. Argon, with no vacancies, does not.

B **Second shell** This shell, which corresponds to the second energy level, has four orbitals—room for a total of eight electrons. Carbon has six electrons: two in the first shell and four in the second. It has four vacancies. Oxygen has two vacancies. Both carbon and oxygen form chemical bonds. Neon, with no vacancies, does not.

A **First shell** A single shell corresponds to the first energy level, which has a single orbital that can hold two electrons. Hydrogen has only one electron in this shell, so it has one vacancy. A helium atom has two electrons (no vacancies), so it does not form bonds.

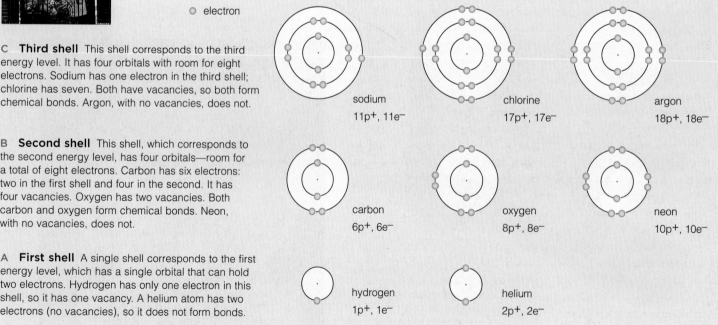

sodium
11p+, 11e−

chlorine
17p+, 17e−

argon
18p+, 18e−

carbon
6p+, 6e−

oxygen
8p+, 8e−

neon
10p+, 10e−

hydrogen
1p+, 1e−

helium
2p+, 2e−

Figure 2.5 Animated Shell models, which help us check for vacancies in atoms. Each circle, or shell, represents all orbitals at one energy level. Atoms with vacancies in the outermost shell tend to form bonds. Remember, atoms do not look anything like these flat diagrams.

atomic apartment building. We draw an atom's shells by filling them with electrons (represented as dots or balls) from the innermost shell out, until there are as many electrons as the atom has protons.

If an atom's outermost shell is full of electrons, it has no vacancies. Atoms of such elements are chemically inactive; they are most stable as single atoms. Helium, neon, and the other inert gases in the right-hand column of the periodic table are like this.

If an atom's outermost shell has room for an extra electron, it has a vacancy. Atoms with vacancies tend to interact with other atoms; they give up, acquire, or share electrons until they have no vacancies in their outermost shell. Any atom is in its most stable state when it has no vacancies.

Atoms and Ions The negative charge of an electron cancels the positive charge of a proton, so an atom is uncharged only when it has as many electrons as protons. An atom with different numbers of electrons and protons is called an **ion**. An ion carries a charge; either it acquired a positive charge by losing an electron, or it acquired a negative charge by pulling an electron away from another atom.

Electronegativity is a measure of an atom's ability to pull electrons from other atoms. Whether the pull is strong or weak depends on the atom's size and how many vacancies it has; it is not a measure of charge.

As an example, when a chlorine atom is uncharged, it has 17 protons and 17 electrons. Seven electrons are in its outer (third) shell, which can hold eight (Figure 2.6). It has one vacancy. An uncharged chlorine atom is highly electronegative—it can pull an electron away from another atom and fill its third shell. When that happens, the atom becomes a chloride ion (Cl^-) with 17 protons, 18 electrons, and a net negative charge.

As another example, an uncharged sodium atom has 11 protons and 11 electrons. This atom has one electron in its outer (third) shell, which can hold eight. It has seven vacancies. An uncharged sodium atom is weakly electronegative, so it cannot pull seven electrons from other atoms to fill its third shell. Instead, it tends to lose the single electron in its third shell. When that happens, two full shells—and no vacancies—remain. The atom has now become a sodium ion (Na^+), with 11 protons, 10 electrons, and a net positive charge.

From Atoms to Molecules Atoms do not like to have vacancies, and try to get rid of them by interacting with other atoms. A **chemical bond** is an attractive force that arises between two atoms when their electrons interact. A **molecule** forms when two or more atoms

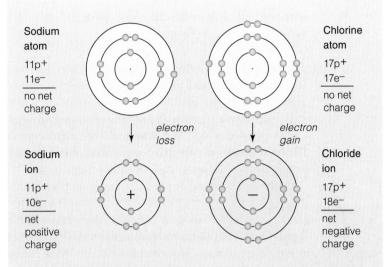

Sodium atom
11p+
11e−
no net charge

Chlorine atom
17p+
17e−
no net charge

electron loss

electron gain

Sodium ion
11p+
10e−
net positive charge

Chloride ion
17p+
18e−
net negative charge

A A sodium atom becomes a positively charged sodium ion (Na^+) when it loses the electron in its third shell. The atom's full second shell is now the outermost, and the atom has no vacancies.

B A chlorine atom becomes a negatively charged chloride ion (Cl^-) when it gains an electron and fills the vacancy in its third, outermost shell.

Figure 2.6 Animated Ion formation.

of the same or different elements join in chemical bonds. The next section explains the main types of bonds in biological molecules.

Compounds are molecules that consist of two or more different elements in proportions that do not vary. Water is an example. All water molecules have one oxygen atom bonded to two hydrogen atoms. Whether water is in the seas, a waterfall, a Siberian lake, or anywhere else, its molecules have twice as many hydrogen as oxygen atoms. By contrast, in a **mixture**, two or more substances intermingle, and their proportions can vary because the substances do not bond with each other. For example, you can make a mixture by swirling sugar into water. The sugar dissolves, but no chemical bonds form.

Always two H for every O

Take-Home Message

Why do atoms interact?

■ An atom's electrons are the basis of its chemical behavior.

■ Shells represent all electron orbitals at one energy level in an atom. When the outermost shell is not full of electrons, the atom has a vacancy.

■ Atoms tend to get rid of vacancies by gaining or losing electrons (thereby becoming ions), or by sharing electrons with other atoms.

■ Atoms with vacancies can form chemical bonds. Chemical bonds connect atoms into molecules.

2.4 What Happens When Atoms Interact?

■ The characteristics of a bond arise from the properties of the atoms that take part in it.

The same atomic building blocks, arranged in different ways, make different molecules. For example, carbon atoms bonded one way form layered sheets of a soft, slippery mineral known as graphite. The same carbon atoms bonded a different way form the rigid crystal lattice of diamond—the hardest mineral. Bond oxygen and hydrogen atoms to carbon and you get sugar.

Although bonding applies to a range of interactions among atoms, we can categorize most bonds into distinct types based on their different properties. Three types—ionic, covalent, and hydrogen bonds—are most common in biological molecules. Which type forms depends on the vacancies and electronegativity of the atoms that take part in it. Table 2.1 compares different ways to represent molecules and their bonds.

Ionic Bonding

Remember from Figure 2.6, a strongly electronegative atom tends to gain electrons until its outermost shell is full. Then it is a negatively charged ion. A weakly electronegative atom tends to lose electrons until its outermost shell is full. Then it is a positively charged ion. Two atoms with a large difference in electronegativity may stay together in an **ionic bond**, which is a strong mutual attraction of two oppositely charged ions. Such bonds do not usually form by the direct transfer of an

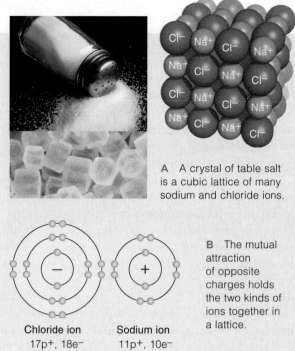

A A crystal of table salt is a cubic lattice of many sodium and chloride ions.

Chloride ion
17p+, 18e−

Sodium ion
11p+, 10e−

B The mutual attraction of opposite charges holds the two kinds of ions together in a lattice.

Figure 2.7 Animated Ionic bonds.

electron from one atom to another; rather, atoms that have already become ions stay close together because of their opposite charges.

Figure 2.7 shows crystals of table salt (sodium chloride, or NaCl). Ionic bonds in such solids hold sodium and chloride ions in an orderly, cubic arrangement.

Covalent Bonding

In a **covalent bond**, two atoms share a pair of electrons. Such bonds typically form between atoms with similar electronegativity and unpaired electrons. By sharing their electrons, each atom's vacancy becomes partially filled (Figure 2.8). Covalent bonds can be stronger than ionic bonds, but they are not always so.

Take a look at the structural formula in Table 2.1. Such formulas show how bonds connect the atoms. A line between two atoms represents a single covalent bond, in which two atoms share one pair of electrons. A simple example is molecular hydrogen (H_2), with one covalent bond between hydrogen atoms (H—H). Two lines between atoms represent a double covalent bond, in which two atoms share two pairs of electrons. Molecular oxygen (O=O) has a double covalent bond linking two oxygen atoms. Three lines indicate a triple covalent bond, in which two atoms share three pairs of electrons. A triple covalent bond links two nitrogen atoms in molecular nitrogen (N≡N).

Table 2.1 Different Ways To Represent the Same Molecule

Common name	Water	Familiar term.
Chemical name	Hydrogen oxide	Systematically describes elemental composition.
Chemical formula	H_2O	Indicates unvarying proportions of elements. Subscripts show number of atoms of an element per molecule. The absence of a subscript means one atom.
Structural formula	H—O—H	Represents each covalent bond as a single line between atoms. The bond angles may also be represented.
Structural model		Shows the positions and relative sizes of atoms.
Shell model		Shows how pairs of electrons are shared in covalent bonds.

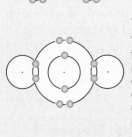

Molecular hydrogen (H—H)

Two hydrogen atoms, each with one proton, share two electrons in a nonpolar covalent bond.

Molecular oxygen (O = O)

Two oxygen atoms, each with eight protons, share four electrons in a double covalent bond.

Water molecule (H—O—H)

Two hydrogen atoms share electrons with an oxygen atom in two polar covalent bonds. The oxygen exerts a greater pull on the shared electrons, so it has a slight negative charge. Each hydrogen has a slight positive charge.

Figure 2.8 Animated Covalent bonds, in which atoms with unpaired electrons in their outermost shell become more stable by sharing electrons. Two electrons are shared in each covalent bond. When sharing is equal, the bond is nonpolar. When one atom exerts a greater pull on the electrons, the bond is polar.

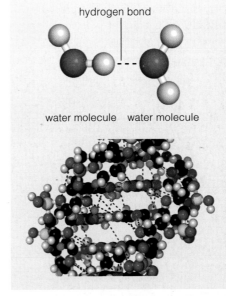

hydrogen bond

water molecule   water molecule

A   A hydrogen (H) bond is an attraction between an electronegative atom and a hydrogen atom taking part in a separate polar covalent bond.

B   Hydrogen bonds are individually weak, but many of them form. Collectively, they are strong enough to stabilize the structures of large biological molecules such as DNA, shown here.

Figure 2.9 Animated Hydrogen bonds. Hydrogen bonds form at a hydrogen atom taking part in a polar covalent bond. The hydrogen atom's slight positive charge weakly attracts an electronegative atom. As shown here, hydrogen (H) bonds can form between molecules or between different parts of the same molecule.

Some covalent bonds are **nonpolar**, meaning that the atoms participating in the bond are sharing electrons equally. There is no difference in charge between the two ends of such bonds. Nonpolar covalent bonds form between atoms with identical electronegativity. The molecular hydrogen (H_2), oxygen (O_2), and nitrogen (N_2) mentioned earlier are examples. These molecules are some of the gases that make up air.

Atoms participating in **polar** covalent bonds do not share electrons equally. Such bonds can form between atoms with a small difference in electronegativity. The atom that is more electronegative pulls the electrons a little more toward its "end" of the bond, so that atom bears a slightly negative charge. The atom at the other end of the bond bears a slightly positive charge.

For example, the water molecule shown in Table 2.1 has two polar covalent bonds (H—O—H). The oxygen atom carries a slight negative charge, but each of the hydrogen atoms carries a slight positive charge. Any such separation of charge into distinct positive and negative regions is called **polarity**. As you will see in the next section, the polarity of the water molecule is very important for the world of life.

Hydrogen Bonding

Hydrogen bonds form between polar regions of two molecules, or between two regions of the same mol-

ecule. A **hydrogen bond** is a weak attraction between a highly electronegative atom and a hydrogen atom taking part in a separate polar covalent bond.

Like ionic bonds, hydrogen bonds form by mutual attraction of opposite charges: The hydrogen atom has a slight positive charge and the other atom has a slight negative charge. However, unlike ionic bonds, hydrogen bonds do not make molecules out of atoms, so they are not chemical bonds.

Hydrogen bonds are weak. They form and break much more easily than covalent or ionic bonds. Even so, many of them form between molecules, or between different parts of a large one. Collectively, they are strong enough to stabilize the characteristic structures of large biological molecules (Figure 2.9).

Take-Home Message

How do atoms interact?

■ A chemical bond forms when the electrons of two atoms interact. Depending on the atoms, the bond may be ionic or covalent.

■ An ionic bond is a strong mutual attraction between ions of opposite charge.

■ Atoms share a pair of electrons in a covalent bond. When the atoms share electrons equally, the bond is nonpolar; when they share unequally, it is polar.

■ A hydrogen bond is an attraction between a highly electronegative atom and a hydrogen atom taking part in a different polar covalent bond.

■ Hydrogen bonds are individually weak, but are collectively strong when many of them form.

2.5 | Water's Life-Giving Properties

■ Water is essential to life because of its unique properties.
■ The unique properties of water are a result of the extensive hydrogen bonding among water molecules.

Life evolved in water. All living organisms are mostly water, many of them still live in it, and all of the chemical reactions of life are carried out in water. What is so special about water?

Polarity of the Water Molecule

The special properties of water begin with the polarity of individual water molecules. In each molecule of water, polar covalent bonds join one oxygen atom with two hydrogen atoms. Overall, the molecule has no charge, but the oxygen pulls the shared electrons a bit more than the hydrogen atoms do. Thus, each of the atoms in a water molecule carries a slight charge: The oxygen atom is slightly negative, and the hydrogen atoms are slightly positive (Figure 2.10a). This separation of charge means a water molecule is polar.

The polarity of each water molecule attracts other water molecules, and hydrogen bonds form between them in tremendous numbers (Figure 2.10b). Extensive hydrogen bonding between water molecules imparts unique properties to water that make life possible.

Water's Solvent Properties

A **solvent** is a substance, usually a liquid, that can dissolve other substances. Dissolved substances are **solutes**. Solvent molecules cluster around ions or molecules of a solute, thereby dispersing them and keeping them separated, or dissolved.

Water is a solvent. Clusters of water molecules form around the solutes in cellular fluids, tree sap, blood, the fluid in your gut, and most other fluids associated with life. When you pour table salt (NaCl) into a cup of water, the crystals of this ionically bonded solid separate into sodium ions (Na^+) and chloride ions (Cl^-). Salt dissolves in water because the negatively charged oxygen atoms of many water molecules pull on each Na^+, and the positively charged hydrogen atoms of many others pull on each Cl^- (Figure 2.11). The collective strength of many hydrogen bonds pulls the ions apart and keeps them dissolved.

Hydrogen bonds also form between water molecules and polar molecules such as sugars, so water easily dissolves polar molecules. Thus, polar molecules are **hydrophilic** (water-loving) substances. Hydrogen bonds do not form between water molecules and nonpolar molecules, such as oils, which are **hydrophobic** (water-dreading) substances. Shake a bottle filled with water and salad oil, then set it on a table. The water gathers together, and the oil clusters at the water's surface as new hydrogen bonds replace the ones broken

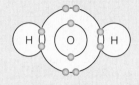

slight negative charge
on the oxygen atom

slight positive charge
on the hydrogen atoms

A The polarity of a water molecule arises because of the distribution of its electrons. The hydrogen atoms bear a slight positive charge, and the oxygen atom bears a slight negative charge.

B Many hydrogen bonds (dashed lines) that form and break rapidly keep water molecules clustered together in liquid water.

C Below 0°C (32°F), the hydrogen bonds hold water molecules rigidly in the three-dimensional lattice of ice. The molecules are less densely packed in ice than in liquid water, so ice floats on water.

The Arctic ice cap is melting because of global warming. It will probably be gone in fifty years, and so will polar bears. Polar bears must now swim farther between shrinking ice sheets, and they are drowning in alarming numbers.

Figure 2.10 Animated Water, a substance that is essential for life.

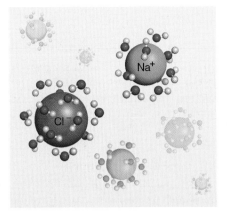

Figure 2.11 Animated Water molecules that surround an ionic solid pull its atoms apart, thereby dissolving them.

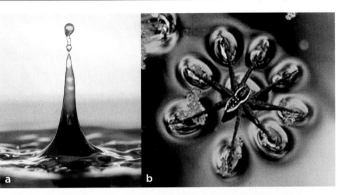

Figure 2.12 Cohesion of water. (**a**) After a pebble hits liquid water, individual molecules do not fly apart. Countless hydrogen bonds keep them together. (**b**) Cohesion keeps fishing spiders from sinking. (**c**) Water rises to the tops of plants because evaporation from leaves pulls cohesive columns of water molecules upward from the roots.

by shaking. The same interactions occur at the thin, oily membrane that separates water inside cells from water outside them. The organization of membranes—and life itself—starts with such interactions. You will read more about membranes in Chapter 5.

Water's Temperature-Stabilizing Effects

All molecules vibrate nonstop, and they move faster as they absorb heat. **Temperature** is a way to measure the energy of this molecular motion. The extensive hydrogen bonding in liquid water restricts the jiggling of water molecules. Thus, compared with other liquids, water absorbs more heat before it becomes measurably hotter. This property means that the temperature of water (and the air around it) stays relatively stable.

When the temperature of water is below its boiling point, hydrogen bonds form as fast as they break. As the water gets hotter, the molecules move faster, and individual molecules at the water's surface begin to escape into the air. By this process—**evaporation**—heat energy converts liquid water to a gas. The energy increase overcomes the attraction between water molecules, which break free.

It takes heat to convert liquid water to a gas, so the surface temperature of water decreases during evaporation. Evaporative water loss can help you and some other mammals cool off when you sweat in hot, dry weather. Sweat, which is about 99 percent water, cools the skin as it evaporates.

Below 0°C (32°F), water molecules do not jiggle enough to break hydrogen bonds, and become locked in the rigid, latticelike bonding pattern of ice (Figure 2.10c). Individual water molecules pack less densely in ice than they do in water, so ice floats on water. During cold winters, ice sheets may form near the surface of ponds, lakes, and streams. Such ice "blankets" insulate liquid water under them, so they help keep fish and other aquatic organisms from freezing.

Water's Cohesion

Another life-sustaining property of water is cohesion. **Cohesion** means that molecules resist separating from one another. You see its effect as surface tension when you toss a pebble into a pond (Figure 2.12a). Although the water ripples and sprays, individual molecules do not fly apart. Its hydrogen bonds collectively exert a continuous pull on the individual water molecules. This pull is so strong that the molecules stay together rather than spreading out in a thin film as other liquids do. Many organisms take special advantage of this unique property (Figure 2.12b).

Cohesion works inside organisms, too. For instance, plants continually absorb water as they grow. Water molecules evaporate from leaves, and replacements are pulled upward from roots (Figure 2.12c). Cohesion makes it possible for columns of liquid water to rise from roots to leaves inside narrow pipelines of vascular tissues. Section 29.3 returns to this topic.

Take-Home Message

Why is water essential to life?

■ Extensive hydrogen bonding among water molecules imparts unique properties to water that make life possible.

■ Water molecules hydrogen-bond with polar (hydrophilic) substances, dissolving them easily. They do not bond with nonpolar (hydrophobic) substances.

■ Ice is less dense than liquid water, so it floats. Ice insulates water beneath it.

■ The temperature of water is more stable than other liquids. Water also stabilizes the temperature of the air near it.

■ Cohesion keeps individual molecules of liquid water together.

■ Hydrogen ions have far-reaching effects because they are chemically active, and because there are so many of them.

■ Link to Homeostasis 1.2

The pH Scale

At any given instant in liquid water, some of the water molecules are separated into hydrogen ions (H^+) and hydroxide ions (OH^-):

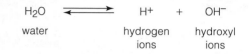

$$H_2O \rightleftarrows H^+ + OH^-$$

water   hydrogen ions   hydroxyl ions

In chemical equations such as this, arrows indicate the direction of the reaction.

pH is a measure of the number of hydrogen ions in a solution. When the number of H^+ ions is the same as the number of OH^- ions, the pH of the solution is 7, or neutral. The pH of pure water (not rainwater or tap water) is like this. The more hydrogen ions, the lower the pH. A one-unit decrease in pH corresponds to a tenfold increase in the amount of H^+ ions, and a one-unit increase corresponds to a tenfold decrease in the amount of H^+ ions. One way to get a sense of the difference is to taste dissolved baking soda (pH 9), distilled water (pH 7), and lemon juice (pH 2). A pH scale that ranges from 0 to 14 is shown in Figure 2.13.

Nearly all of life's chemistry occurs near pH 7. Most of your body's internal environment (tissue fluids and blood) is between pH 7.3 and 7.5.

How Do Acids and Bases Differ?

Substances called **acids** donate hydrogen ions as they dissolve in water. **Bases** accept hydrogen ions. Acidic solutions, such as lemon juice and coffee, contain more H^+ than OH^-, so their pH is below 7. Basic solutions, such as seawater and hand soap, contain more OH^- than H^+. Basic, or alkaline, solutions have a pH greater than 7.

Acids and bases can be weak or strong. Weak acids, such as carbonic acid (H_2CO_3), are stingy H^+ donors. Strong acids give up more H^+ ions. One example is hydrochloric acid (HCl), which separates into H^+ and Cl^- very easily in water:

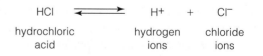

$$HCl \rightleftarrows H^+ + Cl^-$$

hydrochloric acid   hydrogen ions   chloride ions

Inside your stomach, the H^+ from HCl makes gastric fluid acidic (pH 1–2). The acidity activates enzymes that digest proteins in your food.

Acids or bases that accumulate in ecosystems can kill organisms. For instance, fossil fuel emissions and nitrogen-containing fertilizers release strong acids into

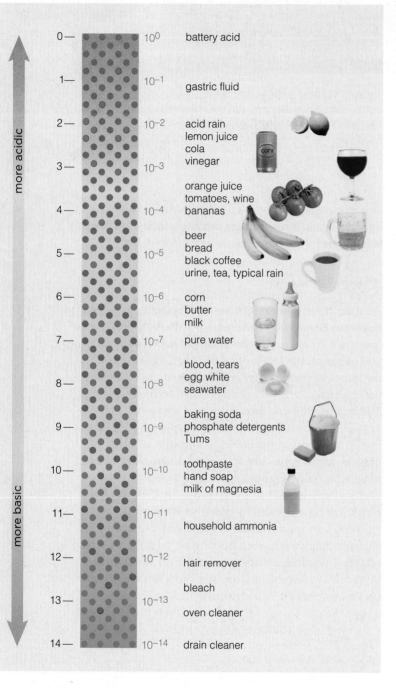

0 —	10^0	battery acid
1 —	10^{-1}	gastric fluid
2 —	10^{-2}	acid rain lemon juice cola vinegar
3 —	10^{-3}	
		orange juice tomatoes, wine bananas
4 —	10^{-4}	
		beer bread black coffee urine, tea, typical rain
5 —	10^{-5}	
6 —	10^{-6}	corn butter milk
7 —	10^{-7}	pure water
		blood, tears egg white seawater
8 —	10^{-8}	
9 —	10^{-9}	baking soda phosphate detergents Tums
10 —	10^{-10}	toothpaste hand soap milk of magnesia
11 —	10^{-11}	household ammonia
12 —	10^{-12}	hair remover
		bleach
13 —	10^{-13}	oven cleaner
14 —	10^{-14}	drain cleaner

more acidic → more basic

Figure 2.13 Animated A pH scale. Here, *red* dots signify hydrogen ions (H^+) and *blue* dots signify hydroxyl ions (OH^-). Also shown are approximate pH values for some common solutions.

This pH scale ranges from 0 (most acidic) to 14 (most basic). A change of one unit on the scale corresponds to a tenfold change in the amount of H^+ ions (*blue* numbers).

Figure It Out: What is the approximate pH of cola?

Answer: 2.5

Figure 2.14 Emissions of sulfur dioxide from a coal-burning power plant. Airborne pollutants such as sulfur dioxide dissolve in water vapor and form acidic solutions. They are a component of acid rain. The far-right photograph shows how acid rain can corrode stone sculptures.

the atmosphere. The acids lower the pH of rain (Figure 2.14). Some ecosystems are being damaged by such acid rain, which changes the composition of water and soil. Organisms in these regions are being harmed by the changes. We return to this topic in Section 48.2.

Salts and Water

A **salt** is a compound that dissolves easily in water and releases ions other than H^+ and OH^-. For example, when dissolved in water, the salt sodium chloride separates into sodium ions and chloride ions:

$$NaCl \longrightarrow Na^+ + Cl^-$$

sodium   sodium   chloride
chloride   ions   ions

Many ions are important components of cellular processes. For example, sodium, potassium, and calcium ions are critical for nerve and muscle cell function. As another example, potassium ions help plants minimize water loss on hot, dry days.

Buffers Against Shifts in pH

Cells must respond quickly to even slight shifts in pH because most enzymes and other biological molecules can function properly only within a narrow pH range. Even a slight deviation from that range can halt cellular processes completely.

Body fluids stay at a consistent pH because they are buffered. A **buffer system** is a set of chemicals, often a weak acid or base and its salt, that can keep the pH of a solution stable. It works because the two chemicals donate and accept ions that contribute to pH.

For example, when a base is added to an unbuffered fluid, the number of OH^- ions increases, so the pH rises. However, when a base is added to a buffered fluid, the acid component of the buffer releases H^+ ions. These combine with the extra OH^- ions, forming water, which does not affect pH. So, the buffered fluid's pH stays the same even when base is added.

Carbon dioxide, a gas that forms in many reactions, takes part in an important buffer system. It becomes carbonic acid when it dissolves in the water component of human blood:

$$H_2O + CO_2 \longrightarrow H_2CO_3$$

carbon dioxide   carbonic acid

The carbonic acid can separate into hydrogen ions and bicarbonate ions:

$$H_2CO_3 \longrightarrow H^+ + HCO_3^-$$

carbonic acid   bicarbonate

This easily-reversed reaction constitutes the buffer system. Any excess OH^- combines with the H^+ to form water, which does not contribute to pH. Any excess H^+ combines with the bicarbonate; thus bonded, the hydrogen does not affect pH:

$$H^+ + HCO_3^- \longrightarrow H_2CO_3$$

bicarbonate   carbonic acid

Together, these reactions keep the blood pH between 7.3 and 7.5—but only up to a point. A buffer system can neutralize only so many ions. Even slightly more than that limit causes the pH to swing widely.

A buffer system failure in a biological system can be catastrophic. In acute respiratory acidosis, carbon dioxide accumulates, and excess carbonic acid forms in blood. The resulting decline in blood pH may cause an individual to enter a coma, a level of unconsciousness that is dangerous. Alkalosis, a potentially lethal rise in blood pH, can also bring on coma. Even an increase to 7.8 can result in tetany, or prolonged muscle spasm.

Take-Home Message

Why are hydrogen ions important in biology?

■ Hydrogen ions contribute to pH. Acids release hydrogen ions in water; bases accept them. Salts release ions other than H^+ and OH^-.

■ Buffer systems keep the pH of body fluids stable. They are part of homeostasis.

What Are You Worth?

Contaminant or nutrient? An average human body contains highly toxic elements such as lead, arsenic, mercury, selenium, nickel, and even a few uranium atoms. The presence of these elements in the body is usually assumed to be the aftermath of environmental pollutants, but occasionally we discover that one of them has a vital function. For example, recently we found that having too little selenium can cause heart problems and thyroid disorders, so it may be part of some biological system we haven't yet unraveled.

The average body contains a substantial amount of fluorine, but as yet we know of no natural metabolic role for this element. Fluorine can substitute for other elements in biological molecules,

How would you vote?

When fluorine replaces calcium in teeth and bones, it changes the structural properties of these body parts. One effect is fewer cavities. Many communities add fluoride to drinking water. Do you want it in yours? See CengageNOW for details, then vote online.

but the substitution tends to make the molecules toxic. Several kinds of predator-deterring plant toxins are simple biological molecules with fluorine substituted for other elements.

Summary

Section 2.1 Most **atoms** have **electrons**, which have a negative **charge**. Electrons move around a **nucleus** of positively charged **protons** and, except in the case of hydrogen, uncharged **neutrons**. Atoms of an **element** have the same number of protons—the **atomic number** (Table 2.2). A **periodic table** lists all of the elements. We refer to **isotopes** of an element by their **mass number**.

Table 2.2 Summary of Players in the Chemistry of Life

Atom	Particles that are basic building blocks of all matter; the smallest unit that retains an element's properties
Element	Pure substance that consists entirely of atoms with the same, characteristic number of protons
Proton (p^+)	Positively charged particle of an atom's nucleus
Electron (e^-)	Negatively charged particle that can occupy a volume of space (orbital) around an atom's nucleus
Neutron	Uncharged particle of an atom's nucleus
Isotope	One of two or more forms of an element, the atoms of which differ in the number of neutrons
Radioisotope	Unstable isotope that emits particles and energy when its nucleus disintegrates
Tracer	Molecule that has a detectable substance (such as a radioisotope) attached
Ion	Atom that carries a charge after it has gained or lost one or more electrons
Molecule	Two or more atoms joined in a chemical bond
Compound	Molecule of two or more different elements in unvarying proportions (for example, water)
Mixture	Intermingling of two or more elements or compounds in proportions that can vary
Solute	Molecule or ion dissolved in a solvent
Acid	Substance that releases H^+ when dissolved in water
Base	Substance that accepts H^+ when dissolved in water
Salt	Substance that releases ions other than H^+ or OH^- when dissolved in water

Section 2.2 Researchers make **tracers** with detectable substances such as **radioisotopes**, which emit particles and energy as they **decay** spontaneously.

■ *Use the animation on CengageNOW to learn how radioisotopes are used in making PET scans.*

Section 2.3 We use **shell models** to view an atom's electron structure. Atoms with different numbers of electrons and protons are **ions**. Atoms with vacancies tend to interact with other atoms by donating, accepting, or sharing electrons. They form different **chemical bonds** depending on their **electronegativity**. A **compound** is a **molecule** of different elements. **Mixtures** are intermingled substances.

■ *Use the animation and interaction on CengageNOW to study electron distribution and the shell model.*

Section 2.4 An **ionic bond** is a very strong association between ions of opposite charge. Two atoms share a pair of electrons in a **covalent bond**, which may be **nonpolar** or **polar** (**polarity** is a separation of charge). **Hydrogen bonds** are weaker than either ionic or covalent bonds.

■ *Use the animation on CengageNOW to compare the types of chemical bonds in biological molecules.*

Section 2.5 **Evaporation** helps liquid water stabilize **temperature**. **Hydrophilic** substances dissolve easily in water; **hydrophobic** substances do not. **Solutes** are substances dissolved in water or another **solvent**. **Cohesion** keeps water molecules together.

■ *Use the animation on CengageNOW to view the structure of the water molecule and properties of liquid water.*

Section 2.6 **pH** reflects the number of hydrogen ions (H^+) in a solution. Typical pH scales range from 0 (most acidic) to 14 (most basic or alkaline). At neutral pH (7), the amounts of H^+ and OH^- ions are the same.

Salts are compounds that release ions other than H^+ and OH^- in water. **Acids** release H^+; **bases** accept H^+. A **buffer system** keeps a solution within a consistent range of pH. Most biological processes are buffered; they work only within a narrow pH range, usually near pH 7.

■ *Use the interaction on CengageNOW to investigate the pH of common solutions.*

Data Analysis Exercise

Living and nonliving things have the same kinds of atoms joined together as molecules, but those molecules differ in their proportions of elements and in how the atoms of those elements are arranged. The three charts in Figure 2.15 compare the proportions of some elements in the human body, Earth's crust, and seawater.

1. Which is the most abundant element in dirt? In the human body? In seawater?

2. What percentage of seawater is oxygen? Hydrogen? How many atoms of hydrogen are there for each atom of oxygen in seawater? In which molecule are hydrogen and oxygen found in that exact proportion?

3. How many atoms of chlorine are there for every atom of sodium in seawater? What common molecule has one atom of chlorine for every atom of sodium?

Human		Earth		Seawater	
Hydrogen	62.0%	Hydrogen	3.1%	Hydrogen	66.0%
Oxygen	24.0	Oxygen	60.0	Oxygen	33.0
Carbon	12.0	Carbon	0.3	Carbon	< 0.1
Nitrogen	1.2	Nitrogen	< 0.1	Nitrogen	< 0.1
Phosphorus	0.2	Phosphorus	< 0.1	Phosphorus	< 0.1
Calcium	0.2	Calcium	2.6	Calcium	< 0.1
Sodium	< 0.1	Sodium	< 0.1	Sodium	0.3
Potassium	< 0.1	Potassium	0.8	Potassium	< 0.1
Chlorine	< 0.1	Chlorine	< 0.1	Chlorine	0.3

Figure 2.15 Comparison of the abundance of some elements in a human, Earth's crust, and typical seawater. Each number is the percent of the total number of atoms in each source. For instance, 120 of every 1,000 atoms in a human body are carbon, compared with only 3 carbon atoms in every 1,000 atoms of dirt.

Self-Quiz

Answers in Appendix III

1. A(n) _____ is a molecule into which a radioisotope has been incorporated.

2. An ion is an atom that has _____ .
   a. the same number of electrons and protons
   b. a different number of electrons and protons
   c. a and b are correct

3. A(n) _____ forms when atoms of two or more elements bond covalently.

4. The measure of an atom's ability to pull electrons away from another atom is called _____ .

5. Atoms share electrons unequally in a(n) _____ bond.

6. Symbols for the elements are arranged according to _____ in the periodic table of the elements.

7. Liquid water has _____ .
   a. tracers
   b. a profusion of hydrogen bonds
   c. cohesion
   d. resistance to increases in temperature
   e. b through d
   f. all of the above

8. A(n) _____ substance repels water.

9. Hydrogen ions (H+) are _____ .
   a. indicated by pH
   b. protons
   c. dissolved in blood
   d. all of the above

10. A(n) _____ is dissolved in a solvent.

11. When dissolved in water, a(n) _____ donates H+.

12. A salt releases ions other than _____ in water.

13. A(n) _____ is a chemical partnership between a weak acid or base and its salt.

14. Match the terms with their most suitable description.
   ___ hydrophilic
   ___ atomic number
   ___ mass number
   ___ temperature
   a. measure of molecular motion
   b. number of protons in nucleus
   c. polar; readily dissolves in water
   d. number of protons and neutrons in nucleus

■ *Visit CengageNOW for additional questions.*

Critical Thinking

1. Alchemists were medieval scholars and philosophers who were the forerunners of modern-day chemists. Many spent their lives trying to transform lead (atomic number 82) into gold (atomic number 79). Explain why they never did succeed in that endeavor.

2. Meats are often "cured," or salted, dried, smoked, pickled, or treated with chemicals that can delay spoilage. Ever since the mid-1800s, sodium nitrite ($NaNO_2$) has been used in processed meat products such as hot dogs, bologna, sausages, jerky, bacon, and ham. Nitrites prevent growth of *Clostridium botulinum*. If ingested, this bacterium can cause a form of food poisoning called botulism.

   In water, sodium nitrite separates into sodium ions (Na^+) and nitrite ions (NO_2^-), which are called nitrites. Nitrites are rapidly converted to nitric oxide (NO), the compound that gives nitrites their preservative qualities. Eating preserved meats increases the risk of cancer, but nitrites may not be at fault. It turns out that nitric oxide has several important functions, including blood vessel dilation (for example, inside a penis during an erection), cell-to-cell signaling, and antimicrobial activities of the immune system. Draw a shell model for nitric oxide and then use it to explain why the molecule is so reactive.

3. Ozone is a chemically active form of oxygen gas. High in Earth's atmosphere, it forms a layer that absorbs about 98 percent of the sun's harmful rays. Oxygen gas consists of two covalently bonded oxygen atoms: O=O. Ozone has three covalently bonded oxygen atoms: O=O—O. Ozone reacts easily with many substances, and gives up an oxygen atom and releases gaseous oxygen (O=O). From what you know about chemistry, why do you suppose ozone is so reactive?

4. David, an inquisitive three-year-old, poked his fingers into warm water in a metal pan on the stove and did not sense anything hot. Then he touched the pan itself and got a nasty burn. Explain why water in a metal pan heats up far more slowly than the pan itself.

5. Some undiluted acids are more corrosive when diluted with water. That is why lab workers are told to wipe off splashes with a towel before washing. Explain.

3 Molecules of Life

IMPACTS, ISSUES | Fear of Frying

The human body requires about one tablespoon of fat each day to remain healthy, but most of us eat far more than that. The average American consumes the equivalent of one stick of butter per day—100 pounds of fat per year—which may be part of the reason why the average American is overweight.

Being overweight increases one's risk for many diseases and health conditions. However, which type of fat we eat may be more important than how much fat we eat. Fats are more than just inert molecules that accumulate in strategic areas of our bodies if we eat too much of them. They are major constituents of cell membranes, and as such they have powerful effects on cell function.

The typical fat molecule has three tails—long carbon chains called fatty acids. Different fats have different fatty acid components. Those with a certain type of double bond in one or more of their fatty acids are called *trans* fats (Figure 3.1). Small amounts of *trans* fats occur naturally in red meat and dairy products, but most of the *trans* fats humans consume come from partially hydrogenated vegetable oil, an artificial food product.

Hydrogenation, a manufacturing process that adds hydrogen atoms to carbons, changes liquid vegetable oils into solid fats. Procter & Gamble Co. developed partially hydrogenated vegetable oil in 1908 as a substitute for the more expensive solid animal fats they were using to make candles. However, the demand for candles began to wane as more households in the United States became wired for electricity, and P & G began to look for another way to sell its proprietary fat. Partially hydrogenated vegetable oil looks a lot like lard, so in 1911 the company began marketing it as a revolutionary new food—a solid cooking fat with a long shelf life, mild flavor, and lower cost than lard or butter.

By the mid-1950s, hydrogenated vegetable oil had become a major part of the American diet. It was (and still is) found in a tremendous buffet of manufactured and fast foods: butter substitutes, cookies, crackers, cakes and pancakes, peanut butter, pies, doughnuts, muffins, chips, granola bars, breakfast bars, chocolate, microwave popcorn, pizzas, burritos, french fries, chicken nuggets, fish sticks, and so on.

For decades, hydrogenated vegetable oil was considered to be a more healthy alternative to animal fats. We now know that *trans* fats in hydrogenated vegetable oils raise the level of cholesterol in our blood more than any other fat, and they directly alter the function of our arteries and veins.

The effects of such changes are serious. Eating as little as 2 grams per day of hydrogenated vegetable oils increases a person's risk of atherosclerosis (hardening of the arteries), heart attack, and diabetes. A single serving of french fries made with hydrogenated vegetable oil contains about 5 grams of *trans* fats.

With this chapter, we introduce you to the chemistry of life. Although every living thing consists of the same basic kinds of molecules—carbohydrates, lipids, proteins, and nucleic acids—small differences in the way those molecules are put together often have big results.

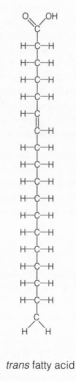

trans fatty acid

See the video! Figure 3.1 *Trans* fats. The arrangement of hydrogen atoms around the carbon–carbon double bond in the middle of a *trans* fatty acid makes it a very unhealthy food. Consider skipping the french fries.

Key Concepts

Structure dictates function

We define cells partly by their capacity to build complex carbohydrates and lipids, proteins, and nucleic acids. All of these organic compounds have functional groups attached to a backbone of carbon atoms. **Sections 3.1, 3.2**

Carbohydrates

Carbohydrates are the most abundant biological molecules. They function as energy reservoirs and structural materials. Different types of complex carbohydrates are built from the same subunits of simple sugars, bonded in different patterns. **Section 3.3**

Lipids

Lipids function as energy reservoirs and as waterproofing or lubricating substances. Some are remodeled into other molecules. Lipids are the main structural component of all cell membranes. **Section 3.4**

Proteins

Structurally and functionally, proteins are the most diverse molecules of life. They include enzymes, structural materials, signaling molecules, and transporters. A protein's function arises directly from its structure. **Sections 3.5, 3.6**

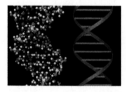

Nucleotides and nucleic acids

Nucleotides have major metabolic roles and are building blocks of nucleic acids. Two kinds of nucleic acids, DNA and RNA, interact as the cell's system of storing, retrieving, and translating information about building proteins. **Section 3.7**

Links to Earlier Concepts

- Having learned about atoms, you are about to enter the next level of organization in nature: the molecules of life. Keep the big picture in mind by reviewing Section 1.1.

- You will be building on your understanding of how electrons are arranged in atoms (2.3) as well as the nature of covalent bonding and hydrogen bonding (2.4).

- Here again, you will consider one of the consequences of mutation in DNA (1.4), this time with sickle-cell anemia as the example.

How would you vote? All packaged foods in the United States now list *trans* fat content, but may be marked "zero grams of *trans* fats" even if a serving contains up to half a gram of it. Should hydrogenated vegetable oils be banned from all food? See CengageNOW for details, then vote online.

Organic Molecules

- All of the molecules of life are built with carbon atoms.
- We can use different models to highlight different aspects of the same molecule.
- Links to Elements 2.1, Covalent bonds 2.4

Carbon—The Stuff of Life

Living things are mainly oxygen, hydrogen, and carbon. Most of the oxygen and hydrogen are in the form of water. Put water aside, and carbon makes up more than half of what is left.

The carbon in living organisms is part of the molecules of life—complex carbohydrates, lipids, proteins, and nucleic acids. These molecules consist primarily of hydrogen and carbon atoms, so they are **organic**. The term is a holdover from a time when such molecules were thought to be made only by living things, as opposed to the "inorganic" molecules that formed by nonliving processes. The term persists, even though we now know that organic compounds were present on Earth long before organisms were, and we can also make them in laboratories.

Carbon's importance to life starts with its versatile bonding behavior. Each carbon atom can form covalent bonds with one, two, three, or four other atoms. Depending on the other elements in the resulting molecule, such bonds may be polar or nonpolar. Many organic compounds have a backbone—a chain of carbon atoms—to which other atoms attach. The ends of a backbone may join so that the carbon chain forms one or more ring structures (Figure 3.2). Such versatility means that carbon atoms can be assembled and remodeled into a variety of organic compounds.

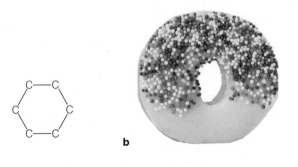

Figure 3.2 Carbon rings. (**a**) Carbon's versatile bonding behavior allows it to form a variety of structures, including rings. (**b**) Carbon rings form the framework of many sugars, starches, and fats, such as those found in doughnuts.

Representing Structures of Organic Molecules

Any molecule's structure can be depicted using different kinds of molecular models. Such models allow us to see different characteristics of the same molecule.

For example, structural models such as the one at *right* show how all the atoms in a molecule connect to one another. In such models, each line indicates one covalent bond. A double line (=) indicates a double bond; a triple line (≡) indicates a triple bond. Some of the atoms or bonds in a molecule may be implied but not shown. Hydrogen atoms bonded to a carbon backbone may also be omitted, and other atoms as well.

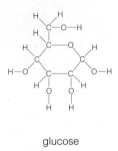

glucose

Carbon ring structures such as the ones that occur in glucose and other sugars are often represented as polygons. If no atom is shown at a corner or at the end of a bond, a carbon atom is implied there:

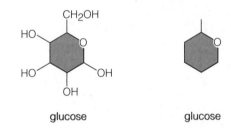

glucose   glucose

Ball-and-stick models such as the one at *right* show the positions of the atoms in three dimensions. Single, double, and triple covalent bonds are all shown as one stick connecting two balls, which represent atoms. Ball size reflects relative size of an atom. Elements are usually coded by color:

glucose

carbon   hydrogen   oxygen   nitrogen   phosphorus

Space-filling models such as the one at *right* show how atoms that are sharing electrons overlap. The elements in space-filling models are coded using the same color scheme as the ones in ball-and-stick models.

glucose

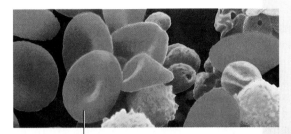

red blood cell

Figure 3.3 shows three different ways to represent the same molecule, hemoglobin, a protein that colors your blood red. Hemoglobin transports oxygen to tissues throughout the body of all vertebrates (animals that have a backbone). A ball-and-stick or space-filling model of such a large molecule can appear very complicated if all of the atoms are included. The space-filling model in Figure 3.3*a* is an example.

To reduce visual complexity, other types of models omit individual atoms. Surface models of large molecules can reveal large-scale features, such as folds or pockets, that can be difficult to see when individual atoms are shown. For example, in the surface model of hemoglobin in Figure 3.3*b*, you can see folds of the molecule that cradle two hemes. Hemes are complex carbon ring structures that often have an iron atom at their center. They are part of many important proteins that you will encounter in this book.

Very large molecules such as hemoglobin are often shown as ribbon models. Such models highlight different features of the structure, such as coils or sheets. In a ribbon model of hemoglobin (Figure 3.3*c*), you can see that the protein consists of four coiled components, each of which folds around a heme.

Such structural details are clues to how a molecule functions. For example, hemoglobin, which is the main oxygen-carrier in vertebrate blood, has four hemes. Oxygen binds at the hemes, so each hemoglobin molecule can carry up to four molecules of oxygen.

Take-Home Message

How are all of the molecules of life alike?

■ Carbohydrates, lipids, proteins, and nucleic acids are organic molecules, which consist mainly of carbon and hydrogen atoms.

■ The structure of an organic molecule starts with its carbon backbone, a chain of carbon atoms that may form a ring.

■ We use different models to represent different characteristics of a molecule's structure. Considering a molecule's structural features gives us insight into how it functions.

A A space-filling model of hemoglobin shows the complexity of the molecule.

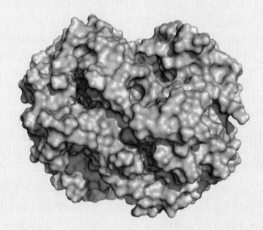

B A surface model of the same molecule reveals crevices and folds that are important for its function. Heme groups, in *red*, are cradled in pockets of the molecule.

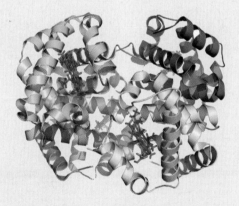

C A ribbon model of hemoglobin shows all four heme groups, also in *red*, held in place by the molecule's coils.

Figure 3.3 Visualizing the structure of hemoglobin, the oxygen-transporting molecule in red blood cells (*top left*). Models that show individual atoms usually depict them color-coded by element. Other models may be shown in various colors, depending on which features are highlighted.

3.2 From Structure to Function

- The function of organic molecules in biological systems begins with their structure.

- Links to Ions 2.3, Polarity 2.4, Acids and bases 2.6

All biological systems are based on the same organic molecules—a legacy of life's common origin—but the details of those molecules can differ among organisms. Remember, depending on the way carbon atoms bond together, they can form diamond, the hardest mineral, or graphite, one of the softest (Section 2.4). Similarly, the building blocks of carbohydrates, lipids, proteins, and nucleic acids bond together in different arrangements to form different molecules.

Functional Groups

An organic molecule that consists only of hydrogen and carbon atoms is called a hydrocarbon. Methane, the simplest hydrocarbon, is one carbon atom bonded to four hydrogen atoms. Most of the molecules of life have at least one **functional group**: a cluster of atoms covalently bonded to a carbon atom of an organic molecule. Functional groups impart specific chemical properties to a molecule, such as polarity or acidity. Figure 3.4 lists a few functional groups that are common in carbohydrates, lipids, proteins, and nucleic acids.

For example, alcohols are a class of organic compounds that have hydroxyl groups (—OH). These polar functional groups can form hydrogen bonds, so alcohols (at least the small ones) dissolve quickly in water. Larger alcohols do not dissolve as easily, because their long, nonpolar hydrocarbon chains repel water. Fatty acids also are like this, which is why lipids that have fatty acid tails do not dissolve easily in water.

Methyl groups impart nonpolar character. Reactive carbonyl groups (—C═O) are part of fats and carbohydrates. Carboxyl groups (—COOH) make amino acids and fatty acids acidic. Amine groups are basic. ATP releases chemical energy when it donates a phosphate group (PO_4) to another molecule. DNA and RNA also contain phosphate groups. Bonds between sulfhydryl

Group	Character	Location	Structure
hydroxyl	polar	amino acids; sugars and other alcohols	—OH
methyl	nonpolar	fatty acids, some amino acids	H—C(H)—H
carbonyl	polar, reactive	sugars, amino acids, nucleotides	—C—H (aldehyde) / —C— (ketone), O
carboxyl	acidic	amino acids, fatty acids, carbohydrates	—C—OH, O / —C—O⁻, O (ionized)
amine	basic	amino acids, some nucleotide bases	—N—H, H, H / —NH⁺, H, H (ionized)
phosphate	high energy, polar	nucleotides (e.g., ATP); DNA and RNA; many proteins; phospholipids	O⁻ / —O—P—O⁻, O / (P) icon
sulfhydryl	forms disulfide bridges	cysteine (an amino acid)	—SH / —S—S— (disulfide bridge)

Figure 3.4 Animated Common functional groups in biological molecules, with examples of where they occur. Because such groups impart specific chemical characteristics to organic compounds, they are an important part of why the molecules of life function as they do.

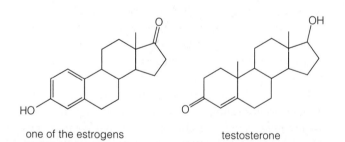

one of the estrogens   testosterone

female wood duck   male wood duck

Figure 3.5 Estrogen and testosterone, sex hormones that cause differences in traits between males and females of many species such as wood ducks (*Aix sponsa*). **Figure It Out:** Which functional groups differ between these hormones? *Answer: The hydroxyl and carbonyl groups differ in position, and testosterone has an extra methyl group.*

Table 3.1 What Cells Do to Organic Compounds

Type of Reaction	What Happens
Condensation	Two molecules covalently bond into a larger one.
Cleavage	A molecule splits into two smaller ones. Hydrolysis is an example.
Functional group transfer	A functional group is transferred from one molecule to another.
Electron transfer	Electrons are transferred from one molecule to another.
Rearrangement	Juggling of covalent bonds converts one organic compound into another.

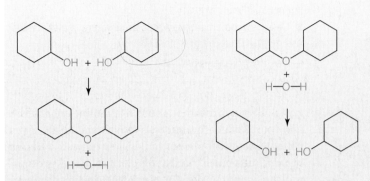

A Condensation. An —OH group from one molecule combines with an H atom from another. Water forms as the two molecules bond covalently.

B Hydrolysis. A molecule splits, then an —OH group and an H atom from a water molecule become attached to sites exposed by the reaction.

Figure 3.6 Animated Two examples of what happens to the organic molecules in cells. (**a**) In condensation, two molecules are covalently bonded into a larger one. (**b**) In hydrolysis, a water-requiring cleavage reaction splits a larger molecule into two smaller molecules.

groups (—SH) stabilize the structure of many proteins. Heat and some kinds of chemicals can temporarily break sulfhydryl bonds in human hair, which is why we can curl straight hair and straighten curly hair.

How much can one functional group do? Consider a seemingly minor difference in the functional groups of two structurally similar sex hormones (Figure 3.5). Early on, an embryo of a wood duck, human, or any other vertebrate is neither male nor female. If it starts making the hormone testosterone, a set of tubes and ducts will become male sex organs and male traits will develop. Without testosterone, those ducts and tubes become female sex organs, and hormones called estrogens will guide the development of female traits.

What Cells Do to Organic Compounds

Metabolism refers to activities by which cells acquire and use energy as they construct, rearrange, and split organic compounds. These activities help each cell stay alive, grow, and reproduce. They require enzymes— proteins that make reactions proceed faster than they would on their own. Some of the most common metabolic reactions are listed in Table 3.1. We will revisit these reactions in Chapter 6. For now, just start thinking about two of them.

With **condensation**, two molecules covalently bond into a larger one. Water usually forms as a product of condensation when enzymes remove an —OH group

from one of the molecules and a hydrogen atom from the other (Figure 3.6a). Some large molecules such as starch form by repeated condensation reactions.

Cleavage reactions split large molecules into smaller ones. One type of cleavage reaction, **hydrolysis**, is the reverse of condensation (Figure 3.6b). Enzymes break a bond by attaching a hydroxyl group to one atom and a hydrogen to the other. The —OH and —H are derived from a water molecule.

Cells maintain pools of small organic molecules— simple sugars, fatty acids, amino acids, and nucleotides. Some of these molecules are sources of energy. Others are used as subunits, or **monomers**, to build larger molecules that are the structural and functional parts of cells. These larger molecules, or **polymers**, are chains of monomers. When cells break down a polymer, the released monomers may be used for energy, or they may reenter cellular pools.

Take-Home Message

How do organic molecules work in living systems?

■ An organic molecule's structure dictates its function in biological systems.

■ Functional groups impart certain chemical characteristics to organic molecules. Such groups contribute to the function of biological molecules.

■ By reactions such as condensation, cells assemble large molecules from smaller subunits of simple sugars, fatty acids, amino acids, and nucleotides.

■ By reactions such as hydrolysis, cells split large organic molecules into smaller ones, and convert one type of molecule to another.

3.3 | Carbohydrates

■ Carbohydrates are the most plentiful biological molecules in the biosphere.

■ Cells use some carbohydrates as structural materials; they use others for stored or instant energy.

■ Link to Hydrogen bonds 2.4

Long-chain hydrocarbons such as gasoline are an excellent source of energy, but cells (which are mostly water) cannot use hydrophobic molecules. Instead, cells use organic molecules that have polar functional groups—molecules that are easily assembled and broken apart inside a cell's watery interior.

Carbohydrates are organic compounds that consist of carbon, hydrogen, and oxygen in a 1:2:1 ratio. Cells use different kinds as structural materials and as sources of instant energy. The three main types of carbohydrates in living systems are monosaccharides, oligosaccharides, and polysaccharides.

Simple Sugars

"Saccharide" is from a Greek word that means sugar. Monosaccharides (one sugar unit) are the simplest of the carbohydrates. Common monosaccharides have a backbone of five or six carbon atoms, one ketone or aldehyde group, and two or more hydroxyl groups. Most monosaccharides are water soluble, so they are easily transported throughout the internal environments of all organisms.

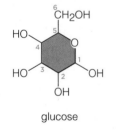

glucose

Sugars that are part of DNA and RNA are monosaccharides with five carbon atoms. Glucose (at *left*) has six carbons. Cells use glucose as an energy source or as a structural material. They also use it as a precursor, or parent molecule, that they remodel into other molecules. For example, vitamin C is derived from glucose.

Short-Chain Carbohydrates

An oligosaccharide is a short chain of covalently bonded monosaccharides (*oligo–* means a few). As examples, disaccharides consist of two sugar monomers. The lactose in milk is a disaccharide, with one glucose and one galactose unit. Sucrose, the most plentiful sugar in nature, has a glucose and a fructose unit (Figure 3.7). Sucrose extracted from sugarcane or sugar beets is our table sugar. Oligosaccharides with three or more sugar units are often attached to lipids or proteins that have important functions in immunity.

Complex Carbohydrates

The "complex" carbohydrates, or polysaccharides, are straight or branched chains of many sugar monomers —often hundreds or thousands. There may be one type or many types of monomers in a polysaccharide. The most common polysaccharides are cellulose, glycogen, and starch. All consist of glucose monomers, but they differ in their chemical properties. Why? The answer begins with differences in patterns of covalent bonding that link their glucose units (Figure 3.8).

For example, the covalent bonding pattern of starch makes the molecule coil like a spiral staircase (Figure 3.8*b*). Starch does not dissolve easily in water, so it resists hydrolysis. This stability is a reason why starch is used to store chemical energy in the watery, enzyme-filled interior of plant cells.

Most plants make much more glucose than they can use. The excess is stored as starch, in roots, stems, and leaves. However, because it is insoluble, starch

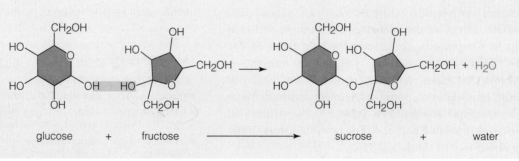

glucose   +   fructose   ⟶   sucrose   +   water

Figure 3.7 Animated The synthesis of a sucrose molecule is an example of a condensation reaction. You are already familiar with sucrose—it is common table sugar.

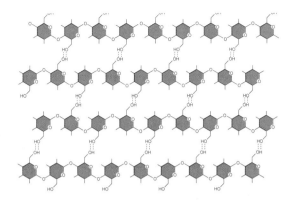

a Cellulose, a structural component of plants. Chains of glucose units stretch side by side and hydrogen bond at many —OH groups. The hydrogen bonds stabilize the chains in tight bundles that form long fibers. Very few types of organisms can digest this tough, insoluble material.

b In amylose, one type of starch, a series of glucose units form a chain that coils. Starch is the main energy reserve in plants, which store it in their roots, stems, leaves, fruits, and seeds (such as coconuts).

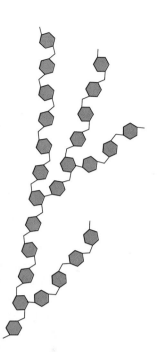

c Glycogen. In animals, this polysaccharide functions as an energy reservoir. It is especially abundant in the liver and muscles of active animals, including people.

Figure 3.8 Structure of (**a**) cellulose, (**b**) starch, and (**c**) glycogen, and their typical locations in a few organisms. All three carbohydrates consist only of glucose units, but the different bonding patterns that link the subunits result in substances with very different properties.

cannot be transported out of cells and distributed to other parts of the plant. When sugars are in short supply, hydrolysis enzymes nibble at the bonds between starch's sugar monomers. Cells make the disaccharide sucrose from the released glucose molecules. Sucrose is soluble and easily transported.

Cellulose, the major structural material of plants, may be the most abundant organic molecule in the biosphere. Glucose chains stretch side by side (Figure 3.8a). Hydrogen bonding between the chains stabilizes them in tight, sturdy bundles. Plant cell walls contain long cellulose fibers. Like steel rods inside reinforced concrete pillars, the tough fibers help tall stems resist wind and other forms of mechanical stress.

In animals, glycogen is the sugar-storage equivalent of starch in plants (Figure 3.8c). Muscle and liver cells store it. When the sugar level in blood falls, the liver cells break down glycogen, and the released glucose subunits enter the blood.

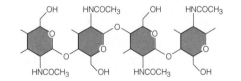

Figure 3.9 Chitin. This polysaccharide strengthens the hard parts of many small animals, such as crabs.

Chitin is a polysaccharide with nitrogen-containing groups on its many glucose monomers (Figure 3.9). Chitin strengthens hard parts of many animals, including the outer cuticle of crabs, beetles, and ticks. It also reinforces the cell wall of many fungi.

Take-Home Message

What are carbohydrates?

■ Subunits of simple carbohydrates (sugars), arranged in different ways, form various types of complex carbohydrates.

■ Cells use carbohydrates for energy, storage, or as structural materials.

■ Lipids function as the body's major energy reservoir, and as the structural foundation of cell membranes.

Lipids are fatty, oily, or waxy organic compounds that are insoluble in water. Many lipids incorporate **fatty acids**: simple organic compounds that have a carboxyl group joined to a backbone of four to thirty-six carbon atoms (Figure 3.10).

Fats

Fats are lipids with one, two, or three fatty acids that dangle like tails from a small alcohol called glycerol. Most neutral fats, such as butter and vegetable oils, are triglycerides. **Triglycerides** are fats with three fatty acid tails linked to the glycerol (Figure 3.11). Triglycerides are the most abundant energy source in vertebrate bodies, and the richest. Gram for gram, triglycerides hold more than twice the energy of glycogen. Triglycerides are concentrated in adipose tissue that insulates and cushions parts of the body.

Figure 3.10 Examples of fatty acids. (**a**) The backbone of stearic acid is fully saturated with hydrogen atoms. (**b**) Oleic acid, with a double bond in its backbone, is unsaturated. (**c**) Linolenic acid, also unsaturated, has three double bonds. The first double bond occurs at the third carbon from the end of the tail, so oleic acid is called an omega-3 fatty acid. Omega-3 and omega-6 fatty acids are "essential fatty acids." Your body does not make them, so they must come from food.

The fatty acid tails of saturated fats have only single covalent bonds. Animal fats tend to remain solid at room temperature because their saturated fatty acid tails pack tightly. Fatty acid tails of unsaturated fats have one or more double covalent bonds. Such rigid bonds usually form kinks that prevent unsaturated fats from packing tightly (Figure 3.12*a*). Most vegetable oils are unsaturated, so they tend to remain liquid at room temperature. Partially hydrogenated vegetable oils are an exception. The double bond in these *trans* fatty acids keeps them straight. *Trans* fats pack tightly, so they are solid at room temperature (Figure 3.12*b*).

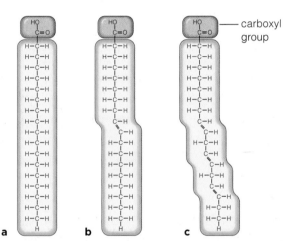

carboxyl group

a   b   c

Figure 3.11 Animated Triglyceride formation by the condensation of three fatty acids with one glycerol molecule. The photograph shows triglyceride-insulated emperor penguins during an Antarctic blizzard.

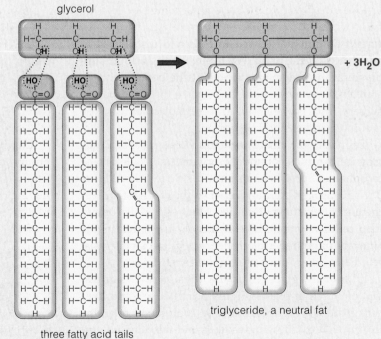

glycerol

+ 3H₂O

triglyceride, a neutral fat

three fatty acid tails

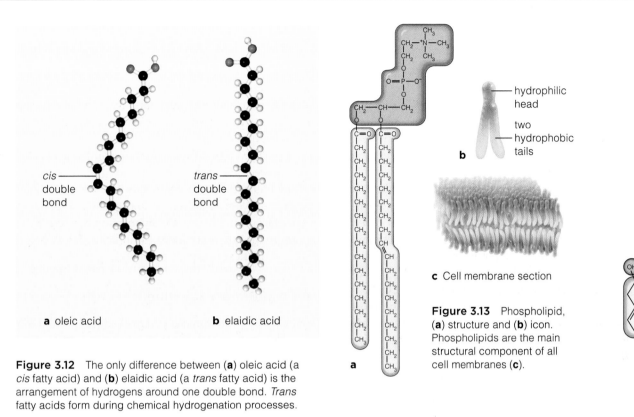

cis double bond

trans double bond

a oleic acid

b elaidic acid

Figure 3.12 The only difference between (**a**) oleic acid (a *cis* fatty acid) and (**b**) elaidic acid (a *trans* fatty acid) is the arrangement of hydrogens around one double bond. *Trans* fatty acids form during chemical hydrogenation processes.

— hydrophilic head

— two hydrophobic tails

b

c Cell membrane section

Figure 3.13 Phospholipid, (**a**) structure and (**b**) icon. Phospholipids are the main structural component of all cell membranes (**c**).

a

Figure 3.14 *Right*, cholesterol. Notice the rigid backbone of four carbon rings.

Phospholipids

Phospholipids have a polar head with a phosphate in it, and two nonpolar fatty acid tails. They are the most abundant lipids in cell membranes, which have two phospholipid layers (Figure 3.13*a–c*). The heads of one layer are dissolved in the cell's watery interior, and the heads of the other layer are dissolved in the cell's fluid surroundings. All of the hydrophilic tails are sandwiched between the heads. You will read about membrane structure and function in Chapters 4 and 5.

Waxes

Waxes are complex, varying mixtures of lipids with long fatty acid tails bonded to long-chain alcohols or carbon rings. The molecules pack tightly, so the resulting substance is firm and water-repellent. Waxes in the cuticle that covers the exposed surfaces of plants help restrict water loss and keep out parasites and other pests. Other types of waxes protect, lubricate,

and soften skin and hair. Waxes, together with fats and fatty acids, make feathers waterproof. Bees store honey and raise new generations of bees inside honeycomb, which they make from beeswax.

Cholesterol and Other Steroids

Steroids are lipids with a rigid backbone of four carbon rings and no fatty acid tails. They differ in the type, number, and position of functional groups. All eukaryotic cell membranes contain steroids. In animal tissues, cholesterol is the most common steroid (Figure 3.14). Cholesterol is remodeled into many molecules, such as bile salts (which help digest fats) and vitamin D (required to keep teeth and bones strong). Steroid hormones are derived from cholesterol. Estrogens and testosterone, hormones that govern reproduction and secondary sexual traits, are examples (Figure 3.5).

Take-Home Message

What are lipids?

■ Lipids are fatty, waxy, or oily organic compounds. They resist dissolving in water. The main classes of lipids are triglycerides, phospholipids, waxes, and steroids.

■ Triglycerides function as energy reservoirs in vertebrate animals.

■ Phospholipids are the main component of cell membranes.

■ Waxes are components of water-repelling and lubricating secretions.

■ Steroids are components of cell membranes, and precursors of many other molecules.

Proteins—Diversity in Structure and Function

- Proteins are the most diverse biological molecule.
- Cells build thousands of different proteins by stringing together amino acids in different orders.

- Link to Covalent bonding 2.4

Proteins and Amino Acids

A **protein** is an organic compound composed of one or more chains of amino acids. An **amino acid** is a small organic compound with an amine group, a carboxyl group (the acid), and one or more atoms called an "R group." Typically, these groups are all attached to the same carbon atom (Figure 3.15). In water, the functional groups ionize: The amine group occurs as —NH_3^+, and the carboxyl group occurs as —COO^-.

Of all biological molecules, proteins are the most diverse. Structural proteins make up spiderwebs and feathers, hooves, hair, and many other body parts. Nutritious types abound in foods such as seeds and

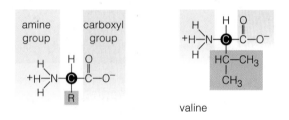

Figure 3.15 Generalized structure of amino acids, and an example. *Green* boxes highlight R groups. Appendix V has models of all twenty of the common amino acids.

eggs. Most enzymes are proteins. Proteins move substances, help cells communicate, and defend the body. Amazingly, cells can synthesize thousands of different proteins from only twenty kinds of amino acids. The complete structures of those twenty amino acids are shown in Appendix V.

Protein synthesis involves bonding amino acids into chains called **polypeptides**. For each type of protein, instructions coded in DNA specify the order in which any of the twenty kinds of amino acids will occur at every place in the chain. A condensation reaction joins the amine group of an amino acid with the carboxyl group of the next in a peptide bond (Figure 3.16).

Levels of Protein Structure

Each type of protein has a unique sequence of amino acids. This sequence is known as the protein's primary structure (Figure 3.17a). Secondary structure emerges as the chain twists, bends, loops, and folds. Hydrogen bonding between amino acids makes stretches of the polypeptide chain form a sheet, or coil into a helix a bit like a spiral staircase (Figure 3.17b). The primary structure of each type of protein is unique, but similar patterns of coils and sheets occur in most proteins.

Much as an overly twisted rubber band coils back on itself, the coils, sheets, and loops of a protein fold up even more into compact domains. A "domain" is a part of a protein that is organized as a structurally stable unit. Such units are a protein's tertiary structure, its third level of organization. Tertiary structure

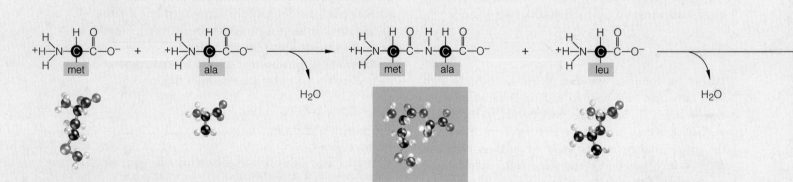

A   DNA encodes the order of amino acids in a new polypeptide chain. Methionine (met) is typically the first amino acid.

B   In a condensation reaction, a peptide bond forms between the methionine and the next amino acid, alanine (ala) in this example. Leucine (leu) will be next. Think about polarity, charge, and other properties of functional groups that become neighbors in the growing chain.

Figure 3.16 Animated Examples of peptide bond formation. Chapter 14 returns to protein synthesis.

makes a protein a working molecule. For instance, the barrel-shaped domains of some proteins function as tunnels through membranes (Figure 3.17c).

Many proteins also have a fourth level of organization, or quaternary structure: They consist of two or more polypeptide chains bonded together or in close association (Figure 3.17d). Most enzymes and many other proteins are globular, with several polypeptide chains folded into shapes that are roughly spherical. Hemoglobin, described shortly, is an example.

Enzymes often attach linear or branched oligosaccharides to polypeptide chains, forming glycoproteins such as those that impart unique molecular identity to a tissue or to a body.

Some proteins aggregate by many thousands into much larger structures, with their polypeptide chains organized into strands or sheets. Some of these fibrous proteins contribute to the structure and organization of cells and tissues. The keratin in your fingernails is an example. Other fibrous proteins, such as the actin and myosin filaments in muscle cells, are part of the mechanisms that help cells and cell parts move.

Take-Home Message

What are proteins?

■ Proteins consist of chains of amino acids. The order of amino acids in a polypeptide chain dictates the type of protein.

■ Polypeptide chains twist and fold into coils, sheets, and loops, which fold and pack further into functional domains.

a Protein primary structure: Amino acids bonded as a polypeptide chain.

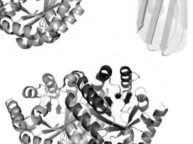

b Protein secondary structure: A coiled (helical) or sheetlike array held in place by hydrogen bonds (*dotted lines*) between different parts of the polypeptide chain.

helix (coil)　　　　sheet

c Protein tertiary structure: A chain's coils, sheets, or both fold and twist into stable, functional domains such as barrels or pockets.

barrel

d Protein quaternary structure: two or more polypeptide chains associated as one molecule.

Figure 3.17 Four levels of a protein's structural organization.

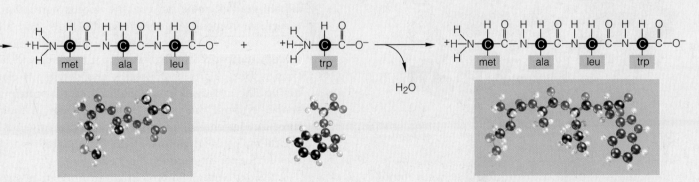

C A peptide bond forms between the alanine and leucine. Tryptophan (trp) will be next. The chain is starting to twist and fold as atoms swivel around some bonds and attract or repel their neighbors.

D The sequence of amino acid subunits in this newly forming peptide chain is now met–ala–leu–trp. The process may continue until there are hundreds or thousands of amino acids in the chain.

3.6 Why Is Protein Structure So Important?

- When a protein's structure goes awry, so does its function.
- Links to Inheritance 1.2, Acids and bases 2.6

Just One Wrong Amino Acid . . .

Sometimes a protein's amino acid sequence changes, with drastic consequences. Let's use hemoglobin as an example. As blood moves through lungs, the hemoglobin inside red blood cells binds oxygen, then gives it up in regions of the body where oxygen levels are low. After giving up oxygen to tissues, the blood circulates back to the lungs, where the hemoglobin inside red blood cells binds more oxygen.

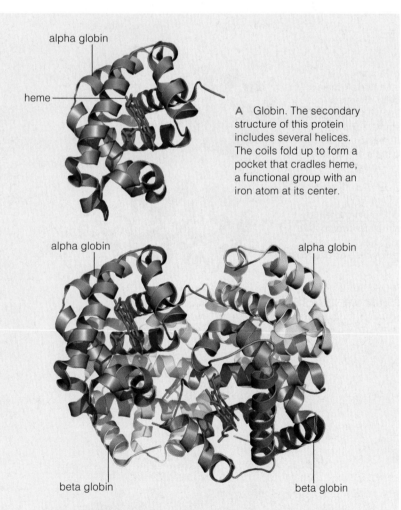

A Globin. The secondary structure of this protein includes several helices. The coils fold up to form a pocket that cradles heme, a functional group with an iron atom at its center.

B Hemoglobin is one of the proteins with quaternary structure. It consists of four globin molecules held together by hydrogen bonds. To help you distinguish among them, the two alpha globin chains are shown here in *green*, and the two beta globin chains are in *brown*.

Figure 3.18 Animated Globin and hemoglobin. (**a**) Globin, a coiled polypeptide chain that cradles heme, a functional group with an iron atom. (**b**) Hemoglobin, an oxygen-transport protein in red blood cells.

Hemoglobin's oxygen-binding properties depend on its structure. Each of the four globin chains in the protein forms a pocket that holds an iron-containing heme group (Figure 3.3 and 3.18). One oxygen molecule can bind to each heme in a hemoglobin protein.

Globin occurs in two slightly different forms, alpha globin and beta globin. In adult humans, two of each form fold up into a hemoglobin molecule. Negatively charged glutamic acid is normally the sixth amino acid in the beta globin chain, but a DNA mutation sometimes puts a different amino acid—valine—in the sixth position (Figure 3.19a,b). Valine is uncharged.

As a result of that one substitution, a tiny patch of the protein changes from polar to nonpolar—which in turn causes the protein's behavior to change slightly. Hemoglobin altered this way is called HbS. Under some conditions, molecules of HbS form large, stable, rod-shaped clumps. Red blood cells containing these clumps become distorted into a sickled shape (Figure 3.19c). Sickled cells tend to clog tiny blood vessels and disrupt blood circulation.

A human has two genes for beta globin, one inherited from each of two parents. (Genes are units of DNA that can encode proteins.) Cells use both genes to make beta globin. If one of a person's genes is normal and the other has the valine mutation, he or she makes enough normal hemoglobin to survive, but not enough to be completely healthy. Someone with two mutated globin genes can make only HbS hemoglobin. The outcome is sickle-cell anemia, a severe genetic disorder (Figure 3.19d).

Proteins Undone—Denaturation

The shape of a protein defines its biological activity: Globin cradles heme, an enzyme speeds a reaction, a receptor responds to some signal. These—and all other proteins—function as long as they stay coiled, folded, and packed in their correct three-dimensional shapes. Heat, shifts in pH, salts, and detergents can disrupt the hydrogen bonds that maintain a protein's shape. Without the bonds that hold them in their three-dimensional shape, proteins and other large biological molecules **denature**—their shape unravels and they no longer function.

Consider albumin, a protein in the white of an egg. When you cook eggs, the heat does not disrupt the covalent bonds of albumin's primary structure. But it destroys albumin's weaker hydrogen bonds, and so the protein unfolds. When the translucent egg white turns opaque, we know albumin has been altered. For a few proteins, denaturation might be reversed if and

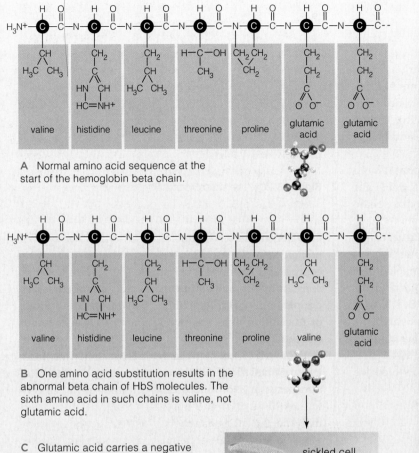

A Normal amino acid sequence at the start of the hemoglobin beta chain.

B One amino acid substitution results in the abnormal beta chain of HbS molecules. The sixth amino acid in such chains is valine, not glutamic acid.

C Glutamic acid carries a negative charge; valine carries no charge. This difference changes the protein so it behaves differently. At low oxygen levels, HbS molecules stick together and form rod-shaped clumps that distort normally rounded red blood cells into sickle shapes. (A sickle is a farm tool that has a crescent-shaped blade.)

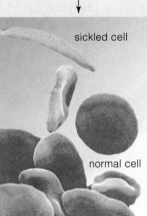

sickled cell

normal cell

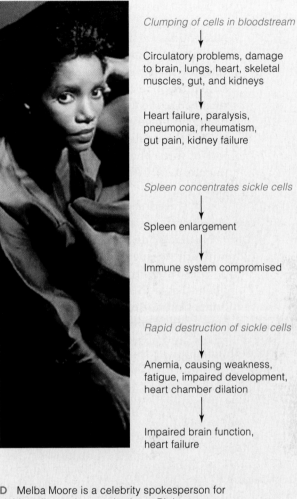

Clumping of cells in bloodstream
↓
Circulatory problems, damage to brain, lungs, heart, skeletal muscles, gut, and kidneys
↓
Heart failure, paralysis, pneumonia, rheumatism, gut pain, kidney failure

Spleen concentrates sickle cells
↓
Spleen enlargement
↓
Immune system compromised

Rapid destruction of sickle cells
↓
Anemia, causing weakness, fatigue, impaired development, heart chamber dilation
↓
Impaired brain function, heart failure

D Melba Moore is a celebrity spokesperson for sickle-cell anemia organizations. *Right,* range of symptoms for a person with two mutated genes for hemoglobin's beta chain.

Figure 3.19 Animated Sickle-cell anemia's molecular basis and symptoms. Section 18.6 explores evolutionary and ecological pressures that maintain this genetic disorder in human populations.

when normal conditions return, but albumin is not one of them. There is no way to uncook an egg.

A protein's structure dictates its function. Enzymes, hormones, transporters, hemoglobin—such proteins are critical for our survival. Their coiled, twisted and folded polypeptide chains form anchors, membrane-spanning barrels, or jaws that attack foreign proteins in the body. Mutations can alter the chains enough to block or enhance an anchoring, transport, or defense function. Sometimes the consequences are awful. Yet such changes also give rise to variation in traits, which

is the raw material of evolution. Learn about protein structure and you are on your way to understanding life's richly normal and abnormal expressions.

Take-Home Message

Why is protein structure important?

■ A protein's function depends on its structure.

■ Mutations that alter a protein's structure may also alter its function.

■ Protein shape unravels if hydrogen bonds are disrupted.

■ Nucleotides are subunits of DNA and RNA. Some have roles in metabolism.

■ Links to Inheritance 1.2, Diversity 1.4, Hydrogen bonds 2.4

Nucleotides are small organic molecules, various kinds of which function as energy carriers, enzyme helpers, chemical messengers, and subunits of DNA and RNA. Each nucleotide consists of a sugar with a five-carbon ring, bonded to a nitrogen-containing base and one or more phosphate groups.

The nucleotide **ATP** (adenosine triphosphate) has a row of three phosphate groups attached to its sugar (Figure 3.20). ATP transfers its outermost phosphate group to other molecules and so primes them to react. You will read about such phosphate-group transfers and their important metabolic role in Chapter 5.

Nucleic acids are polymers—chains of nucleotides in which the sugar of one nucleotide is joined to the phosphate group of the next. An example is **RNA**, or ribonucleic acid, named after the ribose sugar of its component nucleotides. RNA consists of four kinds of nucleotide monomers, one of which is ATP. RNA molecules are important in protein synthesis, which we will discuss in Chapter 14.

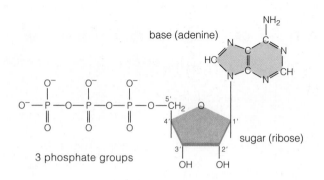

Figure 3.20   The structure of ATP.

DNA, or deoxyribonucleic acid, is another type of nucleic acid named after the deoxyribose sugar of its component nucleotides (Figure 3.21). A DNA molecule consists of two nucleotide chains twisted together as a double helix. Hydrogen bonds between the four kinds of nucleotide hold the two strands of DNA together (Figure 3.22).

Each cell starts out life with DNA inherited from a parent cell. That DNA contains all of the information necessary to build a new cell and, in the case of

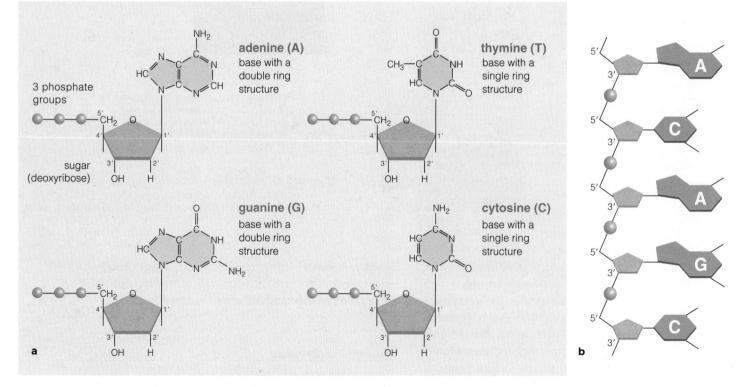

Figure 3.21 Animated (**a**) Nucleotides of DNA. The four kinds of nucleotides in DNA differ only in their component base, for which they are named. The carbon atoms of the sugar rings in nucleotides are numbered as shown. This numbering convention allows us to keep track of the orientation of a chain of nucleotides, as shown in (**b**).

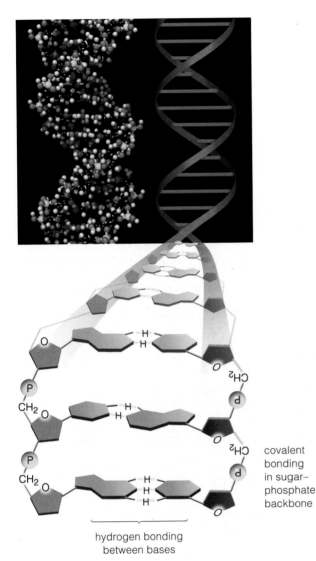

covalent bonding in sugar–phosphate backbone

hydrogen bonding between bases

Figure 3.22 Models of the DNA molecule.

multicelled organisms, an entire individual. The cell uses the order of nucleotide bases in its DNA—the DNA sequence—to construct RNA and proteins. Parts of the sequence are identical or nearly so in all organisms. Other parts are unique to a species, or even to an individual. Chapter 13 returns to DNA structure and function.

Take-Home Message

What are nucleotides and nucleic acids?

■ Different nucleotides are monomers of the nucleic acids DNA and RNA, coenzymes, energy carriers, and messengers.

■ DNA's nucleotide sequence encodes heritable information.

■ Different types of RNA have roles in the processes by which a cell uses the heritable information in its DNA.

Summary

Section 3.1 Under present-day conditions in nature, only living cells make the molecules of life: complex carbohydrates and lipids, proteins, and nucleic acids.

The molecules of life differ, but all of them are **organic** compounds that consist mainly of carbon and hydrogen atoms. Carbon atoms can bond covalently with as many as four other atoms. Long carbon chains or rings form the backbone of the molecules of life.

Section 3.2 **Functional groups** attached to the carbon backbone influence the function of organic compounds. Table 3.2 (next page) summarizes the molecules of life and their functions. By the process of **metabolism**, cells acquire and use energy as they make, rearrange, and break down organic compounds.

Enzymatic reactions that are common in metabolism include **condensation**, which makes **polymers** from smaller **monomers**, and **hydrolysis**, which cleaves molecules into smaller ones.

■ *Use the animation on CengageNOW to explore functional groups, condensation, and hydrolysis.*

Section 3.3 Cells use **carbohydrates** as energy sources, transportable or storable forms of energy, and structural materials. The oligosaccharides and polysaccharides are polymers of monosaccharide monomers.

■ *Use the animation on CengageNOW to see how sucrose forms by condensation of glucose and fructose.*

Section 3.4 **Lipids** are greasy or oily nonpolar molecules, often with one or more **fatty acid** tails, and include **triglycerides** and other **fats**. **Phospholipids** are the main structural component of cell membranes. **Waxes** are part of water-repellent and lubricating secretions; **steroids** are precursors of other molecules.

■ *Use the animation on CengageNOW to see how a triglyceride forms by condensation.*

Section 3.5 **Proteins** are the most diverse molecules of life. Protein structure begins as a linear sequence of **amino acids** called a **polypeptide** chain (primary structure). The chains form sheets and coils (secondary structure), which may pack into functional domains (tertiary structure). Many proteins, including most enzymes, consist of two or more chains (quaternary structure). Fibrous proteins aggregate further into large chains or sheets.

■ *Use the animation on CengageNOW to explore amino acid structure and learn about peptide bond formation.*

■ *Read the InfoTrac article "Protein Folding and Misfolding," David Gossard, American Scientist, September 2002.*

Section 3.6 A protein's structure dictates its function. Sometimes a mutation in DNA results in an amino acid substitution that alters a protein's structure enough to compromise its function. Genetic diseases such as sickle-cell anemia may result.

Shifts in pH or temperature, and exposure to detergent or to salts may disrupt the many hydrogen bonds

IMPACTS, ISSUES REVISITED | Fear of Frying

Several countries are ahead of the United States in restricting the use of *trans* fats in food. In 2004, Denmark passed a law that prohibited importation of foods that contain partially hydrogenated vegetable oils. French fries and chicken nuggets the Danish import from the United States contain almost no *trans* fats; the same foods sold to consumers in the United States contain 5 to 10 grams of *trans* fats per serving.

How would you vote?

New York was the first U.S. city to ban *trans* fats from restaurant food. Should the use of *trans* fats in food be banned entirely? See CengageNOW for details, then vote online.

---

and other molecular interactions that hold a protein in its three-dimensional shape. If a protein unfolds so that it loses its three-dimensional shape (or **denatures**), it also loses its function.

■ *Use the animation on CengageNOW to learn more about hemoglobin structure and sickle-cell mutation.*

Section 3.7 **Nucleotides** are small organic molecules that consist of a sugar bonded to three phosphate groups and a nitrogen-containing base. **ATP** transfers phosphate groups to many kinds of molecules. Other nucleotides are coenzymes or chemical messengers. **DNA** and **RNA** are **nucleic acids**, each composed of four kinds of nucleotides. DNA encodes heritable information about a cell's proteins and RNAs. Different RNAs interact with DNA and with one another to carry out protein synthesis.

■ *Use the animation on CengageNOW to explore DNA.*

Table 3.2 Summary of the Main Organic Molecules in Living Things

Category	Main Subcategories		Some Examples and Their Functions	
CARBOHYDRATES ... contain an aldehyde or a ketone group, and one or more hydroxyl groups	**Monosaccharides** Simple sugars		Glucose	Energy source
	Oligosaccharides Short-chain carbohydrates		Sucrose	Most common form of sugar
	Polysaccharides Complex carbohydrates		Starch, glycogen	Energy storage
			Cellulose	Structural roles
LIPIDS ... are mainly hydrocarbon; generally do not dissolve in water but do dissolve in nonpolar substances, such as alcohols or other lipids	**Glycerides** Glycerol backbone with one, two, or three fatty acid tails (e.g., triglycerides)		Fats (e.g., butter), oils (e.g., corn oil)	Energy storage
	Phospholipids Glycerol backbone, phosphate group, another polar group; often two fatty acids		Lecithin	Key component of cell membranes
	Waxes Alcohol with long-chain fatty acid tails		Waxes in cutin	Conservation of water in plants
	Steroids Four carbon rings; the number, position, and type of functional groups differs		Cholesterol	Component of animal cell membranes; precursor of many steroids, vitamin D
PROTEINS ... are one or more polypeptide chains, each with as many as several thousand covalently linked amino acids	**Mostly fibrous proteins** Long strands or sheets of polypeptide chains; often strong, water-insoluble		Keratin	Structural component of hair, nails
			Collagen	Component of connective tissue
			Myosin, actin	Functional components of muscles
	Mostly globular proteins One or more polypeptide chains folded into globular shapes; many roles in cell activities		Enzymes	Great increase in rates of reactions
			Hemoglobin	Oxygen transport
			Insulin	Control of glucose metabolism
			Antibodies	Immune defense
NUCLEIC ACIDS AND NUCLEOTIDES ... are chains of units (or individual units) that each consist of a five-carbon sugar, phosphate, and a nitrogen-containing base	**Adenosine phosphates**		ATP	Energy carrier
			cAMP	Messenger in hormone regulation
	Nucleotide coenzymes		NAD+, NADP+, FAD	Transfer of electrons, protons (H+) from one reaction site to another
	Nucleic acids Chains of nucleotides		DNA, RNAs	Storage, transmission, translation of genetic information

Data Analysis Exercise

Cholesterol does not dissolve in blood, so it is carried through the bloodstream by lipid–protein aggregates called lipoproteins. Lipoproteins vary in structure. Low-density lipoprotein (LDL) carries cholesterol to body tissues such as artery walls, where it can form health-endangering deposits. LDL is often called "bad" cholesterol. High-density lipoprotein (HDL) carries cholesterol away from tissues to the liver for disposal; it is often called "good" cholesterol.

In 1990, R.P. Mensink and M.B. Katan published a study that tested the effects of different dietary fats on blood lipoprotein levels. Their results are shown in Figure 3.23.

1. In which group was the level of LDL ("bad" cholesterol) highest?

2. In which group was the level of HDL ("good" cholesterol) lowest?

3. An elevated risk of heart disease has been correlated with increasing LDL-to-HDL ratios. In which group was the LDL:HDL ratio highest? Rank the three diets according to their potential effect on cardiovascular health.

	Main Dietary Fats			
	cis-fatty acids	trans-fatty acids	saturated fats	optimal level
LDL	103	117	121	<100
HDL	55	48	55	>40
ratio	1.87	2.43	2.2	<2

Figure 3.23 Effect of diet on lipoprotein levels. Researchers placed 59 men and women on a diet in which 10% of their daily energy intake consisted of cis-fatty acids, trans-fatty acids, or saturated fats. Blood LDL and HDL levels were measured after 3 weeks on the diet; averaged results are shown in mg/dL (milligrams per deciliter). All subjects were tested on each of the diets. The ratio of LDL to HDL is also shown.

Self-Quiz   *Answers in Appendix III*

1. Each carbon atom can share pairs of electrons with up to _____ other atom(s).

2. Sugars are a type of _____ .

3. _____ is a simple sugar (a monosaccharide).
   a. Glucose   c. Ribose   e. both a and b
   b. Sucrose   d. Chitin   f. both a and c

4. Unlike saturated fats, the fatty acid tails of unsaturated fats incorporate one or more _____ .

5. Is this statement true or false? Unlike saturated fats, all of the unsaturated fats are beneficial to health because their fatty acid tails bend and do not pack together.

6. Steroids are among the lipids with no _____ .

7. Which of the following is a class of molecules that encompasses all of the other molecules listed?
   a. triglycerides   c. waxes   e. lipids
   b. fatty acids   d. steroids   f. phospholipids

8. _____ are to proteins as _____ are to nucleic acids.
   a. Sugars; lipids   c. Amino acids; hydrogen bonds
   b. Sugars; proteins   d. Amino acids; nucleotides

9. A denatured protein has lost its _____ .
   a. hydrogen bonds   c. function
   b. shape   d. all of the above

10. _____ consist(s) of nucleotides.
    a. sugars   b. DNA   c. RNA   d. b and c

11. _____ are the richest energy source in the body.
    a. Sugars   b. Proteins   c. Fats   d. Nucleic acids

12. Match each molecule with its most suitable description.
    ___ chain of amino acids   a. carbohydrate
    ___ energy carrier in cells   b. phospholipid
    ___ glycerol, fatty acids, phosphate   c. polypeptide
    ___ two strands of nucleotides   d. DNA
    ___ one or more sugar monomers   e. ATP
    ___ richest source of energy   f. triglycerides

■ *Visit CengageNOW for additional questions.*

Critical Thinking

1. In the following list, identify the carbohydrate, the fatty acid, the amino acid, and the polypeptide:
   a. $^+NH_3$—CHR—COO$^-$ c. (glycine)$_{20}$
   b. $C_6H_{12}O_6$ d. $CH_3(CH_2)_{16}COOH$

2. Lipoproteins are relatively large, spherical clumps of protein and lipid molecules that circulate in the blood of mammals. They are like suitcases that move cholesterol, fatty acid remnants, triglycerides, and phospholipids from one place to another in the body. Given what you know about the insolubility of lipids in water, which of the four kinds of lipids would you predict to be on the outside of a lipoprotein clump, bathed in the fluid portion of blood?

3. In 1976, researchers were developing new insecticides by modifying sugars with chlorine (Cl_2) and other toxic gases. One young member of the team misunderstood instructions to "test" a new molecule. He thought he was supposed to "taste" it. Luckily, the molecule was not toxic, but it was sweet. It became the food additive sucralose.
   Sucralose has three chlorine atoms substituted for three hydroxyl groups of sucrose. The highly electronegative chlorine atoms make sucralose strongly electronegative (Section 2.3). Sucralose binds so strongly to sweet-taste receptors on the tongue that our brain perceives it as 600 times sweeter than sucrose. The body does not recognize sucralose as a carbohydrate. Volunteers ate sucralose labeled with ^{14}C. Analysis of the radioactive molecules in their urine and feces showed that 92.8 percent of the sucralose passed unaltered through the body. Nonetheless, many are worried that the chlorine atoms impart toxicity to sucralose. How would you respond to that concern?

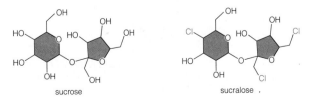

sucrose   sucralose .

Cell Structure and Function

Food for Thought

We find bacteria at the bottom of the ocean, high up in the atmosphere, miles underground—essentially anywhere we look. Mammal intestines typically harbor fantastic numbers of them, but bacteria are not just stowaways there. Intestinal bacteria make vitamins that mammals cannot, and they crowd out more dangerous germs.

Escherichia coli is one of the most common intestinal bacteria of warm-blooded animals. Only a few of the hundreds of types, or strains, of *E. coli*, are harmful. One, O157:H7, makes a potent toxin that can severely damage the lining of the human intestine (Figure 4.1). After ingesting as few as ten O157:H7 cells, a person may become ill with severe cramps and bloody diarrhea that lasts up to ten days. In some people, complications of O157:H7 infection result in kidney failure, blindness, paralysis, and death. About 73,000 people in the United States become infected with *E. coli* O157:H7 each year, and more than 60 die.

E. coli O157:H7 lives in the intestines of other animals —mainly cattle, deer, goats, and sheep—apparently without sickening them. Humans are exposed to the bacteria when they come into contact with feces of animals that harbor it, for example, by eating contaminated ground beef. During slaughter, meat occasionally comes into contact with feces. Bacteria in the feces stick to the meat, then get thoroughly mixed into it during the grinding process. Unless contaminated meat is cooked to at least 71°C (160°F), live bacteria will enter the digestive tract of whoever eats it.

People also become infected by ingesting fresh fruits and vegetables that have come into contact with animal feces. For example, in 2006, at least 205 people became ill and 3 died after eating fresh spinach. The spinach was grown in a field close to a cattle pasture, and water contaminated with manure may have been used to irrigate the field. Washing contaminated produce with water does not remove *E. coli* O157:H7, because the bacteria are sticky.

The economic impact of such outbreaks, which occur with some regularity, extends beyond the casualties. Growers lost $50–100 million dollars recalling fresh spinach after the 2006 outbreak. In 2007, 5.7 million pounds of ground beef were recalled after 14 people were sickened. Food growers and processors are beginning to implement new procedures that they hope will reduce *E. coli* O157:H7 outbreaks. Some meats and produce are now tested for pathogens before sale, and improved documentation should allow a source of contamination to be pinpointed more quickly.

What makes bacteria sticky? Why do people but not cows get sick with *E. coli* O157:H7? You will begin to find answers to these and many more questions that affect your health in this chapter, as you learn about cells and how they work.

See the video! Figure 4.1 *E. coli* O157:H7 bacteria (*above, red*) on intestinal cells (*tan*) of a small child. This type of bacteria can cause a serious intestinal illness in people who eat foods contaminated with it, such as ground beef or fresh produce (*left*).

Key Concepts

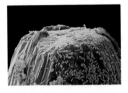

What all cells have in common

Each cell has a plasma membrane, a boundary between its interior and the outside environment. The interior consists of cytoplasm and an innermost region of DNA. **Sections 4.1, 4.2**

Microscopes

Microscopic analysis supports three generalizations of the cell theory: Each organism consists of one or more cells and their products, a cell has a capacity for independent life, and each new cell is descended from another cell. **Section 4.3**

Prokaryotic cells

Archaeans and bacteria are prokaryotic cells, which have few, if any, internal membrane-enclosed compartments. In general, they are the smallest and structurally the simplest cells. **Sections 4.4, 4.5**

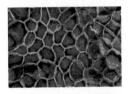

Eukaryotic cells

Cells of protists, plants, fungi, and animals are eukaryotic; they have a nucleus and other membrane-enclosed compartments. They differ in internal parts and surface specializations. **Sections 4.6–4.12**

A look at the cytoskeleton

Diverse protein filaments reinforce a cell's shape and keep its parts organized. As some filaments lengthen and shorten, they move cell structures or the whole cell. **Section 4.13**

Links to Earlier Concepts

- Reflect on the overview of levels of organization in nature in Section 1.1. You will see how the properties of cell membranes emerge from the organization of lipids and proteins (3.4, 3.5).

- What you know about scientific theory (1.6) will help you understand how scientific thought led to the development of the cell theory. This chapter also offers examples of the effects of mutation, and how researchers use tracers (2.2).

- You will consider the cellular location of DNA (3.7) and the sites where carbohydrates (3.2, 3.3) are built and broken apart.

- You will also expand your understanding of the vital roles of proteins in cell functions (3.6), and see how a nucleotide helps control cell activities (3.7).

■ The cell theory, a foundation of modern biology, states that cells are the fundamental units of all life.

■ Link to Theory 1.6

Measuring Cells

Do you ever think of yourself as being about 3/2000 of a kilometer (1/1000 miles) tall? Probably not, yet that is how we measure cells. Use the scale bars in Figure 4.2 like a ruler and you can see that the cells shown are a few micrometers "tall." One micrometer (μm) is one-thousandth of a millimeter, which is one-thousandth of a meter, which is one-thousandth of a kilometer (0.62 miles). The cells are bacteria. Bacteria are among the smallest and structurally simplest cells on Earth. The cells that make up your body are generally larger and more complex than bacteria.

Animalcules and Beasties

Nearly all cells are so small that they are invisible to the naked eye. No one even knew cells existed until after the first microscopes were invented, around the end of the sixteenth century.

The first microscopes were not very sophisticated. Dutch spectacle makers Hans and Zacharias Janssen discovered that objects appear greatly enlarged (mag-nified) when viewed through a series of lenses. The father and son team created the first compound micro-scope (one that uses multiple lenses) in the year 1590, when they mounted two glass lenses inside a tube.

Given the simplicity of their instruments, it is amaz-ing that the pioneers in microscopy observed as much as they did. Antoni van Leeuwenhoek, a Dutch draper, had exceptional skill in constructing lenses and possi-bly the keenest vision. By the mid-1600s, he was spying on the microscopic world of rainwater, insects, fabric, sperm, feces—essentially any sample he could fit into his microscope (Figure 4.3a). He was fascinated by the tiny organisms he saw moving in many of his samples. For example, in scrapings of tartar from his teeth, Leeuwenhoek saw "many very small animalcules, the motions of which were very pleasing to behold." He (incorrectly) assumed that movement defined life, and (correctly) concluded that the moving "beasties" he saw were alive. Perhaps Leeuwenhoek was so pleased to behold his animalcules because he did not grasp the implications of what he was seeing: Our world, and our bodies, teem with microbial life.

Robert Hooke, a contemporary of Leeuwenhoek, added another lens that made the instrument easier to use. Many of the microscopes we use today are still based on his design. Hooke magnified a piece of thinly sliced cork from a mature tree and saw tiny compart-ments (Figure 4.3b). He named them cellulae—a Latin

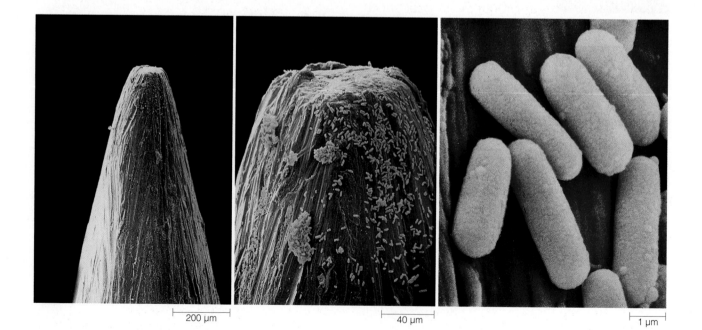

| 200 μm | 40 μm | 1 μm |

Figure 4.2 Rod-shaped bacterial cells on the tip of a household pin, shown at increasingly higher magnifications (enlargements). The "μm" is an abbreviation for micrometers, or 10^{-6} meters.
Figure It Out: About how big are these bacteria? *Answer: About 1 μm wide, and 5 μm long*

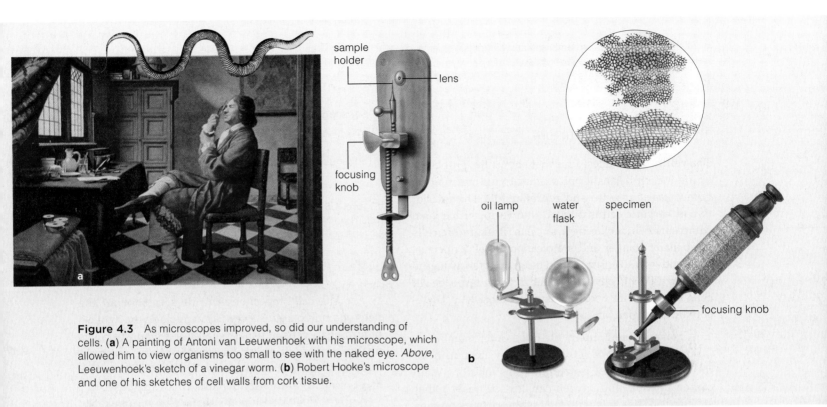

sample holder

lens

focusing knob

oil lamp

water flask

specimen

focusing knob

a

b

Figure 4.3 As microscopes improved, so did our understanding of cells. (**a**) A painting of Antoni van Leeuwenhoek with his microscope, which allowed him to view organisms too small to see with the naked eye. *Above,* Leeuwenhoek's sketch of a vinegar worm. (**b**) Robert Hooke's microscope and one of his sketches of cell walls from cork tissue.

word for the small chambers that monks lived in—and thus coined the term "cell." Actually they were dead plant cell walls, which is what cork consists of, but Hooke did not think of them as being dead because neither he nor anyone else knew cells could be alive. He observed cells "fill'd with juices" in green plant tissues but did not realize they were alive, either.

The Cell Theory Emerges

For nearly 200 years after their discovery, cells were thought to be part of a continuous membrane system in multicelled organisms, not separate entities. By the 1820s, vastly improved lenses brought cells into much sharper focus. Robert Brown, a botanist, was the first to identify a plant cell nucleus. Matthias Schleiden, another botanist, hypothesized that a plant cell is an independent living unit even when it is part of a plant. Schleiden compared notes with the zoologist Theodor Schwann, and both concluded that the tissues of animals as well as plants are composed of cells and their products. Together, the two scientists recognized that cells have a life of their own even when they are part of a multicelled body.

Another insight emerged from physiologist Rudolf Virchow, who studied how cells reproduce—that is, how they divide into descendant cells. Every cell, he realized, had descended from another living cell. These

and many other observations yielded four generalizations that today constitute the **cell theory**:

1. Every living organism consists of one or more cells.

2. The cell is the structural and functional unit of all organisms. A cell is the smallest unit of life, individually alive even as part of a multicelled organism.

3. All living cells come from division of other, preexisting cells.

4. Cells contain hereditary material, which they pass to their offspring during division.

The cell theory, first articulated in 1839 by Schwann and Schleiden and later revised, remains a foundation of modern biology. It was not always so. The theory was a radical new interpretation of nature that underscored life's unity. As with every scientific theory, it has remained (and always will be) open to revision if new data do not support it.

Take-Home Message

What is the cell theory?

- All organisms consist of one or more cells.
- A cell is the smallest unit with the properties of life.
- Each new cell arises from division of another, preexisting cell.
- Each cell passes hereditary material to its offspring.

What Is a Cell?

■ All cells have a plasma membrane and cytoplasm, and all start out life with DNA.

■ Links to Lipid structure 3.4, DNA 3.7

The Basics of Cell Structure

The **cell** is the smallest unit that shows the properties of life, which means it has a capacity for metabolism, homeostasis, growth, and reproduction. The interior of a **eukaryotic cell** is divided into various functional compartments, including a nucleus. **Prokaryotic cells** are usually smaller and simpler; none has a nucleus. Cells differ in size, shape, and activities. Yet, as Figure 4.4 suggests, all cells are similar in three respects. All cells start out life with a plasma membrane, a DNA-containing region, and cytoplasm:

1. A **plasma membrane** is the cell's outer membrane. It separates metabolic activities from events outside of the cell, but does not isolate the cell's interior. Water, carbon dioxide, and oxygen can cross it freely. Other substances cross only with the assistance of membrane proteins. Still others are kept out entirely.

2. All eukaryotic cells start life with a **nucleus**. This double-membraned sac holds a eukaryotic cell's DNA. The DNA inside prokaryotic cells is concentrated in a region of cytoplasm called the **nucleoid**.

3. **Cytoplasm** is a semifluid mixture of water, sugars, ions, and proteins between the plasma membrane and the region of DNA. Cell components are suspended in cytoplasm. For instance, **ribosomes**, structures on which proteins are built, are suspended in cytoplasm.

Diameter (cm)	2	3	6
Surface area (cm²)	12.6	28.2	113
Volume (cm³)	4.2	14.1	113
Surface-to-volume ratio	3:1	2:1	1:1

Figure 4.5 Animated Three examples of the surface-to-volume ratio. This physical relationship between increases in volume and surface area constrains cell size and shape.

Are any cells big enough to be seen without the help of a microscope? A few. They include the "yolks" of bird eggs, cells in watermelon tissues, and the eggs of amphibians and fishes. These cells can be relatively large because they are not very metabolically active. Most of their volume simply acts as a warehouse.

A physical relationship, the **surface-to-volume ratio**, strongly influences cell size and shape. By this ratio, an object's volume increases with the cube of its diameter, but its surface area increases only with the square. The ratio is important because the lipid bilayer can handle only so many exchanges between the cell's cytoplasm and the external environment.

Apply the surface-to-volume ratio to a round cell. As Figure 4.5 shows, when a cell expands in diameter during growth, its volume increases faster than its surface area does. Imagine that a round cell expands until it is four times its original diameter. The volume of the

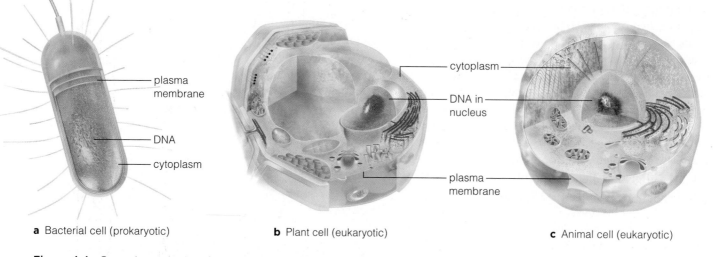

a Bacterial cell (prokaryotic)

plasma membrane

DNA

cytoplasm

b Plant cell (eukaryotic)

cytoplasm

DNA in nucleus

plasma membrane

c Animal cell (eukaryotic)

Figure 4.4 General organization of prokaryotic and eukaryotic cells. If the prokaryotic cell were drawn at the same scale as the other two cells, it would be about this big:

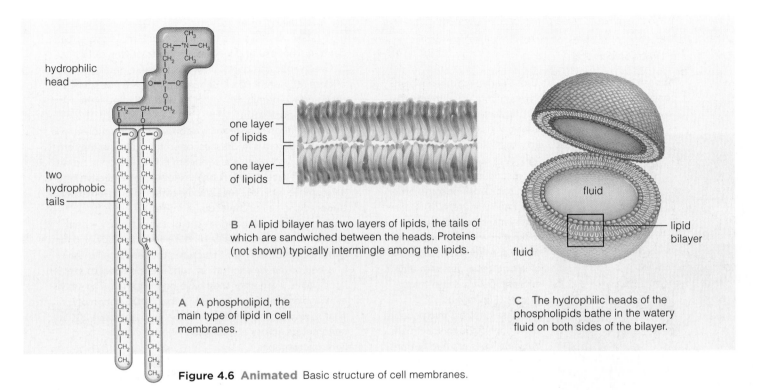

hydrophilic head

two hydrophobic tails

A A phospholipid, the main type of lipid in cell membranes.

one layer of lipids

one layer of lipids

B A lipid bilayer has two layers of lipids, the tails of which are sandwiched between the heads. Proteins (not shown) typically intermingle among the lipids.

fluid

fluid

lipid bilayer

C The hydrophilic heads of the phospholipids bathe in the watery fluid on both sides of the bilayer.

Figure 4.6 Animated Basic structure of cell membranes.

cell has increased 64 times (4^3), but its surface area has increased only 16 times (4^2). Each unit of plasma membrane must now handle exchanges with four times as much cytoplasm. If a cell's circumference gets too big, the inward flow of nutrients and outward flow of wastes will not be fast enough to keep the cell alive.

A big, round cell would also have trouble moving substances through its cytoplasm. Molecules disperse by their own random motions, but they move only so quickly. Nutrients or wastes would not get distributed fast enough to keep up with a large, round, active cell's metabolism. That is why many cells are long and thin, or frilly surfaced with folds that increase surface area. The surface-to-volume ratio of such cells is enough to sustain their metabolism. The amount of raw materials that cross the plasma membrane, and the speed with which they are distributed through cytoplasm, satisfy the cell's needs. Wastes are also removed fast enough to keep the cell from getting poisoned.

Surface-to-volume constraints also affect the body plans of multicelled species. For example, small cells attach end to end in strandlike algae, so each interacts directly with its surroundings. Muscle cells in your thighs are as long as the muscle in which they occur, but each is thin, so it exchanges substances efficiently with fluids in the tissue surrounding it.

Preview of Cell Membranes

The structural foundation of all cell membranes is the **lipid bilayer**, a double layer of lipids organized so that their hydrophobic tails are sandwiched between their hydrophilic heads (Figure 4.6).

Phospholipids are the most abundant type of lipid in cell membranes. Many different proteins embedded in a bilayer or attached to one of its surfaces carry out membrane functions. For example, some proteins form channels through a bilayer; others pump substances across it. In addition to a plasma membrane, many cells also have internal membranes that form channels or enclose sacs. These membranous structures compartmentalize tasks such as building, modifying, and storing substances. Chapter 5 offers a closer look at membrane structure and function.

Take-Home Message

How are all cells alike?

■ All cells start life with a plasma membrane, cytoplasm, and a region of DNA.

■ A lipid bilayer forms the structural framework of all cell membranes.

■ DNA of eukaryotic cells is enclosed by a nucleus. DNA of prokaryotic cells is concentrated in a region of cytoplasm called the nucleoid.

4.3 How Do We See Cells?

■ We use different types of microscopes to study different aspects of organisms, from the smallest to the largest.

■ Link to Tracers 2.2

Modern Microscopes Like those early instruments mentioned in Section 4.1, many types of modern light microscopes still rely on visible light to illuminate objects. All

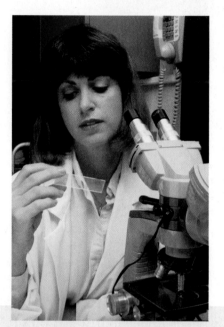

light travels in waves, a property that allows us to focus light with glass lenses. Light microscopes use visible light to illuminate a cell or some other specimen (Figure 4.7a). Curved glass lenses bend the light and focus it as a magnified image of the specimen. Photographs of images enlarged with any microscope are called micrographs (Figure 4.8).

Phase-contrast microscopes shine light through specimens, but most cells are nearly transparent. Their internal details may not be visible unless they

are first stained, or exposed to dyes that only some cell parts soak up. The parts that absorb the most dye appear darkest. The resulting increase in contrast (the difference between light and dark) allows us to see a greater range of detail (Figure 4.8a). Opaque samples are not stained; their surface details are revealed with reflected light microscopes (Figure 4.8b).

With a fluorescence microscope, a cell or a molecule is the light source; it fluoresces, or emits energy in the form of visible light, when a laser beam is focused on it. Some molecules, such as chlorophylls, fluoresce naturally (Figure 4.8c). More typically, researchers attach a light-emitting tracer to the cell or molecule of interest.

The wavelength of light—the distance from the peak of one wave to the peak behind it—limits the power of any light microscope. Why? Structures that are smaller than one-half of the wavelength of light are too small to scatter light waves, even after they have been stained. The smallest wavelength of visible light is about 400 nanometers. That is why structures that are smaller than about 200 nanometers across appear blurry under even the best light microscopes.

Other microscopes can reveal smaller details. For example, electron microscopes use electrons instead

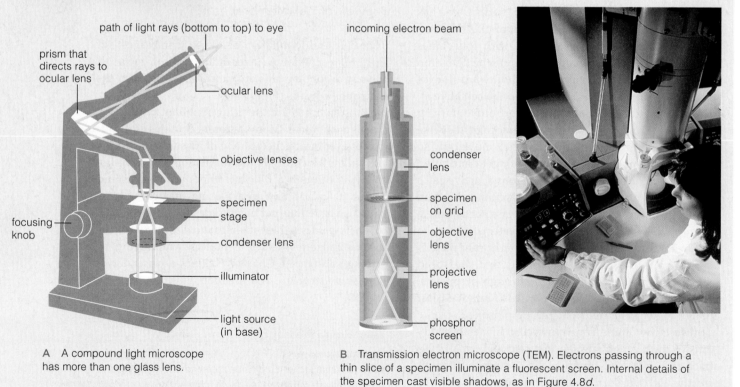

A A compound light microscope has more than one glass lens.

B Transmission electron microscope (TEM). Electrons passing through a thin slice of a specimen illuminate a fluorescent screen. Internal details of the specimen cast visible shadows, as in Figure 4.8d.

Figure 4.7 Animated Examples of microscopes.

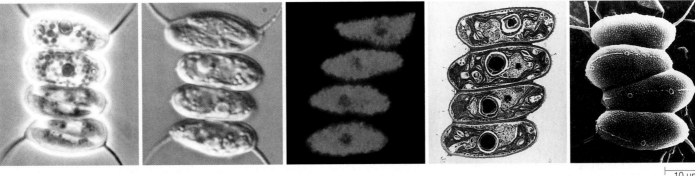

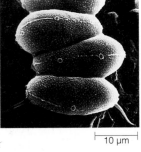

a Light micrograph. A phase-contrast microscope yields high-contrast images of transparent specimens, such as cells.

b Light micrograph. A reflected light microscope captures light reflected from opaque specimens.

c Fluorescence micrograph. The chlorophyll molecules in these cells emitted red light (they fluoresced) naturally.

d A transmission electron micrograph reveals fantastically detailed images of internal structures.

e A scanning electron micrograph shows surface details of cells and structures. Often, SEMs are artificially colored to highlight certain details.

Figure 4.8 Different microscopes can reveal different characteristics of the same aquatic organism—a green alga (*Scenedesmus*). Try estimating the size of one of these algal cells by using the scale bar.

of visible light to illuminate samples (Figure 4.7*b*). Because electrons travel in wavelengths that are much shorter than those of visible light, electron microscopes can resolve details that are much smaller than you can see with light microscopes. Electron microscopes use magnetic fields to focus beams of electrons onto a sample.

With transmission electron microscopes, electrons form an image after they pass through a thin specimen. The specimen's internal details appear on the image as

shadows (Figure 4.8*d*). Scanning electron microscopes direct a beam of electrons back and forth across a surface of a specimen, which has been coated with a thin layer of gold or another metal. The metal emits both electrons and x-rays, which are converted into an image of the surface (Figure 4.8*e*). Both types of electron microscopes can resolve structures as small as 0.2 nanometer.

Figure 4.9 compares the resolving power of light and electron microscopes with that of the unaided human eye.

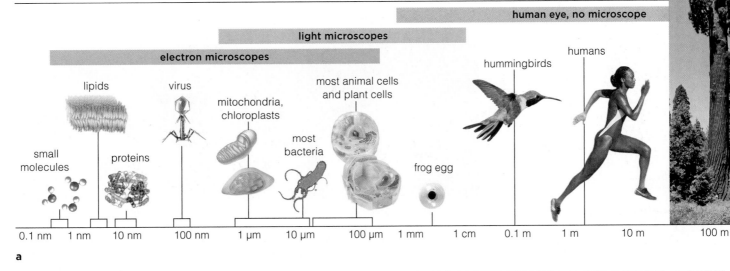

Figure 4.9 (**a**) Relative sizes of molecules, cells, and multicelled organisms. The diameter of most cells is in the range of 1 to 100 micrometers. Frog eggs, one of the exceptions, are 2.5 millimeters in diameter.

The scale shown here is exponential, not linear; each unit of measure is ten times larger than the unit preceding it. (**b**) Units of measure. See also Appendix IX. **Figure It Out:** Which is smallest, a protein, a lipid, or a water molecule?

Answer: A water molecule

1 centimeter (cm)	= 1/100 meter, or 0.4 inch
1 millimeter (mm)	= 1/1000 meter, or 0.04 inch
1 micrometer (µm)	= 1/1,000,000 meter, or 0.00004 inch
1 nanometer (nm)	= 1/1,000,000,000 meter, or 0.00000004 inch

$$1 \text{ meter} = 10^2 \text{ cm} = 10^3 \text{ mm} = 10^6 \text{ µm} = 10^9 \text{ nm}$$

b

4.4 Introducing Prokaryotic Cells

- Bacteria and archaea are the prokaryotes.
- Links to Polysaccharides 3.3, ATP 3.7

The word prokaryote means "before the nucleus," a reminder that the first prokaryotes evolved before the first eukaryotes. Prokaryotes are single-celled (Figure 4.10). As a group, they are the smallest and most metabolically diverse forms of life that we know about. Prokaryotes inhabit nearly all of Earth's environments, including some very hostile places.

Domains Bacteria and Archaea comprise all prokaryotes (Section 1.3). Cells of the two domains are alike in appearance and size, but differ in their structure and metabolic details (Figures 4.11 and 4.12). Some characteristics of archaeans indicate they are more closely related to eukaryotic cells than to bacteria. Chapter 21 revisits prokaryotes in more detail. Here we present an overview of their structure.

Most prokaryotic cells are not much wider than a micrometer. Rod-shaped species are a few micrometers long. None has a complex internal framework, but protein filaments under the plasma membrane impart shape to the cell. Such filaments also act as scaffolding for internal structures.

A rigid **cell wall** surrounds the plasma membrane of nearly all prokaryotes. Dissolved substances easily cross this permeable layer on the way to and from the plasma membrane. The cell wall of most bacteria consists of peptidoglycan, which is a polymer of cross-linked peptides and polysaccharides. The wall of most archaeans consists of proteins. Some types of eukaryotic cells (such as plant cells) also have a wall, but it is structurally different from a prokaryotic cell wall.

Sticky polysaccharides form a slime layer, or capsule, around the wall of many types of bacteria. The sticky capsule helps these cells adhere to many types of surfaces (such as spinach leaves and meat), and it also protects them from predators and toxins. A capsule can protect pathogenic (disease-causing) bacteria from host defenses.

Projecting past the wall of many prokaryotic cells are one or more **flagella** (singular, flagellum): slender cellular structures used for motion. A bacterial flagellum moves like a propeller that drives the cell through fluid habitats, such as a host's body fluids. It differs from a eukaryotic flagellum, which bends like a whip and has a distinctive internal structure.

Protein filaments called **pili** (singular, pilus) project from the surface of some bacterial species (Figure 4.12a). Pili help cells cling to or move across surfaces. One kind, a "sex" pilus, attaches to another bacterium and then shortens. The attached cell is reeled in, and genetic material is transferred from one cell to the other through the pilus.

The plasma membrane of all bacteria and archaeans selectively controls which substances move to and from the cytoplasm, as it does for eukaryotic cells. The plasma membrane bristles with transporters and receptors, and it also incorporates proteins that carry out important metabolic processes.

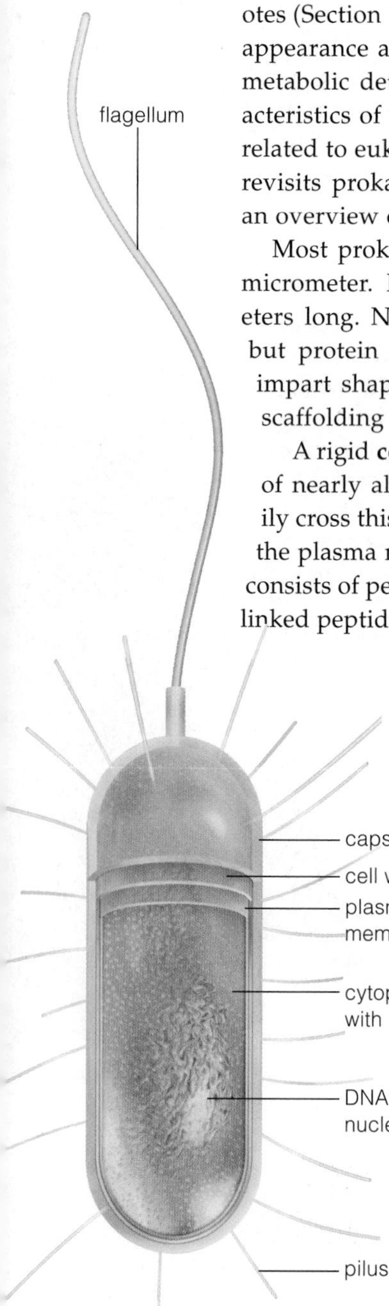

flagellum

capsule
cell wall
plasma membrane
cytoplasm, with ribosomes
DNA in nucleoid
pilus

Figure 4.10 Animated
Generalized body plan of a prokaryote.

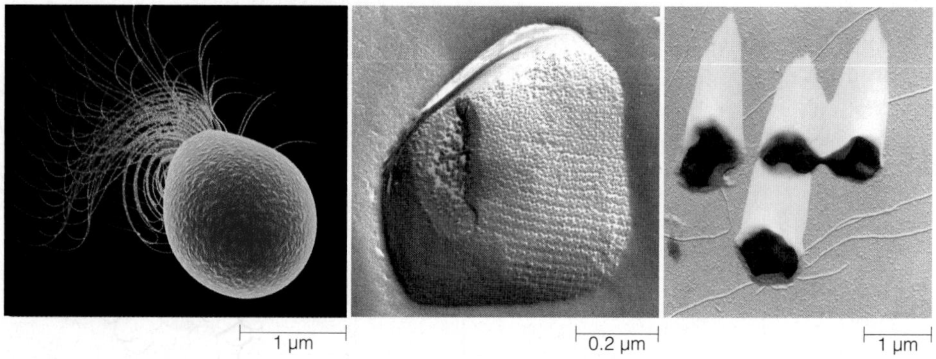

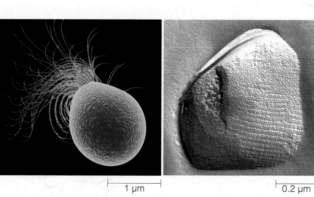

1 μm   0.2 μm   1 μm

a *Pyrococcus furiosus* was discovered in ocean sediments near an active volcano. It lives best at 100°C (212°F), and it makes a rare kind of enzyme that contains tungsten atoms.

b *Ferroglobus placidus* prefers superheated water spewing from the ocean floor. The unique composition of archaean lipid bilayers keeps these membranes intact at extreme heat and pH.

c *Metallosphaera prunae*, discovered in a smoking pile of ore at a uranium mine, prefers high temperatures and low pH. (*White shadows are an artifact of electron microscopy.*)

Figure 4.11 Some like it hot: many archaeans inhabit extreme environments. The cells in this example live without oxygen.

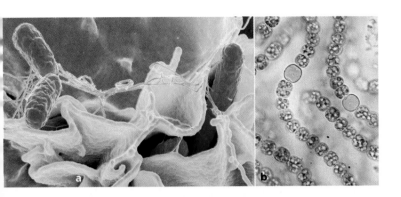

Figure 4.12 Bacteria. (**a**) Protein filaments, or pili, anchor bacterial cells to one another and to surfaces. Here, *Salmonella typhimurium* cells (*red*) use their pili to invade a culture of human cells. (**b**) Ball-shaped *Nostoc* cells stick together in a sheath of their own secretions. *Nostoc* are photosynthetic cyanobacteria. Other types of bacteria are shaped like rods or corkscrews.

For example, the plasma membrane of photosynthetic bacteria has arrays of proteins that capture light energy and convert it to the chemical energy of ATP (Section 3.7). The ATP is then used to build sugars. Similar metabolic processes occur in eukaryotes, but they take place at specialized internal membranes, not the plasma membrane.

The cytoplasm of prokaryotes contains thousands of ribosomes, structures upon which polypeptides are assembled. A prokaryotic cell's single chromosome, a circular DNA molecule, is located in an irregularly shaped region called the nucleoid. Most nucleoids are not enclosed by a membrane. Many prokaryotes also have plasmids in the cytoplasm. These small circles of DNA carry a few genes (units of inheritance) that can confer advantages, such as resistance to antibiotics.

One more intriguing point: There is evidence that all protists, plants, fungi, and animals evolved from a few ancient types of prokaryotes. For example, part of the plasma membrane of cyanobacteria folds into the cytoplasm. Pigments and other molecules that carry out photosynthesis are embedded in the membrane, just as they are in the inner membrane of chloroplasts—structures specialized for photosynthesis in eukaryotic cells. Section 20.4 returns to this topic.

Take-Home Message

What do all prokaryotic cells have in common?

■ All prokaryotes are single-celled organisms with no nucleus. These organisms inhabit nearly all regions of the biosphere.

■ Bacteria and archaeans are the only prokaryotes. Most kinds have a cell wall around their plasma membrane.

■ Prokaryotes have a relatively simple structure, but they are a diverse group of organisms.

4.5 | Microbial Mobs

■ Although prokaryotes are all single-celled, few live alone.

■ Link to Glycoproteins 3.5

Bacterial cells often live so close together that an entire community shares a layer of secreted polysaccharides and glycoproteins. Such communal living arrangements, in which single-celled organisms live in a shared mass of slime, are called **biofilms**. In nature, a biofilm typically consists of multiple species, all entangled in their own mingled secretions. It may include bacteria, algae, fungi, protists, and archaeans. Such associations allow cells living in a fluid to linger in a particular spot rather than be swept away by currents.

The microbial inhabitants of a biofilm benefit each other. Rigid or netlike secretions of some species serve as permanent scaffolding for others. Species that break down toxic chemicals allow more sensitive ones to thrive in polluted habitats that they could not withstand on their own. Waste products of some serve as raw materials for others.

Like a bustling metropolitan city, a biofilm organizes itself into "neighborhoods," each with a distinct microenvironment that stems from its location within the biofilm and the particular species that inhabit it (Figure 4.13). For example, cells that reside near the middle of a biofilm are very crowded and do not divide often. Those at the edges divide repeatedly, expanding the biofilm.

The formation and continuation of a biofilm is not random. Free-living cells sense the presence of other cells. Those that encounter a biofilm with favorable conditions switch their metabolism to support a more sedentary, communal lifestyle, and join in. Flagella disassemble, and sex pili form. If conditions become less favorable, the cells can revert to a free-living mode and swim away to find more hospitable accommodations.

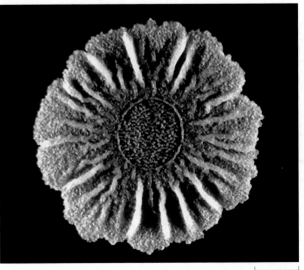

0.2 cm

Figure 4.13 Biofilms. A single species of bacteria, *Bacillus subtilis*, formed this biofilm. Note the distinct regions.

Introducing Eukaryotic Cells

■ Eukaryotic cells carry out much of their metabolism inside organelles enclosed by membranes.

All eukaryotic cells start out life with a nucleus. *Eu–* means true; and *karyon*, meaning kernel, refers to the nucleus. A nucleus is a type of **organelle**: a structure that carries out a specialized function inside a cell. Many organelles, particularly those in eukaryotic cells, are bounded by membranes. Like all cell membranes, those around organelles control the types and amounts of substances that cross them. Such control maintains a special internal environment that allows an organelle to carry out its particular function. That function may be isolating a toxic or sensitive substance from the rest of the cell, transporting some substance through the cytoplasm, maintaining fluid balance, or providing a favorable environment for a reaction that could not occur in the cytoplasm. For example, a mitochondrion makes ATP after concentrating hydrogen ions inside its membrane system.

Much as interactions among organ systems keep an animal body running, interactions among organelles keep a cell running. Substances shuttle from one kind of organelle to another, and to and from the plasma membrane. Some metabolic pathways take place in a series of different organelles.

Table 4.1 lists common components of eukaryotic cells. These cells all start out life with certain kinds of organelles such as a nucleus and ribosomes. They also have a cytoskeleton, a dynamic "skeleton" of proteins (*cyto–* means cell). Specialized cells contain additional

Table 4.1 Organelles of Eukaryotic Cells

Name	Function
Organelles with membranes	
Nucleus	Protecting, controlling access to DNA
Endoplasmic reticulum (ER)	Routing, modifying new polypeptide chains; synthesizing lipids; other tasks
Golgi body	Modifying new polypeptide chains; sorting, shipping proteins and lipids
Vesicles	Transporting, storing, or digesting substances in a cell; other functions
Mitochondrion	Making ATP by sugar breakdown
Chloroplast	Making sugars in plants, some protists
Lysosome	Intracellular digestion
Peroxisome	Inactivating toxins
Vacuole	Storage
Organelles without membranes	
Ribosomes	Assembling polypeptide chains
Centriole	Anchor for cytoskeleton

kinds of organelles and structures. Figure 4.14 shows two typical eukaryotic cells.

Take-Home Message

What do all eukaryotic cells have in common?

■ Eukaryotic cells start life with a nucleus and other membrane-enclosed organelles (structures that carry out specific tasks).

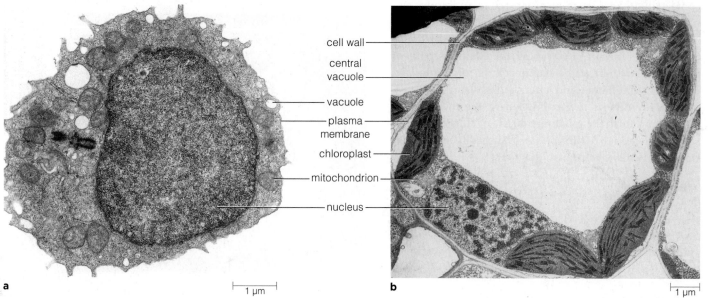

cell wall
central vacuole
vacuole
plasma membrane
chloroplast
mitochondrion
nucleus

a   1 μm

b   1 μm

Figure 4.14 Transmission electron micrographs of eukaryotic cells. (**a**) Human white blood cell. (**b**) Photosynthetic cell from a blade of timothy grass.

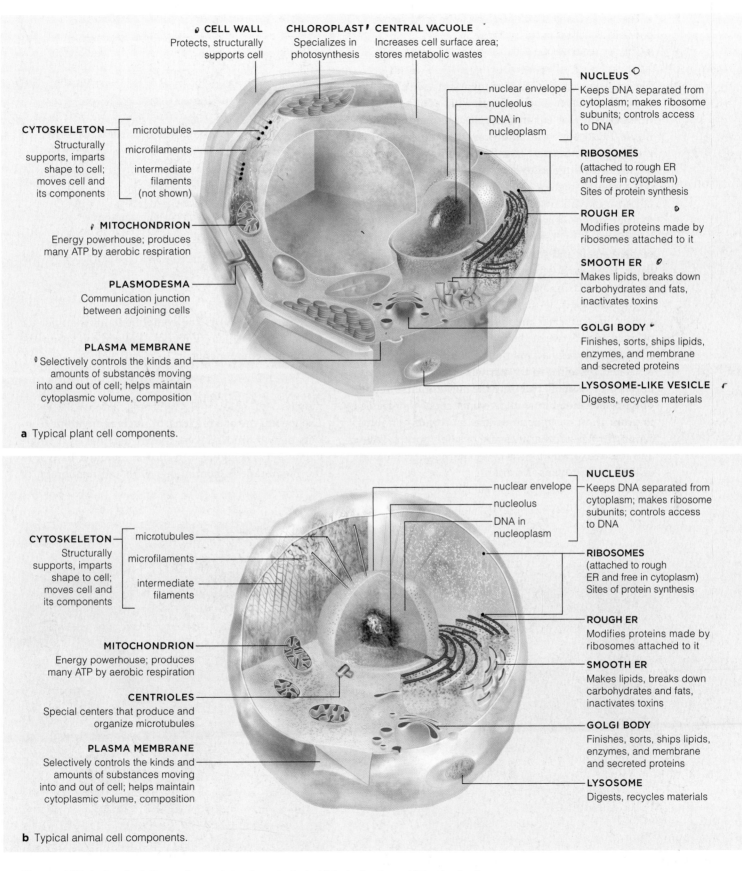

CELL WALL
Protects, structurally supports cell

CHLOROPLAST
Specializes in photosynthesis

CENTRAL VACUOLE
Increases cell surface area; stores metabolic wastes

NUCLEUS
Keeps DNA separated from cytoplasm; makes ribosome subunits; controls access to DNA

nuclear envelope
nucleolus
DNA in nucleoplasm

CYTOSKELETON
Structurally supports, imparts shape to cell; moves cell and its components

microtubules
microfilaments
intermediate filaments (not shown)

RIBOSOMES
(attached to rough ER and free in cytoplasm) Sites of protein synthesis

ROUGH ER
Modifies proteins made by ribosomes attached to it

MITOCHONDRION
Energy powerhouse; produces many ATP by aerobic respiration

SMOOTH ER
Makes lipids, breaks down carbohydrates and fats, inactivates toxins

PLASMODESMA
Communication junction between adjoining cells

GOLGI BODY
Finishes, sorts, ships lipids, enzymes, and membrane and secreted proteins

PLASMA MEMBRANE
Selectively controls the kinds and amounts of substances moving into and out of cell; helps maintain cytoplasmic volume, composition

LYSOSOME-LIKE VESICLE
Digests, recycles materials

a Typical plant cell components.

CYTOSKELETON
Structurally supports, imparts shape to cell; moves cell and its components

microtubules
microfilaments
intermediate filaments

NUCLEUS
Keeps DNA separated from cytoplasm; makes ribosome subunits; controls access to DNA

nuclear envelope
nucleolus
DNA in nucleoplasm

RIBOSOMES
(attached to rough ER and free in cytoplasm) Sites of protein synthesis

MITOCHONDRION
Energy powerhouse; produces many ATP by aerobic respiration

ROUGH ER
Modifies proteins made by ribosomes attached to it

CENTRIOLES
Special centers that produce and organize microtubules

SMOOTH ER
Makes lipids, breaks down carbohydrates and fats, inactivates toxins

PLASMA MEMBRANE
Selectively controls the kinds and amounts of substances moving into and out of cell; helps maintain cytoplasmic volume, composition

GOLGI BODY
Finishes, sorts, ships lipids, enzymes, and membrane and secreted proteins

LYSOSOME
Digests, recycles materials

b Typical animal cell components.

Figure 4.15 Animated Organelles and structures typical of (**a**) plant cells and (**b**) animal cells.

The Nucleus

- The nucleus keeps eukaryotic DNA away from potentially damaging reactions in the cytoplasm.
- The nuclear envelope controls when DNA is accessed.

The nucleus contains all of a eukaryotic cell's DNA. A molecule of DNA is big to begin with, and the nucleus of most kinds of eukaryotic cells has many of them. If you could tease out all of the DNA molecules from the nucleus of a single human cell, unravel them, and stretch them out end to end, you would have a line of DNA about 2 meters (6–1/2 feet) long. That is a lot of DNA for one microscopic nucleus.

The nucleus serves two important functions. First, it keeps a cell's genetic material—its one and only copy of DNA—safe and sound. Isolated in its own compartment, DNA stays separated from the bustling activity of the cytoplasm, and from metabolic reactions that might damage it.

Second, a nuclear membrane controls the passage of molecules between the nucleus and the cytoplasm. For example, cells access their DNA when they make RNA and proteins, so the various molecules involved in this process must pass into the nucleus and out of it. The nuclear membrane allows only certain molecules to cross it, at certain times and in certain amounts. This control is another measure of safety for the DNA, and it is also a way for the cell to regulate the amount of RNA and proteins it makes.

Table 4.2 Components of the Nucleus

Nuclear envelope	Pore-riddled double membrane that controls which substances enter and leave the nucleus
Nucleoplasm	Semifluid interior portion of the nucleus
Nucleolus	Rounded mass of proteins and copies of genes for ribosomal RNA used to construct ribosomal subunits
Chromatin	Total collection of all DNA molecules and associated proteins in the nucleus; all of the cell's chromosomes
Chromosome	One DNA molecule and many proteins associated with it

Figure 4.16 shows the components of the nucleus. Table 4.2 lists their functions. Let's zoom in on the individual components.

The Nuclear Envelope

The membrane of a nucleus, or **nuclear envelope**, consists of two lipid bilayers folded together as a single membrane. As Figure 4.16 shows, the outer bilayer of the membrane is continuous with the membrane of

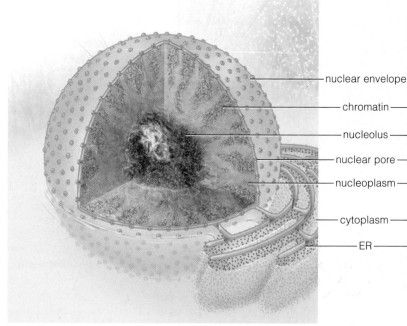
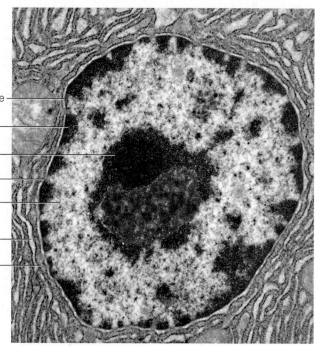

nuclear envelope

chromatin

nucleolus

nuclear pore

nucleoplasm

cytoplasm

ER

1 µm

Figure 4.16 The nucleus. TEM at *right*, nucleus of a mouse pancreas cell.

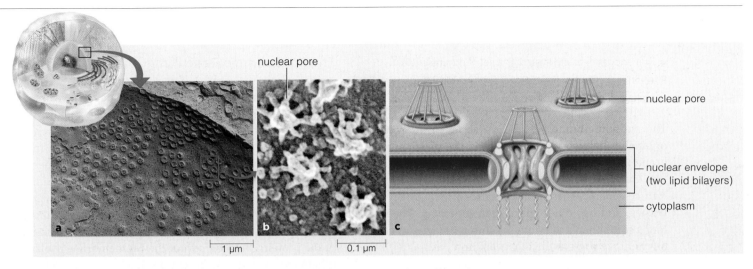

nuclear pore

nuclear pore

nuclear envelope
(two lipid bilayers)

cytoplasm

a

b

c

1 µm

0.1 µm

Figure 4.17 Animated Structure of the nuclear envelope. (**a**) The outer surface of a nuclear envelope was split apart, revealing the pores that span the two lipid bilayers. (**b**) Each nuclear pore is an organized cluster of membrane proteins that selectively allows certain substances to cross it on their way into and out of the nucleus. (**c**) Sketch of the nuclear envelope's structure.

another organelle, the ER. (We will discuss the ER in the next section.)

Different kinds of membrane proteins are embedded in the two lipid bilayers. Some are receptors and transporters; others aggregate into tiny pores that span the membrane (Figure 4.17). These molecules and structures work as a system to transport various molecules across the nuclear membrane. As with all membranes, water and gases cross nuclear membranes freely. All other substances can cross only through transporters and nuclear pores, both of which are selective about which molecules they allow through.

Fibrous proteins that attach to the inner surface of the nuclear envelope anchor DNA molecules and keep them organized. During cell division, these proteins help the cell parcel out the DNA into its offspring.

The Nucleolus

The nuclear envelope encloses **nucleoplasm**, a viscous fluid similar to cytoplasm. The nucleus also contains at least one **nucleolus** (plural, nucleoli), a dense, irregularly shaped region where subunits of ribosomes are assembled from proteins and RNA. The subunits pass through nuclear pores into the cytoplasm, where they join and become active in protein synthesis.

The Chromosomes

Chromatin is the name for all of the DNA, together with its associated proteins, in the nucleus. The genetic material of a eukaryotic cell is distributed among a specific number of DNA molecules. That number is characteristic of the type of organism and the type of cell, but it varies widely among species. For instance, the nucleus of a normal oak tree cell contains 12 DNA molecules; a human body cell, 46; and a king crab cell, 208. Each molecule of DNA, together with its many attached proteins, is called a **chromosome**.

Chromosomes change in appearance over the lifetime of a cell. When a cell is not dividing, its chromatin can appear grainy (as in Figure 4.16). Just before a cell divides, the DNA in each chromosome is copied, or duplicated. Then, during cell division, the chromosomes condense. As they do, they become visible in micrographs. The chromosomes first appear threadlike, then rodlike:

one chromosome
(one unduplicated
DNA molecule)

one chromosome
(one duplicated DNA
molecule, partially
condensed)

one chromosome
(one duplicated DNA
molecule, completely
condensed)

In later chapters, we will look in more detail at the dynamic structure and the function of chromosomes.

Take-Home Message

What is the function of the cell nucleus?

■ A nucleus protects and controls access to a eukaryotic cell's genetic material—its chromosomes.

■ The nuclear envelope is a double lipid bilayer. Proteins embedded in it control the passage of molecules between the nucleus and the cytoplasm.

■ The endomembrane system is a set of organelles that makes, modifies, and transports proteins and lipids.

■ Links to Lipids 3.4, Proteins 3.5

The **endomembrane system** is a series of interacting organelles between the nucleus and the plasma membrane (Figure 4.18). Its main function is to make lipids, enzymes, and proteins for secretion or insertion into cell membranes. It also destroys toxins, recycles wastes, and has other specialized functions. The system's components vary among different types of cells, but here we present the most common ones.

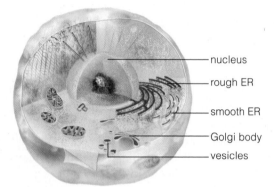

nucleus
rough ER
smooth ER
Golgi body
vesicles

The Endoplasmic Reticulum

Endoplasmic reticulum, or ER, is an extension of the nuclear envelope. It forms a continuous compartment that folds over and over into flattened sacs and tubes. Two kinds of ER are named for their appearance in electron micrographs. Many thousands of ribosomes attach to the outer surface of rough ER (Figure 4.18b). The ribosomes synthesize polypeptide chains, which extrude into the interior of the ER. Inside the ER, the proteins fold and take on their tertiary structure. Some of the proteins become part of the ER membrane itself; others are carried to different destinations in the cell.

Cells that make, store, and secrete a lot of proteins have a lot of rough ER. For example, ER-rich gland cells in the pancreas (an organ) make and secrete enzymes that help digest food in the small intestine.

Smooth ER has no ribosomes, so it does not make proteins (Figure 4.18d). Some of the polypeptides made in the rough ER end up in the smooth ER, as enzymes. These enzymes make most of the cell's membrane lipids. They also break down carbohydrates, fatty acids, and some drugs and poisons. In skeletal muscle cells, a special type of smooth ER called sarcoplasmic reticulum stores calcium ions and has a role in contraction.

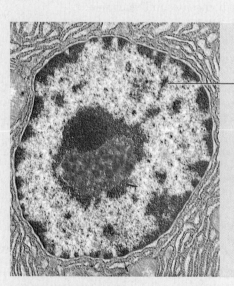

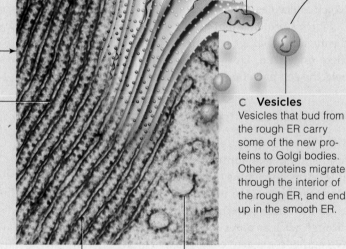

protein

RNA

B Rough ER
Some of the RNA in the cytoplasm is translated into polypeptide chains by ribosomes attached to the rough ER. The chains enter the rough ER, where they are modified into final form.

C Vesicles
Vesicles that bud from the rough ER carry some of the new proteins to Golgi bodies. Other proteins migrate through the interior of the rough ER, and end up in the smooth ER.

A Nucleus
Inside the nucleus, DNA instructions for making proteins are transcribed into RNA, which moves through nuclear pores into the cytoplasm.

ribosome attached to ER

vesicle budding from ER

Figure 4.18 Animated Endomembrane system, where lipids and many proteins are built, then transported to cellular destinations or to the plasma membrane. Chapter 14 describes transcription and translation.

Vesicles

Vesicles are small, membrane-enclosed, saclike organelles. They form in great numbers, and in a variety of types, either on their own or by budding from other organelles or the plasma membrane.

Many types of vesicles transport substances from one organelle to another, or to and from the plasma membrane (Figure 4.18c–f). Other kinds have different roles. For example, **peroxisomes** contain enzymes that digest fatty acids and amino acids. These vesicles form and divide on their own. Peroxisomes have a variety of functions, such as inactivating hydrogen peroxide, a toxic by-product of fatty acid breakdown. Enzymes in the peroxisomes convert hydrogen peroxide to water and oxygen, or they use it in reactions that break down alcohol and other toxins. Drink alcohol, and the peroxisomes in your liver and kidney cells degrade nearly half of it.

Plant and animal cells contain **vacuoles**. Although these vesicles appear empty under a microscope, they serve an important function. Vacuoles are like trash cans; they isolate and dispose of waste, debris, or toxic materials. A central vacuole, described in Section 4.11, helps a plant cell maintain its shape and size.

Golgi Bodies

Many vesicles fuse with and empty their contents into a **Golgi body**. This organelle has a folded membrane that typically looks like a stack of pancakes (Figure 4.18e). Enzymes in a Golgi body put finishing touches on polypeptide chains and lipids that have been delivered from the ER. They attach phosphate groups or sugars, and cleave certain polypeptide chains. The finished products—membrane proteins, proteins for secretion, and enzymes—are sorted and packaged into new vesicles that carry them to the plasma membrane or to lysosomes. **Lysosomes** are vesicles that contain powerful digestive enzymes. They fuse with vacuoles carrying particles or molecules for disposal, such as worn-out cell components. Lysosomal enzymes empty into the other vesicles and digest their contents into bits.

Take-Home Message

What is the endomembrane system?

■ The endomembrane system includes rough and smooth endoplasmic reticulum, vesicles, and Golgi bodies.

■ This series of organelles works together mainly to synthesize and modify cell membrane proteins and lipids.

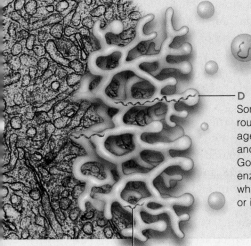

D Smooth ER
Some proteins from the rough ER are packaged into new vesicles and shipped to the Golgi. Others become enzymes of the ER, which assemble lipids or inactivate toxins.

protein in smooth ER

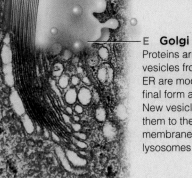

E Golgi body
Proteins arriving in vesicles from the ER are modified into final form and sorted. New vesicles carry them to the plasma membrane or to lysosomes.

F Plasma membrane
Golgi vesicles fuse with the plasma membrane. Lipids and proteins of a vesicle's membrane fuse with the plasma membrane, and the vesicle's contents are released to the exterior of the cell.

4.10 | Lysosome Malfunction

- When lysosomes do not work properly, some cellular materials are not properly recycled, with devastating results.
- Link to Mutations 1.4

Lysosomes serve as waste disposal and recycling centers. Enzymes inside them break large molecules into smaller subunits that the cell can use as building material or eliminate. Different kinds of molecules are broken down by different lysosomal enzymes.

In some people, a genetic mutation causes a deficiency or malfunction in one of the lysosomal enzymes. As a result, molecules that would normally get broken down accumulate instead. The result can be deadly.

For example, cells continually make, use, and break down gangliosides, a kind of lipid. This lipid turnover is especially brisk during early development. In Tay–Sachs disease, the enzyme responsible for ganglioside breakdown misfolds and is destroyed. Most commonly, affected infants seem normal for the first few months. Symptoms begin to appear as gangliosides accumulate to higher and higher levels inside their nerve cells. Within three to six months the child becomes irritable, listless, and may have seizures. Blindness, deafness, and paralysis follow. Affected children usually die by age five (Figure 4.19).

The mutation that causes Tay–Sachs is most prevalent in Jews of Eastern European descent. Cajuns and French Canadians also have a higher than average incidence, but Tay–Sachs occurs in all populations. The mutation can be detected in prospective parents by genetic screening, and in a fetus by prenatal diagnosis.

Researchers continue to explore options for treatment. Potential therapies involve blocking ganglioside synthesis, using gene therapy to deliver a normal version of the missing enzyme to the brain, or infusing normal blood cells from umbilical cords. All treatments are still considered experimental, and Tay–Sachs is still incurable.

Figure 4.19 Conner Hopf was diagnosed with Tay–Sachs disease at age 7–1/2 months. He died at 22 months.

4.11 | Other Organelles

- Eukaryotic cells make most of their ATP in mitochondria.
- Organelles called plastids function in storage and photosynthesis in plants and some types of algae.
- Links to Metabolism 3.2, ATP 3.7

Mitochondria

The **mitochondrion** (plural, mitochondria) is a type of organelle that specializes in making ATP (Figure 4.20). Aerobic respiration, an oxygen-requiring series of reactions that proceeds inside mitochondria, can extract more energy from organic compounds than any other metabolic pathway. With each breath, you are taking in oxygen mainly for the mitochondria in your trillions of aerobically-respiring cells.

Typical mitochondria are between 1 and 4 micrometers in length; a few are as long as 10 micrometers. Some are branched. These organelles can change shape, split in two, and fuse together.

A mitochondrion has two membranes, one highly folded inside the other. This arrangement creates two compartments. Aerobic respiration causes hydrogen ions to accumulate between the two membranes. The buildup causes the ions to flow across the inner membrane, through the interior of membrane transport proteins. That flow drives the formation of ATP.

Nearly all eukaryotic cells have mitochondria, but prokaryotes do not (they make ATP in their cell walls and cytoplasm). The number of mitochondria varies by the type of cell and by the type of organism. For example, a single-celled yeast (a type of fungus) might have only one mitochondrion; a human skeletal muscle cell may have a thousand or more. Cells that have a very high demand for energy tend to have a profusion of mitochondria.

Mitochondria resemble bacteria, in size, form, and biochemistry. They have their own DNA, which is similar to bacterial DNA. They divide independently of the cell, and have their own ribosomes. Such clues led to a theory that mitochondria evolved from aerobic bacteria that took up permanent residence inside a host cell. By the theory of endosymbiosis, one cell was engulfed by another cell, or entered it as a parasite, but escaped digestion. That cell kept its plasma membrane intact and reproduced inside its host. In time, the cell's descendants became permanent residents that offered their hosts the benefit of extra ATP. Structures and functions once required for independent life were no longer needed and were lost over time. Later descendants evolved into mitochondria. We will explore evidence for the theory of endosymbiosis in Section 20.4.

Plastids

Plastids are membrane-enclosed organelles that function in photosynthesis or storage in plants and algal cells. Chloroplasts, chromoplasts, and amyloplasts are common types of plastids.

Photosynthetic cells of plants and many protists contain **chloroplasts**, organelles that are specialized for photosynthesis. Most chloroplasts have an oval or disk shape. Two outer membranes enclose a semifluid interior called the stroma (Figure 4.21). The stroma contains enzymes and the chloroplast's own DNA. Inside the stroma, a third, highly folded membrane forms a single compartment. The folds resemble stacks of flattened disks; the stacks are called grana (singular, granum). Photosynthesis takes place at this membrane, which is called the thylakoid membrane.

The thylakoid membrane incorporates many pigments and other proteins. The most abundant of the pigments are chlorophylls, which appear green. By the process of photosynthesis, the pigments and other molecules harness the energy in sunlight to drive the synthesis of ATP and the coenzyme NADPH. The ATP and NADPH are then used inside the stroma to build carbohydrates from carbon dioxide and water. We will describe the process of photosynthesis in more detail in Chapter 7.

In many ways, chloroplasts resemble photosynthetic bacteria, and like mitochondria they may have evolved by endosymbiosis.

Chromoplasts make and store pigments other than chlorophylls. They have an abundance of carotenoids, a pigment that colors many flowers, leaves, fruits, and roots red or orange. For example, as a tomato ripens, its green chloroplasts are converted to red chromoplasts, and the color of the fruit changes.

Amyloplasts are unpigmented plastids that typically store starch grains. They are notably abundant in cells of stems, tubers (underground stems), and seeds. Starch-packed amyloplasts are dense; in some plant cells, they function as gravity-sensing organelles.

The Central Vacuole

Amino acids, sugars, ions, wastes, and toxins accumulate in the water-filled interior of a plant cell's **central vacuole**. Fluid pressure in the central vacuole keeps plant cells—and structures such as stems and leaves—firm. Typically, the central vacuole takes up 50 to 90 percent of the cell's interior, with cytoplasm confined to a narrow zone in between this large organelle and the plasma membrane. Figure 4.14*b* has an example.

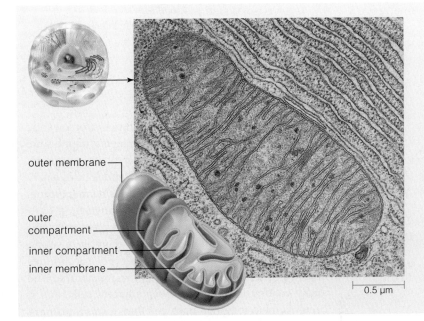

outer membrane
outer compartment
inner compartment
inner membrane
0.5 μm

Figure 4.20 Sketch and transmission electron micrograph of a mitochondrion. This organelle specializes in producing large quantities of ATP.

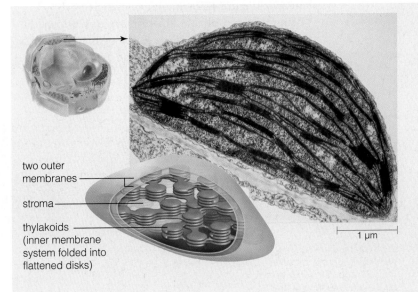

two outer membranes
stroma
thylakoids (inner membrane system folded into flattened disks)
1 μm

Figure 4.21 Animated The chloroplast, a defining character of photosynthetic eukaryotic cells. *Right*, transmission electron micrograph of a chloroplast from a tobacco leaf (*Nicotiana tabacum*). The lighter patches are nucleoids where DNA is stored.

Take-Home Message

What are some other specialized organelles of eukaryotes?

■ Mitochondria are eukaryotic organelles that produce ATP from organic compounds in reactions that require oxygen.

■ Chloroplasts are plastids that carry out photosynthesis.

■ Fluid pressure in a central vacuole keeps plant cells firm.

Cell Surface Specializations

■ A wall or other protective covering often intervenes between a cell's plasma membrane and the surroundings.

■ Link to Tissue 1.1

Eukaryotic Cell Walls

Like most prokaryotic cells, many types of eukaryotic cells have a cell wall around the plasma membrane. The wall is a porous structure that protects, supports, and imparts shape to the cell. Water and solutes easily cross it on the way to and from the plasma membrane. Cells could not live without such exchanges.

Animal cells do not have walls, but plant cells and many protist and fungal cells do. For example, a young plant cell secretes pectin and other polysaccharides onto the outer surface of its plasma membrane. The sticky coating is shared between adjacent cells, and it cements them together. Each cell then forms a **primary wall** by secreting strands of cellulose into the coating. Some of the coating remains as the middle lamella, a sticky layer in between the primary walls of abutting plant cells (Figure 4.22*a,b*).

Being thin and pliable, the primary wall allows the growing plant cell to enlarge. Plant cells with only a thin primary wall can change shape as they develop.

At maturity, cells in some plant tissues stop enlarging and begin to secrete material onto the primary wall's inner surface. These deposits form a firm **secondary wall**, of the sort shown in Figure 4.22*b*. One of the materials deposited is **lignin**, a complex polymer of alcohols that makes up as much as 25 percent of the secondary wall of cells in older stems and roots. Lignified plant parts are stronger, more waterproof, and less susceptible to plant-attacking organisms than younger tissues.

A **cuticle** is a protective body covering made of cell secretions. In plants, a semitransparent cuticle helps protect exposed surfaces of soft parts and limits water loss on hot, dry days (Figure 4.23).

Matrixes Between Cells

Most cells of multicelled organisms are surrounded and organized by **extracellular matrix** (ECM). This nonliving, complex mixture of fibrous proteins and polysaccharides is secreted by cells, and varies with the type of tissue. It supports and anchors cells, separates tissues, and functions in cell signaling.

Primary cell walls are a type of extracellular matrix, which in plants is mostly cellulose. The extracellular matrix of fungi is mainly chitin (Section 3.3). In most

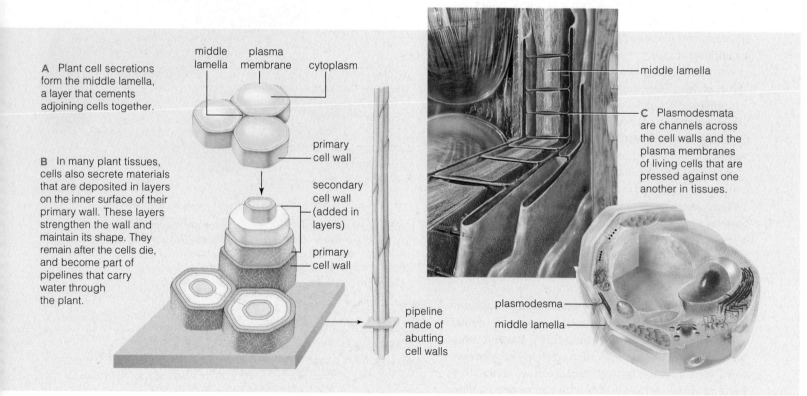

A Plant cell secretions form the middle lamella, a layer that cements adjoining cells together.

middle lamella   plasma membrane   cytoplasm

B In many plant tissues, cells also secrete materials that are deposited in layers on the inner surface of their primary wall. These layers strengthen the wall and maintain its shape. They remain after the cells die, and become part of pipelines that carry water through the plant.

primary cell wall

secondary cell wall (added in layers)

primary cell wall

pipeline made of abutting cell walls

middle lamella

C Plasmodesmata are channels across the cell walls and the plasma membranes of living cells that are pressed against one another in tissues.

plasmodesma

middle lamella

Figure 4.22 Animated Some characteristics of plant cell walls.

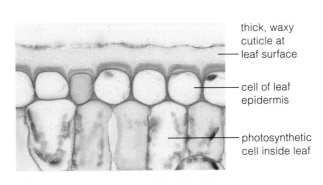

thick, waxy
cuticle at
leaf surface

cell of leaf
epidermis

photosynthetic
cell inside leaf

Figure 4.23 A plant cuticle is a waxy, waterproof covering secreted by living cells.

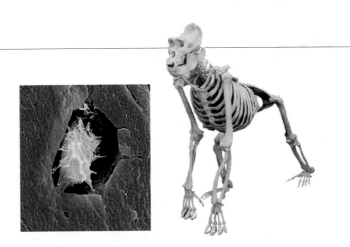

Figure 4.24 A living cell surrounded by hardened bone tissue, the main structural material in the skeleton of most vertebrates.

animals, extracellular matrix consists of various kinds of carbohydrates and proteins; it is the basis of tissue organization, and it provides structural support. For example, bone is mostly extracellular matrix (Figure 4.24). Bone ECM is mostly collagen, a fibrous protein, and it is hardened by mineral deposits.

Cell Junctions

A cell that is surrounded by a wall or other secretions is not isolated; it can still interact with other cells and with the surroundings. In multicelled species, such interaction occurs by way of **cell junctions**, which are structures that connect a cell to other cells and to the environment. Cells send and receive ions, molecules or signals through some junctions. Other kinds help cells recognize and stick to each other and to extracellular matrix.

In plants, channels called plasmodesmata (singular, plasmodesma) extend across the primary wall of two adjoining cells, connecting the cytoplasm of the cells (Figure 4.22c). Substances such as water, ions, nutrients, and signalling molecules can flow quickly from cell to cell through plasmodesmata.

Three types of cell-to-cell junctions are common in most animal tissues: tight junctions, adhering junctions, and gap junctions (Figure 4.25). Tight junctions link cells that line the surfaces and internal cavities of animals. These junctions seal the cells together tightly, so fluid cannot pass between them. Those in your gastrointestinal tract prevent gastric fluid from leaking out of your stomach and damaging your internal tissues. Adhering junctions anchor cells to each other and to extracellular matrix; they strengthen contractile tissues such as heart muscle. Gap junctions are open channels that connect the cytoplasm of adjoining cells; they are similar to plasmodesmata in plants. Gap junctions allow entire regions of cells to respond to a

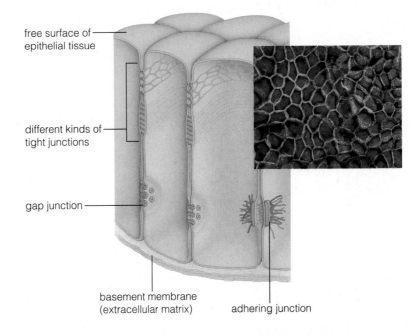

free surface of
epithelial tissue

different kinds of
tight junctions

gap junction

basement membrane
(extracellular matrix)

adhering junction

Figure 4.25 **Animated** Cell junctions in animal tissues. In the micrograph, a continuous array of tight junctions (*green*) seals the abutting surfaces of kidney cell membranes. DNA, which fills each cell's nucleus, appears *red*.

single stimulus. For example, in heart muscle, a signal to contract passes instantly from cell to cell through gap junctions, so all cells contract as a unit.

Take-Home Message

What structures form on the outside of eukaryotic cells?

■ Cells of many protists, nearly all fungi, and all plants, have a porous wall around the plasma membrane. Animal cells do not have walls.

■ Plant cell secretions form a waxy cuticle that helps protect the exposed surfaces of soft plant parts.

■ Cell secretions form extracellular matrixes between cells in many tissues.

■ Cells make structural and functional connections with one another and with extracellular matrix in tissues.

4.13 | The Dynamic Cytoskeleton

- Eukaryotic cells have an extensive and dynamic internal framework called a cytoskeleton.
- Links to Protein structure and function 3.5, 3.6

In between the nucleus and plasma membrane of all eukaryotic cells is a **cytoskeleton**—an interconnected system of many protein filaments. Parts of the system reinforce, organize, and move cell structures, and often the whole cell. Some are permanent; others form only at certain times. Figure 4.26 shows the main types.

Microtubules are long, hollow cylinders that consist of subunits of the protein tubulin. They form a dynamic scaffolding for many cellular processes, rapidly assembling when they are needed, disassembling when they are not. For example, some of the microtubules that assemble before a eukaryotic cell divides separate the cell's duplicated chromosomes, then disassemble. As another example, microtubules that form in the growing end of a young nerve cell support and guide its lengthening in a particular direction.

Microfilaments are fibers that consist primarily of subunits of the globular protein actin. They strengthen or change the shape of eukaryotic cells. Crosslinked, bundled, or gel-like arrays of them make up the **cell cortex**, a reinforcing mesh under the plasma membrane. Actin microfilaments that form at the edge of a cell drag or extend it in a certain direction (Figure 4.26). In muscle cells, microfilaments of myosin and actin interact to bring about contraction.

Intermediate filaments are the most stable parts of a cell's cytoskeleton. They strengthen and maintain cell and tissue structures. For example, some intermediate filaments called lamins form a layer that structurally supports the inner surface of the nuclear envelope.

All eukaryotic cells have similar microtubules and microfilaments. Despite the uniformity, both kinds of elements play diverse roles. How? They interact with accessory proteins, such as the **motor proteins** that can move cell parts in a sustained direction when they are repeatedly energized by ATP.

A cell is like a train station during a busy holiday, with molecules being transported through its interior. Microtubules and microfilaments are like dynamically assembled train tracks. Motor proteins are the freight engines that move along those tracks (Figure 4.27).

Some motor proteins move chromosomes. Others slide one microtubule over another. Some chug along tracks in nerve cells that extend from your spine to your toes. Many engines are organized in series, each moving some vesicle partway along the track before giving it up to the next in line. In plant cells, kinesins drag chloroplasts away from light that is too intense, or toward a light source under low-light conditions.

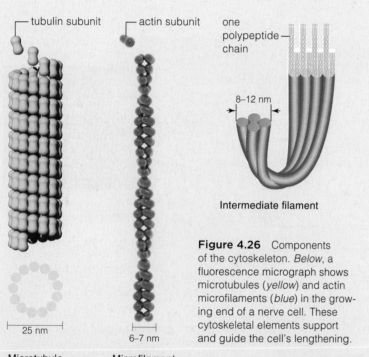

tubulin subunit
actin subunit
one polypeptide chain
8–12 nm
Intermediate filament

Figure 4.26 Components of the cytoskeleton. *Below,* a fluorescence micrograph shows microtubules (*yellow*) and actin microfilaments (*blue*) in the growing end of a nerve cell. These cytoskeletal elements support and guide the cell's lengthening.

25 nm
Microtubule

6–7 nm
Microfilament

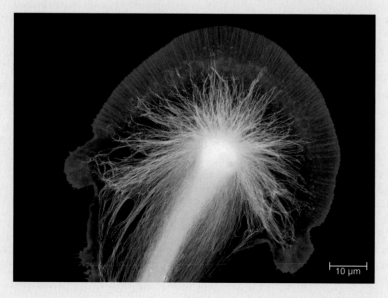

10 µm

Figure 4.27 Animated Kinesin (*tan*), a motor protein dragging cellular freight (in this case, a *pink* vesicle) as it inches along a microtubule.

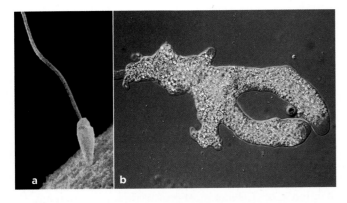

Figure 4.28 (**a**) Flagellum of a human sperm, which is about to penetrate an egg. (**b**) A predatory amoeba (*Chaos carolinense*) extending two pseudopods around its hapless meal: a single-celled green alga (*Pandorina*).

Cilia, Flagella, and False Feet

Organized arrays of microtubules occur in **eukaryotic flagella** (singular, flagellum) and **cilia** (cilium), which are whiplike structures that propel cells such as sperm through fluid (Figure 4.28*a*). Flagella tend to be longer and less profuse than cilia. The coordinated beating of cilia propels motile cells through fluid, and stirs fluid around stationary cells. For example, the coordinated motion of cilia on the thousands of cells lining your airways sweeps particles away from your lungs.

A special array of microtubules extends lengthwise through a flagellum or cilium. This 9+2 array consists of nine pairs of microtubules ringing another pair in the center (Figure 4.29). Protein spokes and links stabilize the array. The microtubules grow from a barrel-shaped organelle called the **centriole**, which remains below the finished array as a basal body.

Amoebas and other types of eukaryotic cells form **pseudopods**, or "false feet" (Figure 4.28*b*). As these temporary, irregular lobes bulge outward, they move the cell and engulf a target such as prey. Elongating microfilaments force the lobe to advance in a steady direction. Motor proteins that are attached to the microfilaments drag the plasma membrane along with them.

Take-Home Message

What is a cytoskeleton?

■ A cytoskeleton of protein filaments is the basis of eukaryotic cell shape, internal structure, and movement.

■ Microtubules organize the cell and help move its parts. Networks of microfilaments reinforce the cell surface. Intermediate filaments strengthen cells and tissues, and maintain their shape.

■ When energized by ATP, motor proteins move along tracks of microtubules and microfilaments. As part of cilia, flagella, and pseudopods, they can move the whole cell.

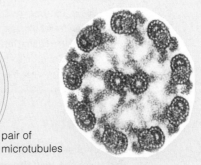

protein spokes
pair of microtubules in a central sheath
plasma membrane
dynein arms
pair of microtubules

A Sketch and micrograph of one eukaryotic flagellum, cross-section. Like a cilium, it contains a 9+2 array: a ring of nine pairs of microtubules plus one pair at its core. Stabilizing spokes and linking elements that connect to the microtubules keep them aligned in this radial pattern.

B Projecting from each pair of microtubules in the outer ring are "arms" of dynein, a motor protein that has ATPase activity. Phosphate-group transfers from ATP cause the dynein arms to repeatedly bind the adjacent pair of microtubules, bend, and then disengage. The dynein arms "walk" along the microtubules. Their motion causes adjacent microtubule pairs to slide past one another

C Short, sliding strokes occur in a coordinated sequence around the ring, down the length of each microtubule pair. The flagellum bends as the array inside bends:

basal body, a microtubule organizing center that gives rise to the 9+2 array and then remains beneath it, inside the cytoplasm

Figure 4.29 Animated Eukaryotic flagella and cilia.

Irradiated meat, poultry, milk, and fruits are now available in supermarkets. By law, irradiated foods must be marked with the symbol on the right. Items that bear this symbol have been exposed to radiation, but are not themselves radioactive. Irradiating fresh foods kills bacteria and prolongs shelf life. However, some worry that the irradiation process may alter the food and produce harmful chemicals.

Whether health risks are associated with consuming irradiated foods is still unknown.

Summary

Sections 4.1–4.3 All organisms consist of one or more cells. By the **cell theory**, the cell is the smallest unit of life, and it is the basis of life's continuity. The **surface-to-volume** ratio limits cell size.

All cells start out life with a **plasma membrane**, a **nucleus** (in **eukaryotic cells**) or **nucleoid** (in **prokaryotic cells**), and **cytoplasm** in which structures such as **ribosomes** are suspended. The **lipid bilayer** is the foundation of all cell membranes. Different types of microscopes use light or electrons to reveal different details of cells.

■ *Use the interactions on CengageNOW to investigate basic membrane structure and the physical limits on cell size.*

■ *Use the animation on CengageNOW to learn how different types of microscopes function.*

Sections 4.4, 4.5 Bacteria and archaeans are prokaryotes (Table 4.3). None has a nucleus. Many have a **cell wall** and one or more **flagella** or **pili**. **Biofilms** are shared living arrangements among bacteria and other microbes.

■ *Use the animation on CengageNOW to view prokaryotic cell structure.*

Sections 4.6–4.11 Eukaryotic cells start out life with a nucleus and other membrane-enclosed **organelles**. The nucleus contains **nucleoplasm** and **nucleoli**. **Chromatin** in the nucleus of a eukaryotic cell is divided into a characteristic number of **chromosomes**. Pores, receptors, and transport proteins in the **nuclear envelope** control the movement of molecules into and out of the nucleus.

The **endomembrane system** includes rough and smooth **endoplasmic reticulum**, **vesicles**, and **Golgi bodies**. This set of organelles functions mainly to make and modify lipids and proteins; it also recycles molecules and particles such as worn-out cell parts, and inactivates toxins.

Mitochondria produce ATP by breaking down organic compounds in the oxygen-requiring pathway of aerobic respiration. **Chloroplasts** are **plastids** that specialize in photosynthesis. Other organelles include **peroxisomes**, **lysosomes**, and **vacuoles** (including **central vacuoles**).

■ *Use the interaction on CengageNOW to survey the major types of eukaryotic organelles.*

■ *Use the animations on CengageNOW to view the nuclear membrane and the endomembrane system.*

■ *Use the animation on CengageNOW to view a chloroplast.*

Section 4.12 Cells of most prokaryotes, protists, fungi, and all plant cells have a wall around the plasma membrane. Older plant cells secrete a rigid, **lignin**-containing **secondary wall** inside their pliable **primary wall**. Many eukaryotic cell types also secrete a **cuticle**. Plasmodesmata connect plant cells. **Cell junctions** connect animal cells to one another and to **extracellular matrix** (ECM).

■ *Study the structure of cell walls and junctions with the animation on CengageNOW.*

Section 4.13 Eukaryotic cells have a **cytoskeleton**. The **cell cortex** consists of **intermediate filaments**. **Motor proteins** that are the basis of movement interact with **microfilaments** in **pseudopods**, or (in **cilia** and **eukaryotic flagella**) **microtubules** that grow from **centrioles**.

■ *Learn more about cytoskeletal elements and their actions with the animation on CengageNOW.*

Self-Quiz   *Answers in Appendix III*

1. The _____ is the smallest unit of life.

2. True or false: Some protists are prokaryotes.

3. Cell membranes consist mostly of _____ .

4. Unlike eukaryotic cells, prokaryotic cells _____ .
   a. have no plasma membrane   c. have no nucleus
   b. have RNA but not DNA   d. a and c

5. Organelles enclosed by membranes are typical features of _____ cells.

6. The main function of the endomembrane system is building and modifying _____ and _____ .

7. Ribosome subunits are built inside the _____ .

8. No animal cell has a _____ .

9. Is this statement true or false? The plasma membrane is the outermost component of all cells. Explain.

10. Enzymes contained in _____ break down worn-out organelles, bacteria, and other particles.

11. Match each cell component with its function.
    ___ mitochondrion   a. protein synthesis
    ___ chloroplast   b. associates with ribosomes
    ___ ribosome   c. ATP by sugar breakdown
    ___ smooth ER   d. sorts and ships
    ___ Golgi body   e. assembles lipids; other tasks
    ___ rough ER   f. photosynthesis

■ *Visit CengageNOW for additional questions.*

Data Analysis Exercise

An abnormal form of the motor protein dynein causes Kartagener syndrome, a genetic disorder characterized by chronic sinus and lung infections. Biofilms form in the thick mucus that collects in the airways, and the resulting bacterial activities and inflammation damage tissues.

Affected men can produce sperm but are infertile (Figure 4.30). Some have become fathers after a doctor injects their sperm cells directly into eggs. Review Figure 4.30, then explain how abnormal dynein could cause the observed effects.

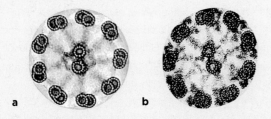

a   b

Figure 4.30 Cross-section of the flagellum of a sperm cell from (**a**) a human male affected by Kartagener syndrome and (**b**) an unaffected male.

Critical Thinking

1. In a classic episode of *Star Trek*, a gigantic amoeba engulfs an entire starship. Spock blows the cell to bits before it reproduces. Think of at least one problem a biologist would have with this particular scenario.

2. Many plant cells form a secondary wall on the inner surface of their primary wall. Speculate on the reason why the secondary wall does not form on the outer surface.

3. A student is examining different samples with a transmission electron microscope. She discovers a single-celled organism swimming in a freshwater pond (*below*).

Which of this organism's structures can you identify? Is it a prokaryotic or eukaryotic cell? Can you be more specific about the type of cell based on what you know about cell structure? Look ahead to Section 22.2 to check your answers.

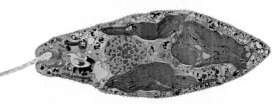

Table 4.3 Summary of Typical Components of Prokaryotic and Eukaryotic Cells

Cell Component	Main Functions	Prokaryotic Bacteria, Archaea	Eukaryotic Protists	Fungi	Plants	Animals
Cell wall	Protection, structural support	✳	✳	✔	✔	—
Plasma membrane	Control of substances moving into and out of cell	✔	✔	✔	✔	✔
Nucleus	Physical separation of DNA from cytoplasm	—*	✔	✔	✔	✔
DNA	Encodes hereditary information	✔	✔	✔	✔	✔
Nucleolus	Assembly of ribosome subunits	—	✔	✔	✔	✔
Ribosome	Protein synthesis	✔	✔	✔	✔	✔
Endoplasmic reticulum (ER)	Synthesis, modification of membrane proteins; lipid synthesis	—	✔	✔	✔	✔
Golgi body	Final modification of membrane proteins; sorting, packaging lipids and proteins into vesicles	—	✔	✔	✔	✔
Lysosome	Intracellular digestion	—	✔	✳	✳	✔
Centriole	Organization of cytoskeletal elements	★	✔	✔	✳	✔
Mitochondrion	ATP formation	—	✔	✔	✔	✔
Chloroplast	Photosynthesis	—	✳	—	✔	—
Central vacuole	Storage	—	—	✳	✔	—
Bacterial flagellum	Locomotion through fluid surroundings	✳	—	—	—	—
Flagellum or cilium with 9+2 microtubule array	Locomotion through or motion within fluid surroundings	—	✳	✳	✳	✔
Cytoskeleton	Cell shape; internal organization; basis of cell movement and, in many cells, locomotion	★	✳	✳	✳	✔

✔ Present in at least part of the life cycle of most or all groups.

✳ Known to be present in cells of at least some groups.

★ Occurs in a form unique to prokaryotes.

* Some planctomycete bacteria have a double membrane around their DNA.

5 | A Closer Look at Cell Membranes

IMPACTS, ISSUES | One Bad Transporter and Cystic Fibrosis

Every cell actively engages in the business of living. Think of how it has to move something as ordinary as water in one direction or the other across its plasma membrane. Water crosses a cell membrane freely. The cell has to be able to take in or send out water at different times in order to keep the cytoplasm from getting too concentrated or too dilute. If all goes well, the cell takes in or sends out water in just the right amounts—not too little, not too much.

Proteins called transporters move ions and molecules, including water, across cell membranes. Different transporters move different substances. One, called CFTR, is a transporter in the plasma membrane of epithelial cells. Sheets of these cells line the passageways and ducts of the lungs, liver, pancreas, intestines, reproductive system, and skin. CFTR pumps chloride ions out of these cells, and water follows the ions. A thin, watery film forms on the surface of the epithelial cell sheets. Mucus slides easily over the wet sheets of cells.

Sometimes a mutation changes the structure of CFTR. When epithelial cell membranes do not have enough working copies of the CFTR protein, chloride ion transport is disrupted. Not enough chloride ions leave the cells, and so not enough water leaves them either. The result is thick, dry mucus that sticks to the epithelial cell sheets.

In the respiratory tract, the mucus clogs airways to the lungs and makes breathing difficult. It is too thick for the ciliated cells lining the airways to sweep out, and bacteria thrive in it. Low-grade infections occur and may persist for years.

These symptoms—outcomes of mutation in the CFTR protein—characterize cystic fibrosis (CF), the most common fatal genetic disorder in the United States. Even with a lung transplant, most CF patients live no longer than thirty years, at which time their lungs usually fail. There is no cure.

More than 10 million people carry a mutated form of the CFTR gene. Some of them have sinus problems, but no other symptoms develop. Most do not know they carry the mutated gene. CF develops when a person inherits a mutated gene from both parents—an unlucky event that occurs in about 1 of 3,300 births (Figure 5.1). Think about it. A startling percentage of the human population can develop severe problems when even one kind of membrane protein does not work.

Your life depends on the functions of thousands of kinds of proteins and other molecules that keep cells working. Each cell functions properly only if it is responsive to conditions in the environments on both sides of its membranes. Cell membranes—these thin boundary layers make the difference between organization and chaos.

Cody, 23

Jeff, 21

Lindsay, 22

Ben, 23

Savannah, 19

Brandon, 18

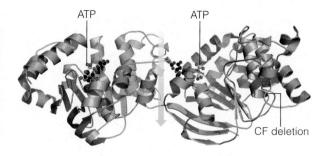
ATP   ATP

CF deletion

See the video! Figure 5.1 Cystic fibrosis. *Left,* a few of the many victims of cystic fibrosis, which occurs most often in people of northern European ancestry. At least one young person dies every day in the United States from complications of this disease.

Above, model of CFTR. The parts shown here are ATP-driven motors that widen or narrow a channel (*gray* arrow) across the plasma membrane. The tiny part of the protein that is lost in most cystic fibrosis mutations is shown on the ribbon in *green.*

Key Concepts

Membrane structure and function

Cell membranes have a lipid bilayer that is a boundary between the outside environment and the cell interior. Diverse proteins embedded in the bilayer or positioned at one of its surfaces carry out most membrane functions. **Sections 5.1, 5.2**

Diffusion and membrane transport

Gradients drive the directional movements of substances across membranes. Transport proteins work with or against gradients to maintain water and solute concentrations. **Sections 5.3, 5.4**

Membrane trafficking

Large packets of substances and engulfed cells move across the plasma membrane by the processes of endocytosis and exocytosis. Membrane lipids and proteins move to and from the plasma membrane during these processes. **Section 5.5**

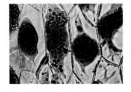

Osmosis

Water tends to diffuse across selectively permeable membranes, including cell membranes, to regions where its concentration is lower. **Section 5.6**

Links to Earlier Concepts

- Reflect again on the road map in Section 1.1. Here you will see how lipids (3.4) and proteins (3.5) become organized in cell membranes (4.2).

- In this chapter, you will consider examples of how a protein's function (3.6) arises from its structure. You will also learn more about the proteins that compose cell junctions (4.12).

- Lipids have both hydrophilic and hydrophobic properties (2.5), a duality that gives rise to the structural organization of all cell membranes.

- Your knowledge of charge (2.1) and the properties of water (2.5) will help you understand the movement of ions and molecules in response to gradients.

- You will revisit the endomembrane system (4.9) as you learn how the cytoskeleton (4.13) is involved in the cycling of membrane lipids and proteins.

- The movement of water into and out of cells is an important part of homeostasis (1.2). A review of what you know about plant cell walls (4.12) will help you understand how this movement affects growth in plants.

5.1 Organization of Cell Membranes

- The basic structure of all cell membranes is the lipid bilayer with many embedded proteins.
- A membrane is a continuous, selectively permeable barrier.
- Links to Emergent properties 1.1, Hydrophilic and hydrophobic 2.5, Lipids 3.4, Membranes 4.2, Tight junctions 4.12

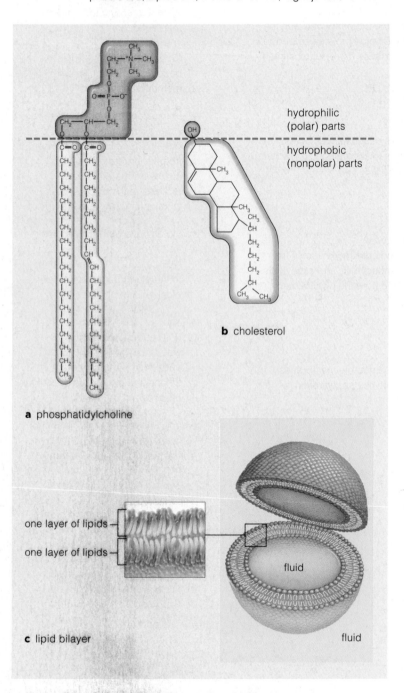

a phosphatidylcholine

b cholesterol

hydrophilic (polar) parts

hydrophobic (nonpolar) parts

one layer of lipids

one layer of lipids

fluid

c lipid bilayer

fluid

Figure 5.2 Cell membrane organization. **(a)** Phosphatidylcholine, the most common phospholipid component of animal cell membranes. **(b)** Cholesterol, the main steroid component of animal cell membranes. Phytosterols are its equivalent in plant cell membranes.

(c) Spontaneous organization of phospholipids into two layers (a lipid bilayer). When mixed with water, phospholipids aggregate into a bilayer, with their hydrophobic tails sandwiched between their hydrophilic heads.

Revisiting the Lipid Bilayer

Properties that are unique to cell membranes emerge when certain lipids—mainly phospholipids—interact. Each phospholipid molecule consists of a phosphate-containing head and two fatty acid tails (Figure 5.2a). The polar head is hydrophilic, which means it interacts with water molecules. The nonpolar tails are hydrophobic, so they do not interact with water molecules. The tails do, however, interact with the tails of other phospholipids. When swirled into water, phospholipids spontaneously assemble into two layers, with all of their nonpolar tails sandwiched between all of their polar heads. Such lipid bilayers are the framework of all cell membranes (Figure 5.2c).

The Fluid Mosaic Model

A **fluid mosaic model** describes the organization of cell membranes. By this model, a cell membrane is a mosaic, a mixed composition of mostly phospholipids, with steroids, proteins, and other molecules dispersed among them (Figure 5.3). The fluid part of the model refers to the behavior of phospholipids in membranes. The phospholipids remain organized as a bilayer, but they also drift sideways, they spin on their long axis, and their tails wiggle.

Variations

Differences in Membrane Composition Membranes differ in composition. The differences reflect their functions in cells. Even the two surfaces of a lipid bilayer are different. For example, carbohydrates attached to certain membrane proteins and lipids project from a plasma membrane but not into the cell. The kinds and numbers of attachments differ from one species to the next, and even among cells of the same body.

Different kinds of cells have different kinds of membrane phospholipids. For example, the fatty acid tails of membrane phospholipids vary in length and saturation. Usually, at least one of the two tails is unsaturated. An unsaturated fatty acid, remember, has one or more double covalent bonds in its carbon backbone (Section 3.4).

Differences in Fluidity We once thought that all proteins in a cell membrane were fixed in place, but key experiments proved otherwise. Two of those experiments are summarized in Figure 5.4. We now know that some proteins stay put, such as those that cluster as rigid pores. Protein filaments of the cytoskeleton

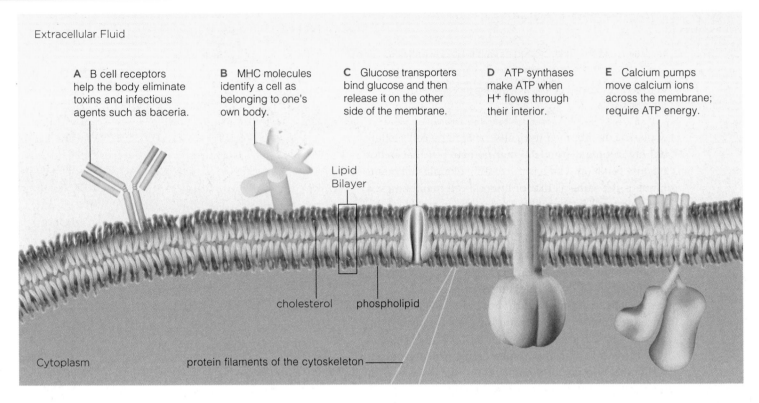

A B cell receptors help the body eliminate toxins and infectious agents such as bacteria.

B MHC molecules identify a cell as belonging to one's own body.

C Glucose transporters bind glucose and then release it on the other side of the membrane.

D ATP synthases make ATP when H+ flows through their interior.

E Calcium pumps move calcium ions across the membrane; require ATP energy.

Lipid Bilayer

cholesterol phospholipid

Cytoplasm protein filaments of the cytoskeleton

Figure 5.3 Animated Fluid mosaic model for the plasma membrane of an animal cell. Section 5.2 presents an overview of the main types of membrane proteins.

lock these and other proteins in place. Tight junctions that link the cytoskeletons of adjacent cells can keep the membrane proteins corraled to the upper or lower surfaces of cells in animal tissues. However, most of the proteins in bacterial and eukaryotic cell membranes drift around very quickly. Part of the reason that the membranes of these organisms are so fluid stems from the composition of the phospholipids in the lipid bilayer.

Archaeans do not build their phospholipids with fatty acids. Instead, they use molecules that have reactive side chains, so the tails of archaean phospholipids form covalent bonds among one another. As a result of this rigid crosslinking, archaean phospholipids do not drift, spin, or wiggle in a bilayer. Thus, the membranes of archaeans are far more rigid than those of bacteria or eukaryotes, a characteristic that may help these cells survive in extreme habitats.

Take-Home Message

What is the function of a cell membrane?

■ A cell membrane is a barrier that selectively controls exchanges between the cell and its surroundings. It is a mosaic of different kinds of lipids and proteins.

■ The foundation of cell membranes is the lipid bilayer—two layers of phospholipids, tails sandwiched between heads.

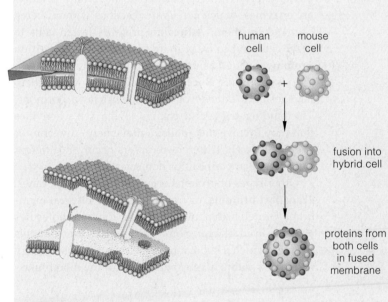

human cell mouse cell

fusion into hybrid cell

proteins from both cells in fused membrane

A Researchers first froze a cell membrane, then they split apart the two layers of its lipid bilayer. Microscopic analysis revealed many proteins embedded within the lipid bilayer.

B Cells of two species were fused into a hybrid cell. In less than one hour, most of the plasma membrane proteins from both species had drifted through the hybrid cell's lipid bilayer and intermingled.

Figure 5.4 Animated Two studies of membrane structure, an observation and an experiment.

5.2 Membrane Proteins

- Cell membrane function begins with the many proteins associated with the lipid bilayer.

- Links to Protein structure 3.5, Protein function 3.6, Cell junctions 4.12

A plasma membrane physically separates a cell's external environment from its internal one, but that is not its only function. The basic structure of a plasma membrane is the same as that of internal cell membranes: a lipid bilayer. The many kinds of proteins in and on the bilayer impart distinct functions to each membrane.

Membrane proteins can be assigned to one of two categories, depending on the way they associate with a membrane. Integral membrane proteins are permanently attached to a lipid bilayer. Some have transmembrane domains—hydrophobic regions that span the entire bilayer. Transmembrane domains anchor the protein in the bilayer, and some form channels all the way through it. Peripheral membrane proteins temporarily attach to one of the bilayer's surfaces by way of interactions with lipids or other proteins.

Each type of protein in a membrane imparts a specific function to it (Figure 5.5). Thus, different cell membranes can have different characteristics. For example, the plasma membrane has proteins that no other cell membrane has. Many peripheral membrane proteins are **enzymes**, which accelerate reactions without being changed by them. **Adhesion proteins** fasten cells to other cells and to ECM in animal tissues. **Recognition proteins** function as unique identity tags for each individual or species. **Receptor proteins** bind to a particular substance outside of the cell, such as a hormone. The binding triggers a change in the cell's activities that may involve metabolism, movement, division, or even cell death. Different receptors occur on different cells, but all are critical for homeostasis.

Other types of proteins occur on all cell membranes. **Transport proteins**, or transporters, are integral membrane proteins that move specific ions or molecules across a lipid bilayer. Some transporters are channels through which a substance diffuses; others use energy to actively pump a substance through the membrane.

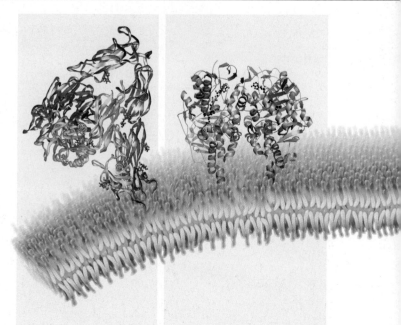

Adhesion Protein

Function Attachment of cells to one another and to extracellular matrix

Occurs only on plasma membranes

Membrane Attachment Integral

Example Integrins, including this one, are also receptors that mediate cell attachment, migration, differentiation, division, and survival.

Example Cadherins are part of adhering junctions between cells.

Example Selectins bind glycoproteins on the surface of cells that function in immunity.

Enzyme

Function Speeds a specific reaction

Membranes provide a relatively stable reaction site for enzymes, particularly those that work in series with other molecules. Arrays of membrane-bound enzymes and other proteins carry out important tasks such as photosynthesis and aerobic respiration.

Membrane Attachment Integral or peripheral

Example The enzyme shown here is a monoamine oxidase of mitochondrial membranes. It catalyzes a hydrolysis reaction that removes an ammonia group (NH_3) from amino acids.

Figure 5.5 Animated Major categories of membrane proteins, with descriptions and examples. You will see the icons above some of the descriptions again in this book.

Transporters span all cell membranes. The other proteins shown are components of plasma membranes. Organelle membranes also incorporate additional kinds of proteins.

Take-Home Message

What do membrane proteins do?

- Various membrane proteins impart functionality to a lipid bilayer.

- A plasma membrane, especially of multicelled species, has receptors and other proteins that function in self-recognition, adhesion, and metabolism.

- All cell membranes have transporters that passively and actively assist specific ions and molecules across the lipid bilayer.

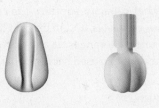

Receptor Protein

Function Binding signaling molecules

Binding causes a change in cell activity, such as gene expression, metabolism, movement, adhesion, division, or cell death.

Membrane Attachment Integral or peripheral

Example The B cell receptor shown here is a protein made only by white blood cells called B lymphocytes. B cell receptors are membrane-bound antibodies. These receptors are vital for immune responses. (We return to immunity in Chapter 38.)

Recognition Protein

Function Identifier of cell type, individual, or species

Membrane Attachment Integral

Example The MHC molecule shown here functions in vertebrate immunity. MHC molecules allow white blood cells called T lymphocytes to identify a cell as *nonself* (foreign) or *self* (belonging to one's own body). Fragments of invading organisms or other nonself particles bound to MHC molecules attract the attention of T lymphocytes. (We return to immunity in Chapter 38.)

Passive Transporter

Function Transport of molecules or ions

Does not require energy

Membrane Attachment Integral

Example On the *left*, a glucose transporter. When glucose binds to this transporter, the protein changes shape, and glucose is released to the other side of the membrane. Passive transporters that change shape are said to be "gated."

Example Other transporters, such as aquaporin, are open channels. Aquaporin transports water.

Example You will see the transporter shown on the *right* several more times in this book. When hydrogen ions flow through a channel in its interior, this molecule synthesizes ATP. Hence its name, ATP synthase.

Active Transporter

Function Transport of molecules or ions

Uses energy (usually in the form of ATP) to pump substances across the membrane

Membrane Attachment Integral

Example The calcium pump shown here uses ATP to pump calcium ions across a membrane.

Example In some contexts, ATP synthase works in reverse, using ATP to pump hydrogen ions across a membrane. In this role, the molecule is an active transporter.

Diffusion, Membranes, and Metabolism

■ Ions and molecules tend to move from one region to another, in response to gradients.

■ Links to Homeostasis 1.2, Charge 2.1, Water 2.5

Membrane Permeability

Any body fluid outside of cells is called extracellular fluid. Many different substances are dissolved in cytoplasm and in extracellular fluid, but the kinds and amounts of solutes in the two fluids differ. The ability of a cell to maintain these differences depends on a property of membranes called **selective permeability**: The membrane allows some substances but not others to cross it. This property helps the cell control which substances and how much of them enter and leave it (Figure 5.6).

Membrane barriers and crossings are vital, because metabolism depends on the cell's capacity to increase, decrease, and maintain concentrations of substances required for reactions. That capacity supplies the cell with raw materials, removes wastes, and maintains the volume and pH within tolerable ranges. It also serves these functions for membrane-enclosed sacs in cells.

Concentration Gradients

Concentration is the number of molecules (or ions) of a substance per unit volume of fluid. A difference in concentration between two adjacent regions is called a **concentration gradient**. Molecules or ions tend to move "down" their concentration gradient, from a region of higher concentration to one of lower concentration.

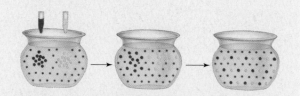

A Dye is dropped into a bowl of water. The dye molecules diffuse until they are evenly dispersed among the water molecules.

B *Red* dye and *yellow* dye are added to a bowl of water. Each substance moves according to its own concentration gradient until all are evenly dispersed.

Figure 5.7 Animated Examples of diffusion.

Why? Like individual atoms, molecules are always in motion. They collide at random and bounce off one another millions of times each second in both regions. However, the more crowded molecules are, the more often they collide. During any interval, more molecules are knocked out of a region of higher concentration than are knocked into it.

Diffusion is the net (or overall) movement of molecules or ions down a concentration gradient. It is an essential way in which substances move into, through, and out of cells. In multicelled species, diffusion also moves substances between cells in different regions of the body or between cells and the body's external environment. For instance, photosynthetic cells inside a leaf produce oxygen. The oxygen diffuses out of the cells and into air spaces inside the leaf, where its concentration is lower. Then it diffuses into the air outside the leaf, where its concentration is lower still.

Any substance tends to diffuse in a direction set by its own concentration gradient, not by the gradients of other solutes that may be sharing the same space. You can observe this tendency by squeezing a drop of dye into water. Dye molecules diffuse slowly into the region where they are less concentrated, regardless of the presence of other solutes (Figure 5.7).

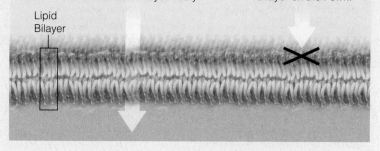

A Gases (such as oxygen and carbon dioxide), small nonpolar molecules, and water cross a bilayer freely.

B Other solutes (molecules and ions) cannot cross a lipid bilayer on their own.

Lipid Bilayer

Figure 5.6 Animated The selectively permeable nature of cell membranes. Small, nonpolar molecules, gases, and water molecules freely cross the lipid bilayer. Polar molecules and ions cross with the help of proteins that span the bilayer.

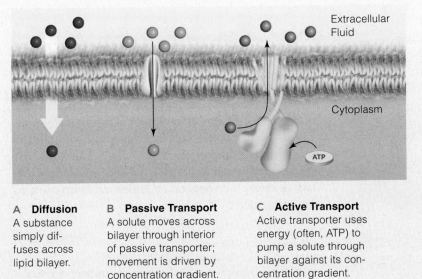

Extracellular Fluid

Cytoplasm

ATP

A Diffusion
A substance simply diffuses across lipid bilayer.

B Passive Transport
A solute moves across bilayer through interior of passive transporter; movement is driven by concentration gradient.

C Active Transport
Active transporter uses energy (often, ATP) to pump a solute through bilayer against its concentration gradient.

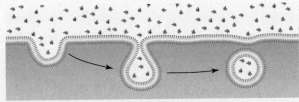

D Endocytosis Vesicle movement brings substances in bulk into cell.

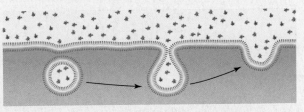

E Exocytosis Vesicle movement ejects substances in bulk from cell.

Figure 5.8 Overview of membrane-crossing mechanisms.

The Rate of Diffusion

How quickly a solute diffuses depends on five factors:

1. Size. It takes less energy to move a smaller molecule, so smaller molecules diffuse faster.

2. Temperature. Molecules move faster at higher temperature, so they collide more often. Rebounds from the collisions propel them away from one another.

3. Steepness of the concentration gradient. The rate of diffusion is higher with steeper gradients. Again, molecules collide more often in a region of greater concentration. So, more molecules bounce out of a region of greater concentration than bounce into it.

4. Charge. Each ion dissolved in a fluid contributes to the fluid's overall electric charge. A difference in charge between two regions can affect the rate and direction of diffusion between them, because opposite charges attract and like charges repel. For example, positively charged substances, such as sodium ions, will move toward a region with a negative charge.

5. Pressure. Diffusion may be affected by a difference in pressure between two adjoining regions. Pressure squeezes molecules together; molecules that are more crowded collide and rebound more frequently.

How Substances Cross Membranes

Selective permeability is a property that arises from a membrane's structure. A lipid bilayer lets gases and nonpolar molecules cross freely, but it is impermeable to ions and large, polar molecules.

A passive transport protein allows a specific solute to follow its gradient across a membrane. The solute binds to the protein, and is released to the other side of the membrane. This process, which is called passive transport or facilitated diffusion, requires no energy input; the movement is driven by the solute's concentration gradient. Some molecules (such as water) that diffuse across a membrane on their own can also move through passive transport proteins.

An active transport protein pumps a specific solute across a membrane against its gradient. This mechanism, active transport, requires energy—typically in the form of ATP.

Other energy-requiring mechanisms move particles in bulk into or out of cells. In endocytosis, a patch of plasma membrane sinks inward, bringing with it molecules on the outside of the cell. In exocytosis, a vesicle in the cytoplasm fuses with the plasma membrane, so that its contents are released outside the cell.

Figure 5.8 shows an overview of these membrane-crossing mechanisms; the sections that follow describe them in detail.

Take-Home Message

What influences the movement of ions and molecules across cell membranes?

■ Diffusion is net movement of molecules or ions into an adjoining region where they are not as concentrated.

■ The steepness of a concentration gradient as well as temperature, molecular size, and electric and pressure gradients affect the rate of diffusion.

■ Substances move across cell membranes by diffusion, passive and active transport, endocytosis, and exocytosis.

5.4 | Passive and Active Transport

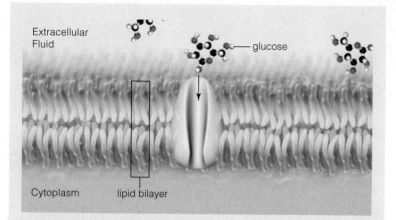

A A glucose molecule (here, in extracellular fluid) binds to a transport protein embedded in the lipid bilayer.

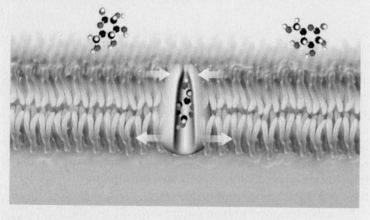

B Binding causes the protein to change shape.

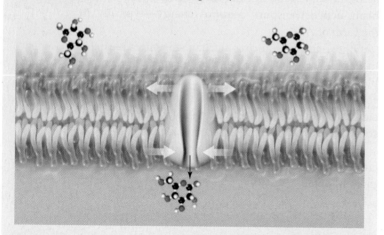

C The glucose molecule detaches from the transport protein on the other side of the membrane (here, in the cytoplasm), and the protein resumes its original shape.

Figure 5.9 Animated Passive transport. This model shows one of the glucose transporters that span the plasma membrane. Glucose crosses in both directions. The net movement of this solute is to the side of the membrane where it is less concentrated.

■ Many types of molecules and ions diffuse across a lipid bilayer only with the help of transport proteins.

Many solutes cross a membrane by associating with transport proteins. Each type of transport protein can move a specific ion or molecule across a membrane. Glucose transporters only transport glucose; calcium pumps only pump calcium; and so on. The specificity means that the amounts and types of substances that cross a membrane depend on which transport proteins are embedded in it.

Passive Transport

In **passive transport**, a concentration gradient drives the diffusion of a solute across a cell membrane, with the assistance of a transport protein. The protein does not require energy to assist the solute's movement; thus, passive transport is also called facilitated diffusion.

Some passive transporters are open channels; others are "gated." A gated transporter changes shape when a molecule binds to it, or in response to a change in electric charge. The protein's shape change moves the solute to the opposite side of the membrane, where it detaches. Then, the transporter reverts to its original shape. Figure 5.9 shows an example, a glucose transporter. Glucose molecules diffuse unassisted across a lipid bilayer, but the transporter increases the rate of diffusion by about 50,000 times.

The net movement of a particular solute through passive transporters tends to be toward the side of the membrane where the solute is less concentrated. This is because molecules or ions simply collide with the transporters more often on the side of the membrane where they are more concentrated.

Passive transport continues until the concentrations on both sides of the membrane are equal. However, such equilibrium rarely occurs in a living system. For example, cells use up glucose as fast as they get it. As soon as a glucose molecule enters a cell, it is broken down for energy or it is used to build other molecules. Thus, there is usually a concentration gradient across the membrane that favors uptake of more glucose.

Active Transport

Solute concentrations shift constantly in the cytoplasm and extracellular fluid. Maintaining a solute's concentration at a certain level often means transporting the solute against its gradient, to the side of a membrane where it is more concentrated. Such pumping does not occur without energy inputs, usually from ATP.

In **active transport**, a transport protein uses energy to pump a solute against its gradient across a cell membrane. Energy, often in the form of a phosphate-group transfer from ATP, changes the shape of the transporter. The change causes the transporter to release the solute to the other side of the membrane.

For example, **calcium pumps** are active transporters that move calcium ions across muscle cell membranes (Figure 5.10). Muscle cells contract when the nervous system causes calcium ions to flood out from a special organelle, the sarcoplasmic reticulum, which is wrapped around the muscle fiber. The flood clears out binding sites on motor proteins that make muscles contract (Section 4.13). Contraction ends after calcium pumps have moved most of the calcium ions back into the sarcoplasmic reticulum, against their concentration gradient. Calcium pumps keep the concentration of calcium in that compartment 1,000 to 10,000 times higher than it is in muscle cell cytoplasm.

The sodium–potassium pump is a **cotransporter**—it moves two substances at the same time (Figure 5.11). Nearly all of the cells in your body have these pumps, which transport sodium and potassium ions in opposite directions across a membrane. Sodium ions (Na^+) in the cytoplasm diffuse into the pump's open channel and bind to its interior. The pump changes shape after it receives a phosphate group from ATP. Its channel opens to the extracellular fluid, and it releases the Na^+. Then, potassium ions (K^+) from extracellular fluid diffuse into the channel and bind to its interior. The transporter releases the phosphate group, then reverts to its original shape. The channel opens to the cytoplasm, and the K^+ is released there.

Bear in mind, the membranes of all cells, not just those of animals, have active transporters. In Section 29.5, for example, you will learn how sugars made in a plant's leaves are pumped into tubes that distribute them through the plant body.

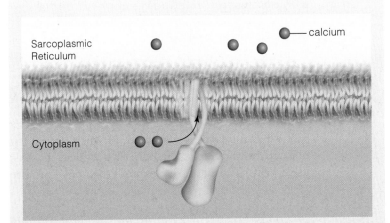

A Calcium ions bind to a calcium transporter (calcium pump).

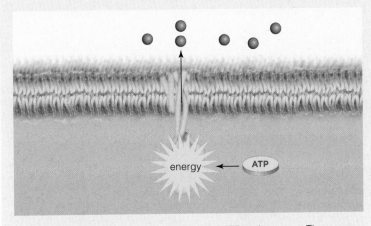

B A phosphate group is transferred from ATP to the pump. The pump changes shape so that it ejects the calcium ions to the opposite side of the membrane, and then resumes its original shape.

Figure 5.10 Animated Active transport. This model shows a calcium transporter. After two calcium ions bind to the transporter, ATP transfers a phosphate group to it, thus providing energy that drives the movement of calcium against its concentration gradient across the cell membrane.

Take-Home Message

If a molecule or ion cannot diffuse through a lipid bilayer, how does it cross a cell membrane?

■ Transport proteins help specific molecules or ions to cross cell membranes. Which substances cross a membrane is mostly determined by the transport proteins embedded in it.

■ In passive transport, a solute binds a protein that releases it on the opposite side of the membrane. No energy is required; the net movement of solute is down its concentration gradient.

■ In active transport, a protein pumps a solute across a membrane, against its concentration gradient. The transporter must be activated, usually by an energy input from ATP.

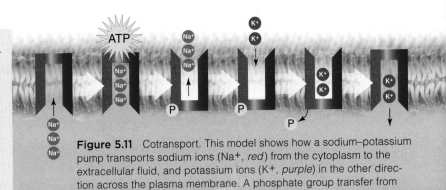

Figure 5.11 Cotransport. This model shows how a sodium–potassium pump transports sodium ions (Na^+, *red*) from the cytoplasm to the extracellular fluid, and potassium ions (K^+, *purple*) in the other direction across the plasma membrane. A phosphate group transfer from ATP provides energy for the transport.

Membrane Trafficking

- By processes of exocytosis and endocytosis, cells take in and expel particles that are too big for transport proteins, as well as substances in bulk.

- Links to Endomembrane system 4.9, Cytoskeleton 4.13

Endocytosis and Exocytosis

Think back on the structure of a lipid bilayer (Figure 5.2). When a bilayer is disrupted, as when part of the plasma membrane pinches off as a vesicle, it seals itself. Why? The disruption exposes the nonpolar fatty acid tails of the phospholipids to their watery surroundings. Remember, in water, phospholipids spontaneously rearrange themselves so that their tails stay together. When a patch of membrane buds, its phospholipid tails are repelled by water on both sides. The water "pushes" the phospholipid tails together, which helps round off the bud as a vesicle, and also seals the rupture in the membrane.

As part of vesicles, patches of membrane constantly move to and from the cell surface (Figure 5.12). The formation and movement of vesicles, which is called membrane trafficking, involves motor proteins and requires ATP (Section 4.13).

By **exocytosis**, a vesicle moves to the cell surface, and the protein-studded lipid bilayer of its membrane fuses with the plasma membrane. As the exocytic vesicle loses its identity, its contents are released to the surroundings (Figure 5.12).

There are three pathways of **endocytosis**, but they all take up substances near the cell's surface. A small patch of plasma membrane balloons inward, and then it pinches off after sinking farther into the cytoplasm. The membrane patch becomes the outer boundary of an endocytic vesicle, which delivers its contents to an organelle or stores them in a cytoplasmic region.

With receptor-mediated endocytosis, molecules of a hormone, vitamin, mineral, or another substance bind to receptors on the plasma membrane. A shallow pit forms in the membrane patch under the receptors. The pit sinks into the cytoplasm and closes back on itself, and in this way it becomes a vesicle (Figure 5.13).

Phagocytosis ("cell eating") is an endocytic pathway. Phagocytic cells such as amoebas engulf microorganisms, cellular debris, or other particles. In animals, macrophages and other phagocytic white blood cells engulf and digest pathogenic viruses and bacteria, cancerous body cells, and other threats.

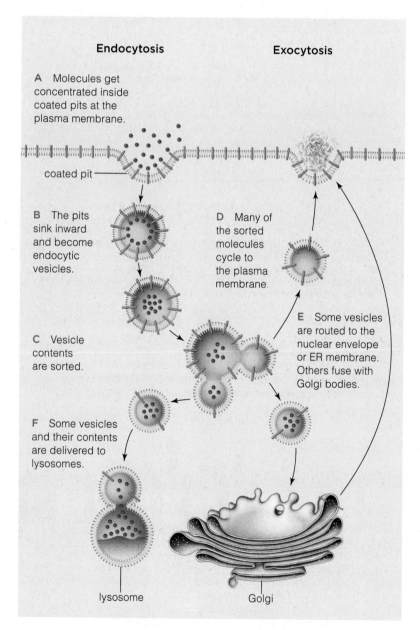

Endocytosis

A Molecules get concentrated inside coated pits at the plasma membrane.

coated pit

B The pits sink inward and become endocytic vesicles.

C Vesicle contents are sorted.

F Some vesicles and their contents are delivered to lysosomes.

lysosome

Exocytosis

D Many of the sorted molecules cycle to the plasma membrane.

E Some vesicles are routed to the nuclear envelope or ER membrane. Others fuse with Golgi bodies.

Golgi

Figure 5.12 Animated Endocytosis and exocytosis.

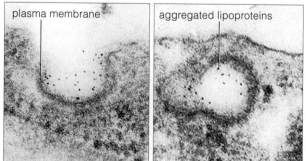

plasma membrane

aggregated lipoproteins

Figure 5.13 Endocytosis of lipoprotein aggregates.

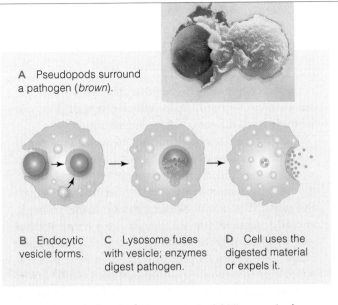

A Pseudopods surround a pathogen (*brown*).

B Endocytic vesicle forms.

C Lysosome fuses with vesicle; enzymes digest pathogen.

D Cell uses the digested material or expels it.

Figure 5.14 Animated Phagocytosis. (**a**) Micrograph of a phagocytic cell with its pseudopods (the extending lobes of cytoplasm) surrounding a pathogen.

(**b–d**) Diagram showing what happens inside a phagocytic cell after pseudopods (the extending lobes of cytoplasm) surround a pathogen. The plasma membrane above the bulging lobes fuses and forms an endocytic vesicle. Inside the cytoplasm, the vesicle fuses with a lysosome, which digests its contents.

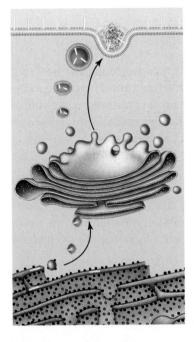

Figure 5.15 How membrane proteins become oriented to the inside or the outside of the cell.

Proteins of the plasma membrane are assembled in the ER, and finished inside Golgi bodies. The proteins become part of vesicle membranes that bud from the Golgi. The membrane proteins automatically become oriented in the proper direction when the vesicles fuse with the plasma membrane.

Figure It Out: What process does the upper arrow represent?

Answer: Exocytosis

We now know that receptor-mediated endocytosis is a misleading name, because receptors also function in phagocytosis. When these receptors bind to a target, they cause microfilaments to assemble in a mesh under the plasma membrane. The microfilaments contract, forcing some cytoplasm and plasma membrane above it to bulge outward as a lobe, or pseudopod (Figures 4.28*b* and 5.14). Pseudopods engulf a target and merge as a vesicle, which sinks into the cytoplasm and fuses with a lysosome (Section 4.9). Enzymes in the lysosome break down the vesicle's contents. Lysosomal enzymes digest the vesicle into fragments and smaller, reusable molecules.

Bulk-phase endocytosis is not as selective. A vesicle forms around a small volume of the extracellular fluid regardless of the kinds of substances dissolved in it.

Membrane Cycling

As long as a cell is alive, exocytosis and endocytosis are continually replacing and withdrawing patches of its plasma membrane, as in Figure 5.12.

The composition of a plasma membrane begins in the ER (Section 4.9). Membrane proteins and lipids are made and modified, and both become part of vesicles that transport them to Golgi bodies for final modification. The finished proteins and lipids are repackaged as new vesicles that travel to the plasma membrane and fuse with it. The lipids and proteins of the vesicle membrane become part of the plasma membrane. This is the process by which new plasma membrane forms.

Figure 5.15 shows what happens when an exocytic vesicle fuses with the plasma membrane. Golgi bodies package membrane proteins facing the inside of a vesicle, so after fusion the proteins face outside the cell.

In a cell that is no longer growing, the total area of the plasma membrane remains more or less constant. Membrane is lost as a result of endocytosis, but it is replaced by membrane arriving as exocytic vesicles.

Take-Home Message

How do cells take in large particles and bulk substances?

■ Exocytosis and endocytosis move materials in bulk across plasma membranes.

■ By exocytosis, a cytoplasmic vesicle fuses with the plasma membrane and releases its contents to the outside of the cell.

■ By endocytosis, a patch of plasma membrane sinks inward and forms a vesicle in the cytoplasm.

■ Receptor-mediated endocytosis and phagocytosis are two endocytic pathways that occur when specific substances bind to receptors. Bulk-phase endocytosis is not specific.

■ Plasma membrane lost during endocytosis is replaced by membrane that surrounds exocytic vesicles.

Which Way Will Water Move?

- Water diffuses across cell membranes by osmosis.
- Osmosis is driven by tonicity, and is countered by turgor.
- Links to Water 2.5, Plant cell walls 4.12

Osmosis

Like any other substance, water molecules tend to diffuse in response to their own concentration gradient. **Osmosis** is the name for this movement. As you read earlier, water crosses cell membranes on its own, and also through transport proteins.

You might be wondering: How can water be more or less concentrated? Think of water's concentration in terms of relative numbers of water molecules and solute molecules. The concentration of water depends on the total number of molecules or ions dissolved in it. The higher the solute concentration, the lower the water concentration.

For example, when you pour some sugar into a container that is partially filled with water, you increase the total volume of liquid. The number of water molecules is unchanged, but they are now dispersed in a larger total volume. As a result of the added solute, the number of water molecules per unit volume—the water concentration—has decreased.

Tonicity

Tonicity refers to the relative concentrations of solutes in two fluids that are separated by a selectively permeable membrane. When the solute concentrations

differ, the fluid with the lower concentration of solutes is said to be **hypotonic**. The other one, with the higher solute concentration, is **hypertonic**. Fluids that are **isotonic** have the same solute concentration.

Tonicity dictates the direction of water movement across membranes: Water diffuses from a hypotonic to a hypertonic fluid. Suppose a container is divided into two sections by a membrane that water, but not sugar, can cross. If you pour water into both compartments and add sugar to just one, you set up a concentration gradient. The sugar solution is hypertonic. By osmosis, water will follow its gradient and diffuse across the membrane into the sugar solution (Figure 5.16).

Now imagine that you have a sheet of a selectively permeable membrane that water, but not sucrose, can cross. You make a bag out of the membrane, then fill it with a 2 percent sucrose solution. If you drop the bag into a solution with 2 percent sucrose (an isotonic solution), the bag stays the same size (Figure 5.17a). If you drop it into a 10 percent sucrose solution (a hypertonic solution), the bag will shrink as water diffuses out of it. If you drop the bag into water with no sucrose in it (which is hypotonic with respect to the solution), it will swell up as water diffuses into it.

A cell is essentially a semipermeable membrane bag of fluid. What happens when the fluid outside of a cell is hypertonic? Water will follow its gradient and cross the membrane to the hypertonic side, and the volume of the cell will decrease as water diffuses out of it. If the outside fluid is very hypotonic, the volume of the cell will increase as water diffuses into it.

Most free-living cells can counter shifts in tonicity by selectively transporting solutes across the plasma membrane. Most cells of multicelled species cannot. In multicelled organisms, maintaining the tonicity of extracellular fluids is part of homeostasis. Thus, tissue fluid is normally isotonic with fluid inside cells (Figure 5.17b). If a tissue fluid were to become hypertonic, the cells would lose water, and they would shrivel (Figure 5.17c). If the fluid were to become hypotonic, too much water would diffuse into the cells, and they would burst (Figure 5.17d).

Effects of Fluid Pressure

Hydrostatic pressure, or as botanists say, **turgor**, often counters osmosis. Both terms refer to pressure that a volume of fluid exerts against a cell wall, membrane, tube, or any other structure that holds it. Cell walls in plants and many protists, fungi, and bacteria resist an increase in the volume of cytoplasm. The walls of

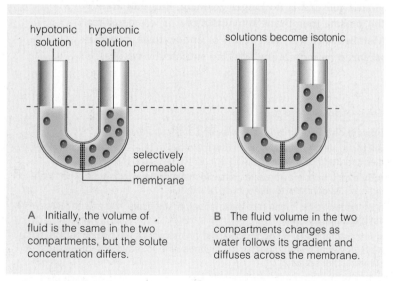

hypotonic solution hypertonic solution

solutions become isotonic

selectively permeable membrane

A Initially, the volume of fluid is the same in the two compartments, but the solute concentration differs.

B The fluid volume in the two compartments changes as water follows its gradient and diffuses across the membrane.

Figure 5.16 Animated Experiment showing a change in fluid volume as an outcome of osmosis. A selectively permeable membrane separates two regions.

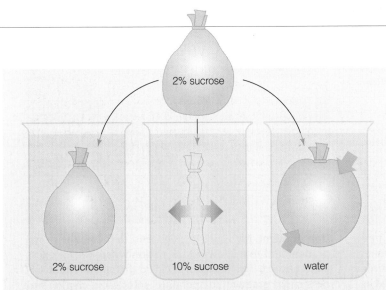

2% sucrose

2% sucrose   10% sucrose   water

A What happens to a semipermeable membrane bag when it is immersed in an isotonic, a hypertonic, or a hypotonic solution?

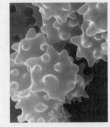

B Red blood cells in an isotonic solution do not change in volume.

C Red blood cells in a hypertonic solution shrivel because water diffuses out of them.

D Red blood cells in a hypotonic solution swell because water diffuses into them.

Figure 5.17 Animated (**a**) A tonicity experiment. (**b–d**) The micrographs show human red blood cells that were immersed in fluids of different tonicity.

Figure 5.18 (**a**) A tomato plant undergoing osmotically induced wilting within thirty minutes after salty water was added to the soil in the pot. (**b**) Cells from an iris petal, plump with water. Their cytoplasm and central vacuole extend to the cell wall. (**c**) Cells from a wilted iris petal. Their cytoplasm and central vacuole shrank, and the plasma membrane moved away from the wall.

blood vessels resist an increase in blood volume. The amount of hydrostatic pressure that can stop water from diffusing into cytoplasmic fluid or other hypertonic solution is called **osmotic pressure**.

As one example, growing plant cells are hypertonic relative to water in soil (the cytoplasmic fluid usually has more solutes than soil water). Water diffusing into a young plant cell by osmosis exerts fluid pressure on the primary wall. The thin, pliable wall expands under pressure, which lets the cytoplasmic volume increase (Section 4.12). Expansion of the wall—and the cell—ends when the osmotic pressure inside the cell builds up enough to prevent the uptake of additional water.

Hydrostatic pressure also supports soft plant parts. When a plant with soft green leaves is growing well in soil that has enough water, hydrostatic pressure keeps the cells plump—and the plant erect. As the soil dries out, the concentration of salt in soil water increases. If the soil water becomes hypertonic with respect to

cytoplasmic fluid, water diffuses out of the plant's cells and hydrostatic pressure in them falls. The cytoplasm shrinks, and the plant wilts. Adding salt to the soil has the same effect. Figure 5.18 shows what happens when you pour salty water into soil around a tomato plant's roots. Within thirty minutes, the plant droops.

Take-Home Message

Why and how does water move into and out of cells?

■ Water moves in response to its own concentration gradient, which is influenced by solute concentration.

■ Osmosis is a net diffusion of water between two solutions that differ in water concentration and are separated by a selectively permeable membrane.

■ Water tends to move osmotically to regions of greater solute concentration (from a hypotonic to a hypertonic solution). No net diffusion occurs between isotonic solutions.

■ Fluid pressure that a solution exerts against a membrane or wall influences the osmotic movement of water.

CFTR is an active transporter of chloride ions. In about 90 percent of CF patients, loss of a single amino acid from the protein causes the disorder. The mutated CFTR proteins are functional, but enzymes destroy them before they reach the plasma membrane. Thus, cystic fibrosis is most often a result of impaired membrane trafficking of the CFTR protein.

How would you vote?

Lung tissue of a baby with cystic fibrosis; white patches are mucus. Should we screen prospective parents for CF mutations? See CengageNow for details, then vote online.

Summary

Sections 5.1, 5.2 A cell membrane is a **selectively permeable** barrier that separates an internal environment from an external one. Each is a mosaic of lipids (mainly phospholipids) and proteins. The lipids are organized as a double layer in which the nonpolar tails of both layers are sandwiched between the polar heads. Membranes of bacteria and eukaryotic cells can be described by a **fluid mosaic** model; those of archaeans are not fluid.

Proteins that are transiently or permanently associated with a membrane carry out most membrane functions. All membranes have **transport proteins**. Plasma membranes also incorporate **receptor proteins**, **adhesion proteins**, **enzymes**, and **recognition proteins** (Table 5.1).

■ *Use the animation on CengageNOW to learn about membrane structure and the experiments that elucidated it.*

■ *Use the animation on CengageNOW to familiarize yourself with the functions of receptor proteins.*

Section 5.3 A difference in the **concentration** of a substance between adjoining regions of fluid is a **concentration gradient**. Molecules or ions tend to follow their own gradent and move toward the region where they are less concentrated. This behavior is called **diffusion**. The steepness of the gradient, temperature, solute size, charge, and pressure influence the diffusion rate.

Gases, water, and small nonpolar molecules diffuse across a membrane. Most other molecules and ions cross only with the help of transport proteins.

■ *Use the interaction on CengageNOW to investigate diffusion across membranes.*

Section 5.4 Transport proteins move specific solutes across membranes. The types of transport proteins in a membrane determine which substances cross it. **Active transport** proteins such as **calcium pumps** use energy, usually from ATP, to move a solute against its concentration gradient. **Passive transport** proteins work without an energy input; solute movement is driven by the concentration gradient. **Cotransporters** move solutes in different directions across a membrane.

■ *Use the animation on CengageNOW to compare the processes of passive and active transport.*

Section 5.5 **Exocytosis**, **endocytosis**, and **phagocytosis** move bulk substances and large particles across plasma membranes. With exocytosis, a cytoplasmic vesicle fuses with the plasma membrane, and its contents are released to the outside of the cell. The vesicle's membrane lipids and proteins become part of the plasma membrane. With endocytosis, a patch of plasma membrane balloons into the cell, and forms a vesicle that sinks into the cytoplasm. Plasma membrane lost by endocytosis is replaced by exocytic vesicles.

■ *Use the animations on CengageNOW to discover how membrane components are cycled, and to explore phagocytosis.*

Section 5.6 **Osmosis** is the diffusion of water across a selectively permeable membrane, from the region with a lower solute concentration (**hypotonic**) toward the region with a higher solute concentration (**hypertonic**). There is no net movement of water between **isotonic** solutions. **Osmotic pressure** is the amount of **turgor** or **hydrostatic pressure** (fluid pressure against a cell membrane or wall) that stops osmosis.

■ *Use the animation on CengageNOW to explore osmosis.*

Table 5.1 Common Types of Membrane Proteins

Category	Function	Examples
Passive transporters	Allow ions or small molecules to cross a membrane to the side where they are less concentrated. Open or gated channels.	Porins; glucose transporter
Active transporters	Pump ions or molecules through membranes to the side where they are more concentrated. Require energy input, as from ATP.	Calcium pump; serotonin transporter
Receptors	Initiate change in a cell's activity by responding to an outside signal (e.g., by binding to a signaling molecule).	Insulin receptor; B cell receptor
Cell adhesion molecules	Help cells stick to one another and to extracellular matrix.	Integrins; cadherins
Recognition proteins	Identify cells as self (belonging to one's own body or tissue)	Histocompatibility molecules
Enzymes	Speed reactions without being altered by them.	Diverse hydrolases

Data Analysis Exercise

In most individuals with cystic fibrosis, the 508th amino acid of the CFTR protein (a phenylalanine) is missing. A CFTR protein with this change is synthesized correctly, and it can transport ions correctly, but it never reaches the plasma membrane to do its job.

Sergei Bannykh and his coworkers developed a procedure to measure the relative amounts of the CFTR protein localized in different regions of a cell. They compared the pattern of CFTR distribution in normal cells with the pattern in CFTR-mutated cells. A summary of their results is shown in Figure 5.19.

1. Which organelle contains the least amount of CFTR protein in normal cells? In CF cells? Which contains the most?

2. In which organelle is the amount of CFTR protein in CF cells closest to the amount in normal cells?

3. Where is the mutated CFTR protein getting held up?

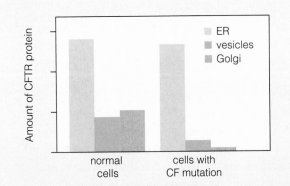

Figure 5.19 Comparison of the amounts of CFTR protein associated with endoplasmic reticulum (*blue*), vesicles traveling from ER to Golgi (*green*), and Golgi bodies (*orange*). The patterns of CFTR distribution in normal cells, and cells with the most common cystic fibrosis mutation, were compared.

Self-Quiz   *Answers in Appendix III*

1. Cell membranes consist mainly of a _____ .
  a. carbohydrate bilayer and proteins
  b. protein bilayer and phospholipids
  c. lipid bilayer and proteins

2. In a lipid bilayer, _____ of all the lipid molecules are sandwiched between all the _____ .
  a. hydrophilic tails; hydrophobic heads
  b. hydrophilic heads; hydrophilic tails
  c. hydrophobic tails; hydrophilic heads
  d. hydrophobic heads; hydrophilic tails

3. By the _____ model, cell membranes are flexible structures composed of a mixture of many different types of molecules.

4. Most membrane functions are carried out by _____ .
  a. proteins   c. nucleic acids
  b. phospholipids   d. hormones

5. Organelle membranes incorporate _____ .
  a. transport proteins   c. recognition proteins
  b. adhesion proteins   d. all of the above

6. Some _____ proteins are also receptors.

7. Diffusion is the movement of ions or molecules from a region where they are _____ (more/less) concentrated to another where they are _____ (more/less) concentrated.

8. Name one molecule that can readily diffuse across a lipid bilayer.

9. Some sodium ions cross a cell membrane through transport proteins that first must be activated by an energy boost. This is an example of _____ .
  a. passive transport   c. facilitated diffusion
  b. active transport   d. a and c

10. Immerse a living cell in a hypotonic solution, and water will tend to _____ .
  a. move into the cell   c. show no net movement
  b. move out of the cell d. move in by endocytosis

11. Fluid pressure against a wall or cell membrane is called _____ .

12. Vesicles form by _____ .
  a. endocytosis   d. halitosis
  b. exocytosis   e. a and c
  c. phagocytosis   f. a through c

13. Put the following structures in order according to an exocytic trafficking pathway.
  a. plasma membrane   c. endoplasmic reticulum
  b. Golgi bodies   d. post-Golgi vesicles

14. Match the term with its most suitable description.
  ___ phagocytosis   a. identity protein
  ___ passive transport   b. basis of diffusion
  ___ recognition protein c. important in membranes
  ___ active   d. one cell engulfs another
      transport   e. requires energy boost
  ___ phospholipid   f. docks for signals and
  ___ gradient   substances at cell surface
  ___ receptors   g. no energy boost required
                              to move solutes

■ *Visit CengageNOW for additional questions.*

Critical Thinking

1. Water moves osmotically into *Paramecium*, a single-celled aquatic protist. If unchecked, the influx would bloat the cell and rupture its plasma membrane, and the cell would die. An energy-requiring mechanism that involves contractile vacuoles (*right*) expels excess water. Water enters the vacuole's tubelike extensions and collects inside. A full vacuole contracts and squirts water out of the cell through a pore. Are *Paramecium*'s surroundings hypotonic, hypertonic, or isotonic?

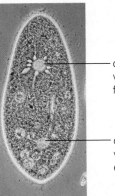

contractile vacuole filled

contractile vacuole empty

Ground Rules of Metabolism

A Toast to Alcohol Dehydrogenase

The next time someone asks you to have a drink, stop for a moment and think about the cells in your body that detoxify alcohol. It makes no difference whether you drink a bottle of beer, a glass of wine, or 1–1/2 ounces of vodka. Each holds the same amount of alcohol or, more precisely, ethanol (CH_3CH_2OH). Ethanol molecules move quickly from the stomach and small intestine into the bloodstream. Almost all of the ethanol someone drinks ends up in the liver, which has impressive numbers of alcohol-metabolizing enzymes. One of those enzymes, alcohol dehydrogenase, helps rid the body of ethanol and other toxic alcohols (Figure 6.1).

Detoxifying alcohol is hard on liver cells. It causes a slow-down in protein and glucose synthesis, and disrupts lipid and carbohydrate breakdown. Mitochondria use oxygen in ethanol metabolism—oxygen that normally would take part in the breakdown of fatty acids. Fatty acids accumulate as large fat globules in the tissues of heavy drinkers. As liver cells die of oxygen starvation, there are fewer and fewer cells for detoxification. One possible outcome is alcoholic hepatitis, a common disease characterized by inflammation and destruction of liver tissue. Alcoholic cirrhosis, another possibility, leaves the liver permanently scarred. (The word cirrhosis is from the Greek word *kirros*, or orange-colored, after the abnormal skin color of people with the disease.) Eventually, the liver just stops working, with dire health consequences.

The liver is the largest gland in the human body, weighing about 1.4 kg (3 pounds). It lies in the upper right side of the abdominal cavity. The liver has many important functions that affect the entire body. It helps digest fats and regulate the body's blood sugar level, and it breaks down many toxic compounds, not just ethanol. It also makes plasma proteins that circulate in blood. Plasma proteins are essential for blood clotting, immune function, and maintaining the solute balance of body fluids.

Now think about a self-destructive behavior known as binge drinking. The idea is to consume large amounts of alcohol in a brief period of time. Binge drinking is currently the most serious drug problem on college campuses throughout the United States. For example, a 2006 study showed that almost half of 4,580 undergraduate students surveyed are binge drinkers, meaning they consumed five or more alcoholic drinks in a two-hour period at least once during the year before the survey.

Binge drinking does far more than damage the liver. Aside from the related 500,000 injuries from accidents, the 600,000 assaults by intoxicated students, 100,000 cases of date rape, and 400,000 incidences of (whoops) unprotected sex among students, binge drinking kills upwards of 1,400 students every year. With this sobering example, we turn to metabolism, the cell's capacity to acquire and use energy.

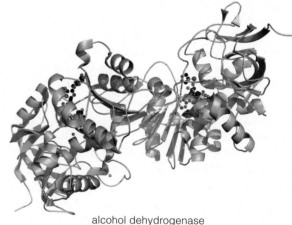

alcohol dehydrogenase

See the video! Figure 6.1 Alcohol dehydrogenase. This enzyme, which helps the body break down ethanol and other toxic alcohols, makes it possible for humans to drink beer, wine, and other alcoholic beverages.

Key Concepts

Energy flow in the world of life

Energy tends to disperse spontaneously. Each time energy is transferred, some of it disperses. Organisms maintain their organization only by continually harvesting energy. ATP couples reactions that release usable energy with reactions that require energy. **Sections 6.1, 6.2**

How enzymes work

Enzymes tremendously increase the rate of metabolic reactions. Environmental factors such as temperature, salt, and pH influence enzyme function. **Section 6.3**

The nature of metabolism

Metabolic pathways are energy-driven sequences of enzyme-mediated reactions. They concentrate, convert, or dispose of materials in cells. Controls over enzymes that govern key steps in metabolic pathways can shift cell activities fast. **Section 6.4**

Metabolism everywhere

Knowledge about metabolism, including how enzymes work, can help you interpret some natural phenomena. **Section 6.5**

Links to Earlier Concepts

- In this chapter, you will gain insight into how organisms tap into a one-way flow of energy to maintain their organization (1.2).

- Your knowledge of chemical bonding (2.4) and carbohydrates (3.3) will help you understand how cells store and retrieve energy in chemical bonds. You will also see how ATP (3.7) connects energy-requiring processes of metabolism (3.2) with energy-releasing ones.

- This chapter revisits the relationship between protein structure and function (3.5, 3.6), this time in the context of enzymes (1.2) and how they work. Factors such as temperature (2.5) and pH (2.6) affect enzyme function.

- You will start thinking about how cells harvest energy from organic molecules in sequences of electron (2.3) transfers, and the membrane proteins (5.2) that carry out such reactions.

- You will see an example of how scientists harness metabolic reactions to make tracers (2.2), and how such tracers help us better understand natural phenomena such as biofilms (4.5).

■ Assembly of the molecules of life starts with energy input into living cells.

■ Links to Life's organization 1.2, Chemical bonding 2.4, Carbohydrates 3.3

Energy Disperses

Energy, remember, is the capacity to do work, but this definition is not perfect. Even the best physicists cannot say, exactly, what energy is. However, even without a perfect definition, we can understand energy just by thinking of familiar kinds such as light, electricity, pressure, heat, and motion (Figure 6.2).

We also understand that one form of energy can be converted to another form. For example, a light bulb can change electricity into light, and an automobile can change the chemical energy of gasoline into the energy of motion. What may not be obvious is that the total amount of energy in every such conversion stays the same. Energy does not appear from nowhere, and it does not vanish into nothing, a concept that is called the **first law of thermodynamics**.

Another concept describes the way energy behaves: It tends to disperse spontaneously. For example, heat flows from a hot pan to air in a cool kitchen until the temperature of both is the same. We never see cool air raising the temperature of a hot pan. Each form of energy—not just heat—tends to disperse until no part of a system holds more than another part.

Entropy is a measure of how much the energy of a particular system has become dispersed. Let's use the hot pan in a cool kitchen as an example of a system. As heat flows from the pan into the air, the entropy of the system increases. Entropy continues to increase until the heat is uniformly distributed throughout the kitchen, and there is no longer a net flow of heat from one area to another. Our system has now reached its maximum entropy with respect to heat (Figure 6.3).

When we say that energy disperses, we mean that a system tends to change toward a state of maximum entropy. The concept that entropy increases spontaneously is the **second law of thermodynamics**. If we see a decrease in entropy, we can expect that some energy change occurred to make it happen.

Biologists use the concept of entropy as it applies to chemical bonding, because energy flow in the world of life occurs primarily by the making and breaking of chemical bonds. How is entropy related to chemical bonding? Think about it just in terms of motion. Two unbound atoms can vibrate, spin, and rotate in every direction: They are at high entropy with respect to motion. A covalent bond between the atoms constrains them, so they move in fewer ways than they did before bonding. Thus, the entropy of two atoms decreases when a bond forms between them.

Entropy changes are part of the reason why some reactions occur spontaneously, and others require an energy input, as you will see in the next section.

Figure 6.2 Demonstration of a familiar type of energy—the energy of motion.

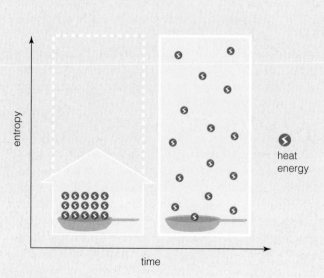

Figure 6.3 Entropy is the "mixedupness" of energy. Entropy tends to increase, but the total amount of energy always stays the same.

Figure 6.4 It takes more than 10,000 pounds of soybeans and corn to raise one 1,000-pound steer. Where do the other 9,000 pounds go? The animal's body breaks down molecules in the food to access energy stored in chemical bonds. Only about 10% of that energy goes toward building body mass. Some of the rest is used for activities (such as movement), but most is lost during energy conversions.

The One-Way Flow of Energy

Work occurs as energy is transferred from one place to another, and such energy transfers often involve the conversion of one form of energy to another. To a biologist, this statement means that all organisms use energy they harvest from the environment to drive cellular work. For example, photosynthetic cells of producers capture light energy from the sun by converting it to chemical energy stored in the bonds of carbohydrates. Most organisms access chemical energy stored in the bonds of carbohydrates by breaking those bonds. Both processes involve many energy transfers.

Some energy escapes with each transfer, usually in the form of heat. This is another way to interpret the second law: Energy transfers are never completely efficient. For example, the typical incandescent light bulb converts about 5% of the energy of electricity into light. The remaining energy, about 95% of it, ends up as heat that radiates from the bulb. Dispersed heat is not very useful for doing work, and it is not easily converted to a more useful form of energy (such as electricity). Because some of the energy in every transfer disperses as heat, and heat is not useful for doing work, we can say that the total amount of energy available for doing work in the universe is always decreasing.

Is life an exception to this depressing flow? An organized body is hardly dispersed. Energy becomes concentrated in each new organism as the molecules of life form and organize into cells. Even so, the second law applies. Living things constantly use energy to grow, to move, to acquire nutrients, to reproduce, and so on. Inevitable losses occur during the energy transfers that maintain life (Figure 6.4). Unless those losses are replenished with energy from another source, the complex organization of life will end. Most of the energy that fuels life on Earth is energy that has been lost from the sun, which has been losing energy since it formed 4.5 billion years ago.

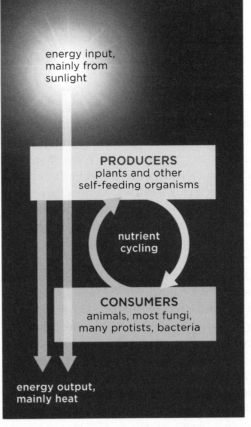

ENERGY IN

Sunlight energy reaches environments on Earth. Producers of nearly all ecosystems secure some and convert it to stored forms of energy. They and all other organisms convert stored energy to forms that can drive cellular work.

ENERGY OUT

With each conversion, there is a one-way flow of a bit of energy back to the environment. Nutrients cycle between producers and consumers.

Figure 6.5 A one-way flow of energy into living organisms compensates for a one-way flow of energy out of them. Energy inputs drive a cycling of materials among producers and consumers.

In our world, energy flows from the sun, through producers, then consumers (Figure 6.5). During this journey, the energy changes form and changes hands many times. Each time, some energy escapes as heat until, eventually, all of it is irrevocably dispersed. However, the second law does not say how quickly the dispersal has to happen. Energy's spontaneous dispersal is resisted by chemical bonds. Think of all the bonds in the countless molecules that make up your skin, heart, liver, fluids, and other body parts. Those bonds hold the molecules, and you, together—at least for the time being.

Take-Home Message

What is energy?

■ Energy is the capacity to do work. It can be converted from one form to another, but it cannot be created or destroyed.

■ Energy tends to spread, or disperse, spontaneously.

■ Organisms can maintain their complex organization only as long as they replenish themselves with energy they harvest from someplace else.

Energy in the Molecules of Life

- All cells store and retrieve energy in chemical bonds of the molecules of life.

- Links to Bonding 2.4, Carbohydrates 3.3, Nucleotides 3.7

Energy In, Energy Out

You already know how chemical bonds join atoms into molecules. When molecules interact, chemical bonds can break, form, or both. A **reaction** is the process by which such chemical change occurs. During a chemical reaction, one or more **reactants** (molecules that enter a reaction) change into one or more **products** (molecules that remain at the reaction's end). A chemical reaction is typically shown as an equation (Figure 6.6).

Every chemical bond holds energy. The amount of energy that a particular bond holds depends on which elements are taking part in it. For example, the covalent bond between an oxygen and hydrogen atom in any water molecule always holds the same amount of energy. That is the amount of energy required to break the bond, and it is also the amount of energy released when the bond forms.

Bond energy and entropy both contribute to a molecule's **free energy**, which is the amount of energy that is available (free) to do work.

In most reactions, the free energy of the reactants differs from the free energy of the products. Reactions in which the reactants have less free energy than the products require a net energy input to proceed. Such reactions are **endergonic**, which means "energy in" (Figure 6.7a).

Cells store energy by running endergonic reactions. For example, energy (in the form of light) drives the overall reactions of photosynthesis, which convert car-

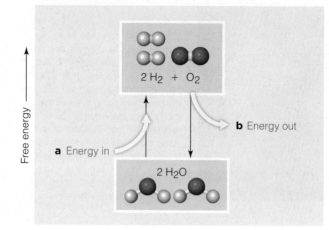

Figure 6.7 Energy inputs and outputs in chemical reactions.

(**a**) Endergonic reactions require an energy input because they convert molecules with lower free energy to molecules with higher free energy.

(**b**) Exergonic reactions end with an energy output because they convert molecules with higher free energy to molecules with lower free energy.

Figure It Out: Which law of thermodynamics explains energy inputs and outputs in chemical reactions?

Answer: The first law

bon dioxide and water to glucose and oxygen. Unlike light, glucose can be stored inside of a cell.

In other reactions, the reactants have greater free energy than the products. Such reactions are **exergonic**, which means "energy out," because they end with a net release of energy (Figure 6.7b). Cells access the free energy of molecules by running exergonic reactions. An example is the overall process of aerobic respiration, which converts glucose and oxygen to carbon dioxide and water for a net energy output.

Why the World Does Not Go Up in Flames

The molecules of life release energy when they combine with oxygen. For example, think of how a spark ignites tinder-dry wood in a campfire. Wood is mostly cellulose, which is a carbohydrate that consists of long chains of repeating glucose units (Section 3.3). A spark initiates a reaction that converts cellulose and oxygen to water and carbon dioxide. The reaction is exergonic, and it releases enough energy to initiate the same reaction with other cellulose and oxygen molecules. That is why a campfire keeps burning once it has been lit.

Earth is rich in oxygen—and in potential exergonic reactions. Why doesn't it burst into flames? Luckily, it takes energy to break the chemical bonds of reactants, even in an exergonic reaction. **Activation energy** is the minimum amount of energy that will get a chemical

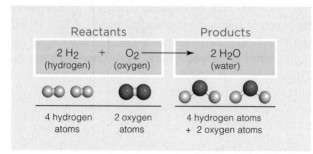

Figure 6.6 Chemical bookkeeping. In equations that represent chemical reactions, reactants are written to the left of an arrow that points to the products. A number before a formula indicates the number of molecules.

Atoms shuffle around in a reaction, but they never disappear: The same number of atoms that enter a reaction remain at the reaction's end.

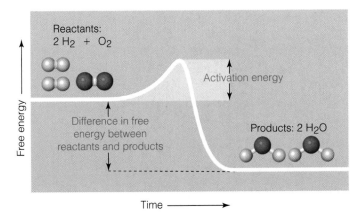

Reactants:
2 H₂ + O₂

Free energy

Activation energy

Difference in free
energy between
reactants and products

Products: 2 H₂O

Time

Figure 6.8 Activation energy. Most reactions will not proceed without an input of activation energy, which is shown here as a bump in an energy hill. In this example, the reactants have more free energy than the products. Activation energy keeps such exergonic reactions from running spontaneously.

reaction started (Figure 6.8). It is independent of any energy difference between reactants and products.

Both endergonic and exergonic reactions have activation energy, but the amount varies with the reaction. For example, guncotton, or nitrocellulose, is a highly explosive derivative of cellulose. Christian Schönbein accidentally discovered a way to manufacture it when he used a cotton apron to wipe up a nitric acid spill on his kitchen table, then hung it up to dry next to his oven. The apron exploded. Being a chemist in the 1800s, Schönbein had immediate hopes that he could market guncotton as a firearm explosive, but it proved to be too unstable. So little activation energy is needed to make guncotton react with oxygen that it explodes spontaneously. The substitute? Gunpowder, which has a higher activation energy for a reaction with oxygen.

ATP—The Cell's Energy Currency

Cells pair reactions that require energy with reactions that release energy. ATP is part of that process for many reactions in cells. **ATP**, or adenosine triphosphate, is an energy carrier: It accepts energy released by exergonic reactions, and delivers energy to endergonic reactions. ATP is the main currency in a cell's energy economy, so we use a cartoon coin to symbolize it.

ATP is a nucleotide with three phosphate groups (Figure 6.9a). The bonds that link those phosphate groups hold a lot of energy. When a phosphate group is transferred from ATP to another molecule, energy is transferred along with it. That energy contributes to the "energy in" part of an endergonic reaction. A phosphate-group transfer is called **phosphorylation**.

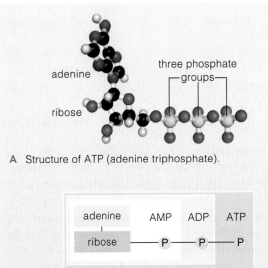

adenine

three phosphate
groups

ribose

A Structure of ATP (adenine triphosphate).

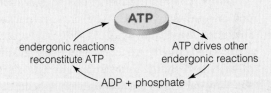

adenine   AMP   ADP   ATP

ribose — P — P — P

B The molecule is called ATP when it has three phosphate groups. After it loses one phosphate group, the molecule is called ADP (adenosine diphosphate); after losing two phosphate groups it is called AMP (adenosine monophosphate).

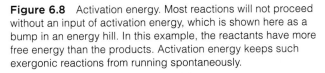

ATP

endergonic reactions
reconstitute ATP

ATP drives other
endergonic reactions

ADP + phosphate

C ATP forms when an endergonic reaction drives the covalent bonding of ADP and phosphate. ATP energy is transferred to another molecule along with a phosphate group, and ADP forms again. Energy from such transfers drives endergonic reactions that are the stuff of cellular work, such as active transport and muscle contraction.

Figure 6.9 **Animated** ATP, the energy currency of all cells.

Cells constantly use up ATP to drive endergonic reactions, so they constantly replenish it. When ATP loses a phosphate, ADP (adenosine diphosphate) forms (Figure 6.9b). ATP forms again when ADP binds phosphate in an endergonic reaction. The cycle of using and replenishing ATP is called the **ATP/ADP cycle** (Figure 6.9c).

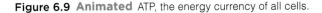

Take-Home Message

How do cells use energy?

■ Cells store and retrieve energy by making and breaking chemical bonds.

■ Activation energy is the minimum amount of energy required to start a chemical reaction.

■ Endergonic reactions cannot run without a net input of energy. Exergonic reactions end with a net release of energy.

■ ATP, the main energy carrier in all cells, couples reactions that release energy with reactions that require energy.

■ Enzymes make specific reactions occur much faster than they would on their own.

■ Links to Temperature 2.5, pH 2.6, Protein structure 3.5, Denaturation 3.6

How Enzymes Work

Centuries might pass before sugar would break down to carbon dioxide and water on its own, yet that same conversion takes just a few seconds inside your cells. Enzymes make the difference. **Enzymes** are catalysts, which are molecules that make chemical reactions proceed much faster than they would on their own. Most enzymes are proteins, but some are RNAs.

Most enzymes are not consumed or changed by participating in a reaction; they can work again and again. Each kind recognizes and alters specific reactants, or **substrates**. For instance, the enzyme thrombin cleaves a specific peptide bond in a protein called fibrinogen.

The polypeptide chains of enzymes are folded into one or more **active sites**. The sites are pockets where substrates bind and where reactions proceed (Figure 6.10). The active site is complementary in shape, size, polarity, and charge to the substrate. That fit is the reason why each enzyme acts only on specific substrates.

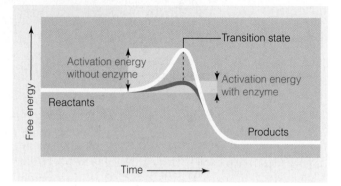

Figure 6.11 **Animated** An enzyme enhances the rate of a reaction by lowering its activation energy. **Figure It Out:** Is this reaction endergonic or exergonic? *Answer: Exergonic*

Activation energy is a bit like a hill that reactants must climb before they can run down the other side to products. When we talk about activation energy, we are really talking about the energy required to break the bonds of the reactants. Depending on the reaction, that energy may force substrates close together, redistribute their charge, or cause some other change. The change brings on the **transition state**, when substrate bonds reach their breaking point and the reaction will run spontaneously to product. Enzymes can help bring on the transition state by lowering activation energy (Figure 6.11). They do this by the following four mechanisms, which work alone or in combination.

Helping Substrates Get Together The closer substrate molecules are to each other, the more likely they are to react. Binding at an active site is as effective as bringing substrates 10 millionfold closer together.

Orienting Substrates In Positions That Favor Reaction On their own, substrates collide from random directions. By contrast, binding at an active site positions substrates so they align appropriately for a reaction.

Inducing a Fit Between Enzyme and Substrate By the **induced-fit model**, a substrate is not quite complementary to an active site. The enzyme restrains the substrate, stretching or squeezing it into a shape that often puts it next to a reactive group or to another molecule. By forcing a substrate to fit into the active site, the enzyme ushers in the transition state.

Shutting Out Water Molecules Metabolism occurs in water-based fluids, but water molecules can interfere with certain reactions. The active sites of some enzymes repel water, and keep it away from the reactions.

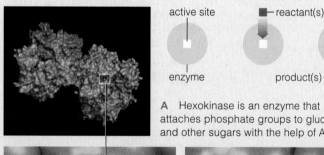

active site ■ reactant(s)

enzyme product(s)●

A Hexokinase is an enzyme that attaches phosphate groups to glucose and other sugars with the help of ATP.

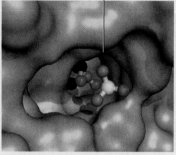 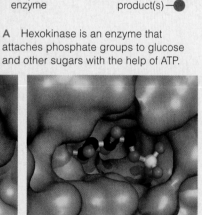

B A glucose and a phosphate meet in hexokinase's active site, the microenvironment of which encourages these molecules to react.

C The glucose has bonded with the phosphate. The product of this reaction, glucose-6-phosphate, is shown leaving the active site.

Figure 6.10 The active site of an enzyme.

Effects of Temperature, pH, and Salinity

Adding energy in the form of heat boosts free energy, which is one reason why molecular motion increases with temperature (Section 2.5). The greater the free energy of reactants, the closer a reaction is to its activation energy. Thus, the rate of an enzymatic reaction typically increases with temperature, but only up to a point. An enzyme denatures above a characteristic temperature. Then, the reaction rate falls sharply as the shape of the enzyme changes and it stops functioning (Figure 6.12). For example, body temperatures above 42°C (107.6°F) adversely affect many of your enzymes, which is why such severe fevers are dangerous.

The pH tolerance of enzymes varies. In the human body, most enzymes work best at pH 6–8. For instance, the hexokinase molecule in Figure 6.10 is most active in areas of the small intestine where the pH is around 8. Some enzymes, like pepsin, work outside the typical range of pH. Pepsin functions only in stomach fluid, where it breaks down proteins in food. The fluid is very acidic, with a pH of about 2 (Figure 6.13).

An enzyme's activity is also influenced by the amount of salt in the surrounding fluid. Too much or too little salt can interfere with the hydrogen bonds that hold an enzyme in its three-dimensional shape.

Help From Cofactors

Cofactors are atoms or molecules (other than proteins) that associate with enzymes and are necessary for their function. Some are metal ions. Organic cofactors are called **coenzymes**. Almost all vitamins are coenzymes or precursors of them.

We can use an enzyme called catalase as an example of how cofactors work. Like hemoglobin (Section 3.6), catalase has four hemes. The iron atom at the center of each heme is a cofactor. Iron, like other metal atoms, affects electrons in nearby molecules. Catalase works by holding a substrate molecule close to one of its iron atoms. The iron pulls on the substrate's electrons, which brings on the transition state.

Catalase is an **antioxidant**, which means it neutralizes free radicals—atoms or molecules with one or more unpaired electrons. These dangerous leftovers of metabolic reactions attack the structure of biological molecules. Free radicals accumulate as we age, in part because the body makes fewer catalase molecules.

Some coenzymes are tightly bound to an enzyme. Others, such as NAD^+ and $NADP^+$, can diffuse freely through the cytoplasm. Unlike enzymes, many coenzymes become modified during a reaction.

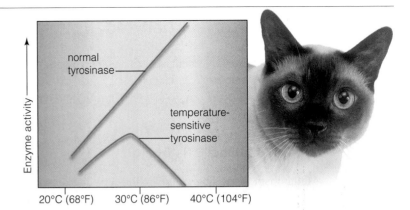

Figure 6.12 The enzyme tyrosinase is involved in the production of melanin, a black pigment in skin cells. Normally, tyrosinase activity increases with temperature between 20°C and 40°C. Mutations can cause tyrosinase activity to plummet at normal body temperatures. The Siamese mutation causes it to be inactive in warmer parts of a cat's body, which end up with less melanin, and lighter fur.

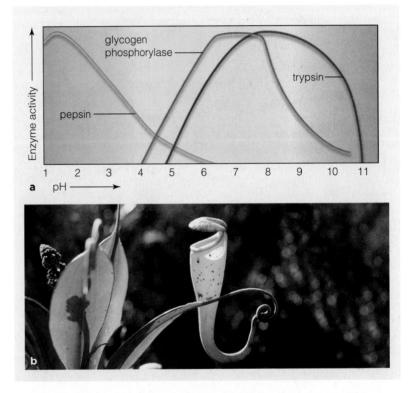

Figure 6.13 Enzymes and pH. (**a**) How pH values affect three enzymes. (**b**) Carnivorous plants of genus *Nepenthes* grow in nitrogen-poor habitats. They secrete acids and protein-digesting enzymes into fluid in a cup made of a modified leaf. The enzymes release nitrogen from small prey, such as insects, that are attracted to odors from the fluid and then drown in it. One of these pepsin-like enzymes functions best at pH 2.6.

Take-Home Message

How do enzymes work?

- Enzymes greatly enhance the rate of specific reactions. Binding at an enzyme's active site causes a substrate to reach its transition state. In this state, the substrate's bonds are at the breaking point.
- Each enzyme works best at certain temperatures, pH, and salt concentration.
- Cofactors associate with enzymes and assist their function.

■ ATP, enzymes, and other molecules interact in organized pathways of metabolism.

■ Links to Electrons 2.3, Metabolism 3.2, Amino acids 3.5, Membrane proteins 5.2

Types of Metabolic Pathways

Metabolism, remember, refers to the activities by which cells acquire and use energy (Section 3.2). Any series of enzyme-mediated reactions by which a cell builds, rearranges, or breaks down an organic substance is called a **metabolic pathway**. Pathways that build molecules from smaller ones are biosynthetic, or anabolic. Other pathways that break molecules apart are degradative, or catabolic.

Many metabolic pathways are linear, a straight line from reactants to products. Others are branched: Their intermediates can continue in more than one sequence of reactions. Still others are cyclic; the last step regenerates a reactant for the first step. For example, a cyclic pathway occurs during the second stage of photosynthesis. The entry point for the reactions is a molecule called RuBP; the last reaction of the pathway converts an intermediate to another molecule of RuBP.

Controls Over Metabolism

Enzymatic reactions do not only run from reactants to products. Many also run in reverse at the same time, with some of the products being converted back to reactants. The rates of the forward and reverse reactions often depend on the concentrations of reactants and products: A high concentration of reactants pushes the reaction in the forward direction. A high concentration of products pushes it in the reverse direction.

Cells conserve energy and resources by making what they need—no more, no less—at any given moment. How does a cell adjust the types and amounts of molecules it produces? Feedback mechanisms help a cell maintain, raise, or lower its production of thousands of different substances. Some of these mechanisms adjust how fast enzyme molecules are made. Others activate or inhibit enzymes that have already been built.

In some cases, molecules that bind to an enzyme directly activate or inhibit it. Such regulatory molecules bind not at the active site, but rather at an allosteric site on the enzyme. An **allosteric** site is a region of an enzyme other than the active site that can bind regulatory molecules (*allo*– means other; *steric* means structure). Binding of an allosteric regulator alters the shape of the enzyme in a way that enhances or inhibits its function (Figure 6.14).

Allosteric effects can cause **feedback inhibition**, in which the end product of a series of enzymatic reactions inhibits the first enzyme in the series (Figure 6.15). For example, isoleucine inhibits its own synthesis, so cells make more of this amino acid when its concentration in the cytoplasm declines. Cells use up their stores of isoleucine and other amino acids—the building blocks of proteins—during protein synthesis (Section 3.5). When protein synthesis slows, less isoleucine gets incorporated into proteins, so the amino acid accumulates. Unused isoleucine binds to an allosteric site on an enzyme in its own synthesis pathway. The

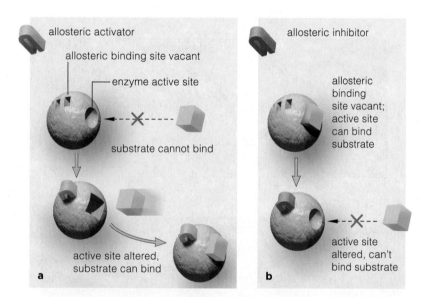

Figure 6.14 Animated Examples of allosteric control. (**a**) An active site becomes functional when an activator binds to an allosteric site. (**b**) An active site stops working when an inhibitor binds to an allosteric site.

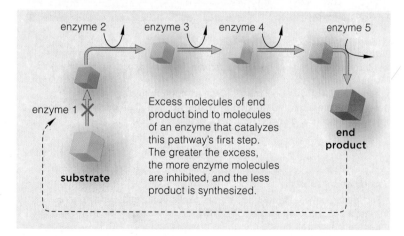

Figure 6.15 Animated Feedback inhibition. In this example, five kinds of enzymes act in sequence to convert a substrate to a product, which inhibits the activity of the first enzyme.

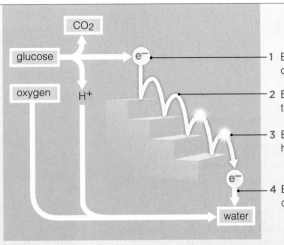

A Glucose and oxygen react when exposed to a spark. Energy is released all at once as CO_2 and water form.

B The same overall reaction occurs in small steps with an electron transfer chain. Energy is released in amounts that cells can harness for cellular work, such as muscle contraction or active transport.

1 Energy input splits glucose into carbon dioxide, electrons, and hydrogen ions (H^+).

2 Electrons lose energy as they move through an electron transfer chain.

3 Energy released by electrons is harnessed for cellular work.

4 Electrons, protons, and oxygen combine to form water.

Figure 6.16 Animated Uncontrolled versus controlled energy release.

binding changes the enzyme's shape, so less isoleucine forms. When the cell starts to make proteins again, it uses up the accumulated isoleucine until the allosteric sites on the enzyme molecules are freed up. Then, isoleucine synthesis begins again.

Redox Reactions

If a glucose molecule breaks apart into water and carbon dioxide all at once, it releases energy explosively (Figure 6.16a). Explosions are not good for cells. The only way cells can capture energy from glucose is to break down the molecule in small, manageable steps. Most of these steps are **oxidation–reduction reactions**. In each of these "redox" reactions, a molecule accepts electrons (it becomes *red*uced) from another molecule (which becomes *ox*idized). To remember what reduced means, think of how the negative charge of an electron "reduces" the charge of a recipient molecule. Think of the x in the word oxidation as a sideways + sign, which represents the increase in charge that occurs when a molecule loses an electron.

Coenzymes are among the many types of molecules that accept electrons in redox reactions, which are also called electron transfers. In the next two chapters, you will learn about the importance of redox reactions in electron transfer chains. An **electron transfer chain** is an organized series of reaction steps in which membrane-bound arrays of enzymes and other molecules give up and accept electrons in turn. Electrons are at a higher energy level when they enter a chain than when they leave. Think of the electrons as descending a staircase and losing a bit of energy at each step (Figure 6.16b).

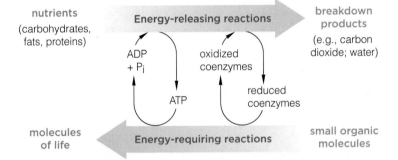

Figure 6.17 ATP forms in energy-releasing reactions, then delivers energy to energy-requiring reactions. Coenzymes (NAD^+, $NADP^+$, and FAD) accept electrons and hydrogen from energy-releasing reactions. The coenzymes (thus reduced to NADH, NADPH, and $FADH_2$) deliver their cargo of electrons and hydrogen to energy-requiring reactions.

Many coenzymes deliver electrons to electron transfer chains in photosynthesis and aerobic respiration. Energy released at certain steps in those chains helps drive the synthesis of ATP. Figure 6.17 is an overview of how ATP and coenzymes connect energy-releasing with energy-requiring pathways. These pathways will occupy our attention in chapters to come.

Take-Home Message

What are metabolic pathways?

■ Metabolic pathways are sequences of enzyme-mediated reactions. Some are biosynthetic; others are degradative.

■ Control mechanisms enhance or inhibit the activity of many enzymes. The adjustments help cells produce only what they require in any given interval.

■ Many metabolic pathways involve electron transfers, or oxidation–reduction reactions. Redox reactions occur in electron transfer chains. The chains are important sites of energy exchange in photosynthesis and aerobic respiration.

6.5 Night Lights

- Bioluminescence is visible evidence of metabolism.

- Links to Tracers 2.2, Biofilms 4.5

Enzymes of Bioluminescence At night, in the warm waters of tropical seas or in the summer air above fields and gardens, you may see shimmers or flashes of light. The light, which is emitted from metabolic reactions in living organisms, is **bioluminescence** (from the Greek *bio–*, for life, and the Latin *lumen*, for shine). In different species, it helps attract mates or prey, or confuse predators.

Bioluminescent organisms emit light when enzymes called luciferases convert chemical bond energy to light energy (luciferase is a generic term that refers to many different enzymes). Figure 6.18 shows firefly luciferase, a temperature-sensitive enzyme that uses ATP to energize a light-emitting pigment molecule. Any substrate of a luciferase is called luciferin:

$$\text{luciferin} + \text{ATP} \rightarrow \text{luciferin-ADP} + P_i$$

Energized by the transfer, the modified luciferin spontanously releases its extra energy as light:

$$\text{luciferin-ADP} + O_2 \rightarrow \text{oxyluciferin} + \text{AMP} + CO_2 + \text{light}$$

Different luciferins emit colors across the spectrum of visible light—from red to orange, yellow, green, blue, and purple. Some even emit infrared or ultraviolet light.

A Research Connection Many species of protists, fungi, bacteria, insects, jellyfishes, and fishes are bioluminescent. Researchers can transfer genes for bioluminescence from one of these species into another, nonluminescent species,

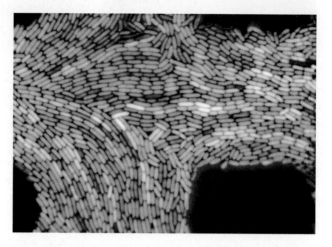

Figure 6.19 Bioluminescent biofilm. These bacteria have been altered to carry bioluminescence genes from a species of jellyfish.

so the recipient organisms light up under certain conditions. In itself, making organisms glow seems like a bizarre thing to do. However, the researchers are using the bioluminescence as a visible tracer in a variety of experiments.

For example, the *Escherichia coli* bacteria in Figure 6.19 are recipients of genes from a type of bioluminescent jellyfish. The bioluminescent light emitted by these cells indicates their metabolic activity. Differences in the intensity of light from individual cells reflect actual differences in metabolic activity among the cells in this biofilm. These bacteria are genetically identical; how could metabolic activity differ among them? The answer must be that each cell's metabolism depends on its location within the biofilm. Such research might help us discover why some bacterial cells, but not others, become resistant to antibiotics and are able to establish long-term infections in humans.

Figure 6.18 Bioluminescence. *Left*, a North American firefly (*Photinus pyralis*) emits a flash from its light organ, which contains peroxisomes packed with luciferase molecules. Firefly flashes may help potential mates find each other in the dark. *Right*, structure of firefly luciferase.

A Toast to Alcohol Dehydrogenase

In the human body, alcohol dehydrogenase (ADH) converts ethanol to acetaldehyde, an organic molecule even more toxic than ethanol and the most likely source of various hangover symptoms:

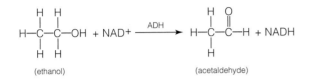

A different enzyme, aldehyde dehydrogenase (ALDH), very quickly converts toxic acetaldehyde to nontoxic acetate:

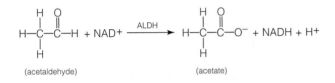

Thus, the overall pathway of ethanol metabolism in humans is:

In the average adult human body, this metabolic pathway can detoxify between 7 and 14 grams of ethanol per hour. The average alcoholic beverage contains between 10 and 20 grams of ethanol, which is why having more than one drink in any two-hour interval may result in a hangover.

Most organisms have alcohol dehydrogenase, which detoxifies the tiny quantities of alcohols that form in some metabolic pathways. In animals, the enzyme also detoxifies alcohols made by gut-inhabiting bacteria, and those in foods such as ripe fruit.

Despite the small amounts of alcohol that humans encounter naturally, our bodies make at least nine different kinds of alcohol dehydrogenase. It is interesting to speculate about why so many of them have evolved.

We do understand how some mutations in ADH affect our alcohol metabolism. For example, some mutations cause one of the ADH enzymes to be overactive, in which case acetaldehyde accumulates more quickly than ALDH can detoxify it:

People who carry such mutations become flushed and feel very ill after drinking even a small amount of alcohol. This unpleasant experience may be part of the reason that these people are less likely to become alcoholic than other people.

Different mutations that result in an underactive ALDH also cause acetaldehyde to accumulate:

These mutations are associated with the same effect—and the same protection from alcoholism—as mutations that cause an ADH to be overactive. Both types of mutations are common in people of Asian descent. For this reason, the alcohol flushing reaction is often called "Asian flush."

Mutations that disrupt the activity of an ADH enzyme have the opposite effect. Such mutations result in slowed alcohol metabolism, and people who carry them may not feel the ill effects of drinking alcoholic beverages as much as other people do. When these people drink alcohol, they have a tendency to become alcoholics. The study mentioned in the chapter opener showed that one-quarter of the undergraduate students who binged also had other signs of alcoholism.

Alcoholics will continue to drink despite the knowledge that doing so has tremendous negative consequences. In the United States, alcohol abuse is the leading cause of cirrhosis of the liver. The liver becomes so scarred, hardened, and filled with fat that it loses its function (Figure 6.20). It stops making the protein albumin, so the solute balance of body fluids is disrupted, and the legs and abdomen swell with watery fluid. It cannot remove drugs and other toxins from the blood, so they accumulate in the brain—which impairs mental functioning and alters personality. Restricted blood flow through the liver causes veins to enlarge and rupture, so internal bleeding is a risk. The damage to the body results in a heightened susceptibility to diabetes and liver cancer. Once cirrhosis has been diagnosed, a person has about a 50% chance of death within 10 years.

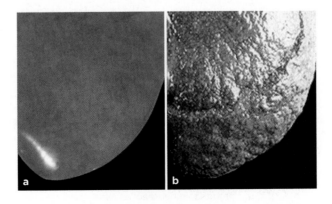

Figure 6.20 Alcoholic liver disease. (**a**) Normal human liver. (**b**) Enlarged, cirrhotic liver of an alcoholic. As few as 2 drinks per day can cause this disease.

Summary

Section 6.1 **Energy** is defined as a capacity to do work. Energy cannot be created or destroyed (**first law of thermodynamics**), but it can be converted from one form to another and thus transferred between objects or systems. Energy tends to disperse spontaneously (**second law of thermodynamics**). A bit disperses at each energy transfer, usually in the form of heat.

All living things maintain their organization only as long as they harvest energy from someplace else. Energy flows in one direction through the biosphere, starting mainly from the sun, then into and out of ecosystems. Producers and then consumers use energy to assemble, rearrange, and break down organic molecules that cycle among organisms throughout ecosystems.

Section 6.2 Cells store and retrieve **free energy** by making and breaking chemical bonds in metabolic **reactions**, in which **reactants** are converted to **products** (Table 6.1). **Activation energy** is the minimum energy required to start a reaction. **Endergonic** reactions require a net energy input. **Exergonic** reactions end with a net energy release.

ATP is an energy carrier between reaction sites in cells. It has three phosphate bonds; when a phosphate is transferred to another molecule, the energy of the bond is transferred along with it. Phosphate-group transfers (**phosphorylations**) to and from ATP couple reactions that release energy with reactions that require energy. Cells regenerate ATP by the **ATP/ADP cycle**.

■ *Use the animation on CengageNOW to learn about energy changes in chemical reactions and the role of ATP.*

Section 6.3 **Enzymes** are proteins or RNAs that greatly enhance the rate of a chemical reaction. Enzymes lower a reaction's activation energy by boosting local concentrations of **substrates**, orienting substrates in positions that favor reaction, inducing the fit between a substrate and the enzyme's **active site** (**induced-fit model**), and sometimes excluding water; all of which bring on a substrate's **transition state**. Each type of enzyme works best within a characteristic range of temperature, salt concentration, and pH. Most enzymes require the assistance of **cofactors**, which are metal ions or organic **coenzymes**. Cofactors in some **antioxidants** help them detoxify free radicals.

■ *Use the animation and interaction on CengageNOW to investigate how enzymes facilitate reactions.*

Section 6.4 Cells concentrate, convert, and dispose of most substances in enzyme-mediated reaction sequences called **metabolic pathways**. **Allosteric** sites are points of control by which a cell adjusts the types and amounts of substances it makes. **Feedback inhibition** is one example of enzyme control. **Oxidation–reduction** (redox) **reactions** in **electron transfer chains** allow cells to harvest energy in manageable increments.

■ *Use the animation on CengageNOW to compare the effects of controlled and uncontrolled energy release, and to observe mechanisms that exert control over enzymes.*

Section 6.5 **Bioluminescence** is light emitted by living organisms. Most bioluminescence is the product of enzyme-mediated reactions that often include ATP.

Self-Quiz
Answers in Appendix III

1. _____ is life's primary source of energy.
   a. Food   b. Water   c. Sunlight   d. ATP

2. Energy _____ .
   a. cannot be created or destroyed
   b. can change from one form to another
   c. tends to disperse spontaneously
   d. all of the above

3. Entropy _____ . (Choose all that are correct.)
   a. disperses   c. always increases, overall
   b. is a measure of disorder   d. is energy

4. If we liken a chemical reaction to an energy hill, then an _____ reaction is an uphill run.
   a. endergonic   c. ATP-assisted
   b. exergonic   d. both a and c

5. If we liken a chemical reaction to an energy hill, then activation energy is like _____ .
   a. a burst of speed   c. coasting downhill
   b. a bump at the hilltop d. both a and b

6. _____ are always changed by participating in a reaction. (Choose all that are correct.)
   a. Enzymes   c. Reactants
   b. Cofactors   d. Intermediates

7. Enzymes _____ .
   a. are proteins, except for a few RNAs
   b. lower the activation energy of a reaction
   c. are changed by the reactions they catalyze
   d. a and b

Table 6.1   Key Players in Metabolic Reactions

Reactant	Substance that enters a metabolic reaction; also called a substrate of an enzyme
Intermediate	Substance that forms in a reaction or pathway between the reactants and products
Product	Substance that remains at the end of a reaction or pathway
Enzyme	Protein or RNA that greatly enhances the rate of a reaction, but is not changed by participating in it
Cofactor	Molecule or ion that assists enzymes; may carry electrons, hydrogen, or functional groups to other reaction sites
Energy carrier	Mainly ATP; couples reactions that release energy with reactions that require energy

Data Analysis Exercise

Ethanol is a toxin, so it makes sense that drinking it can cause various symptoms of poisoning—headache, stomach ache, nausea, fatigue, impaired memory, dizziness, tremors, and diarrhea, among other ailments. All are symptoms of hangover, the common word for what happens as the body is recovering from a bout of heavy drinking.

The most effective treatment for a hangover is to avoid drinking in the first place. Folk remedies (such as aspirin, coffee, bananas, more alcohol, honey, barley grass, pizza, milkshakes, glutamine, raw eggs, charcoal tablets, or cabbage) abound, but few have been studied scientifically. In 2003, Max Pittler and his colleagues tested one of them. The researchers gave 15 participants an unmarked pill containing either artichoke extract or a placebo (an inactive substance) just before or after drinking enough alcohol to cause a hangover. The results are shown in Figure 6.21.

1. How many participants experienced a hangover that was worse with the placebo than with the artichoke extract?

2. How many participants experienced a worse hangover with the artichoke extract?

3. Calculate the numbers you counted in questions 1 and 2 as a percentage of the total number of participants. How much difference is there between the percentages?

4. Does this data support the hypothesis that artichoke extract is an effective hangover treatment? Why or why not?

Participant (Age, Gender)	Severity of Hangover	
	Artichoke Extract	Placebo
1 (34, F)	1.9	3.8
2 (48, F)	5.0	0.6
3 (25, F)	7.7	3.2
4 (57, F)	2.4	4.4
5 (34, F)	5.4	1.6
6 (30, F)	1.5	3.9
7 (33, F)	1.4	0.1
8 (37, F)	0.7	3.6
9 (62, M)	4.5	0.9
10 (36, M)	3.7	5.9
11 (54, M)	1.6	0.2
12 (37, M)	2.6	5.6
13 (53, M)	4.1	6.3
14 (48, F)	0.5	0.4
15 (32, F)	1.3	2.5

Figure 6.21 Results of a study that tested artichoke extract as a hangover preventive. All participants were tested once with the placebo and once with the extract, with a week interval between. Each rated the severity of 20 hangover symptoms on a scale of 0 (not experienced) to 5 ("as bad as can be imagined"). The 20 ratings were averaged as a single, overall rating, which is listed here.

8. Which of the following statements is not correct? A metabolic pathway _____ .
   - a. is a sequence of enzyme-mediated reactions
   - b. may be biosynthetic or degradative
   - c. generates heat
   - d. can include an electron transfer chain
   - e. none of the above

9. A molecule that donates electrons becomes _____ , and the one that accepts electrons becomes _____ .
   - a. reduced; oxidized
   - b. reduced; reduced
   - c. oxidized; reduced
   - d. oxidized; oxidized

10. A free radical is an atom or molecule that _____ .
   - a. carries no charge
   - b. has too many electrons
   - c. has an unpaired electron
   - d. has too few electrons

11. An antioxidant is a molecule that _____ .
   - a. detoxifies free radicals
   - b. degrades toxins
   - c. balances charge
   - d. oxidizes free radicals

12. Match each term with its most suitable description.
   - ___ reactant
   - ___ enzyme
   - ___ entropy
   - ___ product
   - ___ redox reaction
   - ___ cofactor
   - ___ first law
   - a. assists enzymes
   - b. there at reaction's end
   - c. enters a reaction
   - d. increases spontaneously
   - e. energy cannot be created or destroyed
   - f. a form of give and take
   - g. usually unchanged by participating in a reaction

■ *Visit CengageNOW for additional questions.*

Critical Thinking

1. Beginning physics students are often taught the basic concepts of thermodynamics with two phrases: First, you can't win. Second, you can't break even. Explain.

2. Dixie Bee wanted to make JELL-O shots for her next party, but felt guilty about encouraging her guests to consume alcohol. She tried to compensate for the toxicity of the alcohol by adding pieces of healthy fresh pineapple to the shots, but when she did, the JELL-O never set up. What happened? Hint: JELL-O is mainly sugar and collagen, a protein.

3. Free radicals are atoms or molecules that are like ions with the wrong number of electrons. They form in many enzyme-catalyzed reactions, such as the digestion of fats and amino acids. They slip out of electron transfer chains. They form when x-rays and other kinds of ionizing radiation strike water and other molecules. Free radicals react easily with the molecules of life, and can destroy them.

Hydrogen peroxide (H_2O_2) forms in most organisms as a by-product of aerobic respiration. This toxic molecule can easily become an even more dangerous free radical, so cells must dispose of it fast or risk being damaged. One molecule of catalase can inactivate about 6 million hydrogen peroxide molecules per minute by combining them two at a time. Catalase also inactivates other toxins, including ethanol. Given that its active site specifically binds hydrogen peroxide, how can this enzyme act on other substances?

4. Catalase combines two hydrogen peroxide molecules ($H_2O_2 + H_2O_2$) to make two molecules of water. A gas also forms. What is the gas?

5. Hydrogen peroxide bubbles if dribbled on an open cut but does not bubble on unbroken skin. Explain why.

Where It Starts—Photosynthesis

Biofuels

Plants and other photosynthesizers harvest energy from sunlight, then store it in chemical bonds of organic molecules they make from carbon dioxide and water. That process is photosynthesis, and it feeds them, us, and most other life on Earth. Photosynthesis also satisfies almost all of our uniquely human needs for fuel—for energy that we can use to heat our homes, cook our food, and run our machines. Three hundred million years ago, photosynthesis supported the growth of vast swamp forests. Successive forests slowly decayed, compacted, and became fossil fuels that we now extract from the earth. Coal, petroleum, and natural gas are composed of molecules that were originally assembled by ancient photosynthesizers. As such, their supply is limited.

Where are we going to get our energy once Earth's supply of fossil fuels runs out? The sun gives off a lot of it, but unlike plants, we cannot capture usable energy from the sun in an economically feasible way. Luckily for us, photosynthesizers are still at it. The molecules they make end up in vegetation, agricultural products, and ultimately in animals and animal wastes—all biomass (organic matter that is not fossilized).

A lot of energy is locked up in biomass. We can burn it, but that is an inefficient and messy way to release its energy. We can, however, convert it to oils, gases, or alcohols that burn cleanly and yield more energy per unit volume than burning the biomass itself. Such biofuels are a renewable source of energy: they can be made from crops, weeds, or what we now consider waste. We now make biodiesel from oils that come from algae, soybeans, rapeseed, flaxseed, and even restaurant kitchens. Methane seeps from manure ponds, landfills, and cows; we just need to find an efficient way to collect it.

We can also make ethanol from biomass. First, the carbohydrates in the biomass must be broken down to their component sugars. Making ethanol from the sugars is the easy part: We simply feed them to microorganisms that can convert sugars to ethanol. It is much more difficult to break down the carbohydrates in biomass cost-effectively, and without fossil fuels. The more cellulose in the biomass, the more involved that process is. Cellulose is a tough, insoluble carbohydrate. Breaking the bonds between its component sugars uses a lot of chemicals and energy, which adds cost to the biofuel product.

Today, we make ethanol from sugar-rich food crops such as corn, sugar beets, and sugarcane. These crops can be expensive to grow, and using them to make biofuel competes with our food supply. Thus, researchers are working to find an inexpensive way to break down the abundant cellulose in fast-growing weeds like switchgrass (Figure 7.1), and agricultural wastes such as wood chips, wheat straw, cotton stalks, and rice hulls—all biomass that we now dump in landfills or burn.

See the video! Figure 7.1 Biofuels. (**a**) Switchgrass (*Panicum virgatum*) grows wild in North American prairies. (**b**) Researchers Ratna Sharma and Mari Chinn of North Carolina State University are working to make biofuel production from biomass like switchgrass and agricultural wastes economically feasible.

Key Concepts

The rainbow catchers

The flow of energy through the biosphere starts when chlorophylls and other photosynthetic pigments absorb the energy of visible light. **Sections 7.1, 7.2**

Making ATP and NADPH

Photosynthesis proceeds through two stages in the chloroplasts of plants and many types of protists. In the first stage, sunlight energy is converted to the chemical bond energy of ATP. The coenzyme NADPH forms in a pathway that also releases oxygen. **Sections 7.3–7.5**

Making sugars

The second stage is the "synthesis" part of photosynthesis. Sugars are assembled from CO_2. The reactions use ATP and NADPH that form in the first stage of photosynthesis. Details of the reactions vary among organisms. **Sections 7.6, 7.7**

Evolution and photosynthesis

The evolution of photosynthesis changed the composition of Earth's atmosphere. New pathways that detoxified the oxygen by-product of photosynthesis evolved. **Section 7.8**

Photosynthesis, CO_2, and global warming

Photosynthesis by autotrophs removes CO_2 from the atmosphere; metabolism by all organisms puts it back in. Human activities have disrupted this balance, and so have contributed to global warming. **Section 7.9**

Links to Earlier Concepts

- The concept of energy flow (Section 6.1) helps explain how photosynthesizers can harvest energy from the sun. A review of electron energy levels (2.3) and chemical bonding (2.4) may be useful.

- Photosynthesis is the primary function of chloroplasts (4.11). Surface specializations (4.12) indirectly support photosynthesis in plants.

- You will see how energy carriers such as ATP link energy-releasing metabolic reactions with energy-requiring ones (6.2), and how cells harvest energy with electron transfer chains (6.4).

- What you know about carbohydrates (3.3), membrane proteins (5.2), and concentration gradients (5.3) will help you understand the chemical processes of photosynthesis.

- You will see how free radicals (6.3) influenced the evolution of organisms that changed our atmosphere.

- An example of nutrient cycling (1.2) illustrates one of the ways that photosynthesis connects the biosphere with its inhabitants.

How would you vote? Ethanol and other fuels manufactured from crops cost more than gasoline, so they are renewable energy sources and have fewer emissions. Would you pay a premium to drive a vehicle that uses biofuels? If so, how much? See CengageNOW for details, then vote online.

Sunlight as an Energy Source

■ Photosynthetic organisms use pigments to capture the energy of sunlight.

■ Links to Electrons 2.3, Bonding 2.4, Carbohydrates 3.3, Energy 6.1, Metabolism 6.4

Energy flow through nearly all ecosystems on Earth begins when photosynthesizers intercept energy from the sun. Let's turn now to the details of that process.

Properties of Light

Visible light is part of a spectrum of electromagnetic energy radiating from the sun. Such radiant energy travels in waves, undulating across space a bit like waves moving across an ocean. The distance between the crests of two successive waves of light is called **wavelength**, which we measure in nanometers (nm).

The electromagnetic energy of light is organized in packets called photons. A photon's energy and its wavelength are related, so all photons traveling at the same wavelength have the same amount of energy. Photons with the least amount of energy travel in longer wavelengths; those with the most energy travel in shorter wavelengths.

By the metabolic pathway of **photosynthesis**, organisms can harness the energy of light to build organic molecules from inorganic raw materials. Only light of wavelengths between 380 and 750 nanometers drives photosynthesis. Humans and many other organisms perceive light of all of these wavelengths combined as white, and particular wavelengths as different colors. White light separates into its component colors when it passes through a prism. The prism bends the longer wavelengths more than it bends the shorter ones, so a rainbow of colors forms (Figure 7.2).

Figure 7.2 also shows where visible light falls within the electromagnetic spectrum, which is the range of all wavelengths of radiant energy. Wavelengths of UV (ultraviolet) light, x-rays, and gamma rays are shorter than about 380 nanometers. They are energetic enough to alter or break the chemical bonds of DNA and other biological molecules, so they are a threat to life.

The Rainbow Catchers

Pigments are the molecular bridges between sunlight and photosynthesis. A **pigment** is an organic molecule that selectively absorbs light of specific wavelengths. Wavelengths of light that are not absorbed are reflected, and that reflected light gives each pigment its characteristic color. For example, a pigment that absorbs violet, blue, and green light reflects the remainder of the visible light spectrum: yellow, orange, and red light. This pigment would appear orange to us.

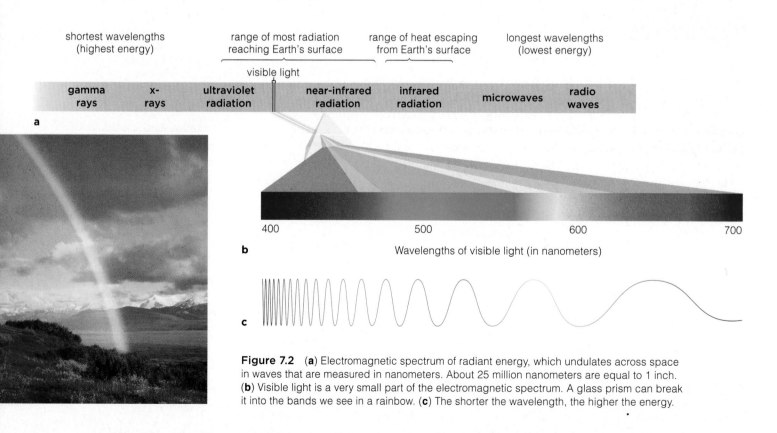

Figure 7.2 (**a**) Electromagnetic spectrum of radiant energy, which undulates across space in waves that are measured in nanometers. About 25 million nanometers are equal to 1 inch. (**b**) Visible light is a very small part of the electromagnetic spectrum. A glass prism can break it into the bands we see in a rainbow. (**c**) The shorter the wavelength, the higher the energy.

Table 7.1 Some Pigments in Photosynthesizers

Pigment	Color	Plants	Protists	Bacteria	Archaeans
		Occurrence in Photosynthetic Organisms			
Chlorophyll *a*	green	X	X	X	
Other chlorophylls	green	X	X	X	
Phycobilins					
phycocyanobilin	blue		X	X	
phycoerythrobilin	red		X	X	
phycoviolobilin	purple		X	X	
phycourobilin	orange		X	X	
Carotenoids					
carotenes					
β-carotene	orange	X	X	X	
α-carotene	orange	X	X	X	
lycopene	red	X	X		
xanthophylls					
lutein	yellow	X	X	X	
zeaxanthin	yellow	X	X	X	
fucoxanthin	orange	X	X		
Anthocyanins	purple	X	X	X	
Retinal	purple				X

Figure 7.3 Structure of two photosynthetic pigments. Both structures are derived from evolutionary remodeling of the same synthesis pathway. The light-catching part of each is the array of single bonds alternating with double bonds.

In chlorophyll, the array is a ring structure almost identical to a heme group. Heme groups are part of hemoglobin, which is also a pigment (Figure 3.3).

chlorophyll *a*   β-carotene

Chlorophyll *a* is by far the most common photosynthetic pigment in plants, photosynthetic protists, and cyanobacteria. Chlorophyll *a* absorbs violet and red light, so it appears green. Accessory pigments absorb additional colors of light for photosynthesis. A few of the 600 or so known accessory pigments are listed in Table 7.1.

Most types of photosynthetic organisms use a mixture of pigments for photosynthesis. In leaves of typical plants, chlorophyll is usually so abundant that it masks the colors of all the other pigments. Thus, most leaves usually appear green. In autumn, however, pigment synthesis slows in many kinds of leafy plants, and chlorophyll breaks down faster than it is replaced. Other pigments tend to be more stable than chlorophyll, so the leaves of such plants turn red, orange, yellow, or purple as their chlorophyll content declines and accessory pigments become visible.

Collectively, photosynthetic pigments absorb nearly all of the wavelengths of visible light. Different kinds cluster in photosynthetic membranes. Together, they can absorb a broad range of wavelengths, like a radio antenna that can pick up different stations.

The light-trapping part of a pigment is an array of atoms in which single bonds alternate with double bonds (Section 2.4 and Figure 7.3). Electrons of these atoms occupy one large orbital that spans all of the atoms. Electrons in such arrays easily absorb photons, so pigment molecules are a bit like antennas that are specialized for receiving light energy.

Absorbing a photon excites electrons. Remember, an energy input can boost an electron to a higher energy level (Section 2.3). The excited electron returns quickly to a lower energy level by emitting the extra energy. As you will see in Section 7.4, photosynthetic cells can capture energy emitted from an electron by bouncing the energy like a super-speed volleyball among a team of photosynthetic pigments. When the energy reaches the team captain—a special pair of chlorophylls—the reactions of photosynthesis begin.

Take-Home Message

How do photosynthetic organisms absorb light?

■ Energy radiating from the sun travels through space in waves and is organized as packets called photons.

■ The spectrum of radiant energy from the sun includes visible light. Humans perceive light of certain wavelengths as different colors. The shorter the wavelength of light, the greater its energy.

■ Pigments absorb specific wavelengths of visible light. Photosynthetic organisms use chlorophyll *a* and other pigments to capture the energy of light. That energy is used to drive the reactions of photosynthesis.

7.2 Exploring the Rainbow

■ Photosynthetic pigments work together to harvest light of different wavelengths.

A   Light micrograph of photosynthetic cells in a strand of *Chladophora*. Engelmann used this green alga to demonstrate that certain colors of light are best for photosynthesis.

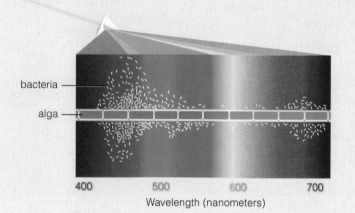

B   Engelmann directed light through a prism so that bands of colors crossed a water droplet on a microscope slide. The water held a strand of *Chladophora* and oxygen-requiring bacteria. The bacteria clustered around the algal cells that were releasing the most oxygen—the ones that were most actively engaged in photosynthesis. Those cells were under red and violet light.

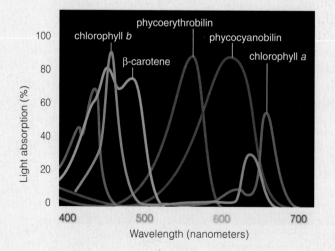

C   Absorption spectra of a few photosynthetic pigments. Line color indicates the characteristic color of each pigment.

At one time, people thought that plants used only the substances in soil to grow. By 1882, a few chemists understood that there were more ingredients in that recipe: water, something in the air, and light. Botanist Theodor Engelmann designed an experiment to test his hypothesis that the color of light affects photosynthesis. It had long been known that photosynthesis releases oxygen, so Engelmann used the amount of oxygen released by photosynthetic cells as a measure of how much photosynthesis was occurring in them.

In his experiment, Engelmann used a prism to divide a ray of light into its component colors, then directed the resulting spectrum across a single strand of photosynthetic alga (Figure 7.4a) suspended in a drop of water. Oxygen-sensing equipment had not yet been invented, so Engelmann used oxygen-requiring bacteria to show him where the oxygen concentration in the water was highest. The bacteria moved through the water and gathered mainly where violet or red light fell across the algal strand (Figure 7.4b). Engelmann concluded that the algal cells illuminated by light of these colors were releasing the most oxygen—a sign that violet and red light are best at driving photosynthesis.

Engelmann's experiment allowed him to correctly identify the colors of light most efficient at driving photosynthesis in *Chladophora*. His results constituted an absorption spectrum—a graph that shows how efficiently the different wavelengths of light are absorbed by a substance. Peaks in the graph indicate wavelengths of light that the substance absorbs best (Figure 7.4c).

Engelmann's absorption spectrum represents the combined spectra of all the photosynthetic pigments in *Chladophora*. Most photosynthetic organisms use a combination of pigments to drive photosynthesis, and the combination differs by species. Why? Different proportions of wavelengths in sunlight reach different parts of Earth. The particular set of pigments in each species is an adaptation that allows an organism to absorb the particular wavelengths of light available in its habitat. For example, water absorbs light between wavelengths of 500 and 600 nm less efficiently than other wavelengths. Algae that live deep underwater have pigments that absorb light in the range of 500–600 nm, which is the range that water does not absorb very well. Phycobilins are the most common pigments in deep-water algae.

Figure 7.4 Animated Discovery that photosynthesis is driven by particular wavelengths of light. Theodor Engelmann used the green alga *Chladophora* (**a**) in an early photosynthesis experiment (**b**). His results constituted one of the first absorption spectra.

(**c**) Absorption spectra of chlorophylls *a* and *b*, β-carotene, and two phycobilins reveal the efficiency with which these pigments absorb different wavelengths of visible light.

Figure It Out: Which are the three main photosynthetic pigments in *Chladophora*?

Answer: chlorophyll a, chlorophyll b, and β-carotene

7.3 Overview of Photosynthesis

■ Chloroplasts are organelles of photosynthesis in plants and other photosynthetic eukaryotes.

The **chloroplast** is an organelle that specializes in photosynthesis in plants and many protists (Figure 7.5a,b). Plant chloroplasts have two outer membranes, and are filled with a semifluid matrix called the **stroma**. Stroma contains the chloroplast's DNA, some ribosomes, and an inner, much-folded **thylakoid membrane**. The folds of a thylakoid membrane typically form stacks of disks (thylakoids) that are connected by channels. The space inside all of the disks and channels is a single, continuous compartment (Figure 7.5b).

Embedded in the thylakoid membrane are many clusters of light-harvesting pigments. These clusters absorb photons of different energies. The membrane also incorporates **photosystems**, which are groups of hundreds of pigments and other molecules that work as a unit to begin the reactions of photosynthesis. Chloroplasts contain two kinds of photosystems, type I and type II, which were named in the order of their discovery. Both types convert light energy into chemical energy.

Often, photosynthesis is summarized by this simple equation, from reactants to products:

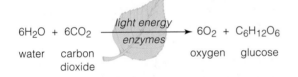

$$6H_2O + 6CO_2 \xrightarrow[\text{enzymes}]{\text{light energy}} 6O_2 + C_6H_{12}O_6$$

water   carbon   oxygen   glucose
         dioxide

However, photosynthesis is actually a series of many reactions that occur in two stages. In the first stage, the **light-dependent reactions**, the energy of light gets converted to the chemical bond energy of ATP. Typically, the coenzyme NADP+ accepts electrons and hydrogen ions, thus becoming NADPH. Oxygen atoms released from the breakdown of water molecules escape from the cell as O_2. The second stage, the **light-independent reactions**, runs on energy delivered by the ATP and NADPH formed in the first stage. That energy drives the synthesis of glucose and other carbohydrates from carbon dioxide and water (Figure 7.5c).

Take-Home Message

What are photosynthesis reactions and where do they occur?

■ In chloroplasts, the first stage of photosynthesis occurs at the thylakoid membrane. In these light-dependent reactions, light energy drives ATP and NADPH formation; oxygen is released.

■ The second stage of photosynthesis occurs in the stroma. In these light-independent reactions, ATP and NADPH drive the synthesis of sugars from water and carbon dioxide.

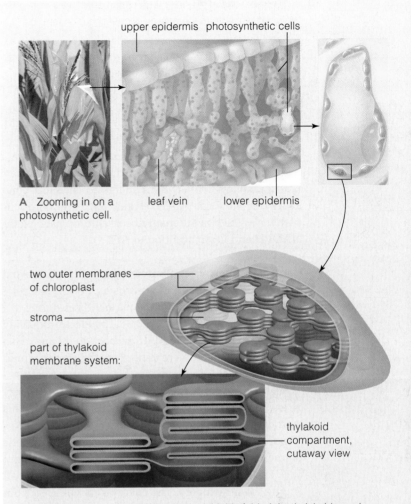

A Zooming in on a photosynthetic cell.

two outer membranes of chloroplast

stroma

part of thylakoid membrane system:

thylakoid compartment, cutaway view

B Chloroplast structure. No matter how highly folded, its thylakoid membrane system forms a single, continuous compartment in the stroma.

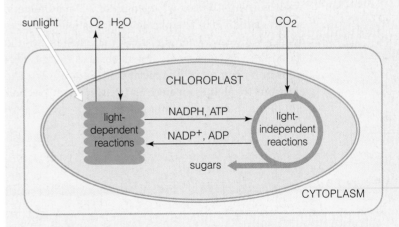

C In chloroplasts, ATP and NADPH form in the light-dependent stage of photosynthesis, which occurs at the thylakoid membrane. The second stage, which produces sugars and other carbohydrates, proceeds in the stroma.

Figure 7.5 Animated Sites of photosynthesis in a typical leafy plant.

7.4 | Light-Dependent Reactions

■ The reactions of the first stage of photosynthesis convert the energy of light to the energy of chemical bonds.

■ Links to Electrons and energy levels 2.3, Chloroplasts 4.11, Membrane proteins 5.2, Membrane properties and gradients 5.3, Energy 6.1, Electron transfer chains 6.4

The first stage of photosynthesis is driven by light, so the collective reactions of this stage are said to be light-dependent. Two different sets of light-dependent reactions constitute a noncyclic and a cyclic pathway. Both pathways convert light energy to chemical bond energy in the form of ATP (Figure 7.6). The noncyclic pathway, which is the main one in chloroplasts, yields NADPH and O_2 in addition to ATP.

Capturing Energy for Photosynthesis Imagine what happens when a pigment absorbs a photon. The photon's energy boosts one of the pigment's electrons to a higher energy level (Section 2.3). The electron quickly emits the extra energy and drops back to its unexcited state. If nothing else were to happen, the energy would be lost to the environment.

In the thylakoid membrane, however, the energy of excited electrons is kept in play. Embedded in this membrane are millions of light-harvesting complexes (Figure 7.7). These circular clusters of photosynthetic pigments and proteins hold on to energy by passing it back and forth, a bit like volleyball players pass a ball among team members. The energy gets volleyed from cluster to cluster until a photosystem absorbs it.

At the center of each photosystem is a special pair of chlorophyll *a* molecules. The pair in photosystem I absorbs energy with a wavelength of 700 nanometers, so it is called P700. The pair in photosystem II absorbs energy with a wavelength of 680 nanometers, so it is called P680. When a photosystem absorbs energy, electrons pop right off of its special pair (Figure 7.8*a*). The electrons then enter an electron transfer chain (Section 6.4) in the thylakoid membrane.

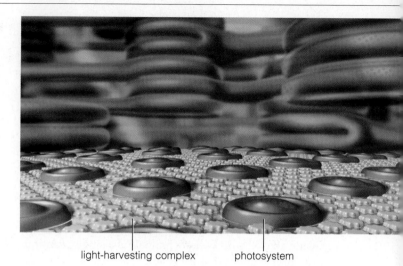

light-harvesting complex   photosystem

Figure 7.7 A view of some of the components of the thylakoid membrane from the stroma.

Replacing Lost Electrons A photosystem can donate only a few electrons to electron transfer chains before it must be restocked with more. Where do the replacements come from? Photosystem II gets more electrons by pulling them off of water molecules. This reaction is so strong that it causes the water molecules to dissociate into hydrogen ions and oxygen (Figure 7.8*b*). The released oxygen diffuses out of the cell as O_2. The process by which any molecule becomes broken down by light energy is called **photolysis**.

Harvesting Electron Energy Light energy is converted to chemical energy when a photosystem donates electrons to an electron transfer chain (Figure 7.8*c*). Light does not take part in chemical reactions. Electrons do. In a series of redox reactions (Section 6.4), they pass from one molecule of the electron transfer chain to the next. With each reaction, the electrons release a bit of their extra energy.

The molecules of the electron transfer chain use the released energy to move hydrogen ions (H^+) across the membrane, from the stroma to the thylakoid compartment (Figure 7.8*d*). Thus, the flow of electrons through electron transfer chains maintains a hydrogen ion gradient across the thylakoid membrane.

This gradient motivates hydrogen ions in the thylakoid compartment to diffuse back into the stroma. However, ions cannot simply diffuse through a lipid bilayer (Section 5.3). H^+ can leave the thylakoid compartment only by flowing through membrane transport proteins called ATP synthases (Section 5.2). Hydrogen ion flow through an ATP synthase causes this protein to attach a phosphate group to ADP (Figure 7.8*g,h*). Thus, a hydrogen ion gradient across a thylakoid membrane drives the formation of ATP in the stroma.

Figure 7.6
Summary of the inputs and outputs of the noncyclic and cyclic light-dependent reactions of photosynthesis.

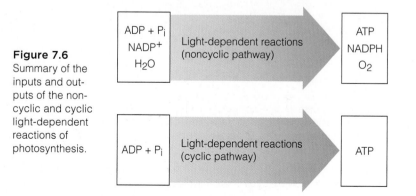

| ADP + P_i
$NADP^+$
H_2O | Light-dependent reactions (noncyclic pathway) → | ATP
NADPH
O_2 |

| ADP + P_i | Light-dependent reactions (cyclic pathway) → | ATP |

The Light-Dependent Reactions of Photosynthesis

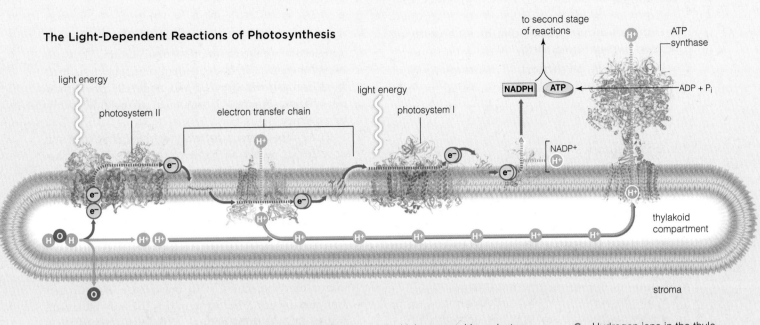

A Light energy drives electrons out of photosystem II.

B Photosystem II pulls replacement electrons from water molecules, which dissociate into oxygen and hydrogen ions (photolysis). The oxygen leaves the cell as O_2.

C Electrons from photosystem II enter an electron transfer chain.

D Energy lost by the electrons as they move through the chain causes H+ to be pumped from the stroma into the thylakoid compartment. An H+ gradient forms across the membrane.

E Light energy drives electrons out of photosystem I, which accepts replacement electrons from electron transfer chains.

F Electrons from photosystem I move through a second electron transfer chain, then combine with NADP+ and H+. NADPH forms.

G Hydrogen ions in the thylakoid compartment are propelled through the interior of ATP synthases by their gradient across the thylakoid membrane.

H H+ flow causes the ATP synthases to attach phosphate to ADP, so ATP forms in the stroma.

Figure 7.8 Animated Noncyclic pathway of photosynthesis. Electrons that travel through two different electron transfer chains end up in NADPH, which delivers them to sugar-building reactions in the stroma. The cyclic pathway (not shown) uses a third type of electron transfer chain.

Accepting Electrons After electrons from a photosystem II have moved through an electron transfer chain, they are accepted by a photosystem I. The photosystem absorbs energy, and electrons pop off of its special pair of chlorophylls (Figure 7.8e). The electrons then enter a second, different electron transfer chain. At the end of this chain, NADP+ accepts the electrons along with H+, so NADPH forms (Figure 7.8f):

$$NADP^+ + 2e^- + H^+ \rightarrow NADPH$$

ATP continues to form as long as electrons continue to flow through transfer chains in the thylakoid membrane. However, when NADPH is not being used, it accumulates in the stroma. The accumulation causes the noncyclic pathway to back up and stall. Then, the cyclic pathway runs independently in type I photosystems. This pathway allows the cell to continue making ATP even when the noncyclic pathway is not running.

The cyclic pathway involves photosystem I and an electron transfer chain that cycles electrons back to it.

The electron transfer chain that acts in the cyclic pathway uses electron energy to move hydrogen ions into the thylakoid compartment. The resulting hydrogen ion gradient drives ATP formation, just as it does in the noncyclic pathway. However, NADPH does not form, because electrons at the end of this chain are accepted by photosystem I, not NADP+. Oxygen (O_2) does not form either, because photosystem I does not rely on photolysis to resupply itself with electrons.

Take-Home Message

What happens in the light-dependent reactions of photosynthesis?

■ In chloroplasts, ATP forms during the light-dependent reactions of photosynthesis, which may occur in a cyclic or a noncyclic pathway.

■ In the noncyclic pathway, electrons flow from water molecules, through two photosystems and two electron transfer chains, and end up in the coenzyme NADPH. This pathway releases oxygen and forms ATP.

■ In the cyclic pathway, electrons lost from photosystem I return to it after moving through an electron transfer chain. ATP forms, but NADPH does not. Oxygen is not released.

Energy Flow in Photosynthesis

■ Energy flow in the light-dependent reactions is an example of how organisms harvest energy.

■ Links to Energy in metabolism 6.1, Redox reactions 6.4

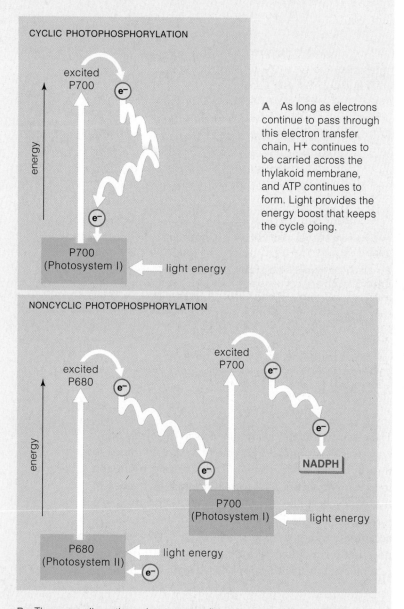

CYCLIC PHOTOPHOSPHORYLATION

A As long as electrons continue to pass through this electron transfer chain, H+ continues to be carried across the thylakoid membrane, and ATP continues to form. Light provides the energy boost that keeps the cycle going.

NONCYCLIC PHOTOPHOSPHORYLATION

B The noncyclic pathway is a one-way flow of electrons from water, to photosystem II, to photosystem I, to NADPH. As long as electrons continue to flow through the two electron transfer chains, H+ continues to be carried across the thylakoid membrane, and ATP and NADPH keep forming. Light provides the energy boosts that keep the pathway going.

Figure 7.9 Animated Energy flow in the light-dependent reactions of photosynthesis. The P700 in photosystem I absorbs photons of a 700-nanometer wavelength. The P680 of photosystem II absorbs photons of a 680-nanometer wavelength. Energy inputs boost P700 and P680 to an excited state in which they lose electrons.

Any light-driven reaction that attaches phosphate to a molecule is called **photophosphorylation**. Thus, the two pathways of light-dependent photosynthesis reactions are also called cyclic and noncyclic photophosphorylation. Figure 7.9 compares energy flow in the two pathways.

The simpler cyclic pathway evolved first, and still operates in nearly all photosynthesizers. Cyclic photophosphorylation yields ATP. No NADPH forms; no oxygen is released. Electrons lost from photosystem I are cycled back to it (Figure 7.9a).

Later, the photosynthetic machinery in some organisms became modified so that photosystem II became part of it. That modification was the beginning of a combined sequence of reactions that removes electrons from water molecules, with the release of hydrogen ions and oxygen. Photosystem II is the only biological system that is strong enough to oxidize—to pull electrons away from—water (Figure 7.9b).

Electrons that leave photosystem II do not return to it. They end up in NADPH, a powerful reducing agent (electron donor). NADPH delivers the electrons to sugar-producing reactions in the stroma.

In both cyclic and noncyclic photophosphorylation, molecules in the electron transfer chains use electron energy to shuttle H+ across the thylakoid membrane. Hydrogen ions accumulate in the thylakoid compartment, forming a gradient that powers ATP synthesis.

Today, the plasma membrane of different species of photosynthetic bacteria incorporates either type I or type II photosystems. Cyanobacteria, plants, and all photosynthetic protists use both types. Which of the two photophosphorylation pathways predominates at any given time depends on the organism's immediate metabolic demands for ATP and NADPH.

Having the alternate pathways is efficient, because cells can direct energy to producing NADPH and ATP or to producing ATP alone. NADPH accumulates when it is not being used. The excess backs up the noncyclic pathway, so the cyclic pathway predominates. The cell still makes ATP, but not NADPH. When sugar production is in high gear, NADPH is being used quickly. It does not accumulate, and the noncyclic pathway is the predominant one.

Take-Home Message

How does energy flow in photosynthesis?

■ Light provides energy inputs that keep electrons flowing through electron transfer chains. Energy lost by electrons as they flow through the chains sets up a hydrogen ion gradient that drives the synthesis of ATP alone, or ATP and NADPH.

7.6 | Light-Independent Reactions: The Sugar Factory

■ The cyclic, light-independent reactions of the Calvin–Benson cycle are the "synthesis" part of photosynthesis.

■ Links to Carbohydrates 3.3, ATP as an energy carrier 6.2

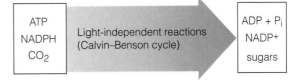

Figure 7.10 Summary of the inputs and outputs of the light-independent reactions of photosynthesis.

The enzyme-mediated reactions of the **Calvin–Benson cycle** build sugars in the stroma of chloroplasts. These reactions are light-independent because light does not power them. Instead, they run on the bond energy of ATP and the reducing power of NADPH—molecules that formed in the light-dependent reactions.

The light-independent reactions build glucose from carbon dioxide (Figure 7.10). Extracting carbon atoms from an inorganic source and incorporating them into an organic molecule is called **carbon fixation**. In most plants, photosynthetic protists, and some bacteria, the enzyme **rubisco** fixes carbon by attaching CO_2 to five-carbon RuBP, or ribulose bisphosphate (Figure 7.11a).

The six-carbon intermediate that forms is unstable. It splits right away into two three-carbon molecules of PGA (phosphoglycerate). The PGAs receive a phosphate group from ATP, and hydrogen and electrons from NADPH (Figure 7.11b). Thus, two molecules of three-carbon PGAL (phosphoglyceraldehyde) form.

Glucose, remember, has six carbon atoms. To make one glucose molecule, six CO_2 must be attached to six RuBP molecules, so twelve PGAL intermediates form. Two of the PGAL combine to form a six-carbon sugar

(Figure 7.11c). The ten remaining PGAL combine and regenerate the six RuBP (Figure 7.11d).

Plants can use the glucose they make in the light-independent reactions as building blocks for other organic molecules, or they can break it down to access the energy held in its bonds. However, most of the glucose is converted at once to sucrose or starch by other pathways that conclude the light-independent reactions. Excess glucose is stored in the form of starch grains inside the stroma of chloroplasts. When sugars are needed in other parts of the plant, the starch is broken down to sugar monomers and exported.

Take-Home Message

What happens during the light-independent reactions of photosynthesis?

■ Driven by ATP energy, the light-independent reactions of photosynthesis use hydrogen and electrons (from NADPH), and carbon and oxygen (from CO_2) to build glucose and other sugars.

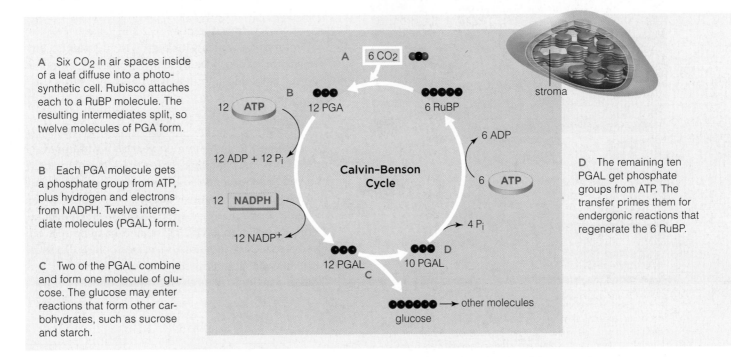

A Six CO_2 in air spaces inside of a leaf diffuse into a photosynthetic cell. Rubisco attaches each to a RuBP molecule. The resulting intermediates split, so twelve molecules of PGA form.

B Each PGA molecule gets a phosphate group from ATP, plus hydrogen and electrons from NADPH. Twelve intermediate molecules (PGAL) form.

C Two of the PGAL combine and form one molecule of glucose. The glucose may enter reactions that form other carbohydrates, such as sucrose and starch.

D The remaining ten PGAL get phosphate groups from ATP. The transfer primes them for endergonic reactions that regenerate the 6 RuBP.

Figure 7.11 Animated Light-independent reactions of photosynthesis which, in chloroplasts, occur in the stroma. The sketch is a summary of six cycles of the Calvin–Benson reactions and their product, one glucose molecule. *Black* balls signify carbon atoms. Appendix VI details the reaction steps.

Adaptations: Different Carbon-Fixing Pathways

- Environments differ, and so do details of photosynthesis.
- Links to Surface specializations 4.12, Controls over metabolic reactions 6.4

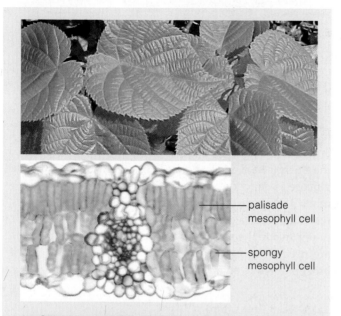

A C3 plant leaves. Chloroplasts are distributed evenly among two kinds of mesophyll cells in leaves of C3 plants such as basswood (*Tilia americana*). The light-dependent and light-independent reactions occur in both cell types.

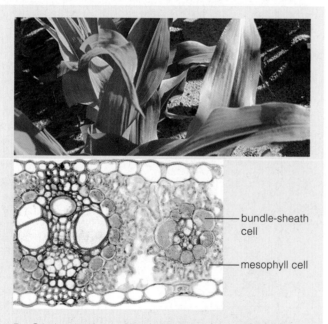

B C4 plant leaves. In C4 plants such as corn (*Zea mays*), carbon is fixed the first time in mesophyll cells, which are near the air spaces in the leaf, but have few chloroplasts. Specialized bundle-sheath cells ringing the leaf veins closely associate with mesophyll cells. Carbon fixation occurs for the second time in bundle-sheath cells, which are stuffed with rubisco-containing chloroplasts.

Rascally Rubisco

Plants that use only the Calvin–Benson cycle to fix carbon are called **C3 plants**, because *three*-carbon PGA is the first stable intermediate to form. Most plants use this pathway, but it can be inefficient in dry weather.

Plant surfaces that are exposed to air typically have a waxy, water-conserving cuticle (Section 4.12). They also have **stomata** (singular, stoma), which are small openings across the epidermal surfaces of leaves and green stems. Stomata close on dry days, which helps the plant minimize evaporative water loss from leaves and stems. However, like water, gases also enter and exit through stomata. When stomata are closed, CO_2 that is required for light-independent reactions cannot diffuse from air into leaves and stems, and the O_2 produced by the light-dependent reactions cannot diffuse out. Thus, when the light-dependent reactions run with stomata closed, oxygen builds up inside the plant. This buildup triggers an alternate pathway that reduces the cell's capacity to build sugars.

Remember that rubisco is the carbon-fixing enzyme of the Calvin–Benson cycle. At high O_2 levels, rubisco attaches oxygen (instead of carbon) to RuBP in a pathway called **photorespiration**. CO_2 is a product of photorespiration, so the cell loses carbon instead of fixing it. In addition, ATP and NADPH are used up to shunt the pathway's intermediates back to the Calvin–Benson cycle. So, sugar production in C3 plants becomes inefficient on dry days (Figures 7.12*a* and 7.13*a*).

Photorespiration can limit growth; C3 plants compensate for rubisco's inefficiency by making a lot of it. Rubisco is the most abundant protein on Earth.

C4 Plants

Over the past 50 to 60 million years, an additional set of reactions that compensates for rubisco's inefficiency evolved independently in many plant lineages. Plants that use the additional reactions also close stomata on dry days, but their sugar production does not decline. Examples are corn, switchgrass, and bamboo. We call these plants **C4 plants** because *four*-carbon oxaloacetate is the first stable intermediate to form in their carbon-fixation reactions (Figures 7.12*b* and 7.13*b*).

Figure 7.12 Different types of plants, different types of cells. Chloroplasts appear as green patches in the cross sections of the leaves. Purple areas are leaf veins.

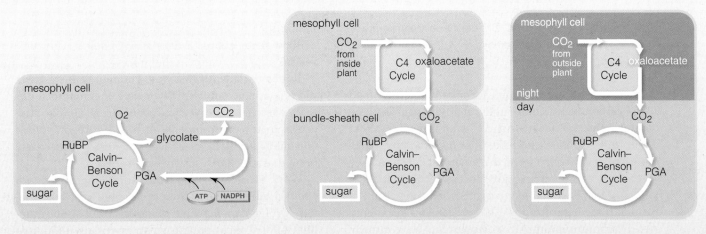

A C3 plants. On dry days, stomata close and oxygen accumulates to high concentration inside leaves. The excess causes rubisco to attach oxygen instead of carbon to RuBP. Cells lose carbon and energy as they make sugars.

B C4 plants. Oxygen also builds up inside leaves when stomata close during photosynthesis. An additional pathway in these plants keeps the CO_2 concentration high enough to prevent rubisco from using oxygen.

C CAM plants open stomata and fix carbon using a C4 pathway at night. When stomata are closed during the day, the organic compounds made during the night are converted to CO_2 that enters the Calvin–Benson cycle.

Figure 7.13 Light-independent reactions in three kinds of plants.

In C4 plants, the first set of light-independent reactions occurs in mesophyll cells. There, carbon is fixed by an enzyme that does not use oxygen even when the oxygen level is high. An intermediate is transported into bundle-sheath cells, where an ATP-requiring reaction converts it to CO_2. Rubisco fixes carbon for a second time as the CO_2 enters the Calvin–Benson cycle in the bundle-sheath cells. The C4 cycle keeps the CO_2 level near rubisco high, so it minimizes photorespiration. C4 plants use more ATP than C3 plants do, but on dry days they can make more sugar.

CAM Plants

Succulents, cactuses, and other **CAM plants** have an alternative carbon-fixing pathway that allows them to conserve water even in regions where the daytime temperatures can be extremely high. CAM stands for Crassulacean Acid Metabolism, after the Crassulaceae family of plants in which this pathway was first studied (Figure 7.14). Like C4 plants, CAM plants use a C4 cycle in addition to the Calvin–Benson cycle, but these two carbon-fixing cycles occur at different times rather than in different cells. The few stomata on a CAM plant open at night, when a C4 cycle fixes carbon from CO_2 in the air. The product of the cycle, a four-carbon acid, is stored in the cell's central vacuole. When the stomata close the next day, the acid moves out of the vacuole and becomes broken down to CO_2, which enters the Calvin–Benson cycle (Figure 7.13c).

Figure 7.14 A CAM plant: *Crassula argentea*, or jade plant.

Take-Home Message

How do carbon-fixing reactions vary?

■ When stomata are closed, oxygen builds up inside leaves of C3 plants. Rubisco then can attach oxygen (instead of carbon dioxide) to RuBP. This reaction, photorespiration, reduces the efficiency of sugar production, so it can limit growth.

■ Plants adapted to dry conditions limit photorespiration by fixing carbon twice. C4 plants separate the two sets of reactions in space; CAM plants separate them in time.

- The evolution of photosynthesis dramatically and permanently changed Earth's atmosphere.

- Link to Free radicals 6.3

Plants are the starting point for nearly all of the food (the carbon-based compounds) that you eat. They are **autotrophs**, or "self-nourishing" organisms. Like other autotrophs, plants can make their own food by securing energy directly from the environment, and they get their carbon from inorganic molecules (such as CO_2). Most bacteria, many protists, all fungi, and all animals are **heterotrophs**. These organisms get energy and carbon from organic molecules that have already been assembled by other organisms, for example by feeding on autotrophs, one another, or organic wastes or remains. *Hetero*– means other, as in "being nourished by others."

Plants are a kind of **photoautotroph**. By the process of photosynthesis, photoautotrophs make sugars from carbon dioxide and water using the energy of sunlight. Each year, plants collectively produce about 220

billion tons of sugar, enough to make about 300 quadrillion sugar cubes. That is a lot of sugar. They also release a lot of oxygen in the process.

It was not always this way. The first cells on Earth did not tap into sunlight. They were **chemoautotrophs** that extracted energy and carbon from simple molecules in the environment, such as hydrogen sulfide and methane. Both gases were plentiful in the nasty brew that was Earth's early atmosphere.

Ways of securing food did not change much for about a billion years. Then cyclic photophosphorylation evolved in the first photoautotrophs, and sunlight offered these organisms an essentially unlimited supply of energy. Not long afterward, the cyclic pathway became modified in some organisms. The new pathway, noncyclic photophosphorylation, split water molecules into hydrogen and oxygen. Molecular oxygen, which had previously been very rare in the atmosphere, began accumulating. From that time on, the world of life would never be the same (Figure 7.15).

The oxygen enrichment of Earth's early atmosphere exerted tremendous selection pressure on life. Oxygen reacts with metals, including enzyme cofactors, and free radicals form during those reactions. Free radicals, remember, are toxic (Section 6.3). Most early cells had no mechanism of detoxifying the oxygen radicals and became extinct. Only a few persisted in deep water, muddy sediments, and other oxygen-free habitats.

New pathways for detoxifying oxygen evolved, and one of them put oxygen's reactive properties to use. Oxygen accepts electrons at the end of electron transfer chains in these ATP-forming reactions, which are collectively called aerobic respiration.

Meanwhile, high in the ancient atmosphere, oxygen molecules were combining into ozone (O_3), a molecule that absorbs short-wavelength ultraviolet radiation in sunlight. The ozone layer that slowly formed in the upper atmosphere eventually shielded life from the sun's dangerous UV radiation. Only then did aerobic species emerge from the deep ocean, out from the mud and sediments, to diversify under the open sky.

Figure 7.15 Then and now—a view of how our atmosphere was irrevocably altered by photosynthesis. Photosynthesis is now the main pathway by which energy and carbon enter the web of life. Plants in this orchard are producing oxygen and carbon-rich parts—apples—at the Jerzy Boyz farm in Chelan, Washington.

Take-Home Message

How did photosynthesis affect Earth's atmosphere?

- The evolution of noncyclic photophosphorylation dramatically changed the oxygen content of Earth's atmosphere.

- In some organisms that survived the change, new pathways evolved for detoxifying oxygen radicals.

- Organisms did not live out in the open until after the ozone layer formed and began absorbing UV light from the sun.

7.9 | A Burning Concern

■ Earth's natural atmospheric cycle of carbon dioxide is out of balance, mainly as a result of human activity.

■ Link to Nutrient cycling 1.2

Have you ever wondered where all the atoms in your body came from? Think about just the carbon atoms. You eat other organisms to get the carbon atoms your body uses for energy and for raw materials. Those atoms may have passed through other heterotrophs before you ate them, but at some point they were part of photoautotrophic organisms. Photoautotrophs strip carbon from carbon dioxide, then use the atoms to build organic compounds. Your carbon atoms—and those of most other organisms— came from carbon dioxide.

Photosynthesis removes carbon dioxide from the atmosphere, and locks its carbon atoms inside organic compounds. When photosynthesizers and other aerobic organisms break down the organic compounds for energy, carbon atoms are released in the form of CO_2, which then reenters the atmosphere. Since photosynthesis evolved, these two processes have constituted a balanced cycle of the biosphere. You will learn more about the carbon cycle in Section 47.7. For now, know that the amount of carbon dioxide that photosynthesis removes from the atmosphere is roughly the same amount that organisms release back into it—at least it was, until humans came along.

As early as 8,000 years ago, humans began burning forests to clear land for agriculture. When trees and other plants burn, most of the carbon locked in their tissues is released into the atmosphere as carbon dioxide. Fires that occur naturally release carbon dioxide the same way.

Today, we are burning a lot more than our ancestors ever did. In addition to wood, we are burning fossil fuels— coal, petroleum, and natural gas—to satisfy our greater and greater demands for energy. As you will see in Section 23.5, fossil fuels are the organic remains of ancient organisms. When we burn these fuels, we release the carbon that has been locked inside of them for hundreds of millions of years back into the atmosphere—as carbon dioxide (Figure 7.16).

Researchers find pockets of our ancient atmosphere in Antarctica. Snow and ice have been accumulating in layers there, year after year, for the last 15 million years. Air and dust trapped in each layer reveal the composition of the atmosphere that prevailed when the layer formed. Thus, we now know that the atmospheric CO_2 level had been relatively stable for about 10,000 years before the industrial revolution. Since 1850, the CO_2 level has been steadily rising. In 2006, it was higher than it had been in *23 million years*.

Our activities have put Earth's atmospheric cycle of carbon dioxide out of balance. We are adding far more CO_2 to the atmosphere than photosynthetic organisms are removing from it. Today, we release around 26 billion tons of carbon dioxide into the atmosphere each year, more than ten times the amount we released in the year 1900. Most of it comes from burning fossil fuels. How do we know? Researchers can determine how long ago the carbon

Figure 7.16 Visible evidence of fossil fuel emissions in the atmosphere: the sky over New York City on a sunny day.

atoms in a sample of CO_2 were part of a living organism by measuring the ratio of different carbon isotopes in it. (You will read more about radioisotope dating techniques in Section 17.6.) These results are correlated with fossil fuel extraction, refining, and trade statistics.

The increase in atmospheric carbon dioxide is having dramatic effects on climate. CO_2 contributes to global warming, as you will read in Section 47.8. We are seeing a warming trend that mirrors the increase in CO_2 levels; Earth is now the warmest it has been for 12,000 years. The climate change is affecting biological systems everywhere. Life cycles are changing: Birds are laying eggs earlier; plants are flowering at the wrong times; mammals are hibernating for shorter periods. Migration patterns and habitats are also changing. The changes may be too fast, and many species may become extinct as a result.

Under normal circumstances, extra carbon dioxide stimulates photosynthesis, which means extra CO_2 uptake. However, the changes that we are already seeing in temperature and moisture patterns as a result of global warming are offsetting this benefit. Such changes are proving harmful to plants and other photosynthesizers.

Much research today targets development of energy sources that are not based on fossil fuels. For example, photosystem II catalyzes photolysis, the most efficient oxidation reaction in nature. Researchers are working to duplicate its catalytic function in artificial systems. If they are successful, then perhaps we too might be able to use light to split water into hydrogen, oxygen, and electrons— all of which can be used as clean sources of energy. Other research is focused on ways to remove carbon dioxide from the atmosphere—for example, by improving the efficiency of the enzyme rubisco in plants.

Corn and sugarcane are currently the top ethanol biofuel crops. These C4 plants flourish in hot, dry regions where photorespiration can limit the growth of C3 plants. They do not grow as well in areas where the growing season temperature averages less than 16°C (60°F), in part because the activity of rubisco decreases at this temperature (Section 6.3). C3 plants can compensate for the lowered activity by making more enzyme. C4 plants cannot. Their carbon-fixing cell specialization means there is less space in the

leaf for rubisco-containing chloroplasts. This space constraint limits the ability of C4 plants to make extra rubisco in cold climates.

Summary

Sections 7.1, 7.2 By metabolic pathways of **photosynthesis**, organisms capture the energy of light and use it to build sugars from water and carbon dioxide. **Pigments** such as **chlorophyll** *a* absorb visible light of particular **wavelengths** for photosynthesis.

■ *Use the animation on CengageNOW to see how Engelmann made an absorption spectrum for a photosynthetic alga.*

Section 7.3 In **chloroplasts**, the **light-dependent reactions** of photosynthesis occur at a much-folded **thylakoid membrane**, which incorporates two types of **photosystems**. The membrane forms a continuous compartment in the chloroplast's semifluid interior (**stroma**) where the **light-independent reactions** occur. The overall reactions of photosynthesis can be summarized as follows:

$$6H_2O + 6CO_2 \xrightarrow[\text{enzymes}]{\text{light energy}} 6O_2 + C_6H_{12}O_6$$

water   carbon   oxygen   glucose
        dioxide

■ *Use the animation on CengageNOW to view the sites where photosynthesis takes place.*

Sections 7.4, 7.5 Light-harvesting complexes in the thylakoid membrane absorb photons and pass the energy to photosystems, which then release electrons.

In noncyclic photophosphorylation, electrons released from photosystem II flow through an electron transfer chain. At the end of the chain, they enter photosystem I. Photon energy causes photosystem I to release electrons, which end up in NADPH. Photosystem II replaces lost electrons by pulling them from water, which then dissociates into H+ and O2 (**photolysis**).

In cyclic **photophosphorylation**, the electrons released from photosystem I enter an electron transfer chain, then cycle back to photosystem I. NADPH does not form.

In both pathways, electrons flowing through electron transfer chains cause H+ to accumulate in the thylakoid compartment, and so a hydrogen ion gradient builds up across the thylakoid membrane. H+ flows back across the membrane through ATP synthases. This flow results in the formation of ATP in the stroma.

■ *Use the animation on CengageNOW to review pathways by which light energy is used to form ATP.*

Section 7.6 **Carbon fixation** occurs in light-independent reactions. Inside the stroma, the enzyme **rubisco** attaches

a carbon from CO2 to RuBP to start the **Calvin–Benson cycle**. This cyclic pathway uses energy from ATP, carbon and oxygen from CO2, and hydrogen and electrons from NADPH to make glucose.

■ *Use the animation on CengageNOW to see how glucose is produced in the light-independent reactions.*

Section 7.7 Environments differ, and so do details of sugar production in the light-independent reactions. On dry days, plants conserve water by closing their **stomata**, but O2 from photosynthesis cannot escape. In **C3 plants**, the resulting high O2 level in leaves causes rubisco to attach O2 instead of CO2 to RuBP. This pathway, called **photorespiration**, reduces the efficiency of sugar production. In **C4 plants**, carbon fixation occurs twice. The first reactions release CO2 near rubisco, and thus limit photorespiration when stomata are closed. **CAM plants** open their stomata and fix carbon at night.

Section 7.8 **Autotrophs** make their own food using energy they get directly from the environment, and carbon from inorganic sources such as CO2. **Heterotrophs** get energy and carbon from molecules that other organisms have already assembled.

Earth's early atmosphere held very little free oxygen, and **chemoautotrophs** were common. When the noncyclic pathway of photosynthesis evolved, oxygen released by **photoautotrophs** permanently changed the atmosphere, and was a selective force that favored evolution of new metabolic pathways, including aerobic respiration.

Section 7.9 Photoautotrophs remove CO2 from the atmosphere; the metabolic activity of most organisms puts it back. Human activities disrupt this cycle by adding more CO2 to the atmosphere than photoautotrophs are removing from it. The resulting imbalance is contributing to global warming.

Self-Quiz   *Answers in Appendix III*

1. Photosynthetic autotrophs use _____ from the air as a carbon source and _____ as their energy source.

2. Chlorophyll *a* absorbs mainly violet and red light, and it reflects mainly _____ light.
    a. violet and red   c. yellow
    b. green   d. blue

Data Analysis Exercise

Most corn is grown intensively in vast swaths, which means farmers who grow it use fertilizers and pesticides, both of which are often made from fossil fuels. Corn is an annual plant, and yearly harvests tend to cause runoff that depletes soil and pollutes rivers.

In 2006, David Tilman and his colleagues published the results of a 10-year study comparing the net energy output of various biofuels. The researchers grew a mixure of native perennial grasses without irrigation, fertilizer, pesticides, or herbicides, in sandy soil that was so depleted by intensive agriculture that it had been abandoned. They measured the usable energy in biofuels made from the grasses, from corn, and from soy. They also measured the energy it took to grow and produce each kind of biofuel (Figure 7.17).

1. About how much energy did ethanol produced from one hectare of corn yield? How much energy did it take to grow and produce that ethanol?

2. Which biofuel tested had the highest ratio of energy output to energy input?

3. Which of the three crops would require the least amount of land to produce a given amount of biofuel energy?

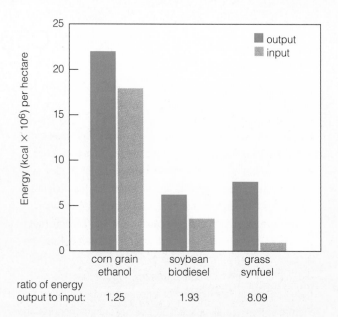

ratio of energy
output to input: 1.25   1.93   8.09

Figure 7.17   Energy inputs and outputs of biofuels from different sources: corn and soy grown on fertile farmland, and grassland plants grown in infertile soil. One hectare is about 2.5 acres.

---

3. Light-dependent reactions in plants occur in the _____ .
a. thylakoid membrane   c. stroma
b. plasma membrane   d. cytoplasm

4. In the light-dependent reactions, _____ .
a. carbon dioxide is fixed   c. CO_2 accepts electrons
b. ATP forms   d. sugars form

5. What accumulates inside the thylakoid compartment during the light-dependent reactions?
a. glucose   b. RuBP   c. hydrogen ions   d. CO_2

6. When a photosystem absorbs light, _____ .
a. sugar phosphates are produced
b. electrons are transferred to ATP
c. RuBP accepts electrons
d. light-dependent reactions begin

7. Light-independent reactions proceed in the _____ .
a. cytoplasm   b. plasma membrane   c. stroma

8. The Calvin–Benson cycle starts when _____ .
a. light is available
b. carbon dioxide is attached to RuBP
c. electrons leave photosystem II

9. What substance is not part of the Calvin–Benson cycle?
a. ATP   d. PGAL
b. NADPH   e. O_2
c. RuBP   f. CO_2

10. A C3 plant absorbs a carbon radioisotope (as $^{14}CO_2$). In which stable, organic compound does the labeled carbon appear first? Which compound forms first if a C4 plant absorbs the same carbon radioisotope?

11. After noncyclic photophosphorylation evolved, its by-product, _____ , accumulated and changed the atmosphere.

12. A cat eats a bird, which ate a caterpillar that chewed on a weed. Which organisms are autotrophs? Heterotrophs?

13. Match each event with its most suitable description.
____PGAL formation   a. rubiscos function
____CO_2 fixation   b. water molecules split
____photolysis   c. ATP, NADPH required
____ATP forms; NADPH   d. electrons cycled back
   does not   to photosystem I

■ Visit *CengageNOW* for additional questions.

Critical Thinking

1. About 200 years ago, Jan Baptista van Helmont wanted to know where growing plants get the materials necessary for increases in size. He planted a tree seedling weighing 5 pounds in a barrel filled with 200 pounds of soil and then watered the tree regularly. After five years, the tree weighed 169 pounds, 3 ounces, and the soil weighed 199 pounds, 14 ounces. Because the tree had gained so much weight and the soil had lost so little, he concluded that the tree had gained all of its additional weight by absorbing the water he had added to the barrel. What really happened?

2. Only about eight classes of pigment molecules are known, but this limited group gets around. For example, photoautotrophs make carotenoids, which move through food webs, as when tiny aquatic snails graze on green algae and then flamingos eat the snails. Flamingos modify the carotenoids. Their cells split beta-carotene to form two molecules of vitamin A. This vitamin is the precursor of retinal, a visual pigment that converts light energy to electric signals in the eyes. Beta-carotene gets dissolved in fat under the skin. Cells that give rise to bright pink feathers take it up. Research another organism to identify sources for pigments that color its surfaces.

8 How Cells Release Chemical Energy

When Mitochondria Spin Their Wheels

In the early 1960s, a Swedish physician, Rolf Luft, mulled over a patient's odd symptoms. The young woman felt weak and hot all the time. Even on the coldest winter days she could not stop sweating, and her skin was always flushed. She was thin, yet had a huge appetite. Luft inferred that his patient's symptoms pointed to a metabolic disorder: Her cells were very active, but much of their activity was being lost as metabolic heat. Luft checked the patient's rate of metabolism, the amount of energy her body was expending. Even while resting, her oxygen consumption was the highest that had ever been recorded! Examination of a tissue sample revealed that the patient's skeletal muscles had plenty of mitochondria, the cell's ATP-producing powerhouses. But there were too many of them, and they were abnormally shaped. The mitochondria were making very little ATP despite working at top speed.

The disorder, now called Luft's syndrome, was the first to be linked to defective mitochondria. The cells of someone with the disorder are like cities that are burning tons of coal in many power plants but not getting much usable energy output.

Skeletal and heart muscles, the brain, and other hardworking body parts with high energy demands are most affected.

More than forty disorders related to defective mitochondria are now known. One of them, Friedreich's ataxia, causes loss of coordination (ataxia), weak muscles, and serious heart problems. Many of those affected die when they are young adults (Figure 8.1).

Like the chloroplasts described in the previous chapter, mitochondria have an internal folded membrane system that allows them to make ATP. By the process of aerobic respiration, electron transfer chains in the mitochondrial membrane set up H+ gradients that power ATP formation.

In Luft's syndrome, electron transfer chains in mitochondria work overtime, but too little ATP forms. In Friedreich's ataxia, a protein called frataxin does not work properly. This protein helps build some of the iron-containing enzymes of electron transfer chains. When it malfunctions, iron atoms that were supposed to be incorporated into the enzymes accumulate inside mitochondria instead.

Oxygen is present in mitochondria. As explained in Section 7.8, toxic free radicals form when oxygen reacts with metals. Too much iron in mitochondria means too many free radicals form, and these destroy the molecules of life faster than the cell can repair or replace them. Eventually, the mitochondria stop working, and the cell dies.

You already have a sense of how cells harvest energy in electron transfer chains. Details of the reactions vary from one type of organism to the next, but all life relies on this ATP-forming machinery. When you consider mitochondria in this chapter, remember that without them, you would not make enough ATP to even read about how they do it.

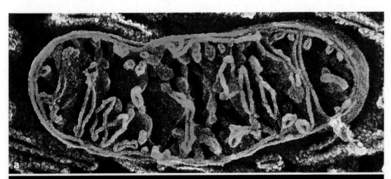

See the video! Figure 8.1 Sister, brother, and broken mitochondria. (**a**) Mitochondria are the body's ATP-producing powerhouses.

(**b**) Friedreich's ataxia, a genetic disorder, prevents mitochondria from making enough ATP. Leah (*left*) started to lose her sense of balance and coordination at age five. Six years later she was in a wheelchair; now she is diabetic and partially deaf. Her brother Joshua (*right*) could not walk by the time he was eleven, and is now blind. Both have heart problems; both had spinal fusion surgery. Special equipment allows them to attend school and work part-time. Leah is a professional model.

Key Concepts

Energy from carbohydrate breakdown

Various degradative pathways convert the chemical energy of glucose and other organic compounds to the chemical energy of ATP. Aerobic respiration yields the most ATP from each glucose molecule. In eukaryotes, it is completed inside mitochondria. **Section 8.1**

Glycolysis

Glycolysis is the first stage of aerobic respiration and of anaerobic fermentation pathways. Enzymes of glycolysis convert glucose to pyruvate. **Section 8.2**

How aerobic respiration ends

The final stages of aerobic respiration break down pyruvate to CO_2. Many coenzymes that become reduced deliver electrons and hydrogen ions to electron transfer chains. Energy released by electrons flowing through the chains is ultimately captured in ATP. Oxygen accepts electrons at the end of the chains. **Sections 8.3, 8.4**

How anaerobic pathways end

Fermentation pathways start with glycolysis. Substances other than oxygen accept electrons at the end of the pathways. Compared with aerobic respiration, the net yield of ATP from fermentation is small. **Sections 8.5, 8.6**

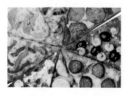

Other metabolic pathways

Molecules other than glucose are common energy sources. Different pathways convert lipids and proteins to substances that may enter glycolysis or the Krebs cycle. **Section 8.7**

Links to Earlier Concepts

- This chapter expands the picture of energy flow through the world of life (Section 6.1). It focuses on metabolic pathways (6.4) that make ATP (6.2) by degrading glucose and other molecules.

- The reactions of carbohydrate breakdown pathways occur either in the cytoplasm (4.2) or inside mitochondria (4.11). You may wish to review the structure of glucose and other carbohydrates (3.3).

- You will come across more examples of electron transfer chains (7.4) and reflect on the global connection between aerobic respiration and photosynthesis (7.8).

How would you vote? Developing new drugs is costly, so pharmaceutical companies tend to ignore Friedreich's ataxia and other disorders that affect relatively few people. Should governments fund private companies to develop treatments for rare disorders? See CengageNOW for details, then vote online.

■ Photoautotrophs make ATP during photosynthesis and use it to synthesize glucose and other carbohydrates.

■ Most organisms, including photoautotrophs, make ATP by breaking down glucose and other organic compounds.

■ Links to Energy 6.1 and 6.2, Metabolism 6.4

Organisms can stay alive only as long as they continue to resupply themselves with energy they get from their environment (Section 6.1). Plants and other photoautotrophs get their energy directly from the sun; heterotrophs eat autotrophs or one another. Regardless of its source, the energy must be converted to a form that can drive the diverse reactions necessary to sustain life. One such form is adenosine triphosphate (ATP), a common currency of energy expenditures in all cells.

Comparison of the Main Pathways

Most organisms make ATP by metabolic pathways that break down carbohydrates and other organic molecules. Some pathways are **aerobic** (they use oxygen); others are **anaerobic** (they occur in the absence of oxygen).

The main pathways by which cells harvest energy from organic molecules are called **aerobic respiration**, and anaerobic **fermentation**. Most types of eukaryotic cells either use aerobic respiration exclusively, or they use it most of the time. Many prokaryotes and protists in anaerobic habitats use alternative pathways. Some prokaryotes have their own unique version of aerobic respiration, but we focus on the pathway as it occurs in eukaryotic cells.

Fermentation and aerobic respiration begin with the same reactions in the cytoplasm. These reactions are **glycolysis**, and they convert one six-carbon molecule of glucose into two **pyruvate**, an organic molecule with a three-carbon backbone. After glycolysis, the pathways of fermentation and aerobic respiration diverge.

Aerobic respiration ends inside mitochondria, where oxygen accepts electrons at the end of electron transfer chains (Figure 8.2). Every breath you take provides your aerobically respiring cells with a fresh supply of oxygen. Fermentation ends in the cytoplasm, where a molecule other than oxygen accepts electrons at the end of the pathway.

Aerobic respiration is much more efficient than the fermentation pathways, which end with a net yield of two ATP per glucose. Aerobic respiration yields about thirty-six ATP per glucose. You and other multicelled organisms could not live without the higher yield of aerobic respiration.

A Carbohydrate breakdown pathways start in the cytoplasm, with glycolysis.

B Fermentation pathways are completed in the semi-fluid matrix of the cytoplasm.

C In eukaryotes, aerobic respiration is completed inside mitochondria.

Figure 8.2 Animated Where the different pathways of carbohydrate breakdown start and end. Aerobic respiration alone can deliver enough ATP to sustain large multicelled organisms such as people, redwoods, and Canada geese.

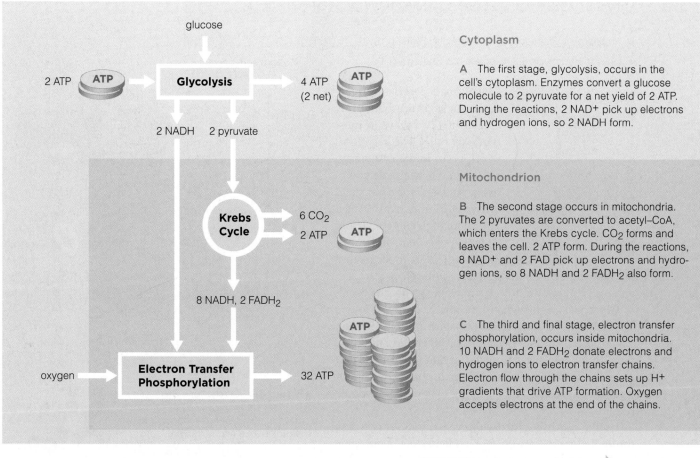

Cytoplasm

A The first stage, glycolysis, occurs in the cell's cytoplasm. Enzymes convert a glucose molecule to 2 pyruvate for a net yield of 2 ATP. During the reactions, 2 NAD+ pick up electrons and hydrogen ions, so 2 NADH form.

Mitochondrion

B The second stage occurs in mitochondria. The 2 pyruvates are converted to acetyl–CoA, which enters the Krebs cycle. CO_2 forms and leaves the cell. 2 ATP form. During the reactions, 8 NAD+ and 2 FAD pick up electrons and hydrogen ions, so 8 NADH and 2 $FADH_2$ also form.

C The third and final stage, electron transfer phosphorylation, occurs inside mitochondria. 10 NADH and 2 $FADH_2$ donate electrons and hydrogen ions to electron transfer chains. Electron flow through the chains sets up H+ gradients that drive ATP formation. Oxygen accepts electrons at the end of the chains.

Figure 8.3 Animated Overview of aerobic respiration. The reactions start in the cytoplasm and end inside mitochondria. The pathway's inputs and outputs are summarized on the *right*. **Figure It Out:** What is aerobic respiration's net yield of ATP?

Answer: 38 − 2 = 36 ATP per glucose

Overview of Aerobic Respiration

This equation summarizes aerobic respiration:

$$C_6H_{12}O_6 \;+\; O_2 \longrightarrow CO_2 \;+\; H_2O$$

glucose   oxygen   carbon dioxide   water

This equation only shows the substances at the start and end of the pathway, but not those at three intermediate stages (Figure 8.3). The first stage, glycolysis, converts glucose to pyruvate. During the second stage, the pyruvate is converted to acetyl–CoA, which enters the Krebs cycle. Carbon dioxide that forms in the second-stage reactions leaves the cell.

Electrons and hydrogen ions released by the reactions of the first two stages are picked up by two coenzymes, NAD+ (or nicotinamide adenine dinucleotide) and FAD (or flavin adenine dinucleotide). When these two coenzymes are carrying electrons and hydrogen, they are reduced (Section 6.4), and we refer to them as NADH and $FADH_2$.

Few ATP form during the first two stages. The big payoff occurs in the third stage after coenzymes give up electrons and hydrogen to electron transfer chains —the machinery of electron transfer phosphorylation. Operation of the transfer chains sets up hydrogen ion (H+) gradients that drive ATP formation. Oxygen in mitochondria accepts electrons and H+ at the end of the transfer chains, so water forms.

Take-Home Message

How do cells access the chemical energy in carbohydrates?

■ Most cells convert the chemical energy of carbohydrates to the chemical energy of ATP by aerobic respiration and fermentation pathways. These pathways start in the cytoplasm, with glycolysis.

■ Fermentation pathways end in the cytoplasm. They do not use oxygen. The net yield per glucose molecule is two ATP.

■ In eukaryotes, aerobic respiration ends in mitochondria. It uses oxygen, and the net yield per glucose molecule is thirty-six ATP.

- An energy investment of ATP starts glycolysis.

- Links to Hydrolysis 3.2, Glucose 3.3, Endergonic 6.2

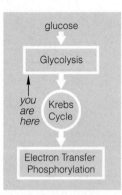

glucose

Glycolysis

you are here | Krebs Cycle

Electron Transfer Phosphorylation

Glycolysis is a series of reactions that begins carbohydrate breakdown pathways in most types of cells. The reactions, which occur in the cytoplasm, convert a glucose molecule to two pyruvates, for a net yield of two ATP and two NADH. The word "glycolysis" (from Greek *glyk–*, sweet; and *–lysis*, loosening) refers to the release of chemical energy from sugars. Different sugars can enter glycolysis, but for clarity we focus on glucose.

Glycolysis begins when a molecule of glucose enters a cell through a membrane transport protein (Section 5.2). The cell invests two ATP in the endergonic reactions that begin the pathway (Section 6.2). In the first reaction, an enzyme transfers a phosphate group from ATP to the glucose, thus forming glucose-6-phosphate (Figure 8.4a).

Unlike glucose, glucose-6-phosphate does not pass through glucose transporters in the plasma membrane, so it is trapped inside the cell. Almost all of the glucose that enters a cell is immediately converted to glucose-6-phosphate. This phosphorylation keeps the glucose concentration in the cytoplasm lower than it is in the fluid outside of the cell. By maintaining this concentration gradient across the plasma membrane, the cell favors uptake of even more glucose.

Glycolysis continues as glucose-6-phosphate accepts a phosphate group from another ATP, then splits in two (Figure 8.4b). PGAL is an abbreviation for the two three-carbon intermediates that result.

Enzymes attach a phosphate to each PGAL, forming two molecules of PGA (Figure 8.4c). In this reaction, two electrons and a hydrogen ion are transferred from each PGAL to NAD^+, so two NADH form. These reduced coenzymes will give up their cargo of electrons and hydrogen ions in reactions that follow glycolysis.

A phosphate group is transferred from each PGA to ADP, so two ATP form (Figure 8.4d). Two more ATP form when a phosphate is transferred from another pair of intermediates to two ADP (Figure 8.4e). Both reactions are **substrate-level phosphorylations**—direct transfers of a phosphate group from a substrate to ADP.

Remember, two ATP were invested to initiate the reactions of glycolysis. A total of four ATP form, so the *net* yield is two ATP per molecule of glucose that enters glycolysis (Figure 8.4f).

glycolysis occurs in the cytoplasm

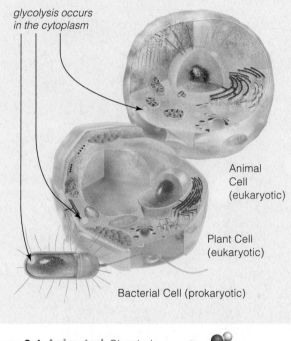

Animal Cell (eukaryotic)

Plant Cell (eukaryotic)

Bacterial Cell (prokaryotic)

Figure 8.4 Animated Glycolysis. This first stage of carbohydrate breakdown starts and ends in the cytoplasm of prokaryotic and eukaryotic cells.

Glucose (*right*) is the reactant in the example shown on the *opposite page*; we track its six carbon atoms (*black* balls). Appendix VI shows the complete structures of the intermediates and the products.

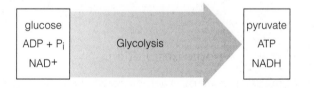

glucose
ADP + P_i
NAD+

Glycolysis

pyruvate
ATP
NADH

Cells invest two ATP to start glycolysis, so the *net* energy yield from one glucose molecule is two ATP. Two NADH also form, and two pyruvate molecules are the end products (*above*).

Glycolysis ends with the formation of two three-carbon pyruvate molecules. These products may now enter the second stage reactions of aerobic respiration or fermentation.

Take-Home Message

What is glycolysis?

- Glycolysis is the first stage of carbohydrate breakdown in both aerobic respiration and fermentation.

- In glycolysis, one molecule of glucose is converted to two molecules of pyruvate, with a net energy yield of two ATP. Two NADH also form. The reactions occur in the cytoplasm.

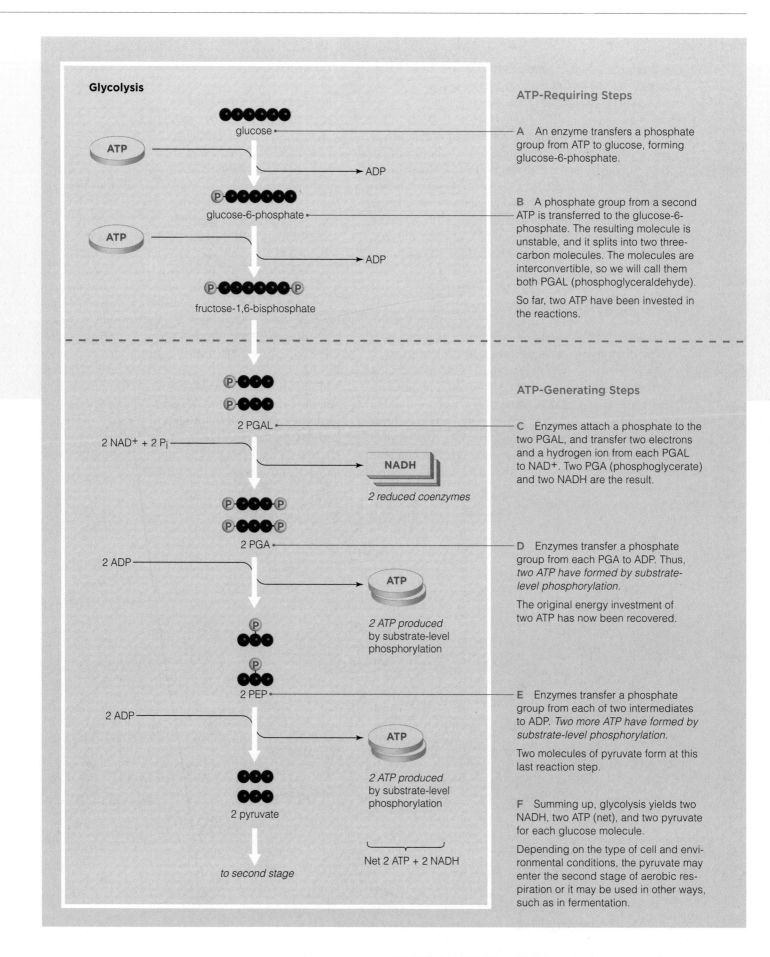

Glycolysis

glucose

ATP

→ ADP

glucose-6-phosphate

ATP

→ ADP

fructose-1,6-bisphosphate

2 PGAL

2 NAD⁺ + 2 Pᵢ

NADH

2 reduced coenzymes

2 PGA

2 ADP

ATP

2 ATP produced by substrate-level phosphorylation

2 PEP

2 ADP

ATP

2 ATP produced by substrate-level phosphorylation

2 pyruvate

Net 2 ATP + 2 NADH

to second stage

ATP-Requiring Steps

A An enzyme transfers a phosphate group from ATP to glucose, forming glucose-6-phosphate.

B A phosphate group from a second ATP is transferred to the glucose-6-phosphate. The resulting molecule is unstable, and it splits into two three-carbon molecules. The molecules are interconvertible, so we will call them both PGAL (phosphoglyceraldehyde).

So far, two ATP have been invested in the reactions.

ATP-Generating Steps

C Enzymes attach a phosphate to the two PGAL, and transfer two electrons and a hydrogen ion from each PGAL to NAD⁺. Two PGA (phosphoglycerate) and two NADH are the result.

D Enzymes transfer a phosphate group from each PGA to ADP. Thus, *two ATP have formed by substrate-level phosphorylation.*

The original energy investment of two ATP has now been recovered.

E Enzymes transfer a phosphate group from each of two intermediates to ADP. *Two more ATP have formed by substrate-level phosphorylation.*

Two molecules of pyruvate form at this last reaction step.

F Summing up, glycolysis yields two NADH, two ATP (net), and two pyruvate for each glucose molecule.

Depending on the type of cell and environmental conditions, the pyruvate may enter the second stage of aerobic respiration or it may be used in other ways, such as in fermentation.

Second Stage of Aerobic Respiration

■ The second stage of aerobic respiration finishes the break-down of glucose that began in glycolysis.

■ Links to Mitochondria 4.11, Metabolism 6.4

The second stage of aerobic respiration occurs inside mitochondria. It includes two sets of reactions, acetyl–CoA formation and the **Krebs cycle**, that break down the pyruvate products of glycolysis (Figure 8.5). All of the carbon atoms that were once part of glucose end up in CO_2, which departs the cell. Only two ATP form. The big payoff is the formation of many reduced coenzymes that drive the third and final stage of aerobic respiration.

Acetyl–CoA Formation

The second-stage reactions start when two pyruvate formed by glycolysis enter the inner compartment of a mitochondrion. An enzyme splits each three-carbon pyruvate into one molecule of CO_2 and a two-carbon acetyl group (Figure 8.6a). The CO_2 diffuses out of the cell, and the acetyl group combines with coenzyme A (abbreviated CoA), forming acetyl–CoA. Electrons and hydrogen ions released by the reaction combine with the coenzyme NAD^+, so NADH forms.

The Krebs Cycle

The Krebs cycle breaks down acetyl–CoA to CO_2. The cycle is not a physical object, such as a wheel. It is a pathway, a sequence of enzyme-mediated reactions. It is called a cycle because the last reaction in the sequence regenerates the substrate of the first (Section 6.4). In the Krebs cycle, the substrate of the first reaction—and the product of the last—is four-carbon oxaloacetate.

Use Figure 8.6 to follow what happens during each cycle of Krebs reactions. First, the two carbon atoms of acetyl–CoA are transferred to four-carbon oxaloacetate, forming citrate, a form of citric acid (Figure 8.6b). The Krebs cycle is also called the citric acid cycle after this first intermediate. In later reactions, two CO_2 form and depart the cell; two NAD^+ accept hydrogen ions and electrons, so two NADH form (Figure 8.6c,d); ATP forms by substrate-level phosphorylation (Figure 8.6e); and FAD and another NAD^+ accept hydrogen ions and electrons (Figure 8.6f,g). The final steps of the pathway regenerate oxaloacetate (Figure 8.6h).

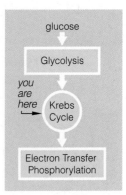

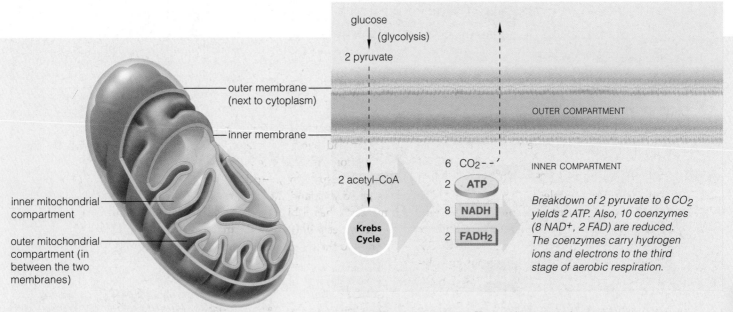

A An inner membrane divides a mitochondrion's interior into two compartments. The second and third stages of aerobic respiration take place at this membrane.

B The second stage starts after membrane proteins transport pyruvate from the cytoplasm to the inner compartment. Six carbon atoms enter these reactions (in two molecules of pyruvate), and six leave (in six CO_2). Two ATP form and ten coenzymes are reduced.

Figure 8.5 Animated Zooming in on aerobic respiration inside a mitochondrion.

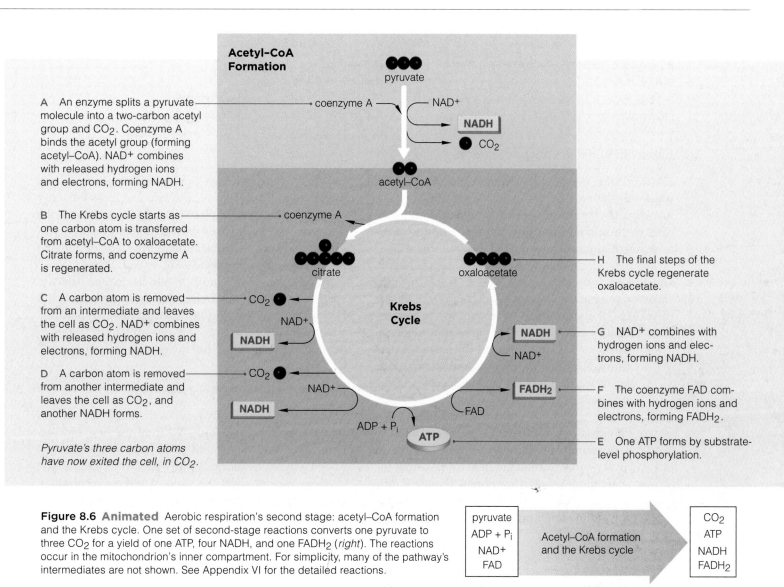

Acetyl–CoA Formation

pyruvate

A An enzyme splits a pyruvate molecule into a two-carbon acetyl group and CO_2. Coenzyme A binds the acetyl group (forming acetyl–CoA). NAD+ combines with released hydrogen ions and electrons, forming NADH.

coenzyme A — NAD+ — NADH — CO_2

acetyl–CoA

B The Krebs cycle starts as one carbon atom is transferred from acetyl–CoA to oxaloacetate. Citrate forms, and coenzyme A is regenerated.

coenzyme A

citrate

oxaloacetate

H The final steps of the Krebs cycle regenerate oxaloacetate.

C A carbon atom is removed from an intermediate and leaves the cell as CO_2. NAD+ combines with released hydrogen ions and electrons, forming NADH.

CO_2 — NAD+ — NADH

Krebs Cycle

NADH — **G** NAD+ combines with hydrogen ions and electrons, forming NADH.

NAD+

D A carbon atom is removed from another intermediate and leaves the cell as CO_2, and another NADH forms.

CO_2 — NAD+ — NADH

FADH2 — **F** The coenzyme FAD combines with hydrogen ions and electrons, forming FADH2.

FAD

ADP + P_i — ATP

E One ATP forms by substrate-level phosphorylation.

Pyruvate's three carbon atoms have now exited the cell, in CO_2.

Figure 8.6 Animated Aerobic respiration's second stage: acetyl–CoA formation and the Krebs cycle. One set of second-stage reactions converts one pyruvate to three CO_2 for a yield of one ATP, four NADH, and one FADH2 *(right)*. The reactions occur in the mitochondrion's inner compartment. For simplicity, many of the pathway's intermediates are not shown. See Appendix VI for the detailed reactions.

pyruvate ADP + P_i NAD+ FAD	Acetyl–CoA formation and the Krebs cycle	CO_2 ATP NADH FADH2

Remember, glycolysis converted one glucose molecule to two pyruvate, and these were converted to two acetyl–CoA when they entered the inner compartment of a mitochondrion. There, the second stage reactions convert the two molecules of acetyl–CoA to six CO_2. At this point in aerobic respiration, one glucose molecule has been broken down completely: Six carbon atoms have left the cell, in six CO_2. Two ATP formed, which adds to the small net yield of glycolysis. However, six NAD+ were reduced to six NADH, and two FAD were reduced to two FADH2.

What is so important about reduced coenzymes? A molecule becomes reduced when it receives electrons (Section 6.4). Electrons carry energy that can be used to drive endergonic reactions. In this case, the electrons picked up by coenzymes during the first two stages of aerobic respiration carry energy that drives the reactions of the third stage.

In total, two ATP form and ten coenzymes (eight NAD+ and two FAD) are reduced during acetyl–CoA formation and the Krebs cycle (the second stage of aerobic respiration). Add in the two NAD+ reduced in glycolysis, and the full breakdown of each glucose molecule has a big potential payoff. Twelve reduced coenzymes will deliver electrons (and the energy they carry) to the third stage of aerobic respiration.

Take-Home Message

What happens during the second stage of aerobic respiration?

■ The second stage of aerobic respiration includes acetyl–CoA formation and the Krebs cycle. The reactions occur in the inner compartment of mitochondria.

■ The pyruvate that formed in glycolysis is converted to acetyl–CoA and carbon dioxide. The acetyl–CoA enters the Krebs cycle, which breaks it down to CO_2.

■ For each two pyruvates broken down in the second-stage reactions, two ATP form, and ten coenzymes (eight NAD+ and two FAD) are reduced.

8.4 | Aerobic Respiration's Big Energy Payoff

■ Many ATP are formed during the third and final stage of aerobic respiration.

■ Links to Cell membranes 5.1, Membrane proteins 5.2, Electron transfer chains 6.4

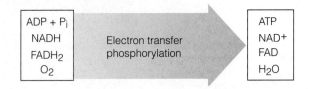

Electron Transfer Phosphorylation

The third stage of aerobic respiration, **electron transfer phosphorylation**, also occurs inside mitochondria. Its name means the flow of electrons through mitochondrial electron transfer chains ultimately results in the attachment of phosphate to ADP, which forms ATP.

The third stage begins with the coenzymes NADH and FADH$_2$, which became reduced in the first two stages of aerobic respiration. These coenzymes donate their cargo of electrons and hydrogen ions to electron transfer chains embedded in the inner mitochondrial membrane (Figure 8.7a). As the electrons pass through the chains, they give up energy little by little (Section 6.4). Some molecules of the transfer chains harness that energy to actively transport hydrogen ions from the inner mitochondrial compartment to the outer one. The ions accumulate in the outer compartment, so an H$^+$ concentration gradient forms across the inner mitochondrial membrane (Figure 8.7b).

This gradient attracts hydrogen ions back toward the inner mitochondrial compartment.

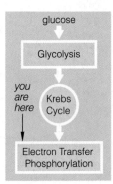

However, hydrogen ions cannot diffuse across a lipid bilayer without assistance. The ions can cross the inner mitochondrial membrane only by flowing through the interior of ATP synthases (Section 5.2 and Figure 8.7d). The flow causes these membrane transport proteins to attach phosphate groups to ADP, so ATP forms.

At the end of the mitochondrial electron transfer chains, oxygen accepts electrons and combines with H$^+$, forming water (Figure 8.7c). Aerobic respiration, which literally means "taking a breath of air," refers to oxygen as the final electron acceptor in this pathway.

Summing Up: The Energy Harvest

Thirty-two ATP typically form in the third stage of aerobic respiration. Add four ATP from the first and second stages, and the overall yield from the breakdown of one glucose molecule is thirty-six ATP (Figure 8.8).

Many factors affect the yield of aerobic respiration. For example, the two NADH from glycolysis cannot cross mitochondrial membranes; they transfer electrons

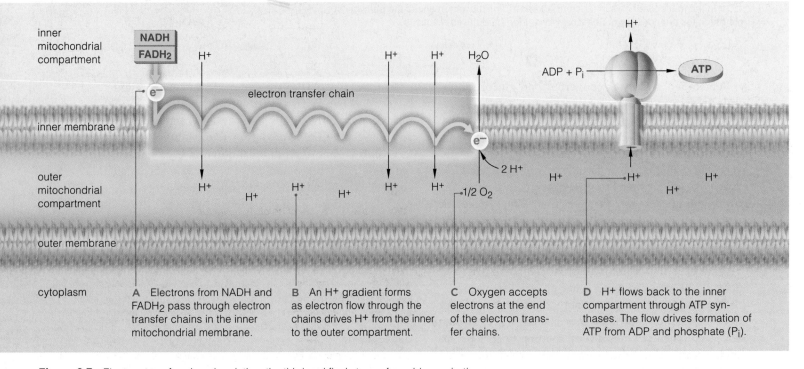

A Electrons from NADH and FADH$_2$ pass through electron transfer chains in the inner mitochondrial membrane.

B An H$^+$ gradient forms as electron flow through the chains drives H$^+$ from the inner to the outer compartment.

C Oxygen accepts electrons at the end of the electron transfer chains.

D H$^+$ flows back to the inner compartment through ATP synthases. The flow drives formation of ATP from ADP and phosphate (P$_i$).

Figure 8.7 Electron transfer phosphorylation, the third and final stage of aerobic respiration.

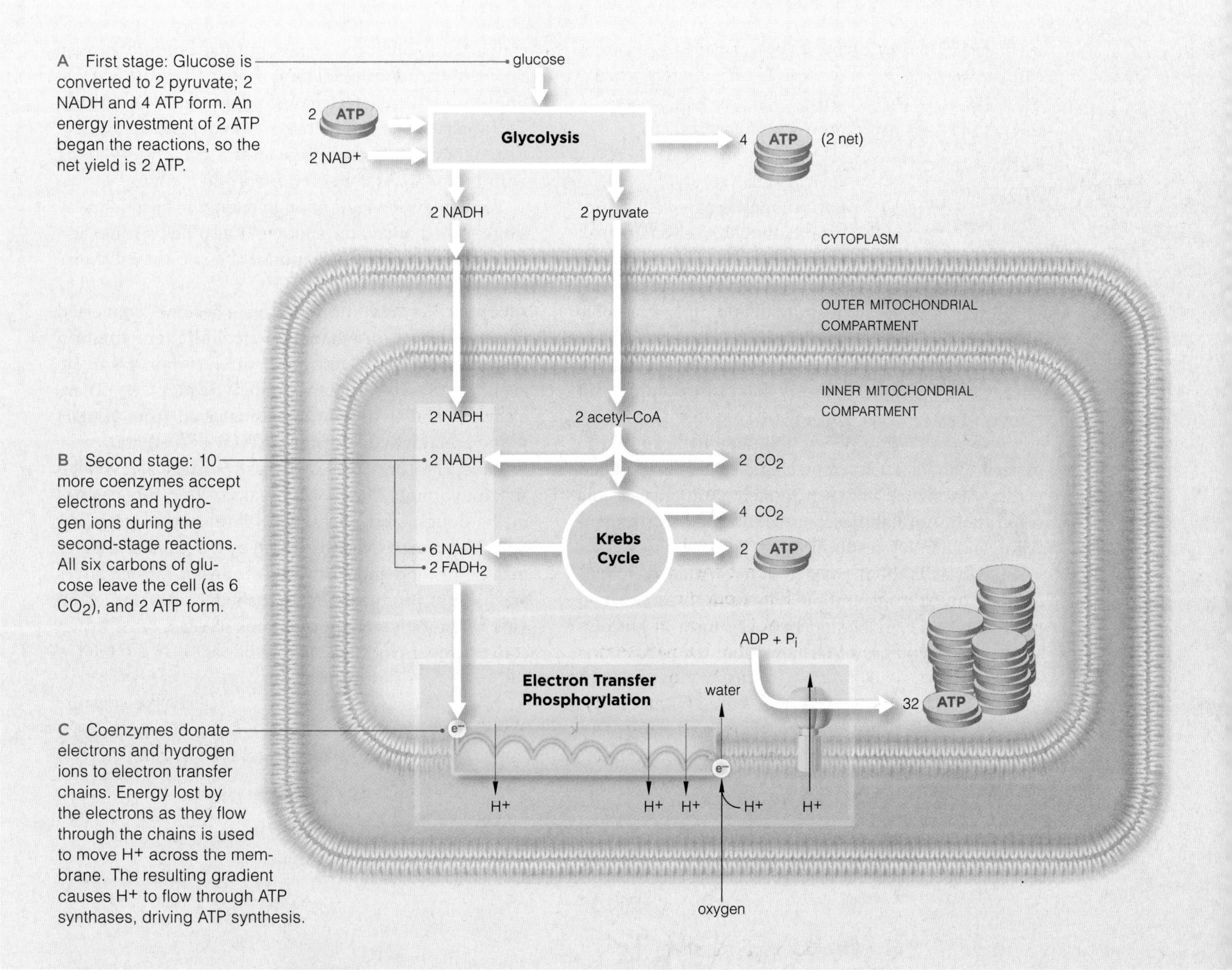

A First stage: Glucose is converted to 2 pyruvate; 2 NADH and 4 ATP form. An energy investment of 2 ATP began the reactions, so the net yield is 2 ATP.

glucose

2 **ATP**

2 NAD+

Glycolysis

4 **ATP** (2 net)

2 NADH

2 pyruvate

CYTOPLASM

OUTER MITOCHONDRIAL COMPARTMENT

INNER MITOCHONDRIAL COMPARTMENT

2 NADH

2 acetyl–CoA

B Second stage: 10 more coenzymes accept electrons and hydrogen ions during the second-stage reactions. All six carbons of glucose leave the cell (as 6 CO_2), and 2 ATP form.

2 NADH

2 CO_2

4 CO_2

6 NADH
2 FADH$_2$

Krebs Cycle

2 **ATP**

ADP + P$_i$

C Coenzymes donate electrons and hydrogen ions to electron transfer chains. Energy lost by the electrons as they flow through the chains is used to move H+ across the membrane. The resulting gradient causes H+ to flow through ATP synthases, driving ATP synthesis.

Electron Transfer Phosphorylation

e−

water

e−

32 **ATP**

H+   H+ H+   H+   H+

oxygen

Figure 8.8 Animated Summary of the steps in aerobic respiration. The typical overall yield of aerobic respiration is 36 ATP per glucose.

and hydrogen ions to molecules that can. After crossing the membranes, the intermediary molecules transfer the electrons to NAD+ or FAD in the inner compartment. This shuttling mechanism, which differs among cells, influences the ATP yield. In brain and skeletal muscle cells, the yield is thirty-eight ATP. In liver, heart, and kidney cells, it is thirty-six.

Remember that some energy dissipates with every transfer (Section 6.1). Even though aerobic respiration is a very efficient way of retrieving energy from carbohydrates, about 60 percent of the energy harvested in this pathway disperses as metabolic heat.

Take-Home Message

What is the third stage of aerobic respiration?

■ In aerobic respiration's third stage, electron transfer phosphorylation, energy released by electrons flowing through electron transfer chains is ultimately captured in the attachment of phosphate to ADP.

■ The reactions begin when coenzymes deliver electrons and hydrogen ions to electron transfer chains in the inner mitochondrial membrane.

■ Energy released by electrons as they pass through electron transfer chains pumps hydrogen ions from the inner mitochondrial compartment to the outer one. Thus, an H+ gradient forms.

■ The gradient drives H+ flow through ATP synthases, which results in ATP formation. The typical net yield of aerobic respiration is thirty-six ATP per glucose.

■ Fermentation pathways break down carbohydrates without using oxygen. The final steps in these pathways regenerate NAD+ but do not produce ATP.

Fermentation Pathways

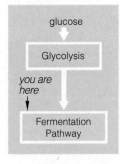

Bacteria and single-celled protists that inhabit sea sediments, animal guts, improperly canned food, sewage treatment ponds, deep mud, and other anaerobic habitats are fermenters. Some of these organisms, including the bacteria that cause botulism, do not tolerate aerobic conditions. They will die when exposed to oxygen. Others, such as the single-celled fungi called yeasts, can switch between fermentation and aerobic respiration. Animal muscle cells can use both fermentation and aerobic respiration.

Glycolysis is the first stage of fermentation, just as it is for aerobic respiration (Figure 8.4). Again, two pyruvate, two NADH, and two ATP form in glycolysis. In the last stages of fermentation, the pyruvate is converted to other molecules, but it is not fully broken down to carbon dioxide and water. Electrons do not flow through transfer chains, so no more ATP forms. The final steps of fermentation only regenerate NAD+. Regenerating this coenzyme allows glycolysis—along with the small ATP yield it offers—to continue.

Fermentation yields enough energy to sustain many single-celled anaerobic species. It also helps some aerobic species produce ATP under anaerobic conditions.

Alcoholic Fermentation Pyruvate becomes converted to ethyl alcohol, or ethanol, in **alcoholic fermentation** (Figures 8.9a and 8.10). First, three-carbon pyruvate is split into two-carbon acetaldehyde and CO_2. Then, electrons and hydrogen are transferred from NADH to the acetaldehyde, forming NAD+ and ethanol.

Bakers mix one species of yeast, *Saccharomyces cerevisiae*, into dough. These cells break down carbohydrates in the dough, and release CO_2 in alcoholic fermentation. The dough expands (rises) as CO_2 forms bubbles in it. Some wild and cultivated strains of *Saccharomyces* are used to produce wine. Crushed grapes are left in vats along with large populations of yeast cells, which convert sugars in the juice to ethanol.

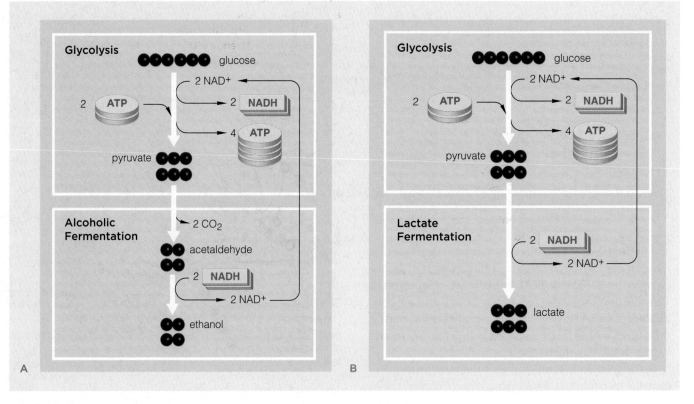

Figure 8.9 Animated (**a**) Alcoholic fermentation. (**b**) Lactate fermentation. In both pathways, the final steps do not produce ATP. They regenerate NAD+. The net yield is two ATP per molecule of glucose (from glycolysis).

Figure 8.10 Animated Alcoholic fermentation in action. (**a**) A vintner examines a fermentation product of *Saccharomyces*. (**b**) A commercial vat of yeast dough rising with the help of yeast cells. (**c**) Scanning electron micrograph of yeast cells.

Lactate Fermentation In **lactate fermentation**, electrons and hydrogen ions are transferred from NADH directly to pyruvate. This reaction converts pyruvate to three-carbon lactate (lactic acid), and also converts NADH to NAD+ (Figure 8.9*b*).

Some lactate fermenters spoil food (Section 21.6), but others preserve it. For instance, we use *Lactobacillus acidophilus*, which digests lactose in milk, to ferment dairy products such as buttermilk, cheese, and yogurt. Other species of yeast ferment and preserve pickles, corned beef, sauerkraut, and kimchi.

Take-Home Message

What is fermentation?

■ ATP can form by carbohydrate breakdown in fermentation pathways, which are anaerobic.

■ The end product of lactate fermentation is lactate. The end product of alcoholic fermentation is ethanol.

■ Both pathways have a net yield of two ATP per glucose molecule. The ATP forms during glycolysis.

■ Fermentation reactions regenerate the coenzyme NAD+, without which glycolysis (and ATP production) would stop.

8.6 | The Twitchers

■ Both lactate fermentation and aerobic respiration yield ATP for muscles that partner with bones.

■ Links to Hemoglobin 3.6, Pigments 7.1

Skeletal muscles, which move bones, consist of cells fused as long fibers. The fibers differ in how they make ATP.

Slow-twitch muscle fibers have many mitochondria and produce ATP by aerobic respiration. They dominate during prolonged activity, such as long runs. Slow-twitch fibers are red because they have an abundance of myoglobin, a pigment related to hemoglobin. Myoglobin stores oxygen in muscle tissue.

Fast-twitch muscle fibers have few mitochondria and no myoglobin; they are pale. The ATP they make by lactate fermentation sustains only short bursts of activity, such as sprints or weight lifting (Figure 8.11). The pathway makes ATP quickly but not for long; it cannot support sustained activity. That is one reason chickens cannot fly very far. The flight muscles of a chicken are mostly fast-twitch fibers, which make up the "white" breast meat. Chickens fly only in short bursts, which they do to escape predators. More often, a chicken walks or runs. Its leg muscles are mostly slow-twitch muscle, the "dark meat."

Would you expect to find more light or dark breast muscles in a migratory duck? An ostrich? An albatross that can skim the ocean surface for months?

Most human muscles are a mix of fast-twitch and slow-twitch fibers, but the proportions vary among muscles and among individuals. Great sprinters tend to have more fast-twitch fibers. Great marathon runners tend to have more slow-twitch fibers. Section 33.5 offers a closer look at energy-releasing pathways in skeletal muscle.

Figure 8.11 Sprinters and lactate fermentation. The micrograph, a cross-section through a human thigh muscle, reveals two types of fibers. Light fibers sustain short, intense bursts of speed; they make ATP by lactate fermentation. Dark fibers contribute to endurance; they make ATP by aerobic respiration. They appear dark because the tissue was stained for the presence of ATP synthase.

- Pathways that break down molecules other than carbohydrates also keep organisms alive.

- Links to Metabolism 3.2, Carbohydrates 3.3, Lipids 3.4, Proteins 3.5, Kilocalories 6.2

The Fate of Glucose at Mealtime and Between Meals

As you (and all other mammals) eat, glucose and other breakdown products of digestion are absorbed across the gut lining, and blood transports these small organic molecules throughout the body. The concentration of glucose in the bloodstream rises, and in response the pancreas (an organ) increases its rate of insulin secretion. The increase causes cells to take up glucose faster. Cells convert the glucose to glucose-6-phosphate, an intermediate of glycolysis (Figure 8.4).

When a cell takes in a lot of glucose, ATP-forming machinery goes into high gear. Unless the ATP is used quickly, its concentration rises in the cytoplasm. The high concentration of ATP causes glucose-6-phosphate to be diverted away from glycolysis and into a biosynthesis pathway that forms glycogen, a polysaccharide (Section 3.3). Liver and muscle cells especially favor the conversion of glucose to glycogen, and these cells maintain the body's largest stores of glycogen.

What happens if you eat too many carbohydrates? When the blood level of glucose gets too high, acetyl–CoA is diverted away from the Krebs cycle and into a pathway that makes fatty acids. That is why excess dietary carbohydrate ends up as fat (Table 8.1).

Between meals, the blood level of glucose declines. If the decline were not countered, that would be bad news for the brain, your body's glucose hog. At any time, your brain is taking up more than two-thirds of the freely circulating glucose. Why? Except in times of starvation, the brain's many nerve cells (neurons) use only this sugar, and they cannot store it.

The pancreas responds to low glucose levels in the blood by secreting glucagon. The hormone causes liver cells to convert stored glycogen to glucose. The cells release glucose into the bloodstream, so the glucose level rises, and brain cells keep working. Thus, hormones control whether cells use glucose as an energy source immediately or save it for later.

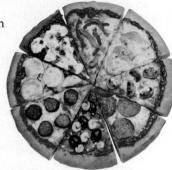

Glycogen makes up about 1 percent of an average adult's total energy reserves, which is the energy equivalent of about two cups of cooked pasta. Unless you eat regularly, you will completely deplete your liver's glycogen stores in less than twelve hours.

Energy From Fats

Of the total energy reserves in a typical adult who eats well, about 78 percent (about 10,000 kilocalories) is stored in body fat, and 21 percent in proteins.

How does a human body access its fat reservoir? A fat molecule, recall, has a glycerol head and one, two, or three fatty acid tails (Section 3.4). The body stores most fats as triglycerides, which have three fatty acid tails. Triglycerides accumulate in fat cells of adipose tissue. This tissue is an energy reservoir. It also insulates and pads the buttocks and other strategic areas of the body.

When the blood glucose level falls, triglycerides are tapped as an energy alternative. Enzymes in fat cells cleave the bonds between glycerol and the fatty acids, and both are released into the bloodstream. Enzymes in liver cells convert the glycerol to PGAL, which is an intermediate of glycolysis (Figure 8.4). Nearly all cells of your body can take up the fatty acids. Inside the cells, enzymes cleave the fatty acid backbones and convert the fragments to acetyl–CoA, which can enter the Krebs cycle (Figures 8.6 and 8.12).

Compared to carbohydrate breakdown, fatty acid breakdown yields more ATP per carbon atom. Between meals or during steady, prolonged exercise, fatty acid breakdown supplies about half of the ATP that muscle, liver, and kidney cells require.

Energy From Proteins

Some enzymes in your digestive system split dietary proteins into their amino acid subunits, which are then absorbed into the bloodstream. Cells use amino acids to build proteins or other molecules. Even so, when you eat more protein than your body needs, the amino

Table 8.1	Disposition of Organic Compounds
During meals	Excess glucose converted to glycogen or fat
Between meals	Glycogen degraded, glucose subunits enter glycolysis
	Fats degraded to fatty acids; some fragments enter glycolysis, others converted to acetyl–CoA
	Proteins degraded to amino acids, fragments become intermediates in Krebs cycle

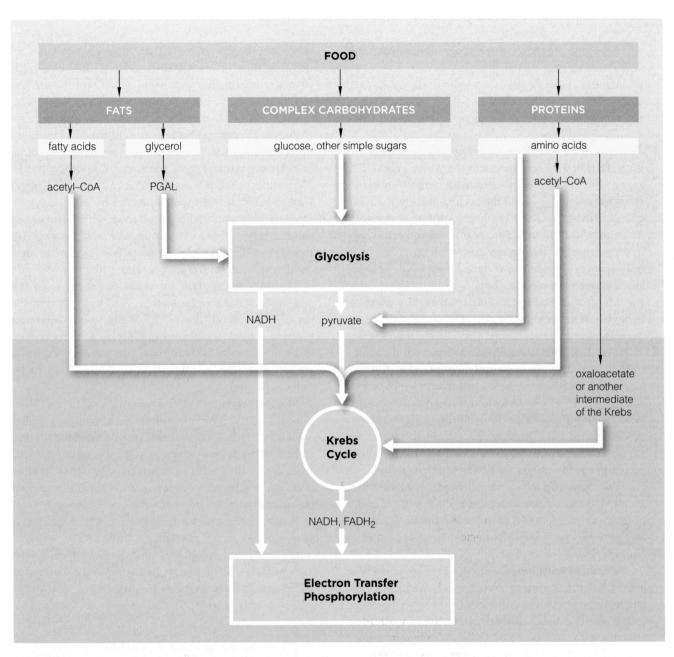

Figure 8.12 Animated Reaction sites where a variety of organic compounds enter aerobic respiration. Such compounds are alternative energy sources in the human body.

In humans and other mammals, complex carbohydrates, fats, and proteins from food do not enter the pathway of aerobic respiration directly. First, the digestive system, and then individual cells, break apart all of these molecules into simpler subunits. We return to this topic in Chapter 40.

acids become broken down further. Their NH_3^+ group is removed, and it becomes ammonia (NH_3). Their carbon backbone is split, and depending on the amino acid, acetyl–CoA, pyruvate, or an intermediate of the Krebs cycle forms. Your cells can divert any of these organic molecules into the Krebs cycle (Figure 8.12).

Maintaining and accessing energy reserves is complicated business. Controlling the use of glucose is important because it is the fuel of choice for the brain.

However, providing all of your cells with energy starts with the kinds and amounts of food you eat.

Take-Home Message

How are molecules other than glucose metabolized?

■ In humans and other mammals, the entrance of glucose or other organic compounds into an energy-releasing pathway depends on the kinds and proportions of carbohydrates, fats, and proteins in the diet.

■ Energy inputs drive the organization of molecules into units called cells.

■ Links to Life's organization 1.1, Water 2.5, Membrane organization 5.1, Selective permeability 5.3, Energy 6.1, Evolution of photosynthesis 7.8

At this point in the book, you may still have difficulty understanding the connections between yourself—a highly intelligent being—and such remote-sounding events as energy flow and the cycling of carbon, hydrogen, and oxygen. Is this really the stuff of humanity?

Think about the structure of a water molecule. Two hydrogen atoms sharing electrons with an oxygen may not seem very close to your daily life. Yet, through that sharing, water molecules have a polarity that makes them hydrogen-bond with one another. The chemical behavior of three simple atoms is a foundation for the organization of lifeless matter into living things.

Water also interacts with other molecules dispersed in it. Remember, phospholipids spontaneously organize into a two-layered film when they are mixed with water. Such lipid bilayers are the structural and functional foundation of all cell membranes.

Cells—and life—arose from such organization, but they continue by processes of metabolic control. With a membrane to contain them, metabolic reactions can proceed independently of conditions in the environment. With molecular functions built into their membranes, cells sense shifts in those conditions. Response mechanisms "tell" the cell what molecules to build or tear apart, and when to do it.

There is no mysterious force that creates proteins in cells. DNA, the double-stranded encyclopedia of inheritance, has a structure—a chemical message—that helps cells make and break down molecules, one generation after the next. Your own DNA strands tell your trillions of cells how to build proteins.

So yes, carbon, hydrogen, oxygen, and other atoms of organic molecules are the stuff of you, and us, and all of life. Yet life is more than molecules. It takes an ongoing flow of energy to turn molecules into cells, cells into organisms, organisms into communities, and so on through the biosphere (Section 1.1).

Photosynthesizers use energy from the sun and raw materials to feed themselves and, indirectly, nearly all other forms of life. Long ago they enriched the whole atmosphere with oxygen, a leftover of photosynthesis. That atmosphere favored aerobic respiration, a novel way to break down food molecules by using oxygen. Photosynthesizers made more food with leftovers of aerobic respiration—carbon dioxide and water. With this connection, the cycling of carbon, hydrogen, and oxygen through living things came full circle.

With few exceptions, infusions of energy from the sun sustain life's organization. And energy, remember, flows through the world of life in one direction (Section 6.1 and Figure 8.13). Only as long as energy lost from ecosystems is replaced with energy inputs—mainly from the sun—can life continue in all of its rich expressions.

In short, each new life is no more and no less than a marvelously complex system for prolonging order. Sustained with energy transfusions from the sun, life continues by its capacity for self-reproduction. With energy and the codes of inheritance in DNA, matter becomes organized, generation after generation. Even as individuals die, life elsewhere is prolonged. With each death, molecules are released and may be cycled as raw materials for new generations.

With this flow of energy and cycling of materials, each birth is affirmation of life's ongoing capacity for organization, each death a renewal.

Take-Home Message

What is the basis of life's unity and diversity?

■ Through biology, we have gained a profound insight into nature: Life's diversity, and its continuity, arise from unity at the level of molecules and energy.

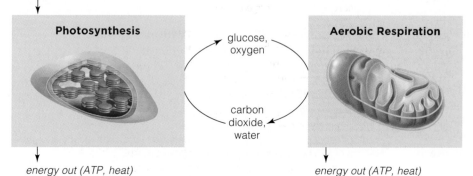

energy in (mainly from sunlight)

Photosynthesis

glucose, oxygen

Aerobic Respiration

carbon dioxide, water

energy out (ATP, heat)

energy out (ATP, heat)

Figure 8.13 Summary of links between photosynthesis (the main energy-requiring process) and aerobic respiration (the main energy-releasing process). Notice the one-way flow of energy.

Figure It Out: What does the middle circle represent?
Answer: The cycling of materials

At least 83 proteins are directly involved in the electron transfer chains of electron transfer phosphorylation in mitochondria. A defect in any one of them—or in any of the thousands of other proteins used by mitochondria, such as frataxin (*right*)—can wreak havoc in the body. About one in 5,000 people suffer from a known mitochondrial disorder. New research is showing that mitochondrial defects may be involved in

many other illnesses such as diabetes, hypertension, Alzheimer's and Parkinson's disease, and even aging.

How would you vote?

Developing new drugs is costly. Should governments fund companies that research treatments for rare diseases? See CengageNOW for details, then vote online.

Summary

Section 8.1 Most organisms convert chemical energy of carbohydrates to the chemical energy of ATP. **Anaerobic** and **aerobic** pathways of carbohydrate breakdown start in the cytoplasm with the same set of reactions, **glycolysis**, which converts glucose and other sugars to **pyruvate**. **Fermentation** pathways end in the cytoplasm, do not use oxygen, and yield two ATP per molecule of glucose. Most eukaryotic cells use **aerobic respiration**, which uses oxygen and yields much more ATP than fermentation. In eukaryotes, it is completed in mitochondria.

■ *Use the animation on CengageNOW for an overview of aerobic respiration.*

Section 8.2 Enzymes of glycolysis use two ATP to convert one molecule of glucose or another six-carbon sugar to two molecules of pyruvate. In the reactions, electrons and hydrogen ions are transferred to two NAD+, which are thereby reduced to NADH. Four ATP also form by **substrate-level phosphorylation**, the direct transfer of a phosphate group from a reaction intermediate to ADP.

The net yield of glycolysis is two pyruvate, two ATP, and two NADH per glucose molecule. The pyruvate may continue in fermentation in the cytoplasm, or it may enter mitochondria and the next steps of aerobic respiration.

■ *Use the animation on CengageNOW for a step-by-step journey through glycolysis.*

Section 8.3 The second stage of aerobic respiration, acetyl–CoA formation and the **Krebs cycle**, takes place in the inner compartment of mitochondria. The first steps convert two pyruvate from glycolysis to two acetyl–CoA and two CO_2. The acetyl–CoA enters the Krebs cycle. It takes two cycles to dismantle the two acetyl–CoA. At this stage, all of the carbon atoms in the glucose molecule that entered glycolysis have left the cell in CO_2.

During these reactions, electrons and hydrogen ions are transferred to NAD+ and FAD, which are thereby reduced to NADH and $FADH_2$. ATP forms by substrate-level phosphorylation.

In total, the second stage of aerobic respiration results in the formation of six CO_2, two ATP, eight NADH, and two $FADH_2$ for every two pyruvates. Adding the yield from glycolysis, the total tally for the first two stages of aerobic respiration is twelve reduced coenzymes and four ATP for each glucose molecule.

■ *Use the animation on CengageNOW to explore a mitochondrion and observe the reactions inside it.*

Section 8.4 Aerobic respiration ends in mitochondria. In the third stage of reactions, **electron transfer phosphorylation**, reduced coenzymes deliver their electrons and H+ to electron transfer chains in the inner mitochondrial membrane. Electrons moving through the chains release energy bit by bit; molecules of the chain use that energy to move H+ from the inner to the outer compartment.

Hydrogen ions that accumulate in the outer compartment form a gradient across the inner membrane. The ions follow the gradient back to the inner compartment through ATP synthases. H+ flow through these transport proteins drives ATP synthesis.

Oxygen combines with electrons and H+ at the end of the transfer chains, thus forming water.

Overall, aerobic respiration typically yields thirty-six ATP for each glucose molecule.

■ *Use the animation on CengageNOW to see how each step in aerobic respiration contributes to a big energy harvest.*

Sections 8.5, 8.6 Fermentation pathways begin with glycolysis and finish in the cytoplasm. They do not use oxygen or electron transfer chains. The final steps oxidize NADH to NAD+, which is required for glycolysis to continue, but produce no ATP. The end product of **lactate fermentation** is lactate. The end product of **alcoholic fermentation** is ethyl alcohol, or ethanol. Both pathways have a net yield of two ATP per glucose (from glycolysis).

Slow-twitch and fast-twitch skeletal muscle fibers can support different activity levels. Aerobic respiration and lactate fermentation proceed in different fibers that make up these muscles.

■ *Use the animation on CengageNOW to compare alcoholic and lactate fermentation.*

Section 8.7 In humans and other mammals, the simple sugars from carbohydrate breakdown, glycerol and fatty acids from fat breakdown, and carbon backbones of amino acids from protein breakdown may enter aerobic respiration at various reaction steps.

■ *Use the interaction on CengageNOW to follow the breakdown of different organic molecules.*

Section 8.8 The diversity and continuity of life arises from its unity at the level of molecules and energy.

Data Analysis Exercise

Tetralogy of Fallot (TF) is a genetic disorder characterized by four major malformations of the heart. The circulation of blood is abnormal, so TF patients have too little oxygen in their blood. Inadequate oxygen levels result in damaged mitochondrial membranes, which in turn cause cells to self-destruct. In 2004, Sarah Kuruvilla and her colleages looked at abnormalities in the mitochondria of heart muscle in TF patients. Some of their results are shown in Figure 8.14.

1. Which abnormality was most strongly associated with TF?

2. Can you make any correlations between blood oxygen content and mitochondrial abnormalities in these patients?

Figure 8.14 Mitochondrial changes in Tetralogy of Fallot (TF). (**a**) Normal heart muscle. Many mitochondria between the fibers provide muscle cells with ATP for contraction. (**b**) Heart muscle from a person with TF shows swollen, broken mitochondria.

(**c**) Mitochondrial abnormalities in TF patients. SPO_2 is oxygen saturation of the blood. A normal value of SPO_2 is 96%. Abnormalities seen are indicated by "+" signs.

Patient (age)	SPO$_2$ (%)	Mitochondrial Abnormalities in TF			
		Number	Shape	Size	Broken
1 (5)	55	+	+	–	–
2 (3)	69	+	+	–	–
3 (22)	72	+	+	–	–
4 (2)	74	+	+	–	–
5 (3)	76	+	+	–	+
6 (2.5)	78	+	+	–	+
7 (1)	79	+	+	–	–
8 (12)	80	+	–	+	–
9 (4)	80	+	+	–	–
10 (8)	83	+	–	+	–
11 (20)	85	+	+	–	–
12 (2.5)	89	+	–	+	–

c

Self-Quiz
Answers in Appendix III

1. Is the following statement true or false? Unlike animals, which make many ATP by aerobic respiration, plants make all of their ATP by photosynthesis.

2. Glycolysis starts and ends in the _____ .
   a. nucleus
   b. mitochondrion
   c. plasma membrane
   d. cytoplasm

3. Which of the following metabolic pathways require molecular oxygen (O_2)?
   a. aerobic respiration
   b. lactate fermentation
   c. alcoholic fermentation
   d. all of the above

4. Which molecule does not form during glycolysis?
   a. NADH   b. pyruvate   c. FADH$_2$ d. ATP

5. In eukaryotes, aerobic respiration is completed in the _____ .
   a. nucleus
   b. mitochondrion
   c. plasma membrane
   d. cytoplasm

6. Which of the following reaction pathways is not part of the second stage of aerobic respiration?
   a. electron transfer phosphorylation
   b. acetyl–CoA formation
   c. Krebs cycle
   d. glycolysis
   e. a and d

7. After the Krebs cycle runs _____ time(s), one glucose molecule has been completely oxidized.
   a. one   b. two   c. three   d. six

8. In the third stage of aerobic respiration, _____ is the final acceptor of electrons from glucose.
   a. water   b. hydrogen   c. oxygen   d. NADH

9. In alcoholic fermentation, _____ is the final acceptor of electrons stripped from glucose.
   a. oxygen
   b. pyruvate
   c. acetaldehyde
   d. sulfate

10. Fermentation makes no more ATP beyond the small yield from glycolysis. The remaining reactions _____ .
    a. regenerate FAD
    b. regenerate NAD$^+$
    c. regenerate NADH
    d. regenerate FADH$_2$

11. Your body cells can use _____ as an alternative energy source when glucose is in short supply.
    a. fatty acids
    b. glycerol
    c. amino acids
    d. all of the above

12. Match the event with its most suitable description.
    ___ glycolysis
    ___ fermentation
    ___ Krebs cycle
    ___ electron transfer phosphorylation

    a. ATP, NADH, FADH$_2$, and CO$_2$ form
    b. glucose to two pyruvates
    c. NAD$^+$ regenerated, little ATP
    d. H$^+$ flows through ATP synthases

■ *Visit CengageNOW for additional questions.*

Critical Thinking

1. At high altitudes, oxygen levels are low. Mountain climbers risk altitude sickness, which is characterized by shortness of breath, weakness, dizziness, and confusion. The early symptoms of cyanide poisoning resemble altitude sickness. Cyanide binds tightly to cytochrome *c* oxidase, a protein complex that is the last component of mitochondrial electron transfer chains. Cytochrome *c* oxidase with bound cyanide can no longer transfer electrons. Explain why cyanide poisoning starts with the same symptoms as altitude sickness.

2. As you learned, membranes impermeable to hydrogen ions are required for electron transfer phosphorylation. Membranes in mitochondria serve this function in eukaryotes. Prokaryotic cells do not have this organelle, but they can make ATP by electron transfer phosphorylation. How do you think they do it, given that they have no mitochondria?

II PRINCIPLES OF INHERITANCE

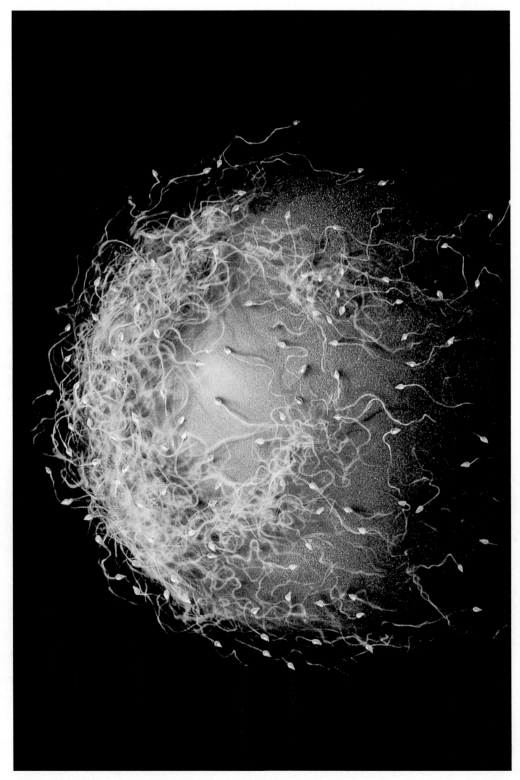

Human sperm, one of which will penetrate this mature egg and so set the stage for the development of a new individual in the image of its parents. This exquisite art is based on a scanning electron micrograph.

How Cells Reproduce

IMPACTS, ISSUES Henrietta's Immortal Cells

Each human starts out as a fertilized egg. By the time of birth, the human body consists of about a trillion cells, all descended from that single cell. Even in an adult, billions of cells divide every day and replace their damaged or aged predecessors. However, human cells cultured in the laboratory tend to divide a few times and die within weeks.

Since the mid-1800s, researchers had been trying to coax human cells to become immortal—to keep dividing outside of the body. Why? Many human diseases propagate only in human cells. Immortal cell lineages, or cell lines, would allow the researchers to study such diseases without experimenting on people. At Johns Hopkins University, George and Margaret Gey were among those researchers. They had been trying to culture human cells for almost thirty years when, in 1951, their assistant Mary Kubicek prepared one last sample of human cancer cells. Mary named the cells *HeLa*, after the first and last names of the patient from whom they were taken.

The HeLa cells began to divide. They divided again and again. Four days later, there were so many cells that the researchers had to transfer part of the population to more culture tubes. The cell populations increased at a phenomenal rate; cells were dividing every twenty-four hours and coating the inside of the tubes within days.

Sadly, cancer cells in the patient were dividing just as fast. Six months after she had been diagnosed with cancer, malignant cells had invaded tissues throughout her body.

Two months after that, Henrietta Lacks, a young woman from Baltimore, was dead.

Although Henrietta passed away, her cells lived on in the Geys' laboratory (Figure 9.1). The Geys used HeLa cells to distinguish among the viral strains that cause polio, which at the time was epidemic. They also shipped the cells to other laboratories all over the world. Researchers still use cell culture techniques developed by the Geys. They also continue to use HeLa cells to investigate cancer, viral growth, protein synthesis, the effects of radiation on cells, and more. Some HeLa cells even traveled into space for experiments on the *Discoverer XVII* satellite.

Henrietta Lacks was thirty-one, a wife and mother of four, when runaway cell divisions killed her. Decades later, her legacy continues to help humans all around the world, through her cells that are still dividing day after day.

Understanding cell division—and, ultimately, how new individuals are put together in the image of their parents—starts with answers to three questions. First, what kind of information guides inheritance? Second, how is that information copied inside a parent cell before being distributed to each of its descendant cells? Third, what kinds of mechanisms parcel out the information to descendant cells?

We will require more than one chapter to survey the nature of inheritance. In this chapter, we introduce the structures and mechanisms that cells use to reproduce.

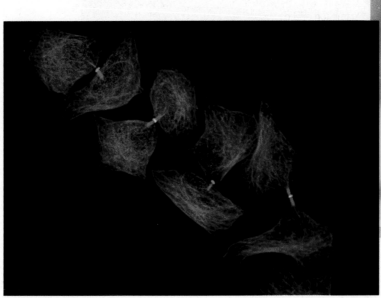

See the video! Figure 9.1 *Left*, dividing HeLa cells—a cellular legacy of Henrietta Lacks, who was a young casualty of cancer. *Right*, Henrietta's son David holds a picture of his parents.

Key Concepts

Chromosomes and dividing cells

Individuals have a characteristic number of chromosomes in each of their cells. The chromosomes differ in length and shape, and they carry different portions of the cell's hereditary information. Division mechanisms parcel out the information into descendant cells. **Section 9.1**

Where mitosis fits in the cell cycle

A cell cycle starts when a new cell forms by division of a parent cell, and ends when the cell completes its own division. A typical cell cycle proceeds through intervals of interphase, mitosis, and cytoplasmic division. **Section 9.2**

Stages of mitosis

Mitosis divides the nucleus, not the cytoplasm. It has four sequential stages: prophase, metaphase, anaphase, and telophase. A bipolar spindle forms and moves the cell's duplicated chromosomes into two parcels, which end up in two genetically identical nuclei. **Section 9.3**

How the cytoplasm divides

After nuclear division, the cytoplasm divides. Typically, one nucleus ends up in each of two new cells. The cytoplasm of an animal cell simply pinches in two. In plant cells, a cross-wall forms in the cytoplasm and divides it. **Section 9.4**

The cell cycle and cancer

Built-in mechanisms monitor and control the timing and rate of cell division. On rare occasions, the surveillance mechanisms fail, and cell division becomes uncontrollable. Tumor formation and cancer are outcomes. **Section 9.5**

Links to Earlier Concepts

- Before you start this chapter, think back on the changing appearance of chromosomes in the nucleus of eukaryotic cells (Section 4.8).

- You may also wish to review the introduction to microtubules and motor proteins (4.13). Doing so will help you understand how the mitotic spindle works.

- A review of plant cell walls (4.12) will help give you a sense of why plant cells do not divide by pinching their cytoplasm into two parcels, as animal cells do.

9.1 | Overview of Cell Division Mechanisms

- Individual cells or organisms produce offspring by the process of reproduction.

- Link to Nucleus 4.8

When a cell reproduces, each of its cellular offspring inherits information encoded in parental DNA along with enough cytoplasm to start up its own operation. DNA contains protein-building instructions. Some of the proteins are structural materials; others are enzymes that speed construction of organic molecules. If a new cell does not inherit all of the information required to build proteins, it will not grow or function properly.

A parent cell's cytoplasm contains all the enzymes, organelles, and other metabolic machinery necessary for life. A descendant cell that inherits a blob of cytoplasm is getting start-up metabolic machinery that will keep it running until it can make its own.

Mitosis, Meiosis, and the Prokaryotes

In general, a eukaryotic cell cannot simply split in two, because only one of its descendant cells would get the nucleus—and thus, the DNA. A cell's cytoplasm splits only after its DNA has been packaged into more than one nucleus by way of mitosis or meiosis.

Mitosis is a nuclear division mechanism that occurs in the somatic cells (body cells) of multicelled eukaryotes. Mitosis and cytoplasmic division are the basis of increases in body size during development, and ongoing replacements of damaged or dead cells. Many species of plants, animals, fungi, and single-celled protists also make copies of themselves, or reproduce asexually, by mitosis (Table 9.1).

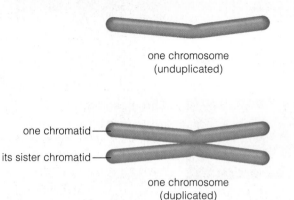

one chromosome
(unduplicated)

one chromatid ———
its sister chromatid ———

one chromosome
(duplicated)

Figure 9.2 A eukaryotic chromosome in the unduplicated state and duplicated state. Eukaryotic cells duplicate their chromosomes before mitosis or meiosis begins. After duplication, each chromosome consists of two sister chromatids.

Meiosis is a nuclear division mechanism that precedes the formation of gametes or spores, and it is the basis of sexual reproduction. In humans and all other mammals, the gametes called sperm and eggs develop from immature reproductive cells. Spores, which protect and disperse new generations, form during the life cycle of fungi, plants, and many kinds of protists.

As you will discover in this chapter and the next, meiosis and mitosis have much in common. Even so, their outcomes differ.

What about prokaryotes—bacteria and archaeans? Such cells reproduce asexually by prokaryotic fission, which is an entirely different mechanism. We consider prokaryotic fission later, in Section 21.5.

Key Points About Chromosome Structure

The genetic information of each eukaryotic species is distributed among some characteristic number of chromosomes that differ in length and shape (Section 4.8). Before a eukaryotic cell enters mitosis or meiosis, each of its chromosomes consists of one double-stranded DNA molecule (Figure 9.2). After the chromosomes are duplicated, each consists of *two* double-stranded DNA molecules. Those two molecules of DNA stay attached as a single chromosome until late in nuclear division. Until they separate, they are called **sister chromatids**.

During the early stages of mitosis and meiosis, each duplicated chromosome coils back on itself again and again, until it is highly condensed. Figure 9.3a shows an example of a duplicated human chromosome when it is most condensed. The structural organization of a chromosome arises from interactions between each DNA molecule and the proteins associated with it.

Table 9.1   Comparison of Cell Division Mechanisms

Mechanisms	Functions
Mitosis, cytoplasmic division	In all multicelled eukaryotes, the basis of: 1. Increases in body size during growth 2. Replacement of dead or worn-out cells 3. Repair of damaged tissues In single-celled and many multicelled species, also the basis of asexual reproduction
Meiosis, cytoplasmic division	In single-celled and multicelled eukaryotes, the basis of sexual reproduction; part of the processes by which gametes and sexual spores form (Chapter 10)
Prokaryotic fission	In bacteria and archaeans alone, the basis of asexual reproduction (Section 21.5)

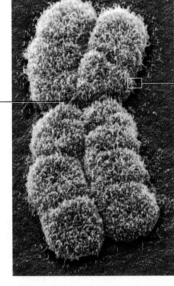

centromere

A Duplicated human chromosome in its most condensed form. If this chromosome were actually the size shown in the micrograph, its two DNA strands would stretch out about 800 meters (0.5 miles).

At regular intervals, a double-stranded DNA molecule winds twice around "spools" of proteins called **histones**. In a micrograph, these DNA–histone spools look like beads on a string (Figure 9.3*d*). Each "bead" is a **nucleosome**, the smallest unit of structural organization in eukaryotic chromosomes (Figure 9.3*e*).

When a duplicated chromosome condenses, its sister chromatids constrict where they attach to one another. This constricted region is called the **centromere** (Figure 9.3*a*). The location of a centromere differs for each type of chromosome. During nuclear division, a kinetochore forms at the centromere. Kinetochores are binding sites for microtubules that attach to chromatids.

What is the point of all this structural organization? It allows a huge amount of DNA to pack into a little nucleus. For example, the DNA from one of your body cells would stretch out to about 2 meters (6.5 feet)! That is a lot of DNA to pack into a nucleus that is typically less than 10 micrometers in diameter. The packing also serves a regulatory purpose. As you will see in Chapter 15, enzymes cannot access DNA that is tightly coiled.

Take-Home Message

What is cell division and why does it happen?

■ When a cell divides, each of its descendant cells receives a required number of chromosomes and some cytoplasm. In eukaryotic cells, the nucleus divides first, then the cytoplasm.

■ Mitosis is a nuclear division mechanism that is the basis of body size increases, cell replacements, and tissue repair in multicelled eukaryotes. Mitosis is also the basis of asexual reproduction in single-celled and some multicelled eukaryotes.

■ In eukaryotes, a nuclear division mechanism called meiosis precedes the formation of gametes and, in many species, spores. It is the basis of sexual reproduction.

multiple levels of coiling of DNA and proteins

B When a chromosome is at its most condensed, the DNA is packed into tightly coiled coils.

fiber

C When the coiled coils unwind, a molecule of chromosomal DNA and its associated proteins are organized as a cylindrical fiber.

beads on a string

D A loosened fiber shows a "beads-on-a-string" organization. The "string" is the DNA molecule; each "bead" is one nucleosome.

DNA double helix

core of histones

nucleosome

E A nucleosome consists of part of a DNA molecule looped twice around a core of histone proteins.

Figure 9.3 Animated Chromosome structure. Several levels of structural organization allow a lot of DNA to pack very tightly into small nuclei.

■ The cell cycle is a sequence of stages through which a cell passes during its lifetime.

■ Link to Microtubules 4.13

The life of a cell passes through a sequence of events between each cell division and the next (Figure 9.4). Interphase, mitosis, and cytoplasmic division are the stages of this **cell cycle**. The length of the cycle is about the same for all cells of the same type, but it differs from one cell type to the next. For example, the stem cells in your red bone marrow divide every 12 hours. Their descendants become red blood cells that replace 2 to 3 million worn-out ones in your blood each second. Cells in the tips of a bean plant root divide every 19 hours. In a sea urchin embryo, which develops rapidly from a fertilized egg, the cells divide every 2 hours.

The Life of a Cell

By a process called **DNA replication**, a cell copies all of its DNA before it divides. This work is completed during interphase, which is typically the longest interval of the cell cycle. **Interphase** consists of three stages, during which a cell increases its mass, roughly doubles the number of its cytoplasmic components, and replicates its DNA:

G1 Interval ("Gap") of cell growth and activity before the onset of DNA replication

S   Time of "Synthesis" (DNA replication)

G2 Second interval (Gap), after DNA replication when the cell prepares for division

Gap intervals were named because outwardly they seem to be periods of inactivity. Actually, most cells going about their metabolic business are in G1. Cells preparing to divide enter S, when they copy their DNA. During G2, they make the proteins that will drive mitosis. Once S begins, DNA replication usually proceeds at a predictable rate and ends before the cell divides.

Control mechanisms work at certain points in the cell cycle. Some function as built-in brakes on the cell cycle. Apply the brakes that work in G1, and the cycle stalls in G1. Lift the brakes, and the cycle runs again.

Such controls are an important part of keeping the body functioning correctly. For example, the neurons (nerve cells) in most parts of your brain remain permanently in G1 of interphase; once they mature, they will never divide again. Experimentally driving them out of G1 causes them to die, not divide, because cells normally self-destruct if their cell cycle goes awry.

Cell suicide is is important because if controls over the cell cycle stop working, the body may be endangered. As you will see shortly, cancer begins this way. Crucial controls are lost, and the cell cycle spins out of control.

Mitosis and the Chromosome Number

After G2, a cell enters mitosis. Identical descendant cells result, each with the same number and kind of chromosomes as the parent. The **chromosome number** is the sum of all chromosomes in a cell of a given type. The body cells of gorillas have 48, those of human cells have 46. Pea plant cells have 14.

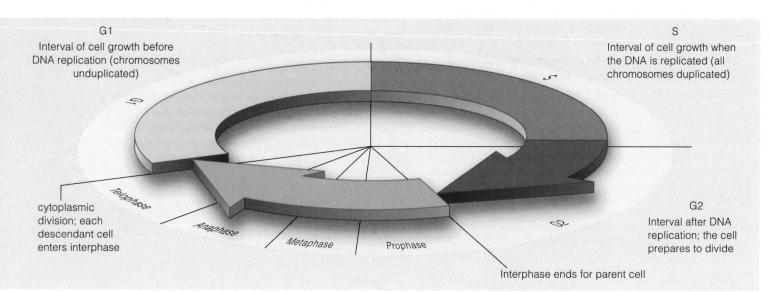

G1
Interval of cell growth before DNA replication (chromosomes unduplicated)

S
Interval of cell growth when the DNA is replicated (all chromosomes duplicated)

cytoplasmic division; each descendant cell enters interphase

telophase

Anaphase

Metaphase

Prophase

Interphase ends for parent cell

G2
Interval after DNA replication; the cell prepares to divide

Figure 9.4 Animated Eukaryotic cell cycle. The length of each interval differs among cells.

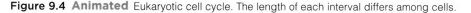

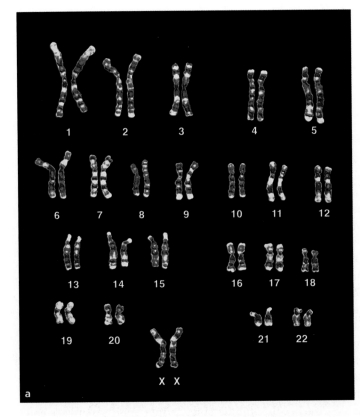

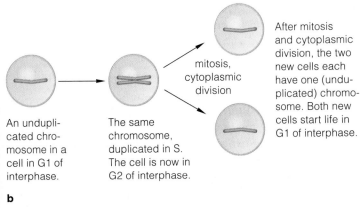

An unduplicated chromosome in a cell in G1 of interphase.

The same chromosome, duplicated in S. The cell is now in G2 of interphase.

mitosis, cytoplasmic division

After mitosis and cytoplasmic division, the two new cells each have one (unduplicated) chromosome. Both new cells start life in G1 of interphase.

b

Figure 9.5 Mitosis maintains the chromosome number.

(a) Human body cells are diploid—they have twenty-three pairs of chromosomes, for a total of forty-six. The last ones in this lineup of human chromosomes are a pair of sex chromosomes: Females have two X chromosomes; males have one X and one Y.

(b) What happens to each one of the forty-six chromosomes? Each time a human body cell undergoes mitosis and cytoplasmic division, its descendant cells end up with a complete set of forty-six chromosomes.

Figure It Out: Were the chromosomes in (a) taken from the cell of a male or a female?   *Answer: Female*

Actually, human body cells have two of each type of chromosome: Their chromosome number is **diploid** ($2n$). The 46 are like two sets of books numbered 1 to 23 (Figure 9.5*a*). You have two volumes of each: a pair. Except for a pairing of sex chromosomes (XY) in males, members of each pair are the same length and shape, and they hold information about the same traits.

Think of them as two sets of books on how to build a house. Your father gave you one set. Your mother had her own ideas about wiring, plumbing, and so on. She gave you an alternate set that says slightly different things about many of those tasks.

With mitosis followed by cytoplasmic division, a diploid parent cell produces two diploid descendant cells. It is not just that each new cell gets forty-six or forty-eight or fourteen chromosomes. If only the total mattered, then one cell might get, say, two pairs of chromosome 22 and no pairs whatsoever of chromosome 9. Neither cell could function like its parent without two of each type of chromosome.

As the next section explains, a dynamic network of microtubules called the **bipolar spindle** forms during nuclear division. The spindle grows into the cytoplasm from opposite ends, or poles, of the cell. During mitosis, some of the spindle's microtubules attach to the duplicated chromosomes. Microtubules from one

pole connect to one chromatid of each chromosome; microtubules from the other pole connect to its sister:

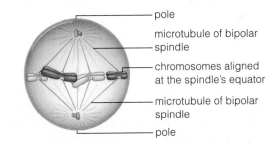

pole

microtubule of bipolar spindle

chromosomes aligned at the spindle's equator

microtubule of bipolar spindle

pole

The microtubules separate sister chromatids and move them to opposite ends of the cell. Two parcels of chromosomes form, and a nuclear membrane forms around each. The cytoplasm divides, and two new cells are the result. Figure 9.5*b* shows a preview of how mitosis maintains the parental chromosome number.

Take-Home Message

What is a cell cycle?

■ A cell cycle is a sequence of stages (interphase, mitosis, and cytoplasmic division) through which a cell passes during its lifetime.

■ During interphase, a new cell increases its mass, doubles the number of its cytoplasmic components, and duplicates its chromosomes. The cycle ends after the cell undergoes mitosis and then divides its cytoplasm.

■ When a nucleus divides by mitosis, each new nucleus has the same chromosome number as the parent cell.

■ There are four main stages of mitosis: prophase, metaphase, anaphase, and telophase.

■ Link to Microtubules and motor proteins 4.13

A cell duplicates its chromosomes during interphase, so by the time mitosis begins, each chromosome consists of two sister chromatids joined at the centromere. During the first stage of mitosis, **prophase**, the chromosomes condense and become visible in micrographs (Figure 9.6a,b). "Mitosis" is from the Greek *mitos*, or thread, for the threadlike appearance of chromosomes during the nuclear division process.

Most animal cells have a centrosome, a region near the nucleus that organizes spindle microtubules while they are forming. The centrosome usually includes two barrel-shaped centrioles, and it is duplicated just before prophase begins. In prophase, one of the two centrosomes (along with its pair of centrioles) moves to the opposite pole of the nucleus. Microtubules that will form the bipolar spindle begin to grow from both centrosomes. (Plant cells do not have centrosomes, but they have other structures that guide spindle growth.) Motor proteins traveling along the microtubules help the spindle grow in the appropriate directions. Motor protein movement is driven by ATP (Section 4.13).

As prophase ends, the nuclear envelope breaks up and spindle microtubules penetrate the nuclear region (Figure 9.6c). Some microtubules from each spindle pole stop growing after they overlap in the middle of the cell. Others continue to grow until they reach a chromosome and attach to it.

One chromatid of each chromosome is tethered by microtubules extending from one spindle pole, and its sister chromatid is tethered by microtubules extending from the other spindle pole. The opposing sets of microtubules begin a tug-of-war by adding and losing tubulin subunits. As the microtubules grow and shrink, they push and pull the chromosomes. Soon, all the microtubules are the same length. At that point, they have aligned the chromosomes midway between the spindle poles (Figure 9.6d). The alignment marks **metaphase** (from ancient Greek *meta*, between).

Anaphase is the interval when sister chromatids of each chromosome separate and move toward opposite spindle poles (Figure 9.6e). Three cell activities bring this about. First, the spindle microtubules attached to each chromatid shorten. Second, motor proteins drag the chromatids along shrinking microtubules toward each spindle pole. Third, the microtubules that overlap midway between spindle poles begin to slide past one

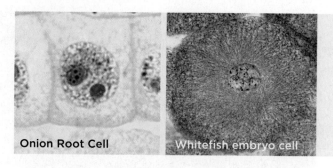

Onion Root Cell   Whitefish embryo cell

Figure 9.6 Animated Mitosis.

Opposite page, micrographs of onion root (plant) cells are shown on the *left*; whitefish embryo (animal) cells on the *right*.

The drawings show a diploid (2*n*) animal cell. For clarity, only two pairs of chromosomes are illustrated, but cells of nearly all eukaryotes have more than two. The two chromosomes of the pair inherited from one parent are coded *purple*; the two chromosomes inherited from the other parent are coded *blue*.

Above, interphase cells are shown for comparison, but interphase is not part of mitosis.

another. Motor proteins drive this movement, which pushes the spindle poles farther apart. Anaphase ends as each chromosome and its duplicate are heading to opposite spindle poles.

Telophase begins when the two clusters of chromosomes reach the spindle poles. Each cluster consists of the parental complement of chromosomes—two of each, if the parent cell was diploid. Vesicles derived from the old nuclear envelope fuse in patches around the clusters as the chromosomes decondense. Patch joins with patch until each set of chromosomes is enclosed by a new nuclear envelope. Thus, two nuclei form (Figure 9.6f). The parent cell in our example was diploid, so each new nucleus is diploid too. Once two nuclei have formed, telophase is over, and so is mitosis.

Take-Home Message

What happens during mitosis?

■ Each chromosome in a cell's nucleus was duplicated before mitosis begins, so each consists of two sister chromatids.

■ In prophase, chromosomes condense and microtubules form a bipolar spindle. The nuclear envelope breaks up. Some of the microtubules attach to the chromosomes.

■ At metaphase, all duplicated chromosomes are aligned midway between the spindle poles.

■ In anaphase, microtubules separate the sister chromatids of each chromosome, and pull them to opposite spindle poles.

■ In telophase, two clusters of chromosomes reach the spindle poles. A new nuclear envelope forms around each cluster.

■ Thus two new nuclei form. Each one has the same chromosome number as the parent cell's nucleus.

 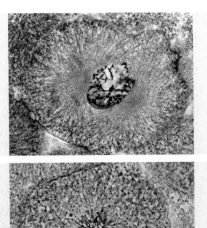

A Early Prophase

Mitosis begins. In the nucleus, the chromatin begins to appear grainy as it organizes and condenses. The centrosome is duplicated.

  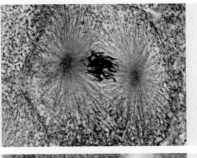

B Prophase

The chromosomes become visible as discrete structures as they condense further. Microtubules assemble and move one of the two centrosomes to the opposite side of the nucleus, and the nuclear envelope breaks up.

 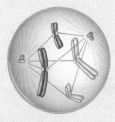

C Transition to Metaphase

The nuclear envelope is gone, and the chromosomes are at their most condensed. Microtubules of the bipolar spindle assemble and attach sister chromatids to opposite spindle poles.

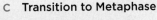 

D Metaphase

All of the chromosomes are aligned midway between the spindle poles. Microtubules attach each chromatid to one of the spindle poles, and its sister to the opposite pole.

 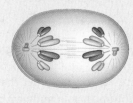 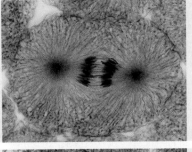

E Anaphase

Motor proteins moving along spindle microtubules drag the chromatids toward the spindle poles, and the sister chromatids separate. Each sister chromatid is now a separate chromosome.

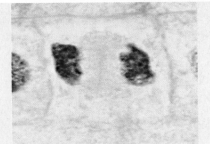 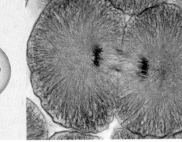

F Telophase

The chromosomes reach the spindle poles and decondense. A nuclear envelope begins to form around each cluster; new plasma membrane may assemble between them. Mitosis is over.

9.4 | Cytoplasmic Division Mechanisms

- In most kinds of eukaryotes, the cell cytoplasm divides between late anaphase and the end of telophase, but the mechanism of division differs.

- Links to Primary wall 4.12, Cytoskeleton 4.13

Division of Animal Cells

A cell's cytoplasm usually divides after mitosis. The process of cytoplasmic division, or **cytokinesis**, differs among eukaryotes. Typical animal cells partition their cytoplasm by pinching in two. The plasma membrane starts to sink inward as a thin indentation between the former spindle poles (Figure 9.7a). The indentation is called a cleavage furrow, and it is the first visible sign that the cytoplasm is dividing. The furrow advances until it extends around the cell. As it does, it deepens along a plane that corresponds to the former spindle equator (midway between the poles).

What is happening? The cell cortex, which is the mesh of cytoskeletal elements just under the plasma membrane, includes a band of actin and myosin filaments that wraps around the cell's midsection. ATP hydrolysis causes these filaments to interact, just as it does in muscle cells, and the interaction results in contraction. The band of filaments, which is called a **contractile ring**, is anchored to the plasma membrane. As it shrinks, the band drags the plasma membrane

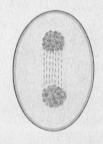 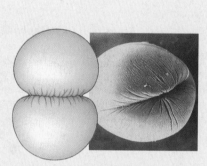 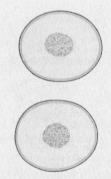

1 Mitosis is completed, and the bipolar spindle is starting to disassemble.

2 At the former spindle equator, a ring of actin filaments attached to the plasma membrane contracts.

3 This contractile ring pulls the cell surface inward as it continues to contract.

4 The contractile ring contracts until the cytoplasm is partitioned and the cell pinches in two.

A   Contractile Ring Formation

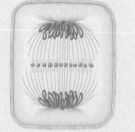 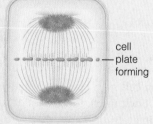 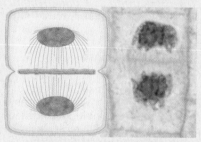 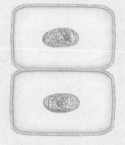

cell plate forming

1 The plane of division (and of the future cross-wall) was established by microtubules and actin filaments that formed and broke up before mitosis began. Vesicles cluster here when mitosis ends.

2 The vesicles fuse with each other and with endocytic vesicles bringing cell wall components and plasma membrane proteins from the cell surface. The fused materials form a cell plate along the plane of division.

3 The cell plate expands outward along the plane of division until it reaches the plasma membrane. When the cell plate attaches to the plasma membrane, it partitions the cytoplasm.

4 The cell plate matures as two new primary cell walls surrounding middle lamella material. The new walls join with the parent cell wall, so each daughter cell becomes enclosed by its own wall.

B   Cell Plate Formation

Figure 9.7 Animated Cytoplasmic division of an animal cell (**a**) and a plant cell (**b**).

inward until the cytoplasm (and the cell) is pinched in two (Figure 9.7a). Two new cells form this way. Each has a nucleus and some of the parent cell's cytoplasm, and each is enclosed in its own plasma membrane.

Division of Plant Cells

Dividing plant cells face a particular challenge. Unlike most animal cells, plant cells remain attached to one another and organized in tissues during development. Thus, plant growth occurs mainly in the direction of cell division, and the orientation of each cell's division is critical to the architecture of the plant.

Accordingly, plants have an extra step in cytokinesis. Microtubules under a plant cell's plasma membrane help orient the cellulose fibers in the cell wall. Before prophase, these microtubules disassemble, and then reassemble in a dense band around the nucleus along the future plane of division. The band, which also includes actin filaments, disappears as microtubules of the bipolar spindle form. An actin-depleted zone is left behind. The zone marks the plane in which cytoplasmic division will occur (Figure 9.7b).

The contractile ring mechanism that works for animal cells would not work for a plant cell. The contractile force of microfilaments is not strong enough to pinch through plant cell walls, which are stiff with cellulose and often lignin.

By the end of anaphase in a plant cell, a set of short microtubules has formed on either side of the division plane. These microtubules now guide vesicles derived from Golgi bodies and the cell's surface to the division plane. There, the vesicles and their wall-building contents start to fuse into a disk-shaped **cell plate**.

The plate grows outward until its edges reach the plasma membrane. It attaches to the membrane, and so partitions the cytoplasm. In time, the cell plate will develop into a primary cell wall that merges with the parent cell's wall. Thus, by the end of division, each of the descendant cells will be enclosed by its own plasma membrane and its own cell wall.

Appreciate the Process!

Take a moment to visualize the cells making up your palms, thumbs, and fingers. Now imagine the mitotic divisions that produced the many generations of cells before them while you were developing inside your mother (Figure 9.8). Be grateful for the precision of the mechanisms that led to the formation of your body parts at the right times, in the proper places. Why? An individual's survival depends on proper timing and

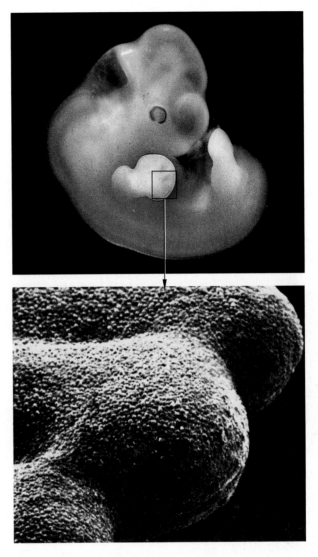

Figure 9.8 The paddlelike structure of a human embryo that develops into a hand by mitosis, cytoplasmic divisions, and other processes. The scanning electron micrograph reveals individual cells.

completion of cell cycle events. If a cell's cycle goes awry, the cell may begin to divide in an uncontrolled manner. Such unchecked divisions can destroy tissues and, ultimately, the individual.

Take-Home Message

How do cells divide?

■ After mitosis, the cytoplasm of the parent cell typically is partitioned into two descendant cells, each with its own nucleus. The process of cytoplasmic division, cytokinesis, differs among different kinds of eukaryotic cells.

■ In animal cells, a contractile ring partitions the cytoplasm. A band of actin filaments that rings the cell midsection contracts and pinches the cytoplasm in two.

■ In plant cells, a cell plate that forms midway between the spindle poles partitions the cytoplasm when it reaches and connects to the parent cell wall.

■ On rare occasions, controls over cell division are lost. Cancer may be the outcome.

■ Links to Receptors 5.2, Enzymes and free radicals 6.3, Ultraviolet radiation 7.1

The Cell Cycle Revisited Every second, millions of cells in your skin, bone marrow, gut lining, liver, and elsewhere are dividing and replacing their worn-out, dead, and dying predecessors. They do not divide at random. Many mechanisms control DNA replication and when cell division begins and ends.

What happens when something goes wrong? Suppose sister chromatids do not separate as they should during mitosis. As a result, one descendant cell ends up with too many chromosomes and the other with too few. Or sup-

pose DNA gets damaged when a chromosome is being duplicated. A cell's DNA can also be damaged by free radicals (Section 6.3), chemicals, or environmental assaults such as ultraviolet radiation (Section 7.1). Such problems are frequent and inevitable, but a cell may not function properly unless they are countered quickly.

The cell cycle has built-in checkpoints that allow problems to be corrected before the cycle advances. Certain proteins, the products of checkpoint genes, can monitor whether a cell's DNA has been copied completely, whether it is damaged, and even whether nutrient concentrations are sufficient to support cell growth. Such proteins interact to advance, delay, or stop the cell cycle (Figure 9.9).

For example, some checkpoint gene products are kinases. This class of enzymes can activate other molecules by transferring a phosphate group to them. When DNA is broken or incomplete, the kinases activate certain proteins in a cascade of signaling events that ultimately stops the cell cycle or causes the cell to die.

As another example, checkpoint gene products called **growth factors** activate genes that stimulate cells to grow and divide. One kind, an epidermal growth factor, activates a kinase by binding to receptors on target cells in epithelial tissues. The binding is a signal to start mitosis.

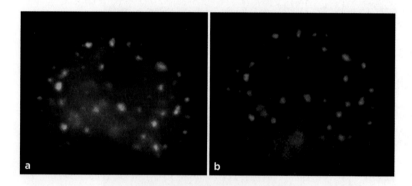

Figure 9.9 Protein products of checkpoint genes in action. A form of radiation damaged the DNA inside this nucleus. (**a**) *Green* dots pinpoint the location of a protein called *53BP1*, and (**b**) *red* dots pinpoint the location of another protein, *BRCA1*. Both proteins have clustered around the same chromosome breaks in the same nucleus. The integrated action of these proteins and others blocks mitosis until the DNA breaks are fixed.

Checkpoint Failure and Tumors Sometimes a checkpoint gene mutates so that its protein product no longer works properly. The result may be that the cell skips interphase, and division occurs over and over with no resting period. Or, DNA that has been damaged may be copied and packaged into descendant cells. In still other cases, the mutation alters signaling mechanisms that make an abnormal cell commit suicide (you will read more about apoptosis, the cellular self-destruct mechanism, in Section 27.6).

When all checkpoint mechanisms fail, a cell loses control over its cell cycle. The cell's descendants may form a **tumor**—an abnormal mass—in the surrounding tissue (Figures 9.10–9.12).

Mutated checkpoint genes are associated with an increased risk of tumor formation, and sometimes they run in families. Usually, one or more checkpoint gene products are missing in tumor cells. Checkpoint gene products that inhibit mitosis are called tumor suppressors because tumors form when they are missing. Checkpoint genes encoding proteins that stimulate mitosis are called proto-oncogenes (from the Greek *onkos*, or tumor); mutations that alter their products or the rate at which they are made can transform a normal cell into a tumor cell.

Moles and other tumors are **neoplasms**—abnormal masses of cells that lost control over how they grow and divide. Ordinary skin moles are among the noncancerous, or benign, neoplasms. They grow very slowly, and their cells retain the surface recognition proteins that keep them in their home tissue (Figure 9.11). Unless a benign neoplasm grows too large or becomes irritating, it poses no threat to the body.

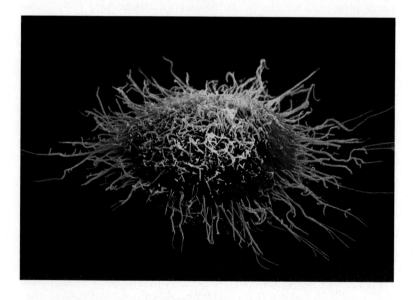

Figure 9.10 Scanning electron micrograph of the surface of a cervical cancer cell, the kind of malignant cell that killed Henrietta Lacks.

Characteristics of Cancer A malignant neoplasm is one that is dangerous to health. **Cancer** occurs when the abnormally dividing cells of a malignant neoplasm disrupt body tissues, physically and metabolically. These typically disfigured cells can break loose from home tissues, slip into and out of blood vessels and lymph vessels, and invade other tissues where they do not belong (Figure 9.11). Cancer cells typically display the following three characteristics:

First, cancer cells grow and divide abnormally. Controls that usually keep cells from getting overcrowded in tissues are lost, so cancer cell populations may reach extremely high densities. The number of small blood vessels, or capillaries, that transport blood to the growing cell mass also increases abnormally.

Second, cancer cells often have an altered plasma membrane and cytoplasm. The membrane may be leaky and have altered or missing proteins. The cytoskeleton may be shrunken, disorganized, or both. The balance of metabolism is often shifted, as in an amplified reliance on ATP formation by fermentation rather than by aerobic respiration.

Third, cancer cells often have a weakened capacity for adhesion. Because their recognition proteins are altered or lost, they do not necessarily stay anchored in their proper tissues, and may break away and establish colonies in distant tissues. Metastasis is the name for this process of abnormal cell migration and tissue invasion.

Unless chemotherapy, surgery, or another procedure eradicates them, cancer cells can put an individual on a painful road to death. Each year, cancers cause 15 to 20 percent of all human deaths in developed countries alone. Cancers are not just a human problem. They are known to occur in most of the animal species studied to date.

Cancer is a multistep process. Researchers already know about many mutations that contribute to it. They are working to identify drugs that target and destroy cancer cells or stop them from dividing.

HeLa cells, for instance, were used in early tests of taxol, a drug that keeps microtubules from disassembling and so hampers mitosis. Frequent divisions of cancer cells make them more vulnerable to this poison than normal cells. Such research may yield drugs that put the brakes on cancer. We return to this topic in later chapters.

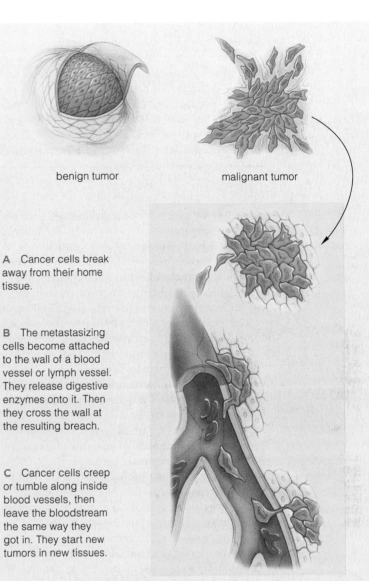

benign tumor   malignant tumor

A Cancer cells break away from their home tissue.

B The metastasizing cells become attached to the wall of a blood vessel or lymph vessel. They release digestive enzymes onto it. Then they cross the wall at the resulting breach.

C Cancer cells creep or tumble along inside blood vessels, then leave the bloodstream the same way they got in. They start new tumors in new tissues.

Figure 9.11 Animated Comparison of benign and malignant tumors. Benign tumors typically are slow-growing and stay in their home tissue. Cells of a malignant tumor migrate abnormally through the body and establish colonies even in distant tissues.

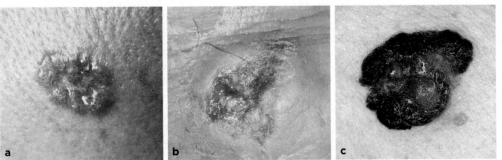

Figure 9.12 Skin cancers. **(a)** A basal cell carcinoma is the most common type. This slow-growing, raised lump is typically uncolored, reddish-brown, or black.

(b) The second most common form of skin cancer is a squamous cell carcinoma. This pink growth, firm to the touch, grows fast under the surface of skin exposed to sun.

(c) Malignant melanoma spreads fastest. Cells form dark, encrusted lumps. They may itch like an insect bite or bleed easily.

HeLa cells divide quickly and indefinitely, so they are notoriously difficult to contain. Even with careful laboratory practice, HeLa cells tend to infest other cell lines grown in the same laboratory, and quickly outgrow the other cells. Most cells appear similar in tissue culture, so the contamination may not be detected. Researchers discovered just how easy it is to propagate HeLa cells in the 1970s, when they found out that dozens of cell lines from various sources—as many as one in three—were not what they were supposed to be. The lines had been completely over-

How would you vote?

HeLa cells are sold worldwide by cell culture firms. Should the family of Henrietta Lacks (*right*) share in the profits of those sales? See CengageNOW for details, then vote online.

grown by HeLa cells. The finding undermined the significance of decades of research that had relied on the contaminated lines.

Summary

Section 9.1 By processes of reproduction, parents produce a new generation of individuals like themselves. Cell division is the bridge between generations. When a cell divides, each descendant cell receives a required number of DNA molecules and some cytoplasm.

Eukaryotic cells undergo mitosis, meiosis, or both. These nuclear division mechanisms partition the duplicated chromosomes of a parent cell into two new nuclei. Cytoplasm divides by a separate mechanism. Prokaryotic cells divide by a different process.

Mitosis followed by cytoplasmic division is the basis of growth, cell replacements, and tissue repair in multicelled species, and also the basis of asexual reproduction in many single-celled and multicelled species.

Meiosis, the basis of sexual reproduction in eukaryotes, precedes the formation of gametes or sexual spores.

A eukaryotic chromosome is a molecule of DNA and many **histones** and other proteins associated with it. The proteins structurally organize the chromosome and affect access to its genes. The smallest unit of organization, the **nucleosome**, is a stretch of double-stranded DNA looped twice around a spool of histones.

When duplicated, a chromosome consists of two **sister chromatids**, each with a kinetochore (an attachment site for microtubules). Sister chromatids remain attached at their **centromere** until late in mitosis (or meiosis).

■ *Use the animation on CengageNOW to explore the structural organization of chromosomes.*

Section 9.2 A **cell cycle** starts when a new cell forms, runs through interphase, and ends when that cell reproduces by nuclear and cytoplasmic division. Most of a cell's activities occur in **interphase**, when it grows, roughly doubles the number of its cytoplasmic components, then duplicates its chromosomes.

Chromosome number is the sum of all chromosomes in cells of a specified type. For example, the chromosome number of human body cells is 46. These cells have two of each kind of chromosome, so they are **diploid**.

■ *Use the interaction on CengageNOW to investigate the stages of the cell cycle.*

Section 9.3 Mitosis is a nuclear division mechanism that maintains the chromosome number. It proceeds in these four sequential stages:

Prophase. Duplicated chromosomes start to condense. Microtubules assemble and form a **bipolar spindle**, and the nuclear envelope breaks up. Some microtubules that extend from one spindle pole harness one chromatid of each chromosome; some that extend from the opposite spindle pole tether its sister chromatid. Other microtubules extend from both poles and grow until they overlap at the spindle's midpoint.

Metaphase. All chromosomes are aligned at the spindle's midpoint.

Anaphase. The sister chromatids of each chromosome detach from each other, and the spindle microtubules start moving them toward opposite spindle poles. Microtubules that overlap at the spindle's midpoint slide past each other, pushing the poles farther apart. Motor proteins drive all of these movements.

Telophase. A cluster of chromosomes that consists of a complete set of chromosomes reaches each spindle pole. A nuclear envelope forms around each cluster, forming two new nuclei. Both nuclei have the parental chromosome number.

■ *Use the animation on CengageNOW to see how mitosis proceeds.*

Section 9.4 Most cells divide in two after their nucleus divides. Mechanisms of **cytokinesis**, or cytoplasmic division differ. In animal cells, a **contractile ring** of microfilaments that is part of the cell cortex pulls the plasma membrane inward until the cytoplasm is pinched in two. In plant cells, a band of microtubules and microfilaments that forms around the nucleus before mitosis marks the site at which a **cell plate** forms. The cell plate expands until it becomes a cross-wall, which partitions the cytoplasm when it fuses to the parent cell wall.

■ *Compare the cytoplasmic division of plant and animal cells with the animation on CengageNOW.*

Section 9.5 Checkpoint gene products such as **growth factors** control the cell cycle. Mutated checkpoint genes can cause **tumors** (**neoplasms**) by disrupting the normal controls. **Cancer** is a multistep process involving altered cells that grow and divide abnormally. Cancer cells may metastasize, or break loose and colonize distant tissues.

■ *Use the animation on CengageNOW to see how cancers spread through the body.*

Data Analysis Exercise

Despite their notorious ability to contaminate other cell lines, HeLa cells continue to be an extremely useful tool in cancer research. One early finding was that HeLa cells vary in chromosome number. The panel of chromosomes in Figure 9.13, originally published in 1989, shows all of the chromosomes in a single metaphase HeLa cell.

1. What is the chromosome number of this HeLa cell?

2. How many extra chromosomes do these cells have, compared to normal human cells?

3. Can you tell these cells came from a female? How?

Figure 9.13 Chromosomes of a HeLa cell line.

Self-Quiz

Answers in Appendix III

1. Mitosis and cytoplasmic division function in _____ .
   a. asexual reproduction of single-celled eukaryotes
   b. growth and tissue repair in multicelled species
   c. gamete formation in prokaryotes
   d. both a and b

2. A duplicated chromosome has _____ chromatid(s).
   a. one   b. two   c. three   d. four

3. The basic unit that structurally organizes a eukaryotic chromosome is the _____ .
   a. higher order coiling   c. nucleosome
   b. bipolar mitotic spindle   d. microfilament

4. The chromosome number is _____ .
   a. the sum of all chromosomes in a cell of a given type
   b. an identifiable feature of each species
   c. maintained by mitosis
   d. all of the above

5. A somatic cell having two of each type of chromosome has a(n) _____ chromosome number.
   a. diploid   b. haploid   c. tetraploid   d. abnormal

6. Interphase is the part of the cell cycle when _____ .
   a. a cell ceases to function
   b. a cell forms its spindle apparatus
   c. a cell grows and duplicates its DNA
   d. mitosis proceeds

7. After mitosis, the chromosome number of the two new cells is _____ the parent cell's.
   a. the same as   c. rearranged compared to
   b. one-half of   d. doubled compared to

8. Name the intervals in the diagram of mitosis *below*.

9. Only _____ is not a stage of mitosis.
   a. prophase   b. interphase   c. metaphase   d. anaphase

10. Which of the following is a subset of the other two?
    a. cancer   b. neoplasm   c. tumor

11. Name one type of checkpoint gene product.

12. Match each stage with the events listed.
    ___metaphase   a. sister chromatids move apart
    ___prophase   b. chromosomes start to condense
    ___telophase   c. new nuclei form
    ___anaphase   d. all duplicated chromosomes are
                        aligned at the spindle equator

▪ *Visit CengageNOW for additional questions.*

Critical Thinking

1. The anticancer drug taxol was first isolated from Pacific yews (*Taxus brevifolia*), which are slow-growing trees (*right*). Bark from about six yew trees provided enough taxol to treat one patient, but removing the bark killed the trees. Fortunately, taxol is now produced using plant cells that grow in big vats rather than in trees. What challenges do you think had to be overcome to get plant cells to grow and divide in laboratories?

2. Suppose you have a way to measure the amount of DNA in one cell during the cell cycle. You first measure the amount at the G1 phase. At what points in the rest of the cycle will you see a change in the amount of DNA per cell?

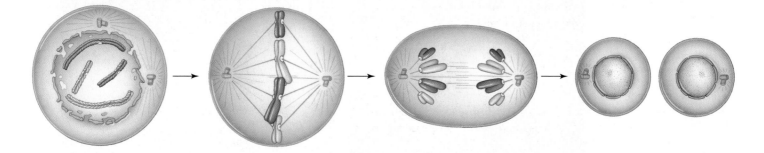

10 | Meiosis and Sexual Reproduction

Why Sex?

If the function of reproduction is the perpetuation of one's genetic material, then an asexual reproducer would seem to win the evolutionary race. In asexual reproduction, all of an individual's genetic information is passed to all of its offspring. Sexual reproduction mixes up genetic information from two parents (Figure 10.1), so only about half of each parent's genetic information is passed to offspring.

So why sex? Variation in forms and combinations of heritable traits is typical of sexually reproducing populations. Remember from Section 1.4 that some forms of traits are more adaptive than others to conditions in the environment. If those conditions change, some of the diverse offspring of sexual reproducers may have forms of traits or combinations of them that help these individuals to survive the change. All offspring of asexual reproducers are adapted the same way to the environment—and equally vulnerable to changes in it.

Other organisms are part of the environment, and they, too, can change. Think of predator and prey—say, foxes and rabbits. If one rabbit is better than others at outrunning the foxes, it has a better chance of escaping, surviving, and passing on the genetic basis for its evasive ability to offspring. Thus, over many generations, the rabbits may get faster. If one fox is better than others at outrunning the faster rabbits, it has a

better chance of eating, surviving, and passing on the genetic basis for its predatory ability to offspring. Thus, over many generations, the foxes may tend to get faster. As one species changes, so does the other—an idea called the Red Queen hypothesis, after Lewis Carroll's book *Through the Looking Glass*. In the book, the Queen of Hearts tells Alice, "It takes all the running you can do, to keep in the same place."

An adaptive trait tends to spread more quickly through a sexually reproducing population than through an asexually reproducing one. Why? In asexual reproduction, new combinations of traits can arise only by mutation. An adaptive trait is perpetuated along with the same set of other traits—adaptive or not—until another mutation occurs. By contrast, sexual reproduction mixes up the genetic information of individuals that often have different forms of traits. It brings together adaptive traits, and separates adaptive traits from maladaptive ones, in far fewer generations than does mutation alone.

However, having a faster pace of achieving genetically diverse populations doesn't mean that sexual reproduction wins the evolutionary race. In terms of numbers of individuals and how long their lineages have endured, the most successful organisms on Earth are bacteria, which reproduce most often by just copying their DNA and dividing.

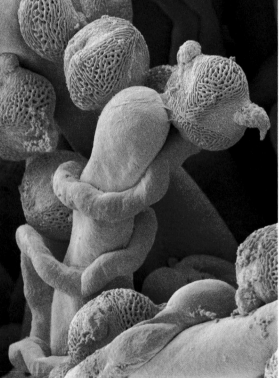

See the video!
Figure 10.1 Moments in the stages of sexual reproduction, a process that mixes up the genetic material of two organisms.

The photo on the *right* shows pollen grains (*orange*) germinating on flower carpels (*yellow*). Pollen tubes with male gametes inside are growing from the grains down into tissues of the ovary, which house the flower's female gametes.

Key Concepts

Sexual versus asexual reproduction

In asexual reproduction, one parent transmits its genetic information (DNA) to offspring. In sexual reproduction, offspring inherit DNA from two parents who usually differ in some number of alleles. Alleles are different forms of the same gene. **Section 10.1**

Stages of meiosis

Meiosis reduces the chromosome number. It occurs only in cells set aside for sexual reproduction. Meiosis sorts a reproductive cell's chromosomes into four haploid nuclei. **Sections 10.2, 10.3**

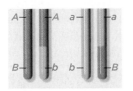

Chromosome recombinations and shufflings

During meiosis, each pair of maternal and paternal chromosomes swaps segments. Then, each chromosome is randomly segregated into one of the new nuclei. Both processes lead to novel combinations of alleles—and traits—among offspring. **Section 10.4**

Sexual reproduction in context of life cycles

Gametes form by different mechanisms in males and females. In most plants, spore formation and other events intervene between meiosis and gamete formation. **Section 10.5**

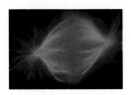

Mitosis and meiosis compared

Meiosis may have originated by evolutionary remodeling of mechanisms that already existed for mitosis and, before that, for repairing damaged DNA. **Section 10.6**

Links to Earlier Concepts

- This chapter returns to the concept of reproduction introduced in Section 1.2. Here, we detail the cellular basis of sexual reproduction, and begin to explore the far-reaching effects of gene shufflings—a process that introduces variations in traits among offspring (1.4).

- You will be revisiting the microtubules that move chromosomes (4.13, 9.3). Be sure you have a clear picture of the structural organization of chromosomes (9.1) and understand chromosome number (9.2).

- You will also draw on your understanding of cytoplasmic division (9.4) and checkpoint gene products (9.5) that monitor and repair chromosomal DNA during the cell cycle.

How would you vote? Japanese researchers successfully created a "fatherless" mouse with genetic material from the eggs of two females. The mouse is healthy and fertile. Should researchers be prevented from trying the same process with human eggs? See CengageNOW for details, then vote online.

10.1 | Introducing Alleles

■ Asexual reproduction produces genetically identical copies of a parent. By contrast, sexual reproduction introduces variation in the combinations of traits among offspring.

■ Link to Diversity 1.4

Each species has a unique set of **genes**: regions in DNA that encode information about traits. An individual's genes collectively contain the information necessary to make a new individual. With **asexual reproduction**, one parent produces offspring, so all of its offspring inherit the same number and kinds of genes. Mutations aside, then, all offspring of asexual reproduction are genetically identical copies of the parent, or **clones**.

Inheritance is far more complicated with **sexual reproduction**, the process involving meiosis, formation of mature reproductive cells, and fertilization. In most sexually reproducing multicelled eukaryotes, the first cell of a new individual has pairs of genes, on pairs of chromosomes. Typically, one chromosome of each pair is maternal and the other is paternal (Figure 10.2).

If the information in every gene of a pair were identical, then sexual reproduction would also produce clones. Just imagine: The entire human population might consist of clones, in which case everybody would look alike. But the two genes of a pair are often not identical. Why not? Inevitably, mutations accumulate in genes and permanently alter the information they carry. Thus, the two genes of any pair might "say" slightly different things about a particular trait. Each different form of a gene is called an **allele**.

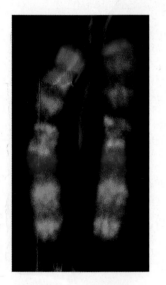

Figure 10.2 A maternal and a paternal chromosome pair. They appear identical in this micrograph, but any gene on one might differ slightly from its partner on the other.

Alleles influence differences in thousands of traits. For instance, whether your chin has a dimple in it or not depends on which alleles you inherited at one chromosome location. One allele says "dimple in the chin." A different allele says "no dimple in the chin." Alleles are one reason individuals of a sexually-reproducing species do not all look alike. The offspring of sexual reproducers inherit new combinations of alleles, which is the basis of new combinations of traits.

Take-Home Message

How does sexual reproduction introduce variation in traits?

■ Alleles are the basis of traits. Sexual reproduction bestows novel combinations of alleles—thus novel combinations of traits—on offspring.

10.2 | What Meiosis Does

■ Meiosis is a nuclear division mechanism that precedes cytoplasmic division of immature reproductive cells. It occurs only in sexually reproducing eukaryotic species.

■ Link to Chromosome number 9.2

Remember, the chromosome number is the total number of chromosomes in a cell of a given type (Section 9.2). A diploid cell has two copies of every chromosome; typically, one of each type was inherited from each of two parents. Except for a pairing of nonidentical sex chromosomes, the chromosomes of a pair are **homologous**, which means they have the same length, shape, and collection of genes (*hom*– means alike).

Mitosis maintains the chromosome number. **Meiosis**, a different nuclear division process, halves the chromosome number. Meiosis occurs in the immature reproductive cells—**germ cells**—of multicelled eukaryotes that reproduce sexually. In animals, meiosis of germ cells results in mature reproductive structures called **gametes**. (Plants have a slightly different process that we will discuss later.) A sperm cell is a type of male gamete; an egg is a type of female gamete. Gametes usually form inside special reproductive structures or organs (Figure 10.3).

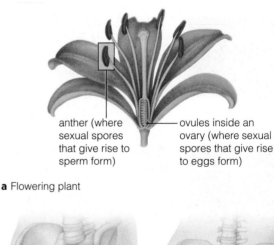

anther (where sexual spores that give rise to sperm form)

ovules inside an ovary (where sexual spores that give rise to eggs form)

a Flowering plant

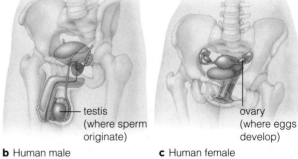

testis (where sperm originate)

ovary (where eggs develop)

b Human male   **c** Human female

Figure 10.3 Examples of reproductive organs, where cells that give rise to gametes originate.

Gametes have a single set of chromosomes, so they are **haploid** (n): their chromosome number is half of the diploid ($2n$) number. Human body cells are diploid, with 23 pairs of homologous chromosomes (Figure 10.4). Meiosis of a human germ cell normally produces gametes with 23 chromosomes: one of each pair. The diploid chromosome number is restored at fertilization, when two haploid gametes (one egg and one sperm) fuse to form a **zygote**, the first cell of a new individual.

Two Divisions, Not One

Meiosis is similar to mitosis in certain respects. A cell duplicates its DNA before the division process starts. The two DNA molecules and associated proteins stay attached at the centromere. For as long as they remain attached, they are sister chromatids (Section 9.1):

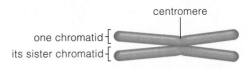

one chromosome in the duplicated state

As in mitosis, the microtubules of a spindle move the chromosomes to opposite poles of the cell. However, meiosis sorts the chromosomes into new nuclei twice. Two consecutive nuclear divisions form four haploid nuclei. There is typically no interphase between the two divisions, which are called meiosis I and II:

Interphase	Meiosis I	Meiosis II
DNA is replicated prior to meiosis I	Prophase I Metaphase I Anaphase I Telophase I	Prophase II Metaphase II Anaphase II Telophase II

In meoisis I, every duplicated chromosome aligns with its partner, homologue to homologue. After they are sorted and arranged this way, each homologous chromosome is pulled away from its partner:

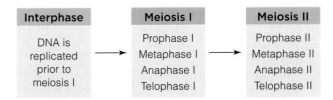

each chromosome in the cell pairs with its homologous partner

then the partners separate

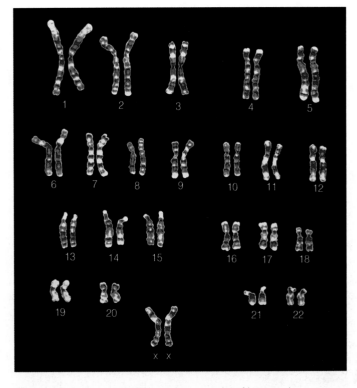

Figure 10.4 Twenty-three homologous pairs of human chromosomes. This example is from a human female, with two X chromosomes. Human males have a different pairing of sex chromosomes (XY).

After homologous chromosomes are pulled apart, each ends up in one of two new nuclei. The chromosomes are still duplicated—the sister chromatids are still attached. During meiosis II, the sister chromatids of each chromosome are pulled apart, so each becomes an individual, unduplicated chromosome:

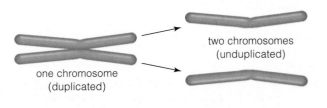

one chromosome (duplicated)

two chromosomes (unduplicated)

Meiosis distributes the duplicated chromosomes of a diploid nucleus ($2n$) into four new nuclei. Each new nucleus is haploid (n), with one unduplicated version of each chromosome. Typically, two cytoplasmic divisions accompany meiosis, so four haploid cells form. Figure 10.5 in the next section shows the chromosomal movements in the context of the stages of meiosis.

Take-Home Message

What is meiosis?

■ Meiosis is a nuclear division mechanism that occurs in immature reproductive cells of sexually reproducing eukaryotes. It halves a cell's diploid ($2n$) chromosome number, to the haploid number (n).

■ Links to Microtubules 4.13, Mitosis 9.2 and 9.3

Meiosis I

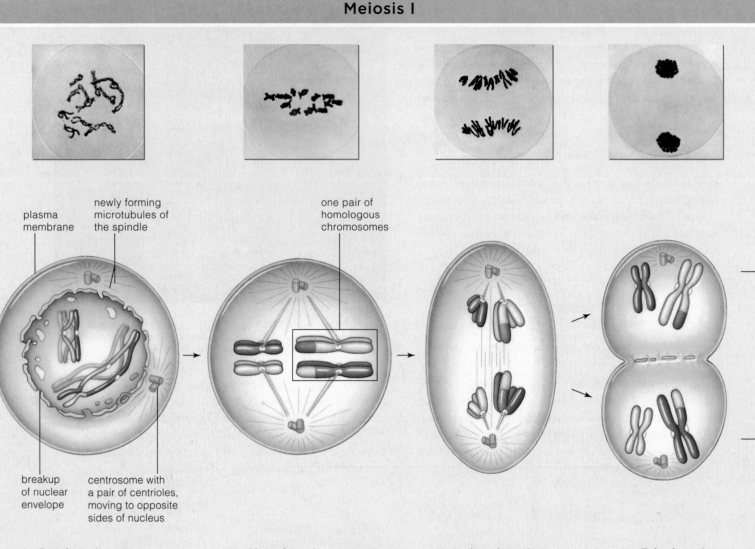

plasma membrane

newly forming microtubules of the spindle

one pair of homologous chromosomes

breakup of nuclear envelope

centrosome with a pair of centrioles, moving to opposite sides of nucleus

A Prophase I

The chromosomes were duplicated in interphase, so every chromosome now consists of two sister chromatids joined at the centromere. The nucleus is diploid (2n)—it contains two sets of chromosomes, one from each parent.

The chromosomes now condense. Homologous chromosomes pair up and swap segments (as indicated by color breaks). A bipolar spindle forms. The centrosome, with its two centrioles, gets duplicated; one centriole pair now moves to the opposite side of the cell as the nuclear envelope breaks up.

B Metaphase I

By the end of prophase I, spindle microtubules had connected the chromosomes to the spindle poles. Each chromosome is now attached to one spindle pole, and its homologue is attached to the other.

The microtubules grow and shrink, pushing and pulling the chromosomes as they do. When all of the microtubules are the same length, the chromosomes are aligned midway between spindle poles. This alignment marks metaphase I.

C Anaphase I

As spindle microtubules shorten, they pull each duplicated chromosome toward one of the spindle poles, so the homologous chromosomes separate.

Which chromosome (maternal or paternal) became attached to a particular spindle pole was random, so either may end up at a particular pole.

D Telophase I

One of each chromosome arrives at each spindle pole.

New nuclear envelopes form around the two clusters of chromosomes as they decondense. There are now two haploid (n) nuclei. The cytoplasm may divide at this point.

Figure 10.5 Animated Meiosis halves the chromosome number. Drawings show a diploid (2n) animal cell. For clarity, only two pairs of chromosomes are illustrated, but cells of nearly all eukaryotes have more than two. The two chromosomes of the pair inherited from one parent are in *purple;* the two inherited from the other parent are in *blue.* Micrographs show meiosis in a lily plant cell (*Lilium regale*).

Figure It Out: Are chromosomes in their duplicated or unduplicated state during metaphase II?

Answer: Duplicated

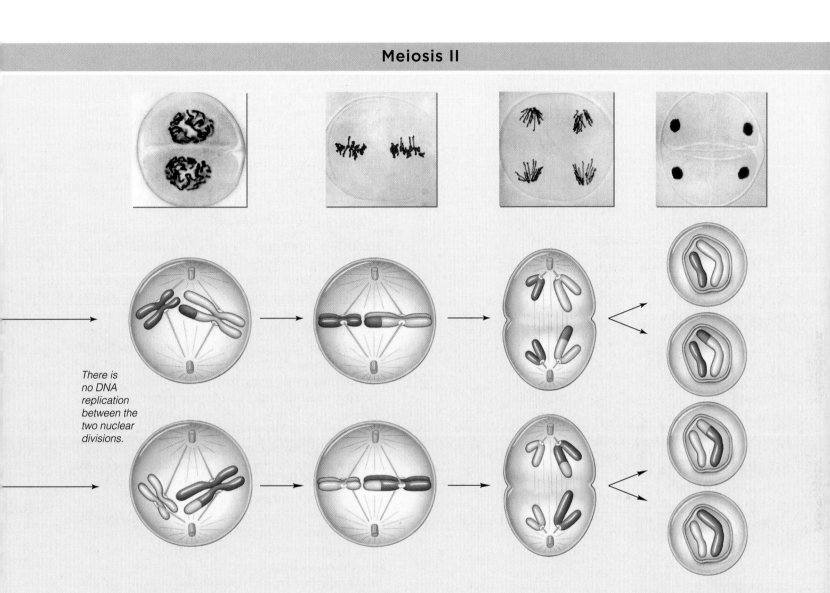

There is no DNA replication between the two nuclear divisions.

E Prophase II

Each nucleus contains one full set of chromosomes. Every chromosome is still duplicated—it consists of two sister chromatids joined at the centromere.

The chromosomes condense as a bipolar spindle forms. One centriole moves to the opposite side of each new nucleus, and the nuclear envelopes break up.

F Metaphase II

By the end of prophase II, spindle microtubules had connected the sister chromatids to the spindle poles. Each chromatid is now attached to one spindle pole, and its sister is attached to the other.

The microtubules grow and shrink, pushing and pulling the chromosomes as they do. When all of the microtubules are the same length, the chromosomes are aligned midway between spindle poles. This alignment marks metaphase II.

G Anaphase II

As spindle microtubules shorten, they pull each sister chromatid toward one of the spindle poles, so the sisters separate.

Which sister chromatid became attached to a particular spindle pole was random, so either may end up at a particular pole.

H Telophase II

Each chromosome now consists of a single, unduplicated molecule of DNA. One of each chromosome arrives at each spindle pole.

New nuclear envelopes form around each cluster of chromosomes as they decondense. There are now four haploid (*n*) nuclei. The cytoplasm may divide.

■ Crossovers and the random sorting of chromosomes in meiosis result in new combinations of traits among offspring.

■ Link to Chromosome structure 9.1

The previous section mentioned briefly that duplicated chromosomes swap segments with their homologous partners during prophase I. It also showed how each chromosome aligns with and then separates from its homologous partner during anaphase I. Both events introduce novel combinations of alleles into gametes. Along with new chromosome combinations that occur in fertilization, these events contribute to the variation in combinations of traits among offspring of sexually reproducing species.

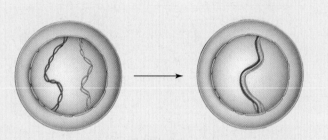

A Two homologous chromosomes, one maternal (*purple*) and one paternal (*blue*) are in their duplicated form: Each is two sister chromatids, joined at the centromere. Homologous chromosomes align and associate tightly during prophase I.

Crossing Over in Prophase I

Figure 10.6*a* illustrates one pair of duplicated chromosomes, early in prophase I of meiosis when they are in the process of condensing. All chromosomes in a germ cell condense this way. When they do, each is drawn close to its homologue. The chromatids of one homologous chromosome become stitched to the chromatids of the other, point by point along their length with little space in between. This tight, parallel orientation favors **crossing over**—the process by which a chromosome and its homologous partner exchange corresponding segments.

Crossing over is a normal and frequent process in meiosis. The rate of crossing over varies among species and among chromosomes; in humans, between 46 and 95 crossovers occur per meiosis, so each chromosome probably crosses over at least once.

B Here, we focus on only two genes. One gene has alleles *A* and *a*; the other has alleles *B* and *b*.

Each crossover event is an opportunity for homologous chromosomes to exchange heritable information. Such swapping would be pointless if genes never varied, but remember, many genes have slightly different forms (alleles). Typically, a number of genes on one chromosome will not be identical to their partners on the homologous chromosome.

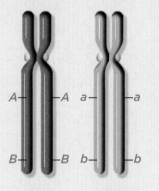

C Close contact between the homologous chromosomes promotes crossing over between nonsister chromatids, so paternal and maternal chromatids exchange segments.

crossover

We will return to the impact of crossing over in later chapters. For now, remember that crossing over introduces novel combinations of alleles in both members of a pair of homologous chromosomes, which results in novel combinations of traits among offspring.

D Crossing over mixes up paternal and maternal alleles on homologous chromosomes.

Figure 10.6 Animated Crossing over. *Blue* signifies a paternal chromosome, and *purple*, its maternal homologue.

For clarity, we show only one pair of homologous chromosomes and one crossover, but more than one crossover may occur in each chromosome pair.

Segregation of Chromosomes into Gametes

Normally, all new nuclei that form in meiosis I receive the same number of chromosomes, but which homologue ends up in which nucleus is random.

The process of chromosome segregation begins in prophase I. Suppose segregation is happening right now in one of your own germ cells. Crossovers have already made genetic mosaics of your chromosomes, but let's put crossing over aside to simplify tracking. Just call the twenty-three chromosomes you inherited from your mother the maternal ones, and the twenty-three you inherited from your father the paternal ones.

By metaphase I, microtubules emanating from both spindle poles have aligned all of the duplicated chromosomes at the spindle equator (Figure 10.5b). Have they attached all of the maternal chromosomes to one pole and all of the paternal chromosomes to the other? Probably not. Spindle microtubules attach to the kinetochores of the first chromosome they contact, regardless of whether it is maternal or paternal. Homologues get attached to opposite spindle poles. Thus, there is no pattern to the attachment of the maternal or paternal chromosomes to a particular spindle pole: Either homologous chromosome can end up at either pole.

Then, in anaphase I, each duplicated chromosome separates from its homologous partner and is pulled toward the pole to which it is attached.

Think of meiosis in a germ cell with just three pairs of chromosomes. By metaphase I, the three pairs would be attached to the spindle poles in one of four possible combinations (Figure 10.7). There would be eight (2^3) possible combinations of maternal and paternal chromosomes in new nuclei that form at telophase I.

By telophase II, each of the two nuclei would have divided and given rise to two new, identical haploid nuclei. (The nuclei would be identical because the sister chromatids of each duplicated chromosome were identical in our hypothetical example.) So, there would be eight possible combinations of maternal and paternal chromosomes in the four haploid nuclei that form by meiosis of that one germ cell.

Cells that give rise to human gametes have twenty-three pairs of homologous chromosomes, not three. Each time a human germ cell undergoes meiosis, the four gametes that form end up with one of 8,388,608 (or 2^{23}) possible combinations of homologous chromosomes! Remember, any number of genes may occur as different alleles on the maternal and paternal homologues. Are you getting an idea of why such fascinating combinations of traits show up among the generations of your own family tree?

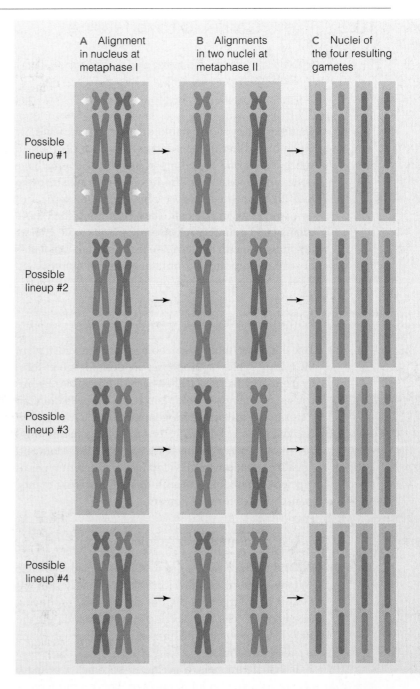

A Alignment in nucleus at metaphase I

B Alignments in two nuclei at metaphase II

C Nuclei of the four resulting gametes

Possible lineup #1

Possible lineup #2

Possible lineup #3

Possible lineup #4

Figure 10.7 **Animated** Hypothetical segregation of three pairs of chromosomes in meiosis I. Which chromosome of each pair gets packaged into which of the two new nuclei is random. *Left:* the four possible metaphase I lineups of three pairs of homologous chromosomes. *Right:* the resulting eight combinations of maternal (*purple*) and paternal (*blue*) chromosomes in the new nuclei.

Take-Home Message

How does meiosis introduce variation in combinations of traits?

■ Crossing over is recombination between nonsister chromatids of homologous chromosomes during prophase I. It makes new combinations of parental alleles.

■ Homologous chromosomes can be attached to either spindle pole in prophase I, so each homologue can be packaged into either one of the two new nuclei. Thus, the random assortment of homologous chromosomes increases the number of potential combinations of maternal and paternal alleles in gametes.

10.5 | From Gametes to Offspring

■ Aside from meiosis, the details of gamete formation and fertilization differ among plants and animals.

■ Link to Cytoplasmic division 9.4

Gametes are typically haploid, but do you know how much they differ in their details? For example, human sperm have one flagellum, roundworm sperm have none, and opossum sperm have two. Crayfish sperm look like pinwheels. Most eggs are microscopic, but an ostrich egg in its shell can weigh more than 2,200 grams (4.85 pounds). A flowering plant's male gamete is simply a sperm nucleus. We leave most of the details of sexual reproduction for later chapters, but you will need to know a few concepts before you get there.

Gamete Formation in Plants

Two kinds of multicelled bodies form in most plant life cycles. Typical **sporophytes** are diploid; spores form by meiosis in their specialized parts (Figure 10.8a). Spores consist of one or a few haploid cells. The cells undergo mitosis and give rise to a **gametophyte**, a multicelled haploid body inside which one or more gametes form. As an example, pine trees are sporophytes. Male and female gametophytes develop inside different types of pine cones that form on each tree. In flowering plants, gametophytes form in flowers.

Gamete Formation in Animals

Diploid germ cells give rise to animal gametes. In male animals, a germ cell develops into a primary spermatocyte. This large cell divides by meiosis, producing four haploid cells that develop into spermatids (Figure 10.9). Each spermatid matures as a male gamete, which is called a **sperm**.

In female animals, a germ cell becomes a primary oocyte, which is an immature egg. This cell undergoes

meiosis and division, as occurs with a primary spermatocyte. However, the cytoplasm of a primary oocyte divides unequally, so the four cells that result differ in size and function (Figure 10.10).

Two haploid cells form when the primary oocyte divides after meiosis I. One of the cells, the secondary oocyte, gets nearly all of the parent cell's cytoplasm. The other cell, a first polar body, is much smaller. Both cells undergo meiosis II and cytoplasmic division. One of the two cells that forms by division of the secondary oocyte develops into a second polar body. The other cell gets most of the cytoplasm and matures into a female gamete, which is called an ovum (plural, ova), or **egg**.

Polar bodies are not nutrient-rich or plump with cytoplasm, and generally do not function as gametes. In time they will degenerate. Their formation simply ensures that the egg will have a haploid chromosome number, and also will get enough metabolic machinery to support early divisions of the new individual.

More Shufflings at Fertilization

At **fertilization**, the fusion of two gametes produces a zygote. Fertilization restores the parental chromosome number. If meiosis did not precede fertilization, the chromosome number would double with every generation. If the chromosome number changes, so does the individual's set of genetic instructions. This set is like a fine-tuned blueprint that must be followed exactly, page by page, in order to build a body that functions normally. Changes in the blueprint can have serious—if not lethal—consequences, particularly in animals.

Fertilization also contributes to the variation that we see among offspring of sexual reproducers. Think about it in terms of human reproduction. The 23 pairs of homologous chromosomes are mosaics of genetic information after prophase I crossovers. Each gamete

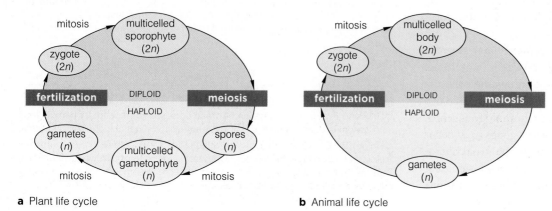

a Plant life cycle

b Animal life cycle

Figure 10.8 (a) Generalized life cycle for most plants. A pine tree is a sporophyte.

(b) Generalized life cycle for animals. The zygote is the first cell to form when the nuclei of two gametes, such as a sperm and an egg, fuse at fertilization.

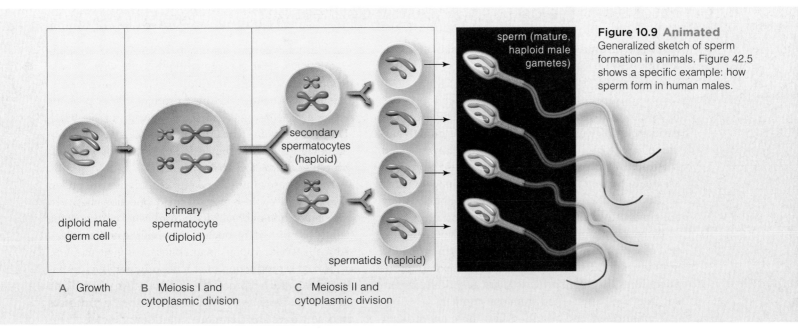

A Growth   B Meiosis I and cytoplasmic division   C Meiosis II and cytoplasmic division

Figure 10.9 Animated Generalized sketch of sperm formation in animals. Figure 42.5 shows a specific example: how sperm form in human males.

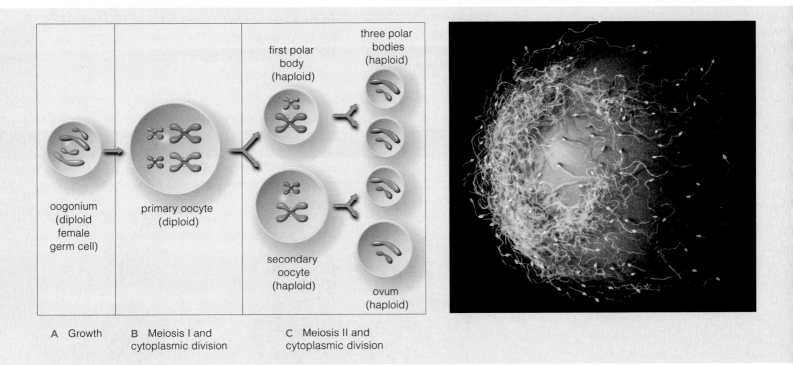

A Growth   B Meiosis I and cytoplasmic division   C Meiosis II and cytoplasmic division

Figure 10.10 Animated Animal egg formation. Eggs are far larger than sperm and larger than the three polar bodies. The painting, based on a scanning electron micrograph, depicts human sperm surrounding an ovum. Figure 42.10 shows how human eggs form.

that forms receives one of millions of possible combinations of those chromosomes. Then, out of all the male and female gametes that form, which two actually get together at fertilization is a matter of chance. The sheer number of ways that parental genetic information can combine at fertilization is staggering!

Take-Home Message

Where does meiosis fit into the life cycle of plants and animals?

■ Meiosis and cytoplasmic division precede the development of haploid gametes in animals and spores in plants.

■ The union of two haploid gametes at fertilization results in a diploid zygote.

10.6 Mitosis and Meiosis—An Ancestral Connection?

■ Though they have different results, mitosis and meiosis are fundamentally similar processes.

■ Links to Mitosis 9.3, Cell cycle controls 9.5

By mitosis and cytoplasmic division, one cell becomes two new cells. This process is the basis of growth and tissue repair in all multicelled species. Single-celled eukaryotes (and some multicelled ones) also reproduce asexually by way of mitosis and cytoplasmic division. Mitotic (asexual) reproduction results in clones, which are genetically identical copies of a parent.

By contrast, meiosis produces haploid parent cells, two of which fuse to form a diploid cell that is a new individual of mixed parentage. Meiotic (sexual) reproduction results in offspring that are genetically different from the parent—and from one another.

Though their end results differ, there are striking parallels between the four stages of mitosis and meiosis II (Figure 10.11). As one example, a bipolar spindle separates chromosomes during both processes. There are more similarities at the molecular level.

Long ago, the molecular machinery of mitosis may have been remodeled into meiosis. For example, cer-tain proteins repair breaks in DNA. These proteins monitor DNA for damage while it is being duplicated prior to mitosis. All modern species, from prokaryotes to mammals, make these proteins. Other proteins repair DNA that gets damaged during mitosis itself. This same set of repair proteins also seals up breaks in homologous chromosomes during crossover events in prophase I of meiosis.

In anaphase of mitosis, sister chromatids are pulled apart. What would happen if the connections between the sisters did not break? Each duplicated chromosome would be pulled to one or the other spindle pole—which is what happens in anaphase I of meiosis.

Sexual reproduction may have originated by mutations that affected processes of mitosis. As you will see in later chapters, the remodeling of existing processes into new ones is a common evolutionary theme.

Take-Home Message

Are the processes of mitosis and meiosis related?

■ Meiosis may have evolved by the remodeling of existing mechanisms of mitosis.

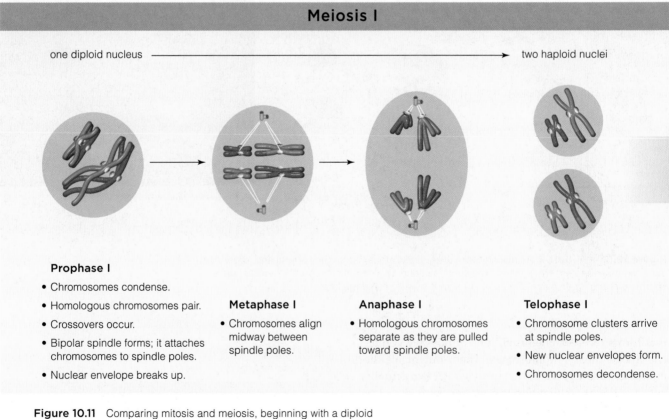

Meiosis I

one diploid nucleus ————————————————→ two haploid nuclei

Prophase I

• Chromosomes condense.

• Homologous chromosomes pair.

• Crossovers occur.

• Bipolar spindle forms; it attaches chromosomes to spindle poles.

• Nuclear envelope breaks up.

Metaphase I

• Chromosomes align midway between spindle poles.

Anaphase I

• Homologous chromosomes separate as they are pulled toward spindle poles.

Telophase I

• Chromosome clusters arrive at spindle poles.

• New nuclear envelopes form.

• Chromosomes decondense.

Figure 10.11 Comparing mitosis and meiosis, beginning with a diploid cell containing two paternal and two maternal chromosomes.

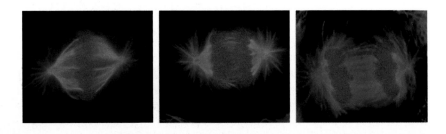

Right, a bipolar spindle at metaphase, anaphase, and telophase of mitosis in a mouse cell. The *green* stain identifies the microtubules of the spindle. The *blue* stain identifies DNA in the cell's chromosomes.

Mitosis

one diploid nucleus ⟶ two diploid nuclei

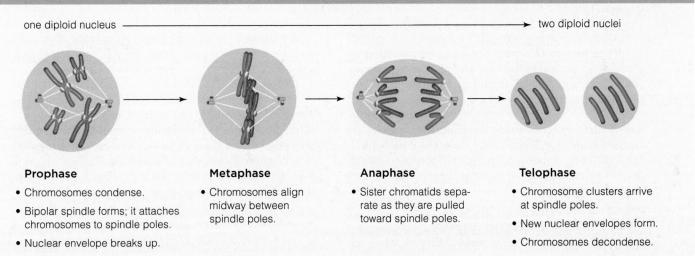

Prophase
- Chromosomes condense.
- Bipolar spindle forms; it attaches chromosomes to spindle poles.
- Nuclear envelope breaks up.

Metaphase
- Chromosomes align midway between spindle poles.

Anaphase
- Sister chromatids separate as they are pulled toward spindle poles.

Telophase
- Chromosome clusters arrive at spindle poles.
- New nuclear envelopes form.
- Chromosomes decondense.

Meiosis II

two haploid nuclei ⟶ four haploid nuclei

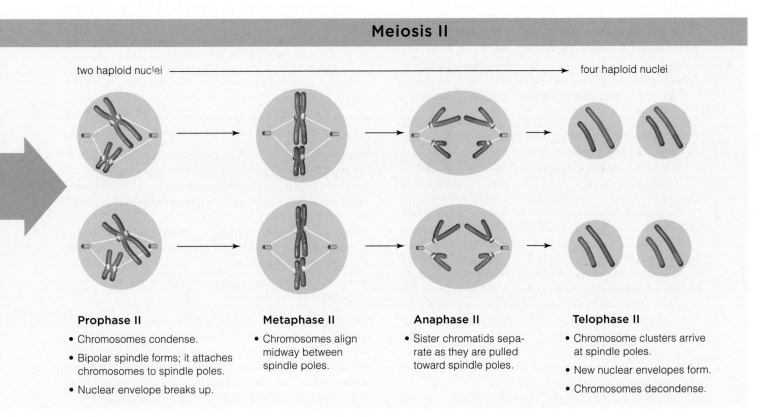

Prophase II
- Chromosomes condense.
- Bipolar spindle forms; it attaches chromosomes to spindle poles.
- Nuclear envelope breaks up.

Metaphase II
- Chromosomes align midway between spindle poles.

Anaphase II
- Sister chromatids separate as they are pulled toward spindle poles.

Telophase II
- Chromosome clusters arrive at spindle poles.
- New nuclear envelopes form.
- Chromosomes decondense.

Why Sex?

There are a few all-female species of fishes, reptiles, and birds in nature, but not mammals. In 2004, researchers fused two mouse eggs in a test tube and made an embryo using no DNA from a male. The embryo developed into Kaguya, the world's first father-less mammal. The mouse grew up healthy, engaged in sex with a male mouse, and gave birth to offspring. The researchers wanted to find out if sperm was required for normal development.

How would you vote?

Researchers made a "fatherless" mouse (*right*) from two eggs. Should they be prevented from trying the process with human eggs? See CengageNOW for details, then vote online.

Summary

Section 10.1 Many eukaryotic life cycles have asexual and sexual phases. Offspring of **asexual reproduction** are genetically identical to their one parent—they are **clones**. The offspring of **sexual reproduction** differ from parents, and often from one another, in the details of shared traits. Meiosis in germ cells, haploid gamete formation, and fertilization occur in sexual reproduction. **Alleles** are different molecular forms of the same **gene**. Each specifies a different version of the gene's product. Meiosis shuffles parental alleles; thus, offspring inherit new combinations of alleles.

Section 10.2 **Meiosis**, a nuclear division mechanism that occurs in eukaryotic **germ cells**, precedes the formation of **gametes**. Meiosis halves the parental chromosome number. The fusion of two **haploid** gamete nuclei during fertilization restores the parental chromosome number in the **zygote**, the first cell of the new individual.

Offspring of most sexual reproducers inherit pairs of chromosomes, one of each pair from the mother and the other from the father. Except in individuals with non-identical sex chromosomes, the members of a pair are **homologous**: They have the same length, the same shape, and the same set of genes. The pairs interact at meiosis.

Section 10.3 All chromosomes are duplicated during interphase, before meiosis. Two divisions, meiosis I and II, halve the parental chromosome number.

In the first nuclear division, meiosis I, each duplicated chromosome lines up with its homologous partner; then the two move apart, toward opposite spindle poles.

Prophase I. Chromosomes condense and align tightly with their homologues. Each pair of homologues typically undergoes crossing over. Microtubules form the bipolar spindle. One of two pairs of centrioles is moved to the other side of the nucleus. The nuclear envelope breaks up, so microtubules growing from each spindle pole can penetrate the nuclear region. The microtubules then attach to one or the other chromosome of each homologous pair.

Metaphase I. A tug-of-war between the microtubules from both poles has positioned all pairs of homologous chromosomes at the spindle equator.

Anaphase I. Microtubules separate each chromosome from its homologue and move both to opposite spindle poles. As anaphase I ends, a cluster of duplicated chromosomes is nearing each spindle pole.

Telophase I. Two nuclei form; typically the cytoplasm divides. All of the chromosomes are still duplicated; each still consists of two sister chromatids.

The second nuclear division, meiosis II, occurs in both nuclei that formed in meiosis I. The chromosomes condense in **prophase II**, and align in **metaphase II**. Sister chromatids of each chromosome are pulled away from each other in **anaphase II**, so each becomes an individual chromosome. By the end of **telophase II**, there are four **haploid** nuclei, each with one set of chromosomes. The chromosomes are unduplicated at this stage.

■ *Use the animation on CengageNOW to explore what happens in the stages of meiosis.*

Section 10.4 Novel combinations of alleles arise by events in prophase I and metaphase I.

The *non*sister chromatids of homologous chromosomes undergo **crossing over** during prophase I: They exchange segments at the same place along their length, so each ends up with new combinations of alleles that were not present in either parental chromosome.

Crossing over during prophase I, and random segregation of maternal and paternal chromosomes into new nuclei, contribute to variation in traits among offspring.

■ *Use the animation on CengageNOW to see how crossing over and metaphase I alignments affect allele combinations.*

Section 10.5 Multicelled diploid and haploid bodies are typical in life cycles of plants and animals. A diploid **sporophyte** is a multicelled plant body that makes haploid spores. Spores give rise to **gametophytes**, or multicelled plant bodies in which haploid gametes form. Germ cells in the reproductive organs of most animals give rise to **sperm** or **eggs**. Fusion of a sperm and egg at **fertilization** results in a zygote.

■ *Use the animation on CengageNOW to see how gametes form.*

Section 10.6 Like mitosis, meiosis requires a bipolar spindle to move and sort duplicated chromosomes, but meiosis occurs only in cells that are set aside for sexual reproduction. Mitosis maintains the parental chromosome number. Meiosis halves the chromosome number, and it introduces new combinations of alleles into offspring. Some mechanisms of meiosis resemble those of mitosis, and may have evolved from them. For example, the same DNA repair enzymes act in both processes.

Data Analysis Exercise

In 1998, researchers at Case Western University were studying meiosis in mouse oocytes when they saw an unexpected and dramatic increase of abnormal meiosis events (Figure 10.12). Improper segregation of chromosomes during meiosis is one of the main causes of human genetic disorders, which we will discuss in Chapter 12.

The researchers discovered that the spike in meiotic abnormalities started immediately after the mouse facility's plastic cages and water bottles were washed in a new, alkaline detergent. The detergent had damaged the plastic, which began to leach bisphenol A (BPA). BPA is a synthetic chemical that mimics estrogen, a hormone. BPA is used to manufacture polycarbonate plastic items (including baby bottles and water bottles) and epoxies (including the coating on the inside of metal cans of food).

1. What percentage of mouse oocytes displayed abnormalities of meiosis with no exposure to damaged caging?

2. Which group of mice showed the most meiotic abnormalities in their oocytes?

3. What is abnormal about metaphase I as it is occurring in the oocytes shown in Figure 10.12*b*, *c*, and *d*?

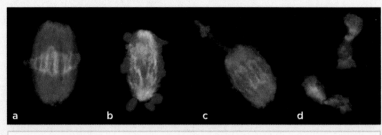

Caging materials	Total number of oocytes	Abnormalities
Control: New cages with glass bottles	271	5 (1.8%)
Damaged cages with glass bottles		
Mild damage	401	35 (8.7%)
Severe damage	149	30 (20.1%)
Damaged bottles	197	53 (26.9%)
Damaged cages with damaged bottles	58	24 (41.4%)

Figure 10.12 Meiotic abnormalities associated with exposure to damaged plastic caging. Fluorescent micrographs show nuclei of single mouse oocytes in metaphase I. (**a**) Normal metaphase; (**b–d**) examples of abnormal metaphase. Chromosomes are *red*; spindle fibers are *green*.

Self-Quiz

Answers in Appendix III

1. Meiosis and cytoplasmic division function in _____ .
   a. asexual reproduction of single-celled eukaryotes
   b. growth and tissue repair
   c. sexual reproduction
   d. both b and c

2. Sexual reproduction requires _____ .
   a. meiosis   c. spore formation
   b. fertilization   d. a and b

3. What is the name for alternative forms of the same gene?

4. Generally, a pair of homologous chromosomes _____ .
   a. carry the same genes   c. are the same length, shape
   b. interact at meiosis   d. all of the above

5. Sister chromatids are joined at the _____ .
   a. kinetochore   c. centriole
   b. spindle   d. centromere

6. Meiosis _____ the parental chromosome number.
   a. doubles   c. maintains
   b. halves   d. mixes up

7. Meiosis ends with the formation of _____ .
   a. two cells   c. four cells
   b. two nuclei   d. four nuclei

8. Sister chromatids of each duplicated chromosome separate during _____ .
   a. prophase I   d. anaphase II
   b. prophase II   e. both b and c
   c. anaphase I

9. How does meiosis contribute to variation in traits among offspring of sexual reproducers?

10. The cell shown at *right* is in anaphase II. I know this because _____ .

11. Match each term with its description.
   ___ chromosome number
   ___ alleles
   ___ metaphase I
   ___ interphase

   a. different molecular forms of the same gene
   b. maybe none between meiosis I, II
   c. all chromosomes aligned at spindle equator
   d. all chromosomes in a given type of cell

■ *Visit CengageNOW for additional questions.*

Critical Thinking

1. Explain why you can predict that meiosis gives rise to genetic differences between parent cells and descendant cells in fewer cell divisions than mitosis does.

2. Assume you can measure the amount of DNA in the nucleus of a primary oocyte, and then in the nucleus of a primary spermatocyte. Each gives you a mass *m*. What mass of DNA would you expect to find in the nucleus of each mature gamete (each egg and sperm) that forms after meiosis? What mass of DNA will be (1) in the nucleus of a zygote that forms at fertilization and (2) in that zygote's nucleus after the first DNA duplication?

3. The diploid chromosome numbers for the somatic cells of several eukaryotic species are listed at *right*. What is the number of chromosomes that normally ends up in gametes of each species? What would that number be after three generations if meiosis did not occur before gamete formation?

Fruit fly, *Drosophila melanogaster*	8
Garden pea, *Pisum sativum*	14
Frog, *Rana pipiens*	26
Earthworm, *Lumbricus terrestris*	36
Human, *Homo sapiens*	46
Amoeba, *Amoeba*	50
Dog, *Canis familiaris*	78
Vizcacha rat, *Tympanoctomys barrerae*	102
Horsetail, *Equisetum*	216

Observing Patterns in Inherited Traits

IMPACTS, ISSUES | The Color of Skin

One of the most visible human traits is the color of skin, which can range from very pale to very dark brown. The color arises from melanosomes, organelles in skin cells that make red and brownish-black pigments called melanins. Most people have about the same number of melanosomes in their skin cells. Skin color variation occurs because the kinds and amounts of melanins made by the melanosomes varies among people.

Dark skin would have been adaptive under the intense sunlight of the African savannas where humans first evolved. Melanin protects skin cells exposed to sunlight because it absorbs ultraviolet (UV) radiation, which damages DNA and other biological molecules. Melanin-rich dark skin acts as a natural sunscreen, so it reduces the risk of certain cancers and other serious problems caused by overexposure to sunlight.

Early human groups that migrated to regions with colder climates were exposed to less sunlight. In these regions, lighter skin would have been adaptive. Why? UV radiation stimulates skin cells to make a molecule the body converts to essential vitamin D. Where sunlight exposure is minimal, UV radiation damage is less of a risk than vitamin D deficiency, which has serious health consequences for developing fetuses and children. People with dark, UV-shielding skin have a high risk of this deficiency in regions where sunlight exposure is minimal.

Like most other human traits, skin color has a genetic basis (Figure 11.1). More than 100 gene products affect the synthesis and deposition of melanin. Mutations in at least some of these genes may have contributed to adaptive variations of human skin color. For example, the *SLC24A5* gene on chromosome 15 encodes a membrane transport protein in melanosomes. Nearly all people of native African, American, or East Asian descent have the same version (allele) of this gene. By contrast, nearly all people of native European descent carry a particular mutation in the gene. The European allele results in less melanin, and lighter skin color, than the unmutated version.

Such genetic patterns offer clues about the past. For example, Chinese and Europeans do not share any skin pigmentation allele that does not also occur in other populations. However, most people of Chinese descent carry a particular allele of the *DCT* gene, the product of which helps convert tyrosine to melanin. Few people of European or African descent have this allele. Taken together, the distribution of the *SLC24A5* and *DCT* genes suggests that (1) an African population was ancestral to both the Chinese and Europeans, and (2) Chinese and European populations separated before their pigmentation genes mutated and their skin color changed.

Skin color is only one of many human traits that can vary because of mutations in single genes. The small scale of such differences is a reminder that all of us share a genetic legacy of common ancestry.

Figure 11.1 Skin color. Variations in skin color may have evolved as a balance between vitamin production and protection against harmful UV radiation.

Variation in skin color and in most other human traits begins with differences in alleles inherited from parents. Fraternal twin girls Kian and Remee were born in 2006 to parents Kylie (*left*) and Remi (*right*). Both Kylie's and Remi's mothers are of European descent, and have pale skin. Both of their fathers are of African descent, and have dark skin.

More than 100 genes affect skin color in humans. Kian and Remee inherited different alleles of some of those genes.

Key Concepts

Where modern genetics started

Gregor Mendel gathered the first experimental evidence of the genetic basis of inheritance. His meticulous work gave him clues that heritable traits are specified in units. The units, which are distributed into gametes in predictable patterns, were later identified as genes. **Section 11.1**

Insights from monohybrid experiments

Some experiments yielded evidence of gene segregation: When one chromosome separates from its homologous partner during meiosis, the alleles on those chromosomes also separate and end up in different gametes. **Section 11.2**

Insights from dihybrid experiments

Other experiments yielded evidence of independent assortment: Genes are typically distributed into gametes independently of other genes. **Section 11.3**

Variations on Mendel's theme

Not all traits appear in Mendelian inheritance patterns. An allele may be partly dominant over a nonidentical partner, or codominant with it. Multiple genes may influence a trait; some genes influence many traits. The environment also influences gene expression. **Sections 11.4–11.7**

Links to Earlier Concepts

- Before starting this chapter, be sure you can generally define genes, alleles, and diploid versus haploid chromosome numbers (Sections 10.1 and 10.2).

- You may want to scan the sections that introduce protein structure (3.5), enzymes (6.3), and pigments (7.1).

- As you read, refer back to the visual road map of the stages of meiosis (10.3).

- You will be considering experimental evidence of two major topics that were introduced earlier—the effects that crossing over and metaphase I alignments have on inheritance (10.4).

How would you vote? Traditionally, humans have been assigned to race categories based on physical attributes such as skin color, which have a genetic basis. Are twins such as Kian and Remee of different races? See CengageNOW for details, then vote online.

11.1 Mendel, Pea Plants, and Inheritance Patterns

■ Recurring inheritance patterns are observable outcomes of sexual reproduction.

■ Links to Genes and alleles 10.1, Diploid and haploid 10.2

By the 1850s, most people had an idea that two parents contribute hereditary material to their offspring, but few suspected that the material is organized as units, or genes. Some thought that hereditary material must be fluid, with fluids from both parents blending at fertilization like milk into coffee.

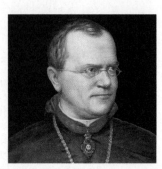

Figure 11.2 Gregor Mendel, the founder of modern genetics.

carpel   anther

A Garden pea flower, cut in half. Sperm form in pollen grains, which originate in male floral parts (anthers). Eggs develop, fertilization takes place, and seeds mature in female floral parts (carpels).

B Pollen from a plant that breeds true for purple flowers is brushed onto a floral bud of a plant that breeds true for white flowers. The white flower had its anthers snipped off. Artificial pollination is one way to ensure that a plant will not self-fertilize.

C Later, seeds develop inside pods of the cross-fertilized plant. An embryo in each seed develops into a mature pea plant.

D Each new plant's flower color is indirect but observable evidence that hereditary material has been transmitted from the parent plants.

Figure 11.3 Animated Garden pea plant (*Pisum sativum*), which can self-fertilize or cross-fertilize. Experimenters can control the transfer of its hereditary material from one flower to another.

The idea of "blending inheritance" failed to explain the obvious. For example, many children who differ in their eye or hair color have the same two parents. If parental fluids blended, then the color would be some blended shade of the parental colors. If neither parent had freckles, freckled children would never pop up. A white horse bred with a black horse should always produce gray offspring, but offspring of such matings are not always gray. Blending inheritance did not explain the variation in traits that people could see with their own eyes.

Charles Darwin did not accept the idea of blending inheritance. However, though inheritance was central to his theory of natural selection, he could not quite see how it works. He saw that forms of traits often vary among individuals in a population. He realized that variations that help individuals survive and reproduce tend to appear more frequently in a population over generations. However, neither he nor anyone else at the time knew that hereditary material is divided into discrete units (genes). That insight is crucial to understanding how heredity works.

Even before Darwin presented his theory of natural selection, someone had been gathering evidence that would support it. Gregor Mendel, an Austrian monk (Figure 11.2), had been carefully breeding thousands of pea plants. By documenting how certain traits are passed from plant to plant, generation after generation, Mendel had been collecting indirect but observable evidence of how inheritance works.

Mendel's Experimental Approach

Mendel spent most of his adult life in Brno, a city near Vienna that is now part of the Czech Republic. He was not a man of narrow interests who just stumbled onto dazzling principles. He lived in a monastery close to European cities that were centers of scientific inquiry. Having been raised on a farm, Mendel was aware of agricultural principles and their applications. He kept abreast of current literature on breeding experiments. He was a dedicated member of an agricultural society, and he won awards for developing improved varieties of fruits and vegetables.

Just after Mendel entered the monastery at Brno, he took courses in mathematics, physics, and botany at the University of Vienna. Few scholars of his time were trained in both plant breeding and mathematics.

Shortly after his university training ended, Mendel started to study *Pisum sativum*, the garden pea plant. This plant is self-fertilizing: Its flowers produce both male and female gametes (sperm and eggs) that can

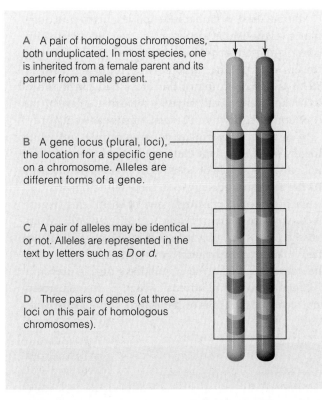

A A pair of homologous chromosomes, both unduplicated. In most species, one is inherited from a female parent and its partner from a male parent.

B A gene locus (plural, loci), the location for a specific gene on a chromosome. Alleles are different forms of a gene.

C A pair of alleles may be identical or not. Alleles are represented in the text by letters such as *D* or *d*.

D Three pairs of genes (at three loci on this pair of homologous chromosomes).

Figure 11.4 Animated A few genetic terms. Like other species with a diploid chromosome number, garden pea plants have pairs of genes, on pairs of homologous chromosomes.

Most genes come in slightly different forms called alleles. Different alleles may result in different versions of a trait. An allele at any given location on a chromosome may or may not be identical with its partner on the homologous chromosome.

come together and give rise to a new plant. Pea plants can "breed true" for certain traits such as white flowers. Breeding true for a trait means that, rare mutations aside, all offspring have the same form of the trait as the parent(s), generation after generation. For example, all offspring of pea plants that breed true for white flowers also have white flowers.

Breeders such as Mendel cross-fertilize plants when they transfer pollen from the flower of one plant to the flower of another. (Pollen grains are structures in which sperm develop. They form in anthers, which are the male parts of a flower.) For example, a breeder may open a flower bud of a true-breeding, white-flowered plant and snip out its anthers. Removing the anthers prevents the flower from fertilizing itself. The breeder then brushes the female parts of the flower with pollen from another plant, perhaps one that breeds true for purple flowers (Figure 11.3). Mendel discovered that the traits of the offspring of such cross-fertilized plants appear in predictable patterns.

Terms Used in Modern Genetics

In Mendel's time, no one knew about genes, meiosis, or chromosomes. As we follow his thinking, we will clarify the picture by substituting some modern terms, as stated here and in Figure 11.4.

1. **Genes** are heritable units of information about traits. Parents transmit genes to offspring. Each gene occurs at a specific location (**locus**) on a specific chromosome.

2. Cells with a diploid chromosome number ($2n$) have pairs of genes, on pairs of homologous chromosomes.

3. A **mutation** is a permanent change in a gene. It may cause a trait to change, as when a gene for flower color specifies purple and a mutated form specifies white. Such alternative forms of a gene are alleles.

4. All members of a lineage that breeds true for a specific trait have identical alleles for that trait. The offspring of a cross, or mating, between two individuals that breed true for different forms of a trait are **hybrids**. A hybrid has nonidentical alleles for the trait.

5. An individual with nonidentical alleles of a gene is **heterozygous** for the gene. An individual with identical alleles of a gene is **homozygous** for the gene.

6. An allele is **dominant** if its effect masks the effect of a **recessive** allele paired with it. Capital letters such as *A* signify dominant alleles; lowercase letters such as *a* signify recessive ones.

7. A **homozygous dominant** individual has a pair of dominant alleles (*AA*). A **homozygous recessive** individual has a pair of recessive alleles (*aa*). A heterozygous individual has a pair of nonidentical alleles (*Aa*). Heterozygotes are hybrids.

8. **Gene expression** is the process by which information in a gene is converted to a structural or functional part of a cell or body. Expressed genes determine traits.

9. Two terms help keep the distinction clear between genes and the traits they specify: **Genotype** refers to the particular alleles an individual carries; **phenotype** refers to an individual's traits.

10. F_1 stands for first-generation offspring of parents (P); F_2 for second-generation offspring. F is an abbreviation for filial (offspring).

Take-Home Message

What contribution did Gregor Mendel make to modern biology?
■ Mendel collected clues about how inheritance works among sexual reproducers by tracking observable traits through generations of pea plants.

11.2 Mendel's Law of Segregation

- Garden pea plants inherit two "units" of information (genes) for a trait, one from each parent.

- Link to Sampling error 1.8

A **testcross** is a method of determining genotype. An individual of unknown genotype is crossed with one that is known to be homozygous recessive. The traits of the offspring may indicate that the individual is heterozygous or homozygous for a dominant trait.

Monohybrid experiments are testcrosses that check for a dominance relationship between two alleles at a single locus. Individuals with different alleles of a gene are crossed (or self-fertilized); traits of the offspring of such a cross may indicate whether one of the alleles is dominant over the other. A typical monohybrid experiment is a cross between individuals that are identically heterozygous at one gene locus ($Aa \times Aa$).

Mendel used monohybrid experiments to find dominance relationships among seven pea plant traits. For example, he crossed plants that bred true for purple flowers with plants that bred true for white flowers. All of the F_1 offspring of this cross had purple flowers. When he crossed those F_1 offspring, some of the F_2 offspring had white flowers! What was going on?

In pea plants, one gene governs purple and white flower color. Any plant that carries the dominant allele (A) will have purple flowers. Only plants homozygous for the recessive allele (a) will have white flowers.

Each gamete carries only one of the alleles (Figure 11.5). If plants homozygous for different alleles are crossed ($AA \times aa$), only one outcome is possible: All of the F_1 offspring are heterozygous (Aa). All of them carry the dominant allele A, so all will have purple flowers.

Mendel crossed hundreds of such F_1 heterozygotes, and recorded the traits of thousands of their offspring.

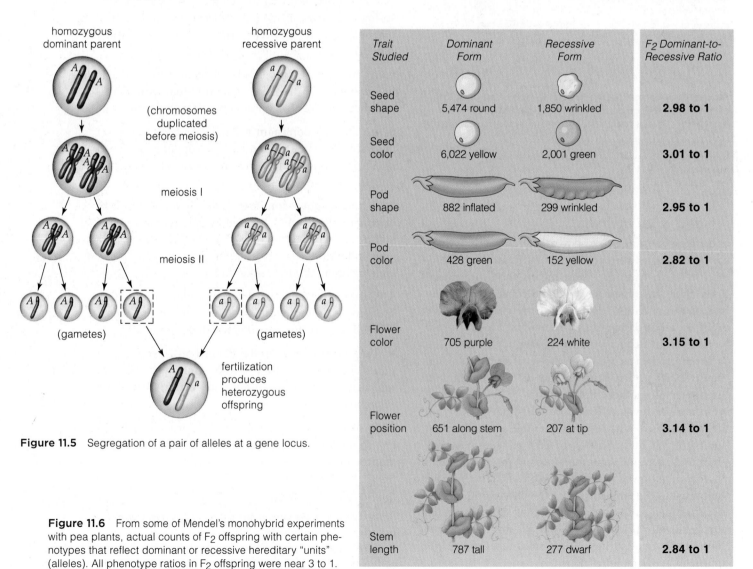

Figure 11.5 Segregation of a pair of alleles at a gene locus.

Trait Studied	Dominant Form	Recessive Form	F_2 Dominant-to-Recessive Ratio
Seed shape	5,474 round	1,850 wrinkled	**2.98 to 1**
Seed color	6,022 yellow	2,001 green	**3.01 to 1**
Pod shape	882 inflated	299 wrinkled	**2.95 to 1**
Pod color	428 green	152 yellow	**2.82 to 1**
Flower color	705 purple	224 white	**3.15 to 1**
Flower position	651 along stem	207 at tip	**3.14 to 1**
Stem length	787 tall	277 dwarf	**2.84 to 1**

Figure 11.6 From some of Mendel's monohybrid experiments with pea plants, actual counts of F_2 offspring with certain phenotypes that reflect dominant or recessive hereditary "units" (alleles). All phenotype ratios in F_2 offspring were near 3 to 1.

About three of every four F₂ plants had the dominant trait, and about one of every four had the recessive trait (Figure 11.6).

Mendel's predictable results hinted that fertilization is a chance event with a finite number of possible outcomes. Mendel knew about **probability**, which is a measure of the chance that a particular outcome will occur. That chance depends on the total number of possible outcomes. For example, if you cross two Aa heterozygotes, the two types of gametes (A and a) can meet four different ways at fertilization:

Possible Event	Probable Outcome
Sperm A meets egg A	1 out of 4 offspring AA
Sperm A meets egg a	1 out of 4 offspring Aa
Sperm a meets egg A	1 out of 4 offspring Aa
Sperm a meets egg a	1 out of 4 offspring aa

Each of the offspring of this cross has 3 chances in 4 of inheriting at least one dominant A allele (and purple flowers). It has 1 chance in 4 of inheriting two recessive a alleles (and white flowers). Thus, the probability that an offspring of this cross will have purple or white flowers is 3 purple to 1 white, which we represent as a ratio of 3:1. We use grids called **Punnett squares** to calculate the probability of genotypes (and phenotypes) that will occur in offspring (Figure 11.7).

Mendel's observed ratios were not exactly 3:1, but he knew that deviations can arise from sampling error (Section 1.8). For example, if you flip a coin, it is just as likely to end up heads as tails (a probability of 1:1). But often it ends up heads, or tails, several times in a row. Thus, if you flip the coin only a few times, the observed ratio might differ greatly from the predicted ratio of 1:1. If you flip it many times, you are more likely to see that ratio. Mendel minimized his sampling error by maximizing his sample sizes.

The results from Mendel's monohybrid experiments became the basis of his law of **segregation**, which we state here in modern terms: Diploid cells have pairs of genes, on pairs of homologous chromosomes. The two genes of each pair are separated from each other during meiosis, so they end up in different gametes.

Take-Home Message

What is Mendel's law of segregation?

■ Diploid cells have pairs of genes, on pairs of homologous chromosomes. The two genes of each pair are separated from each other during meiosis, so they end up in different gametes.

■ Mendel discovered patterns of inheritance in pea plants by recording and analyzing the results of many testcrosses.

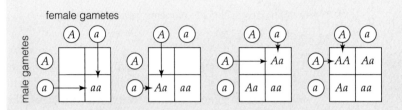

A From left to right, step-by-step construction of a Punnett square. Circles signify gametes, and letters signify alleles: A is dominant; a is recessive. The genotypes of the resulting offspring are inside the squares.

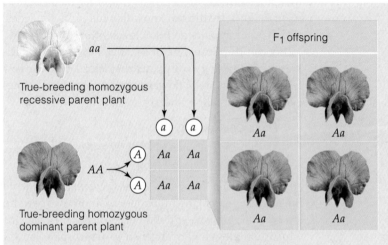

B A cross between two plants that breed true for different forms of a trait produces F₁ offspring that are identically heterozygous.

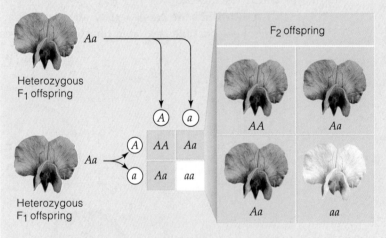

C A cross between the F₁ offspring is the monohybrid experiment. The phenotype ratio of F₂ offspring in this example is 3:1 (3 purple to 1 white).

Figure 11.7 Animated (a) Punnett-square method of predicting probable outcomes of genetic crosses. (b,c) One of Mendel's monohybrid experiments. On average, the ratio of dominant-to-recessive phenotypes among second-generation (F₂) plants of a monohybrid experiment is 3:1.

Figure It Out: How many possible genotypes are there in the F₂ generation?

Answer: *Three: AA, Aa, and aa*

11.3 Mendel's Law of Independent Assortment

- Many genes sort into gametes independently.
- Link to Meiosis 10.3

Dihybrid experiments test for dominance relationships between alleles at two loci. Individuals with different alleles of two genes are crossed (or self-fertilized); the ratio of traits in offspring offer clues about the alleles. Mendel analyzed the numerical results from dihybrid experiments, but given the prevailing understanding of heredity, he could only hypothesize that "units" specifying one trait (such as flower color) sort into gametes independently of "units" specifying other traits (such as plant height). He did not know that the units are genes, which occur in pairs on homologous chromosomes.

We can duplicate one of Mendel's dihybrid experiments by crossing two plants that breed true for two traits. Here, we track flower color (*A*, purple; *a*, white) and height (*B*, tall; *b*, short):

True-breeding parents:	*AABB*	× *aabb*
Gametes:	(*AB*, *AB*)	(*ab*, *ab*)
F$_1$ offspring, all dihybrids:	*AaBb*	

All F$_1$ offspring from this cross are tall with purple flowers. Remember, homologous chromosomes become attached to opposite spindle poles during meiosis, but which homologue gets attached to which pole is random (Section 10.3). Thus, four combinations of alleles are possible in the gametes of *AaBb* dihybrids: *AB*, *Ab*, *aB*, and *ab* (Figure 11.8).

If two dihybrids are crossed, their alleles can combine in sixteen possible ways at fertilization (four types of gametes in one individual × four types of gametes in the other). In our example (*AaBb* × *AaBb*), the sixteen combinations result in four different phenotypes (Figure 11.9). Nine of the sixteen are tall with purple flowers, three are short with purple flowers, three are tall with white flowers, and one is short with white flowers. The ratio of these phenotypes is 9:3:3:1. With more gene pairs, more combinations are possible. If the parents differ in, say, twenty gene pairs, 3.5 billion genotypes are possible!

Mendel published his results in 1866, but apparently his work was read by few and understood by no one. In 1871 he became abbot of his monastery, and his pioneering experiments ended. He died in 1884, never to know that his experiments would be the starting point for modern genetics.

In time, Mendel's hypothesis became known as the law of **independent assortment**. In modern terms, the law states that genes are sorted into gametes independently of other genes. The law holds for many genes in most organisms. However, it requires qualification because there are exceptions. For example, genes that are relatively close together on the same chromosome tend to stay together during meiosis (we will return to this topic in Section 11.5).

Take-Home Message

What is Mendel's law of independent assortment?

- A gene is distributed into gametes independently of how other genes are distributed.

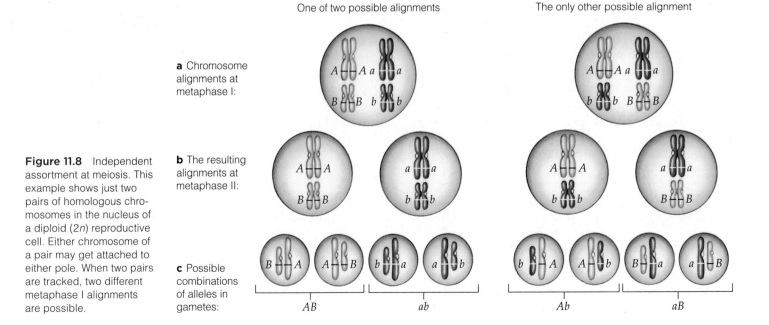

Figure 11.8 Independent assortment at meiosis. This example shows just two pairs of homologous chromosomes in the nucleus of a diploid (2*n*) reproductive cell. Either chromosome of a pair may get attached to either pole. When two pairs are tracked, two different metaphase I alignments are possible.

One of two possible alignments

The only other possible alignment

a Chromosome alignments at metaphase I:

b The resulting alignments at metaphase II:

c Possible combinations of alleles in gametes:

AB   *ab*   *Ab*   *aB*

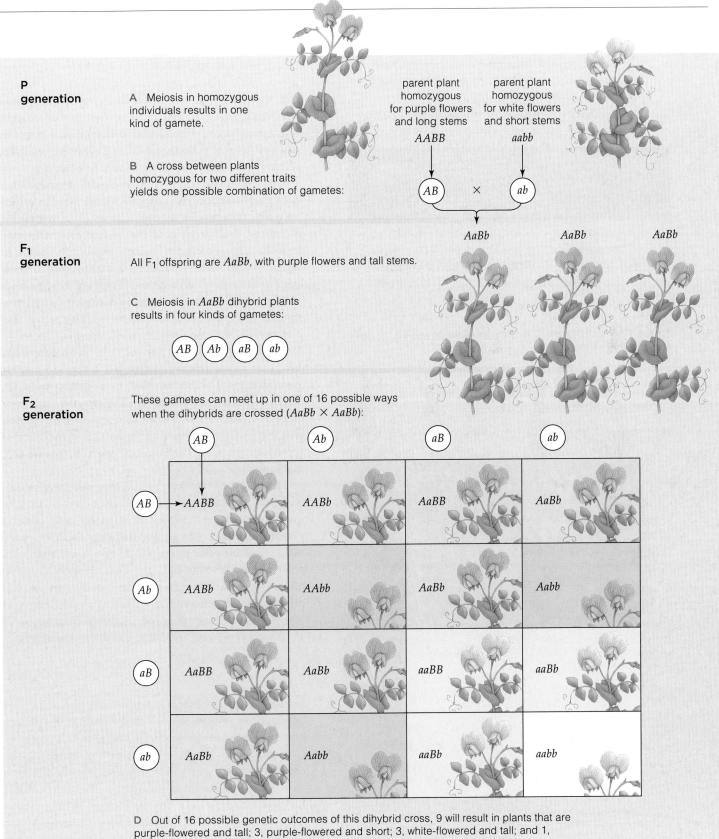

P generation

A Meiosis in homozygous individuals results in one kind of gamete.

B A cross between plants homozygous for two different traits yields one possible combination of gametes:

parent plant homozygous for purple flowers and long stems

AABB

parent plant homozygous for white flowers and short stems

aabb

AB × ab

AaBb   *AaBb*   *AaBb*

F₁ generation

All F₁ offspring are *AaBb*, with purple flowers and tall stems.

C Meiosis in *AaBb* dihybrid plants results in four kinds of gametes:

AB Ab aB ab

F₂ generation

These gametes can meet up in one of 16 possible ways when the dihybrids are crossed (*AaBb* × *AaBb*):

AB Ab aB ab

	AB	Ab	aB	ab
AB	*AABB*	*AABb*	*AaBB*	*AaBb*
Ab	*AABb*	*AAbb*	*AaBb*	*Aabb*
aB	*AaBB*	*AaBb*	*aaBB*	*aaBb*
ab	*AaBb*	*Aabb*	*aaBb*	*aabb*

D Out of 16 possible genetic outcomes of this dihybrid cross, 9 will result in plants that are purple-flowered and tall; 3, purple-flowered and short; 3, white-flowered and tall; and 1, white-flowered and short. The ratio of phenotypes of this dihybrid cross is 9:3:3:1.

Figure 11.9 Animated One of Mendel's dihybrid experiments. Here, *A* is an allele for purple flowers; *a*, white flowers; *B*, tall plants; *b*, short plants. **Figure It Out:** What do the flowers inside the boxes represent?

Answer: Phenotypes of the offspring

11.4 | Beyond Simple Dominance

■ Mendel focused on traits based on clearly dominant and recessive alleles. However, the expression patterns of genes for some traits are not as straightforward.

■ Links to Fibrous proteins 3.5, Pigments 7.1

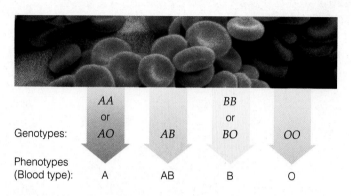

Genotypes: AA or AO | AB | BB or BO | OO

Phenotypes (Blood type): A | AB | B | O

Figure 11.10 **Animated** Combinations of alleles that are the basis of ABO blood typing.

homozygous parent (*RR*) × homozygous parent (*rr*) → heterozygous F₁ offspring (*Rr*)

A Cross a red-flowered with a white-flowered plant, and all of the F₁ offspring will be pink.

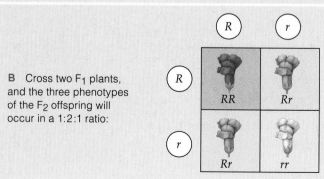

B Cross two F₁ plants, and the three phenotypes of the F₂ offspring will occur in a 1:2:1 ratio:

Figure 11.11 Incomplete dominance in snapdragons.

Codominance in ABO Blood Types

With **codominance**, two nonidentical alleles of a gene are both fully expressed in heterozygotes, so neither is dominant or recessive. Codominance may occur in **multiple allele systems**, in which three or more alleles of a gene persist among individuals of a population.

ABO blood typing is a method of determining an individual's genotype at the *ABO* gene locus, a multiple allele system. The method checks for a membrane glycolipid that helps give the body's cells a unique identity. This glycolipid occurs in slightly different forms. Which form a person has begins with a gene, *ABO*, that encodes an enzyme. There are three alleles of this gene. The *A* and *B* alleles encode different versions of the enzyme. The *O* allele has a mutation that prevents its enzyme product from becoming active.

The alleles you carry for the *ABO* gene determine your blood type (Figure 11.10). The *A* and the *B* allele are codominant when paired. If your genotype is *AB*, then you have both versions of the enzyme, and your blood is type AB. The *O* allele is recessive when paired with either *A* or *B*. If you are *AA* or *AO*, your blood is type A. If you are *BB* or *BO*, it is type B. If you are *OO*, it is type O.

The reason for blood typing is that receiving incompatible blood cells in a transfusion is dangerous. The immune system attacks any red blood cells bearing glycolipids that are foreign to the body. Such an attack can cause the cells to clump or burst, a transfusion reaction with potentially fatal consequences. Type O blood is compatible with all other blood types, so people who have it are called universal blood donors. If you have AB blood, you can receive a transfusion of any blood type; you are called a universal recipient.

Incomplete Dominance

In **incomplete dominance**, one allele of a pair is not fully dominant over its partner, so the heterozygote's phenotype is somewhere between the two homozygotes.

A cross between two true-breeding snapdragons, one red and one white, reveals incomplete dominance: all F₁ offspring are pink-flowered (Figure 11.11). Why? Red snapdragons have two alleles that let them make a lot of red pigment. White snapdragons have two mutated alleles; they do not make any pigment at all, so their flowers are colorless. Pink snapdragons have a "red" allele and a "white" allele; such heterozygotes make only enough pigment to color the flowers pink. Cross two pink F₁ plants and you can expect to see red, pink, and white flowers in a 1:2:1 ratio in F₂ offspring.

Epistasis

Some traits are affected by interactions among different gene products, an effect called **epistasis**. Typically, one gene product suppresses the effect of another, so the resulting phenotype is somewhat unexpected. Epistatic interactions between two genes in chickens cause dramatic variations in their combs (Figure 11.12).

As another example, several genes affect Labrador retriever coat color, which can be black, yellow, or brown (Figure 11.13). A dog's coat color depends on how products of alleles at more than one locus make a dark pigment, melanin, and deposit it in tissues. Allele *B* (black) is dominant to *b* (brown). At a different locus, allele *E* promotes the deposition of melanin in fur, but two recessive alleles (*ee*) reduce it. A dog with two *e* alleles has yellow fur regardless of which alleles it has at the *B* locus.

Single Genes With a Wide Reach

One gene may influence multiple traits, an effect called **pleiotropy**. Genes encoding products used throughout the body are the ones most likely to be pleiotropic. For example, long fibers of fibrillin impart elasticity to the tissues of the heart, skin, blood vessels, tendons, and other body parts. Mutations in the fibrillin gene cause a genetic disorder called Marfan syndrome, in which tissues form with defective fibrillin or none at all. The largest blood vessel leading from the heart, the aorta, is particularly affected. In Marfan syndrome, muscle cells in the aorta's thick wall do not function very well, and the wall itself is not as elastic as it should be. The aorta expands under pressure, so the lack of elasticity eventually makes it thin and leaky. Calcium deposits accumulate inside. Inflamed, thinned, and weakened, the aorta can rupture abruptly during exercise.

Marfan syndrome can be very difficult to diagnose. Affected people are often tall, thin, and loose-jointed, but there are plenty of tall, thin, loose-jointed people without the disorder. Symptoms may not be apparent, so many people are not aware that they have Marfan. Until recently, it killed most of them before the age of fifty. Haris Charalambous was one (Figure 11.14).

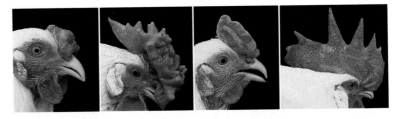

Figure 11.12 Variation in combs, the fleshy, red crest on the head of chickens, is an outcome of interactions among products of alleles at two gene loci.

	(EB)	(Eb)	(eB)	(eb)
(EB)	EEBB black	EEBb black	EeBB black	EeBb black
(Eb)	EEBb black	EEbb chocolate	EeBb black	Eebb chocolate
(eB)	EeBB black	EeBb black	eeBB yellow	eeBb yellow
(eb)	EeBb black	Eebb chocolate	eeBb yellow	eebb yellow

Figure 11.13 *Left to right*, black, chocolate, and yellow Labrador retrievers. Epistatic interactions among products of two gene pairs affect coat color.

Figure 11.14 Rising basketball star Haris Charalambous, who died suddenly when his aorta burst during warmup exercises at the University of Toledo in 2006. He was 21.

Charalambous was very tall and lanky, with long arms and legs—traits that are valued in professional athletes such as basketball players. These traits are also associated with Marfan syndrome.

Like many other people, Charalambous did not know he had Marfan syndrome. An estimated 1 in 5,000 people are affected by Marfan worldwide.

Take-Home Message

Are all alleles clearly dominant or recessive?

■ An allele may be fully dominant, incompletely dominant, or codominant with its partner on a homologous chromosome.

■ In epistasis, two or more gene products influence a trait. In pleiotropy, one gene product influences two or more traits.

■ The farther apart two genes are on a chromosome, the more often crossing over occurs between them.

■ Links to Meiosis 10.3, Crossing over 10.4

As you learned in Section 11.3, alleles of genes on different chromosomes assort independently into gametes. What about genes on the same chromosome? Mendel studied seven genes in pea plants, which have seven pairs of chromosomes. Was he lucky enough to choose one gene on each of those seven chromosome pairs? Some surmised that if he had studied more genes, he would have discovered an exception to his law of independent assortment.

As it turns out, some of the genes Mendel studied *are* on the same chromosome. The genes are far enough apart that crossing over occurs between them very frequently—so frequently that they tend to assort

into gametes independently, just as if they were on different chromosomes. By contrast, genes that are very close together on a chromosome do not tend to assort independently, because crossing over does not happen very often between them. Thus, gametes usually receive parental combinations of alleles of such genes. Genes that do not assort independently are said to be linked (Figure 11.15).

Alleles of some linked genes stay together during meiosis more than others do. The effect is simply due to the relative distance between genes: Genes that are closer together on a chromosome get separated less frequently by crossovers. For example, if genes *A* and *B* are twice as far apart on a chromosome as genes *C* and *D*, then we can expect crossovers to separate alleles of genes *A* and *B* more frequently than they separate alleles of genes *C* and *D*:

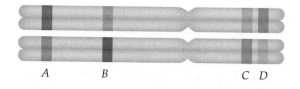

Generalizing from this example, we can say the probability that a crossover event will separate alleles of two genes is proportional to the distance between those genes. In other words, the closer together any two genes are on a chromosome, the more likely gametes will be to receive parental combinations of alleles of those genes. Genes are said to be tightly linked if the distance between them is relatively small.

All genes on one chromosome are called a **linkage group**. Peas have 7 chromosomes, so they have 7 linkage groups. Humans have 23 chromosomes, so they have 23 linkage groups.

Human gene linkages were identified by tracking inheritance in families over several generations. One thing became clear: Crossovers are not at all rare, and may even be required in order for meiosis to run to completion. In many eukaryotes, at least two crossover events occur between every pair of homologous chromosomes during prophase I of meiosis.

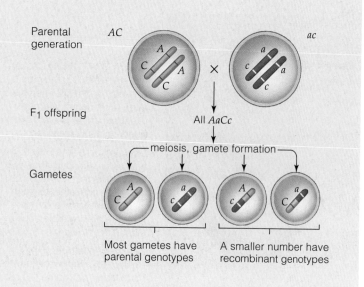

Mendel's blind luck in the genetics game?

Take-Home Message

What is the effect of crossing over on inheritance?

■ All genes on a chromosome are part of a linkage group.

■ Crossing over disrupts linkage groups. The farther apart two genes are on a chromosome, the more often crossing over occurs between them.

Figure 11.15 Animated Linkage and crossing over. Alleles of two genes on the same chromosome stay together when there is no crossover between them, and recombine when there is a crossover between them.

11.6 Genes and the Environment

- The environment can influence gene expression.
- Link to Enzyme function 6.3

Variations in traits are not always the result of differences in alleles. For example, in Section 6.3, you read about a heat-sensitive enzyme, tyrosinase, that affects the coat color of Siamese cats. This enzyme catalyzes one step in the synthesis of melanin, but it only works in cooler body regions, such as the legs, tail, and ears. Tyrosinase also affects coat color in Himalayan rabbits. The rabbits are homozygous for the c^h allele, which encodes a form of tyrosinase that stops working when the temperature in cells exceeds 33°C, or 91°F. Metabolic heat keeps the more massive body parts warm enough to deactivate the enzyme, so the fur is light on their surfaces. The ears and other slender appendages lose metabolic heat faster, so they are cooler, and melanin darkens them (Figure 11.16).

Yarrow plants offer another example of how environment influences phenotype. Yarrow is useful for experiments because it grows from cuttings. All cuttings of a plant have the same genotype, so experimenters know that genes are not the basis for any phenotypic differences among them. In one study, genetically identical yarrow plants had different phenotypes when grown at different altitudes (Figure 11.17).

Invertebrates, too, show phenotypic variation with environmental conditions. For instance, daphnias are microscopic freshwater relatives of shrimps. Aquatic insects prey on them. *Daphnia pulex* living in ponds with few predators have rounded heads, but those in ponds with many predators have more pointed heads (Figure 11.18). *Daphnia*'s predators emit chemicals that trigger the different phenotype.

The environment also affects human genes. One of our genes encodes a protein that transports serotonin across the membrane of brain cells. Serotonin lowers anxiety and depression during traumatic times. Some mutations in the serotonin transporter gene can reduce the ability to cope with stress. It is as if some of us are bicycling through life without an emotional helmet. Only when we crash does the mutation's phenotypic effect—depression—appear. Other human genes affect emotional state, but mutations in this one reduce our capacity to snap out of it when bad things happen.

Take-Home Message

Does the environment influence gene expression?

- Expression of some genes is affected by environmental factors such as temperature. The result may be variation in traits.

Figure 11.16 **Animated** Observable effect of the environment on gene expression. A Himalayan rabbit is homozygous for an allele that encodes a heat-sensitive form of an enzyme required for melanin synthesis. Cooler body parts, such as ears, are dark. The main body mass is warmer, and light. A patch of one rabbit's white fur was shaved off. An ice pack was tied above the fur-free patch. Fur grew back, but it was dark. The ice pack had cooled the patch enough for the enzyme to work, and melanin was produced.

a Mature cutting at high elevation (3,060 meters above sea level)

b Mature cutting at mid-elevation (1,400 meters above sea level)

c Mature cutting at low elevation (30 meters above sea level)

Figure 11.17 Experiment showing environmental effects on phenotype in yarrow (*Achillea millefolium*). Cuttings from the same parent plant were grown in the same kind of soil at three different elevations.

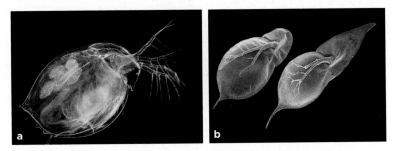

Figure 11.18 (**a**) Light micrograph of a living daphnia. (**b**) Phenotypic effects of the presence of insects that prey on daphnias. The body form at the *left* develops when predators are absent or few. The form at the *right* develops when water contains chemicals emitted by the daphnia's insect predators. It has a longer tail spine and a pointed spine at the head.

11.7 Complex Variations in Traits

■ Individuals of most species vary in some of their shared traits. Many traits show a continuous range of variation.

Continuous Variation

The individuals of a species typically vary in many of their shared traits. Some of those traits appear in two or three forms; others occur in a range of small differences that is called **continuous variation**. Continuous variation is an outcome of polygenic inheritance, in which multiple genes affect a single trait. The more genes and environmental factors that influence a trait, the more continuous is its variation.

Consider eye color. The colored part is the iris, a doughnut-shaped, pigmented structure. Several gene products contribute to iris color by making and distributing melanins. The more melanin deposited in the iris, the less light is reflected from it. Irises that are nearly black have dense melanin deposits that absorb almost all light, and reflect almost none. Melanin deposits are not as extensive in brown or hazel eyes, which reflect some incident light. Green, gray, and blue eyes have the least amount of melanin, so they reflect the most light.

How do we determine whether a trait varies continuously? First,

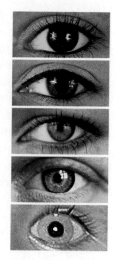

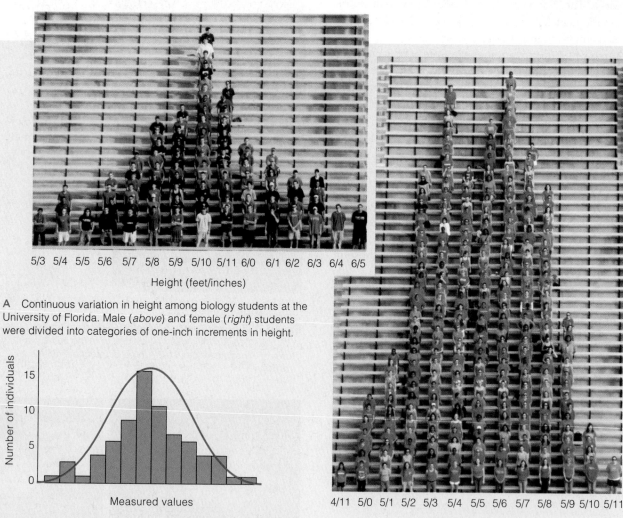

5/3 5/4 5/5 5/6 5/7 5/8 5/9 5/10 5/11 6/0 6/1 6/2 6/3 6/4 6/5

Height (feet/inches)

A Continuous variation in height among biology students at the University of Florida. Male (*above*) and female (*right*) students were divided into categories of one-inch increments in height.

Number of individuals

15

10

5

0

Measured values

B Bar graph showing the data from the male biology students in (**a**). Notice the rough bell-shaped distribution.

4/11 5/0 5/1 5/2 5/3 5/4 5/5 5/6 5/7 5/8 5/9 5/10 5/11

Height (feet/inches)

Figure 11.19 Animated Continuous variation. These examples show continuous variation in body height, one of the traits that help characterize human populations.

we divide the total range of phenotypes into measurable categories, such as inches of height. Next, we count how many individuals of a group fall into each category; this count gives the relative frequencies of phenotypes across our range of measurable values. Finally, we plot the data as a bar chart (Figure 11.19). In such charts, the shortest bars are categories with the fewest individuals, and the tallest bars are the categories with the most. A graph line around the top of the bars shows the distribution of values for the trait. If the line is a bell-shaped curve, or **bell curve**, the trait shows continuous variation.

Regarding the Unexpected Phenotype

Nearly all of the traits Mendel studied appeared in predictable ratios because the gene pairs happened to be on different chromosomes or far apart on the same chromosome. They tended to segregate independently. However, there is often far more variation in phenotypes, and not all of it is a result of crossing over.

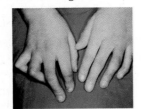

For example, certain mutations cause camptodactyly, in which finger shape and movement are abnormal. Any or all fingers on the left hand, right hand, or both hands may be bent and immobile (*right*).

What causes complex variation? Most organic molecules are synthesized in metabolic pathways that involve many enzymes. Genes encoding those enzymes can mutate in any number of ways, so their products may function within a spectrum of activity that ranges from excessive to not at all. Thus, the end product of a metabolic pathway can be produced within a range of concentration and activity. Environmental factors often add further variations on top of that.

Thus, phenotype results from complex interactions among gene products and the environment. We return to this topic in Chapter 18, as we consider some evolutionary consequences of variation in phenotype.

Take-Home Message

How does phenotype vary?

■ Some traits have a range of small differences, or continuous variation. The more genes and other factors that influence a trait, the more continuous the distribution of phenotype.

■ Enzymes and other gene products control steps of most metabolic pathways. Mutations, interactions among genes, and environmental conditions can affect one or more steps, and thus contribute to variation in phenotypes.

Summary

Section 11.1 By experimenting with pea plants, Mendel gathered evidence of patterns by which parents transmit genes to offspring. **Genes** are units of DNA that hold information about traits. Each has its own **locus**, or location, along the length of a chromosome. **Mutations** give rise to different forms of a gene (alleles).

Individuals that carry two identical alleles of a gene are **homozygous** for the gene: (*AA*) or (*aa*). Individuals that breed true for a trait are homozygous for alleles that affect the trait. Offspring of a cross between individuals homozygous for different alleles of a gene are **hybrids**, or **heterozygous**, with two nonidentical alleles (*Aa*).

A **dominant** allele masks the effect of a **recessive** allele partnered with it on the homologous chromosome. An individual with two dominant alleles (*AA*) is **homozygous dominant**. An individual with two recessive alleles is **homozygous recessive** (*aa*).

The alleles at any or all gene loci constitute an individual's **genotype**. **Gene expression** results in **phenotype**, which refers to an individual's observable traits.

■ *Learn how Mendel crossed garden pea plants, and the definitions of important genetic terms, on CengageNOW.*

Section 11.2 Crossing individuals that breed true for two forms of a trait ($AA \times aa$) yields identically heterozygous F_1 offspring (*Aa*). A cross between such F_1 offspring is a **monohybrid experiment**, which can reveal dominance relationships among the alleles. We use **Punnett squares** to calculate the **probability** of seeing certain phenotypes in F_2 offspring of such **testcrosses**:

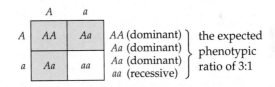

Mendel's monohybrid experiment results led to his law of **segregation** (stated here in modern terms): Diploid organisms have pairs of genes, on pairs of homologous chromosomes. During meiosis, the genes of each pair separate, so each gamete gets one or the other gene.

■ *Use the interaction on CengageNOW to carry out monohybrid experiments and a testcross.*

Section 11.3 Crossing individuals that breed true for two forms of two traits ($AABB \times aabb$) yields identically heterozygous F_1 offspring (*AaBb*). A cross between these F_1 offspring is called a **dihybrid experiment**. The phenotype ratios in F_2 offspring of such testcrosses may reveal dominance relationships among the alleles. Mendel saw a 9:3:3:1 phenotype ratio in his dihybrid experiments:

9 dominant for both traits
3 dominant for *A*, recessive for *b*
3 dominant for *B*, recessive for *a*
1 recessive for both traits

These results led to the law of **independent assortment** (stated in modern terms): Meiosis assorts gene pairs on homologous chromosomes independently of other gene

A person of mixed ethnicity may make gametes that contain different mixtures of alleles for dark and light skin. It is fairly rare that one of those gametes contains all of the alleles for dark skin, or all of the alleles for light skin, but it happens, as evidenced by twins Kian and Remee.

In the case of the *SLC24A5* gene, a mutation that occurred between 6,000 and 10,000 years ago changed the 111th amino

acid of its protein product from alanine to threonine. This small change resulted in the European version of white skin.

---

pairs on the other chromosomes. The random attachment of homologous chromosomes to opposite spindle poles during prophase I is the basis of this outcome.

- *Use the interactions on CengageNOW to observe the results of a dihybrid cross.*

Section 11.4 Inheritance patterns frequently vary. With **incomplete dominance**, an allele is not fully dominant over its partner on a homologous chromosome, and both are expressed. The combination of alleles gives rise to an intermediate phenotype.

Codominant alleles are both expressed at the same time in heterozygotes, as in the **multiple allele system** underlying ABO blood typing. In **epistasis**, interacting products of one or more genes often affect the same trait. A **pleiotropic** gene affects two or more traits.

- *Use the interactions on CengageNOW to explore patterns of non-Mendelian inheritance.*

Section 11.5 The farther apart two genes are on a chromosome, the greater the frequency of crossing over between them. Genes that are relatively close to each other on a chromosome tend to stay together during meiosis because few crossover events occur between them. Genes that are relatively far apart tend to assort independently into gametes. All of the genes on one chromosome constitute a **linkage group**.

Section 11.6 Various environmental factors may affect gene expression in individuals.

- *Use the interactions on CengageNOW to see how the environment can affect phenotype.*

Section 11.7 A trait that is influenced by the products of multiple genes often occurs in a range of small increments of phenotype (**continuous variation**).

- *Use the interaction on CengageNOW to plot the continuous distribution of height for a class.*

Self-Quiz
Answers in Appendix III

1. Alleles are _____ .
   a. different molecular forms of a gene
   b. different phenotypes
   c. self-fertilizing, true-breeding homozygotes

2. A bell curve indicates _____ in a trait.

3. A heterozygote has a _____ for a trait being studied.
   a. pair of identical alleles
   b. pair of nonidentical alleles
   c. haploid condition, in genetic terms

4. The observable traits of an organism are its _____ .
   a. phenotype   c. genotype
   b. sociobiology   d. pedigree

5. Second-generation offspring of a cross between parents who are homozygous for different alleles are the _____ .
   a. F_1 generation   c. hybrid generation
   b. F_2 generation   d. none of the above

6. F_1 offspring of the cross $AA \times aa$ are _____ .
   a. all AA c. all Aa
   b. all aa d. 1/2 AA and 1/2 aa

7. Refer to question 5. Assuming complete dominance, the F_2 generation will show a phenotypic ratio of _____ .
   a. 3:1   b. 9:1   c. 1:2:1   d. 9:3:3:1

8. A testcross is a way to determine _____ .
   a. phenotype   b. genotype   c. both a and b

9. Assuming complete dominance, crosses between two dihybrid F_1 pea plants, which are offspring from a cross $AABB \times aabb$, result in F_2 phenotype ratios of _____ .
   a. 1:2:1   b. 3:1   c. 1:1:1:1   d. 9:3:3:1

10. The probability of a crossover occurring between two genes on the same chromosome _____ .
    a. is unrelated to the distance between them
    b. decreases with the distance between them
    c. increases with the distance between them

11. Two genes that are close together on the same chromosome are _____ .
    a. linked   c. homologous   e. all of the
    b. identical alleles   d. autosomes   above

12. Match each example with the most suitable description.
    ___dihybrid experiment   a. *bb*
    ___monohybrid experiment   b. *AABB* × *aabb*
    ___homozygous condition   c. *Aa*
    ___heterozygous condition   d. *Aa* × *Aa*

- *Visit CengageNOW for additional questions.*

Genetics Problems
Answers in Appendix III

1. Assuming that independent assortment occurs during meiosis, what type(s) of gametes will form in individuals with the following genotypes?
   a. *AABB*   b. *AaBB*   c. *Aabb*   d. *AaBb*

Data Analysis Exercise

A 2000 study measured average skin color of people native to more than fifty regions, and correlated them to the amount of UV radiation received in those regions. Some of their results are shown in Figure 11.20.

1. Which country receives the most UV radiation? The least?

2. The people native to which country have the darkest skin? The lightest?

3. According to this data, how does the skin color of indigenous peoples correlate with the amount of UV radiation incident in their native regions?

Country	Skin Reflectance	UVMED
Australia	19.30	335.55
Kenya	32.40	354.21
India	44.60	219.65
Cambodia	54.00	310.28
Japan	55.42	130.87
Afghanistan	55.70	249.98
China	59.17	204.57
Ireland	65.00	52.92
Germany	66.90	69.29
Netherlands	67.37	62.58

Figure 11.20
Skin color of indigenous peoples and regional incident UV radiation. Skin reflectance measures how much light of 685 nanometers wavelength is reflected from skin; UVMED is the annual average UV radiation received at Earth's surface.

2. Refer to problem 1. Determine the frequencies of each genotype among offspring from the following matings:
- a. $AABB \times aaBB$
- b. $AaBB \times AABb$
- c. $AaBb \times aabb$
- d. $AaBb \times AaBb$

3. Refer to problem 2. Assume a third gene has alleles C and c. For each genotype listed, what allele combinations will occur in gametes, assuming independent assortment?
- a. $AABBCC$
- b. $AaBBcc$
- c. $AaBBCc$
- d. $AaBbCc$

4. Sometimes the gene for tyrosinase mutates so its product is not functional. An individual who is homozygous recessive for such a mutation cannot make melanin. Albinism, the absence of melanin, results. Humans and many other organisms can have this phenotype (*right*). In the following situations, what are the probable genotypes of the father, the mother, and their children?

- a. Both parents have normal phenotypes; some of their children are albino and others are unaffected.
- b. Both parents are albino and have albino children.
- c. The woman is unaffected, the man is albino, and they have one albino child and three unaffected children.

5. Certain genes are vital for development. When mutated, they are lethal in homozygous recessives. Even so, heterozygotes can perpetuate recessive, lethal alleles. The allele *Manx* (M^L) in cats is an example. Homozygous cats ($M^L M^L$) die before birth. In heterozygotes ($M^L M$), the spine develops abnormally, and the cats end up with no tail (*right*).

Two $M^L M$ cats mate. What is the probability that any one of their surviving kittens will be heterozygous?

6. Several alleles affect traits of roses, such as plant form and bud shape. Alleles of one gene govern whether a plant will be a climber (dominant) or shrubby (recessive). All F_1 offspring from a cross between a true-breeding climber and a shrubby plant are climbers. If an F_1 plant is crossed with a shrubby plant, about 50 percent of the offspring will be shrubby; 50 percent will be climbers. Using symbols A and a for the dominant and recessive alleles, make a Punnett-square diagram of the expected genotypes and phenotypes in F_1 offspring and in offspring of a cross between an F_1 plant and a shrubby plant.

7. Mendel crossed a true-breeding pea plant with green pods and a true-breeding pea plant with yellow pods. All the F_1 plants had green pods. Which color is recessive?

8. Suppose you identify a new gene in mice. One of its alleles specifies white fur, another specifies brown. You want to see if the two interact in simple or incomplete dominance. What sorts of genetic crosses would give you the answer?

9. In sweet pea plants, an allele for purple flowers (P) is dominant to an allele for red flowers (p). An allele for long pollen grains (L) is dominant to an allele for round pollen grains (l). Bateson and Punnett crossed a plant having purple flowers/long pollen grains with one having white flowers/round pollen grains. All F_1 offspring had purple flowers and long pollen grains. Among the F_2 generation, the researchers observed the following phenotypes:

296 purple flowers/long pollen grains
19 purple flowers/round pollen grains
27 red flowers/long pollen grains
85 red flowers/round pollen grains

What is the best explanation for these results?

10. Red-flowering snapdragons are homozygous for allele R^1. White-flowering snapdragons are homozygous for a different allele (R^2). Heterozygous plants ($R^1 R^2$) bear pink flowers. What phenotypes should appear among first-generation offspring of the crosses listed? What are the expected proportions for each phenotype?
- a. $R^1 R^1 \times R^1 R^2$
- b. $R^1 R^1 \times R^2 R^2$
- c. $R^1 R^2 \times R^1 R^2$
- d. $R^1 R^2 \times R^2 R^2$

(Incompletely dominant alleles are usually designated by superscript numerals, as shown, not by uppercase letters for dominance and lowercase letters for recessiveness.)

11. A single mutant allele gives rise to an abnormal form of hemoglobin (Hb^S, not Hb^A). Homozygotes ($Hb^S Hb^S$) develop sickle-cell anemia (Section 3.6). Heterozygotes ($Hb^A Hb^S$) have few symptoms. A couple who are both heterozygous for the Hb^S allele plan to have children. For each of the pregnancies, state the probability that they will have a child who is:
- a. homozygous for the Hb^S allele
- b. homozygous for the Hb^A allele
- c. heterozygous: $Hb^A Hb^S$

Chromosomes and Human Inheritance

Strange Genes, Tortured Minds

"This man is brilliant." That was the extent of a letter of recommendation from Richard Duffin, a professor of mathematics at Carnegie Mellon University. Duffin wrote the line in 1948 on behalf of John Forbes Nash, Jr. Nash was twenty years old at the time and applying for admission to Princeton University's graduate school. Over the next ten years, Nash made his reputation as one of the foremost mathematicians. He was socially awkward, but so are many highly gifted people. Nash showed no symptoms warning of the paranoid schizophrenia that eventually would debilitate him.

Full-blown symptoms emerged in his thirtieth year. Nash had to abandon his position at the Massachusetts Institute of Technology. Two decades passed before he was able to return to his pioneering work in mathematics.

Of every 100 people, 1 is affected by schizophrenia. This neurobiological disorder (NBD) is characterized by delusions, hallucinations, disorganized speech, and abnormal social behavior. Exceptional creativity often accompanies schizophrenia. It also accompanies other NBDs, including autism, chronic depression, and bipolar disorder, which manifests itself as jarring swings in mood and social behavior.

Certainly not every person with a high IQ has a neurobiological disorder, but a higher percentage of creative geniuses have NBDs than nongeniuses. In fact, emotionally healthy, highly creative people have more personality traits in common with people affected by NBDs than with individuals closer to the norm. For instance, both tend to be hypersensitive to environmental stimuli. Some may be on the edge of mental instability. Those who develop NBDs become part of a crowd that includes physicist Sir Isaac Newton, philosopher Socrates, composer Ludwig von Beethoven, painter Vincent van Gogh, psychiatrist Sigmund Freud, politician Winston Churchill, poets Edgar Allan Poe and Lord Byron, writers James Joyce and Ernest Hemingway, and many others (Figure 12.1).

We have not yet identified all the interactions among genes and environment that might tip such individuals one way or the other. But we do know about several mutations that predispose them to develop NBDs.

Indeed, NBDs tend to run in families. Many of the same families that produce creative geniuses also produce people affected by an NBD—sometimes more than one type. For decades, neuroscientists have been studying these families. They have identified several gene loci that, when mutated, are associated with a spectrum of neurobiological disorders.

With this intriguing connection, we invite you into the study of the chromosomal basis of human inheritance.

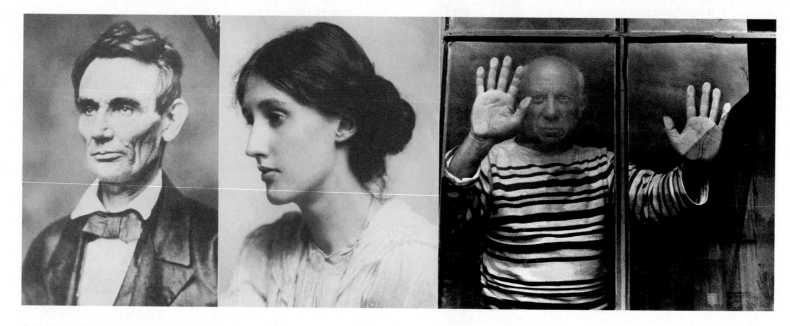

See the video! Figure 12.1 NBDs and creativity. Abraham Lincoln (*left*) suffered from chronic depression even as he changed the course of U.S. history. *Center*, Virginia Woolf's suicide after a mental breakdown is a tragic example of creative writers, who are, as a group, eighteen times more suicidal, ten times more likely to be depressed, and twenty times more likely to have bipolar disorder than the average person. Pablo Picasso (*right*) suffered from depression, and perhaps schizophrenia.

Key Concepts

16 17 18

22 X Y

Autosomes and sex chromosomes

Animals have chromosomes called autosomes. The members of autosome pairs are identical in length, shape, and which genes they carry. In sexually reproducing species, the members of a pair of sex chromosomes differ between females and males. **Section 12.1**

Autosomal inheritance

Many genes on autosomes are expressed in Mendelian patterns of simple dominance. Some dominant or recessive alleles result in genetic disorders. **Sections 12.2, 12.3**

Sex-linked inheritance

Some traits are affected by genes on the X chromosome. Inheritance patterns of such traits differ between males and females. **Section 12.4**

Changes in chromosome structure or number

On rare occasions, a chromosome may undergo a large-scale, permanent change in its structure, or the number of autosomes or sex chromosomes may change. In humans, such changes usually result in a genetic disorder. **Sections 12.5, 12.6**

Human genetic analysis

Various analytical and diagnostic procedures often reveal genetic disorders. What an individual, and society at large, should do with the information raises ethical questions. **Sections 12.7, 12.8**

Links to Earlier Concepts

- In this chapter, you will be drawing upon your knowledge of chromosome structure (Section 9.1), mitosis (9.3), meiosis (10.3, 10.4), and gamete formation (11.3).

- Before you start, make sure you understand dominant and recessive alleles, and homozygous and heterozygous conditions (11.1).

- Sampling error (1.8), carbohydrates (3.3), protein structure (3.6), the cell cortex (4.13), pigments (7.1), glycolysis (8.2), and nutrition (8.7) will all turn up again, this time in the context of different genetic disorders.

Human Chromosomes

■ In humans, two sex chromosomes are the basis of sex. All other human chromosomes are autosomes.

■ Links to Chromosome structure 9.1 and 9.3, Meiosis 10.3

Most animals, including humans, normally are either female or male. They also have a diploid chromosome number (2*n*), with pairs of homologous chromosomes in their body cells. Typically, all except one pair of those chromosomes are **autosomes**, which carry the same genes in both females and males. Autosomes of a pair have the same length, shape, and centromere location. Members of a pair of **sex chromosomes** differ between females and males. The differences determine an individual's sex.

The sex chromosomes of humans are called X and Y. Body cells of human females contain two X chromosomes (XX); those of human males contain one X and one Y chromosome (XY). The X and Y chromosomes differ in length, shape, and which genes they carry, but they interact as homologues during prophase I.

XX females and XY males are the rule among fruit flies, mammals, and many other animals, but there are other patterns. In butterflies, moths, birds, and certain fishes, males have two identical sex chromosomes, not females. Environmental factors (not sex chromosomes) determine sex in some species of invertebrates, turtles, and frogs. As an example, the temperature of the sand in which sea turtle eggs are buried determines the sex of the hatchlings.

Sex Determination

In humans, a new individual inherits a combination of sex chromosomes that dictates whether it will become a male or a female. All eggs made by a human female have one X chromosome. One-half of the sperm cells made by a male carry an X chromosome; the other half carry a Y chromosome. If an X-bearing sperm fertilizes an X-bearing egg, the resulting zygote will develop into a female. If the sperm carries a Y chromosome, the zygote will develop into a male (Figure 12.2*a*).

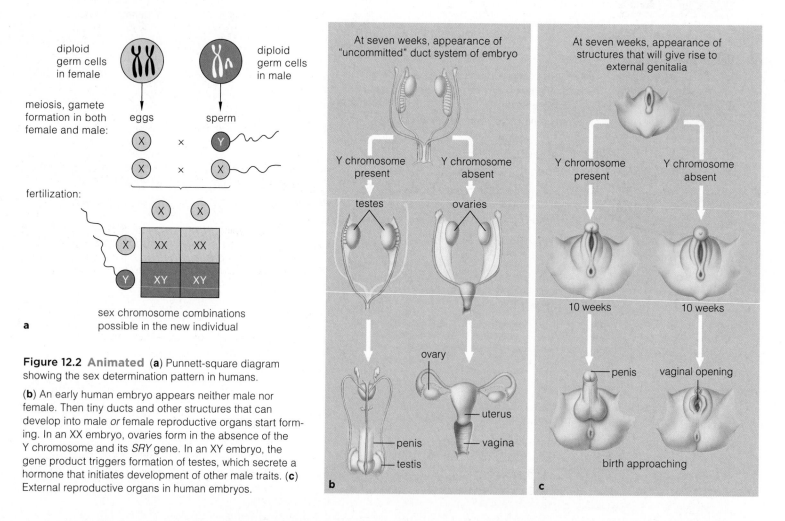

Figure 12.2 Animated (**a**) Punnett-square diagram showing the sex determination pattern in humans.

(**b**) An early human embryo appears neither male nor female. Then tiny ducts and other structures that can develop into male *or* female reproductive organs start forming. In an XX embryo, ovaries form in the absence of the Y chromosome and its *SRY* gene. In an XY embryo, the gene product triggers formation of testes, which secrete a hormone that initiates development of other male traits. (**c**) External reproductive organs in human embryos.

The human Y chromosome carries only 307 genes, but one of them is the *SRY* gene—the master gene for male sex determination. Its expression in XY embryos triggers the formation of testes, which are male gonads (Figure 12.2b). Some of the cells in these primary male reproductive organs make testosterone, a sex hormone that controls the emergence of male secondary sexual traits such as facial hair, increased musculature, and a deep voice. How do we know *SRY* is the male sex master gene? Mutations in this gene cause XY individuals to develop external genitalia that appear female.

An XX embryo has no Y chromosome, no *SRY* gene, and much less testosterone, so primary female reproductive organs (ovaries) form instead of testes. Ovaries make estrogens and other sex hormones that will govern the development of female secondary sexual traits, such as enlarged, functional breasts, and fat deposits around the hips and thighs.

The human X chromosome carries 1,336 genes. Some of those genes are associated with sexual traits, such as the distribution of body fat and hair. However, most of the genes on the X chromosome govern nonsexual traits such as blood clotting and color perception. Such genes are expressed in both males and females. Males, remember, also inherit one X chromosome.

Karyotyping

Sometimes the structure of a chromosome can change during mitosis or meiosis. Chromosome number can change also. A diagnostic tool called karyotyping helps us determine an individual's diploid complement of chromosomes (Figure 12.3). With this procedure, cells taken from an individual are put into a fluid growth medium that stimulates mitosis. The growth medium contains colchicine, a poison that binds tubulin and so interferes with assembly of mitotic spindles. The cells enter mitosis, but the colchicine prevents them from dividing, so their cell cycle stops at metaphase.

The cells and the medium are transferred to a tube. Then, the cells are separated from the liquid medium, and a hypotonic solution is added (Section 5.6). The cells swell up with water, so the chromosomes inside of them move apart. The cells are spread on a microscope slide and stained so the chromosomes become visible with a microscope.

The microscope reveals metaphase chromosomes in every cell. A micrograph of a single cell is digitally rearranged so the images of the chromosomes are lined up by centromere location, and arranged according to size, shape, and length. The finished array constitutes the individual's **karyotype**, which is compared with a

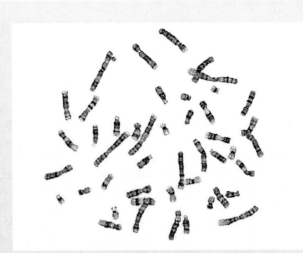

A The chromosomes of one body cell are isolated, then stained to reveal differences in banding patterns.

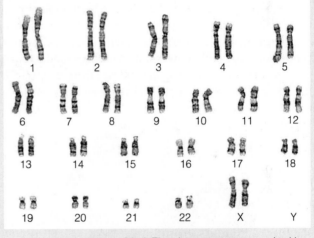

B The image is reassembled: The chromosomes are paired by size, centromere position, and other characteristics.

Figure 12.3 Animated Karyotyping, a diagnostic tool that reveals an image of a single cell's diploid complement of chromosomes. This human karyotype shows 22 pairs of autosomes and a pair of X chromosomes.
Figure It Out: Was this cell taken from a male or female? *Answer: Female*

normal standard. The karyotype shows whether there are extra or missing chromosomes. Some other kinds of structural abnormalities are also visible.

Take-Home Message

What is the basis of sex determination in humans?
- Members of a pair of sex chromosomes differ between males and females.
- Other human chromosomes are autosomes—the same in males and females.

12.2 Examples of Autosomal Inheritance Patterns

■ Many human traits can be traced to autosomal dominant or recessive alleles that are inherited in Mendelian patterns. Some of those alleles cause genetic disorders.

■ Links to Carbohydrates 3.3, Nutrition and aerobic respiration 8.7, Terms in genetics 11.1

Autosomal Dominant Inheritance

A dominant allele on an autosome (an autosomal dominant allele) is expressed in homozygotes and heterozygotes, so any trait it specifies tends to appear in every generation. When one parent is heterozygous, and the other is homozygous for the recessive allele, each of their children has a 50 percent chance of inheriting the dominant allele—and displaying the trait associated with it (Figure 12.4a).

An autosomal dominant allele causes achondroplasia, a genetic disorder that affects about 1 out of 10,000 people. Adult heterozygotes average about four feet, four inches tall, and have abnormally short arms and legs relative to other body parts (Figure 12.4a). While they were still embryos, the cartilage model on which a skeleton is constructed did not form properly. Most homozygotes die before or not long after birth.

A different autosomal dominant allele is responsible for Huntington's disease. With this genetic disorder, involuntary muscle movements increase as the nervous system slowly deteriorates. Typically, symptoms do not start until after age thirty; affected people die during their forties or fifties. The mutation that causes Huntington's alters a protein necessary for brain cell development. It is an expansion mutation, in which three nucleotides are duplicated many times. Hundreds of thousands of other expansion repeats occur harmlessly in and between genes on human chromosomes. This one alters the function of a critical gene product.

A dominant allele that causes severe problems can persist because expression of the allele does not interfere with reproduction. Achondroplasia is an example. With Huntington's and other late-onset disorders, people tend to reproduce before symptoms appear, so the allele may be passed unknowingly to children.

Autosomal Recessive Inheritance

Because autosomal recessive alleles are expressed only in homozygotes, traits associated with them may skip generations. Heterozygotes are carriers: They do not have the trait. Any child of two carriers has a 25 percent chance of inheriting the allele from both parents and being a homozygote with the trait (Figure 12.4b). All children of homozygous parents will be homozygous.

Galactosemia is a heritable metabolic disorder that affects about 1 in every 50,000 newborns. This case of autosomal recessive inheritance involves an allele for an enzyme that helps digest the lactose in milk or in milk products. The body normally converts lactose to glucose and galactose. Then, a series of three enzymes converts the galactose to glucose-6-phosphate (Figure 12.5). This intermediate can enter glycolysis or it can be converted to glycogen (Sections 3.3 and 8.7).

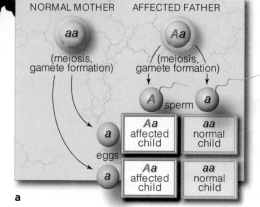
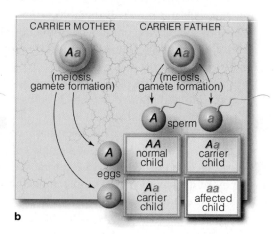

Figure 12.4 Animated (a) Example of autosomal dominant inheritance. A dominant allele (*red*) is fully expressed in heterozygotes. Achondroplasia, an autosomal dominant disorder, affects the three men shown above. At center, Verne Troyer (Mini Me in the Mike Myers spy movies), stands two feet, eight inches tall.

(b) An autosomal recessive pattern. In this example, both parents are heterozygous carriers of the recessive allele (*red*).

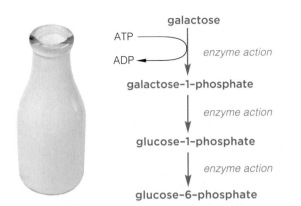

galactose

ATP ——┐
ADP ◄——┘ *enzyme action*

↓

galactose-1-phosphate

enzyme action

↓

glucose-1-phosphate

enzyme action

↓

glucose-6-phosphate

Figure 12.5 How galactose is normally converted to a form that can enter the breakdown reactions of glycolysis. A mutation that affects the second enzyme in the conversion pathway gives rise to galactosemia.

People with galactosemia do not make one of these three enzymes; they are homozygous recessive for a mutated allele. Galactose-1-phosphate accumulates to toxic levels in their body, and it can be detected in the urine. The condition leads to malnutrition, diarrhea, vomiting, and damage to the eyes, liver, and brain.

When they do not receive treatment, galactosemics typically die young. When they are quickly placed on a diet that excludes all dairy products, the symptoms may not be as severe.

What About Neurobiological Disorders?

Most of the neurobiological disorders mentioned in the chapter introduction do not follow simple patterns of Mendelian inheritance. In most cases, mutations in one gene do not give rise to depression, schizophrenia, or bipolar disorder. Multiple genes and environmental factors contribute to these outcomes. Nonetheless, it is useful to search for mutations that make some people more vulnerable to NBDs.

For example, researchers who conducted extensive family and twin studies have predicted that mutations in specific genes on chromosomes 1, 3, 5, 6, 8, 11–15, 18, and 22 increase an individual's chance of developing schizophrenia. Similarly, mutations in specific genes have been linked to bipolar disorder and depression.

Take-Home Message

How do we link traits to alleles on autosomal chromosomes?

■ Many traits, including some genetic disorders, can be traced to dominant or recessive alleles on autosomes because they are inherited in simple Mendelian patterns.

12.3 | Too Young to be Old

■ Progeria, genetic disorder that results in accelerated aging, is caused by mutations in an autosome.

Imagine being ten years old with a mind trapped in a body that is getting a bit more shriveled, more frail—old—every day. You are barely tall enough to peer over the top of a table. You weigh less than thirty-five pounds. Already you are bald and have a wrinkled nose. Possibly you have a few more years to live. Would you, like Mickey Hays and Fransie Geringer, still be able to laugh (Figure 12.6)?

On average, of every 8 million newborn humans, one will grow old far too soon. On one of its autosomes, that rare individual carries a mutated allele that gives rise to Hutchinson–Gilford progeria syndrome. While that new individual was still an embryo inside its mother, billions of DNA replications and mitotic cell divisions distributed the information encoded in that gene to each newly formed body cell. Its legacy will be an accelerated rate of aging and a sharply reduced life span.

The disorder arises by spontaneous mutation of a gene for lamin, a protein that normally makes up intermediate filaments in the nucleus (Section 4.13). The altered lamin is not processed properly. It builds up on the inner nuclear membrane and distorts the nucleus. How this buildup causes the symptoms of progeria is not yet known.

Those symptoms start before age two. Skin that should be plump and resilient starts to thin. Skeletal muscles weaken. Limb bones that should lengthen and grow stronger soften. Premature baldness is inevitable. Affected people do not usually live long enough to reproduce, so progeria does not run in families.

Most progeriacs can expect to die in their early teens as a result of strokes or heart attacks. These final insults are brought on by a hardening of the wall of arteries, a condition typical of advanced age. Fransie was seventeen when he died. Mickey died at age twenty.

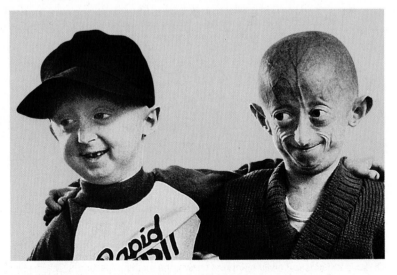

Figure 12.6 Mickey (*left*) and Fransie (*right*) met at a gathering of progeriacs at Disneyland, California. They were not yet ten years old.

12.4 Examples of X-Linked Inheritance Patterns

- X chromosome alleles give rise to phenotypes that reflect Mendelian patterns of inheritance. Some of those alleles cause genetic disorders.

- Links to Cell cortex 4.13, Pigments 7.1

The X chromosome carries over 6 percent of all human genes. Mutations on this sex chromosome are known to cause or contribute to over 300 genetic disorders.

A recessive allele on an X chromosome (an X-linked recessive allele) leaves certain clues when it causes a genetic disorder. First, more males than females are affected by such X-linked recessive disorders. This is because heterozygous females have a second X chromosome that carries the dominant allele, which masks the effects of the recessive one. Heterozygous males have only one X chromosome, so they are not similarly protected (Figure 12.7). Second, an affected father cannot pass his X-linked recessive allele to a son because all children who inherit their father's X chromosome are female. Thus, a heterozygous female is the bridge between an affected male and his affected grandson.

X-linked dominant alleles that cause disorders are rarer than X-linked recessive ones, probably because they tend to be lethal in male embryos. Females, with two X chromosomes, most often have one functional allele that can dampen the effects of a mutated one.

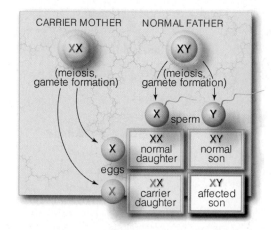

Figure 12.7 Animated X-linked recessive inheritance. In this case, the mother carries a recessive allele on one of her X chromosomes (*red*).

Hemophilia A

Hemophilia A is an X-linked recessive disorder that interferes with blood clotting. Most of us have a blood clotting mechanism that quickly stops bleeding from minor injuries. The mechanism involves protein products of genes on the X chromosome. Bleeding is prolonged in males who carry a mutated form of one of

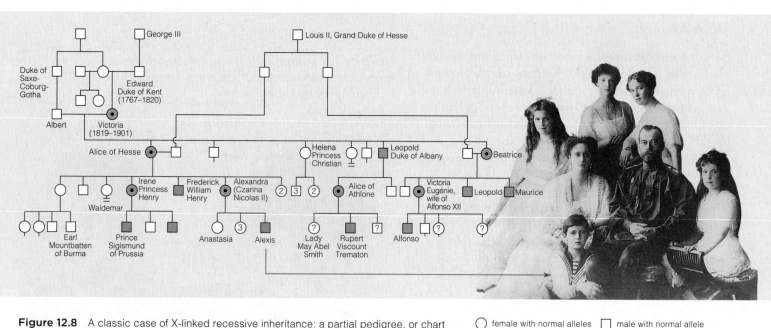

Figure 12.8 A classic case of X-linked recessive inheritance: a partial pedigree, or chart of genetic connections, of the descendants of Queen Victoria of England. At one time, the recessive X-linked allele that resulted in hemophilia was present in eighteen of Victoria's sixty-nine descendants, who sometimes intermarried.

Of the Russian royal family members shown, the mother (Alexandra Czarina Nicolas II) was a carrier. Through her obsession with the vulnerability of her son Alexis, a hemophiliac, she became involved in political intrigue that helped trigger the Russian Revolution of 1917.

- ◯ female with normal alleles
- ◉ female carrier
- ③ three females
- ☐ male with normal allele
- ◼ affected male
- ?/? status unknown

Figure It Out: How many of Alexis' siblings were affected by hemophilia A? Answer: None

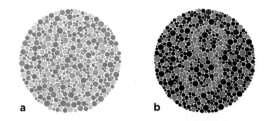

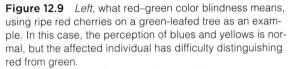

Figure 12.9 *Left,* what red–green color blindness means, using ripe red cherries on a green-leafed tree as an example. In this case, the perception of blues and yellows is normal, but the affected individual has difficulty distinguishing red from green.

Above, two of many Ishihara plates, which are standardized tests for different forms of color blindness. (**a**) You may have one form of red–green color blindness if you see the numeral 7 instead of 29 in this circle. (**b**) You may have another form if you see a 3 instead of an 8.

these X-linked genes, or in females who are homozygous for a mutation (clotting time is close to normal in heterozygous females). Affected people tend to bruise easily, and internal bleeding causes problems in their muscles and joints.

In the nineteenth century, the incidence of hemophilia A was relatively high in royal families of Europe and Russia, probably because the common practice of inbreeding kept the harmful allele circulating in their family trees (Figure 12.8). Today, about 1 in 7,500 people is affected, but that number may be rising because the disorder is now treatable. More affected people are living long enough to transmit the allele to children.

Red–Green Color Blindness

The pattern of X-linked recessive inheritance shows up among individuals who have some degree of color blindness. The term refers to a range of conditions in which an individual cannot distinguish among some or all colors in the spectrum of visible light. Mutated genes result in altered function of the photoreceptors (light-sensitive receptors) in the eyes.

Normally, humans can sense the differences among 150 colors. A person who is red–green color blind sees fewer than 25 colors: Some or all of the receptors that respond to red and green wavelengths are weakened or absent. Some people confuse red and green colors. Others see green as shades of gray, but perceive blues and yellows quite well (Figure 12.9). Two sections of a standard set of tests for color blindness are shown in Figure 12.9*a,b.*

Duchenne Muscular Dystrophy

Duchenne muscular dystrophy (DMD) is one of several X-linked recessive disorders that is characterized by muscle degeneration. DMD affects about 1 in 3,500 people, almost all of them boys.

A gene on the X chromosome encodes dystrophin, which is a protein that structurally supports the fused cells in muscle fibers by anchoring the cell cortex to the plasma membrane. When dystrophin is abnormal or absent, the cell cortex weakens and muscle cells die. The cell debris that remains in the tissues triggers chronic inflammation.

DMD is typically diagnosed in boys between the ages of three and seven. The rapid progression of this disorder cannot be stopped. When an affected boy is about twelve years old, he will begin to use a wheelchair. His heart muscles will start to break down. Even with the best of care, he will probably die before he is thirty years old, from a heart disorder or from respiratory failure (suffocation).

Take-Home Message

How do we link traits to alleles on sex chromosomes?

■ Mutated alleles on sex chromosomes cause or contribute to more than 300 genetic disorders. They are inherited in characteristic patterns.

■ A female heterozygous for one of those alleles may not show symptoms.

■ Males (XY) transmit an X-linked allele only to daughters, not to their sons.

12.5 Heritable Changes in Chromosome Structure

- On rare occasions, a chromosome's structure changes. Many of the alterations have severe or lethal outcomes.
- Links to Protein structure 3.6, Meiosis 10.3

Large-scale changes in the structure of a chromosome may give rise to a genetic disorder. Such changes are rare, but they do occur spontaneously in nature. They can also be induced by exposure to certain chemicals or radiation. Either way, such alterations may be detected by karyotyping. Large-scale changes in chromosome structure include duplications, deletions, inversions, and translocations.

Duplication Even normal chromosomes have DNA sequences that are repeated two or more times. These repetitions are called **duplications**:

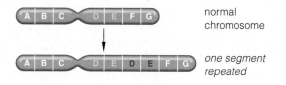

normal chromosome

one segment repeated

Duplications can occur through unequal crossovers at prophase I. Homologous chromosomes align side by side, but their DNA sequences misalign at some point along their length. The probability of misalignment is greater in regions where DNA has long repeats of the same sequence of nucleotides. A stretch of DNA gets deleted from one chromosome and is spliced into the partner chromosome. Some duplications, such as the expansion mutations that cause Huntington's, cause genetic abnormalities or disorders. Others have been evolutionarily important.

Deletion A **deletion** is the loss of some portion of a chromosome:

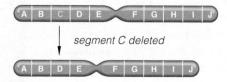

segment C deleted

In mammals, deletions usually cause serious disorders and are often lethal. The loss of genes results in the disruption of growth, development, and metabolism. For instance, a small deletion in chromosome 5 causes mental impairment and an abnormally shaped larynx. Affected infants tend to make a sound like the meow of a cat, hence the name of the disorder, cri-du-chat, which is French for "cat's cry" (Figure 12.10).

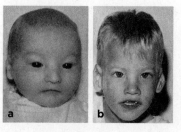

Figure 12.10 Cri-du-chat syndrome. (**a**) This infant's ears are low relative to his eyes. (**b**) Same boy, four years later. The high-pitched monotone of cri-du-chat children may persist into their adulthood.

Inversion With an **inversion**, part of the sequence of DNA within the chromosome becomes oriented in the reverse direction, with no molecular loss:

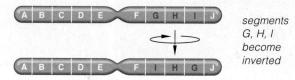

segments G, H, I become inverted

An inversion may not affect a carrier's health if it does not disrupt a gene region. However, it may affect an individual's fertility. Crossovers in an inverted region during meiosis may result in deletions or duplications that affect the viability of forthcoming embryos. Some carriers do not know that they have an inversion until they are diagnosed with infertility and their karyotype is tested.

Translocation If a chromosome breaks, the broken part may get attached to a different chromosome, or to a different part of the same one. This structural change is a **translocation**. Most translocations are reciprocal, or balanced; two chromosomes exchange broken parts:

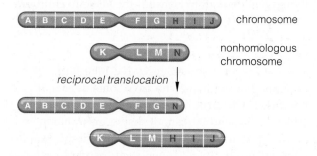

chromosome

nonhomologous chromosome

reciprocal translocation

A reciprocal translocation that does not disrupt a gene may have no adverse effect on its bearer. Many people do not even realize they carry a translocation until they have difficulty with fertility. The two translocated chromosomes pair abnormally with their non-translocated counterparts during meiosis. They segregate improperly about half of the time, so about half of the resulting gametes will carry major duplications or deletions. If one of these gametes unites with a normal gamete at fertilization, the resulting embryo almost always dies.

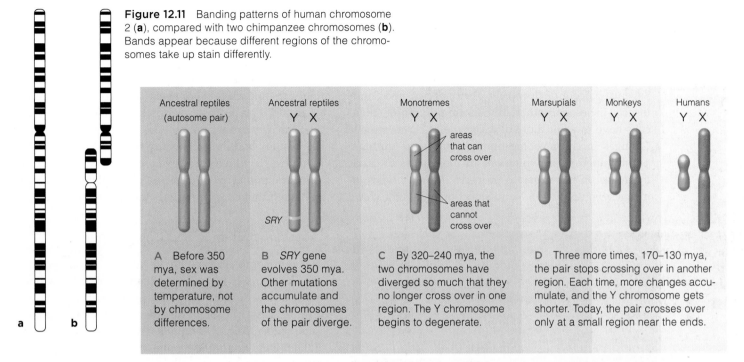

Figure 12.11 Banding patterns of human chromosome 2 (**a**), compared with two chimpanzee chromosomes (**b**). Bands appear because different regions of the chromosomes take up stain differently.

Ancestral reptiles (autosome pair)	Ancestral reptiles Y X	Monotremes Y X	Marsupials Y X	Monkeys Y X	Humans Y X

SRY

areas that can cross over

areas that cannot cross over

A Before 350 mya, sex was determined by temperature, not by chromosome differences.

B *SRY* gene evolves 350 mya. Other mutations accumulate and the chromosomes of the pair diverge.

C By 320–240 mya, the two chromosomes have diverged so much that they no longer cross over in one region. The Y chromosome begins to degenerate.

D Three more times, 170–130 mya, the pair stops crossing over in another region. Each time, more changes accumulate, and the Y chromosome gets shorter. Today, the pair crosses over only at a small region near the ends.

Figure 12.12 Evolution of the Y chromosome. Mya stands for million years ago.

Does Chromosome Structure Evolve?

As you see, alterations in chromosome structure may reduce fertility; individuals who are heterozygous for multiple changes may not be able to produce offspring at all. However, accumulation of multiple alterations in homozygous individuals can be the start of a new species. It may seem as if this outcome would be rare, but it can and does occur over generations in nature. Karyotyping studies show that structural alterations have been built into the DNA of nearly all species.

For example, certain duplications have allowed one copy of a gene to mutate while a different copy carries out its original function. The multiple and strikingly similar globin chain genes of humans and other primates apparently evolved by this process. Four globin chains associate in each hemoglobin molecule (Section 3.6). Which version of the globin chains participate in the association influences the oxygen-binding behavior of the resulting protein.

Some chromosome structure alterations contributed to differences among closely related organisms, such as apes and humans. Body cells of humans have twenty-three pairs of chromosomes, but those of chimpanzees, gorillas, and orangutans have twenty-four. Thirteen human chromosomes are almost identical with chimpanzee chromosomes. Nine more are similar, except for some inversions. One human chromosome matches up with two in chimpanzees and the other great apes

(Figure 12.11). During human evolution, two chromosomes fused end to end and formed our chromosome 2. How do we know? The fused region contains remnants of a telomere, which is a special DNA sequence that caps the ends of chromosomes.

As another example, X and Y chromosomes were once homologous autosomes in reptilelike ancestors of mammals. In those organisms, ambient temperature probably determined sex, as it still does in turtles and some other modern reptiles. Then, about 350 million years ago, one of the two chromosomes underwent a structural change that interfered with crossing over in meiosis, and the two homologues diverged over evolutionary time. Eventually, the chromosomes became so different that they no longer crossed over at all in the changed region, which by that time held the *SRY* gene on the Y chromosome. One of the genes on the modern X chromosome is similar to *SRY*; it probably evolved from the same ancestral gene (Figure 12.12).

Take-Home Message

Does chromosome structure change?

■ A segment of a chromosome may be duplicated, deleted, inverted, or translocated. Such a change is usually harmful or lethal, but may be conserved in the rare circumstance that it has a neutral or beneficial effect.

12.6 Heritable Changes in the Chromosome Number

■ Occasionally, new individuals end up with the wrong chromosome number. Consequences range from minor to lethal.

■ Links to Sampling error 1.8, Mitosis 9.3, Meiosis 10.3

Seventy percent of flowering plant species, and some insects, fishes, and other animals are **polyploid**—their cells have three or more of each type of chromosome. Changes in chromosome number may arise through **nondisjunction**, in which one or more pairs of chromosomes do not separate properly during mitosis or meiosis (Figure 12.13). Nondisjunction affects the chromosome number at fertilization. For example, suppose that a normal gamete fuses with an $n+1$ gamete (one that has an extra chromosome). The new individual will be trisomic ($2n+1$), having three of one type of chromosome and two of every other type. As another example, if an $n-1$ gamete fuses with a normal n gamete, the new individual will be $2n-1$, or monosomic.

Trisomy and monosomy are types of **aneuploidy**, a condition in which cells have too many or too few copies of a chromosome. Autosomal aneuploidy is usually fatal in humans, and it causes many miscarriages.

Autosomal Change and Down Syndrome

A few trisomic humans are born alive, but only those that have trisomy 21 will reach adulthood. A newborn with three chromosomes 21 will develop Down syndrome. This autosomal disorder is the most common type of aneuploidy in humans; it occurs once in 800 to 1,000 births and affects more than 350,000 people in the United States alone. Figure 12.13*a* shows a karyotype for a trisomic 21 female. The affected individuals have upward-slanting eyes, a fold of skin that starts at the inner corner of each eye, a deep crease across the sole of each palm and foot, one (instead of two) horizontal furrows on their fifth fingers, slightly flattened facial features, and other symptoms.

Not all of the outward symptoms develop in every individual. That said, trisomic 21 individuals tend to have moderate to severe mental impairment and heart problems. Their skeleton grows and develops abnormally, so older children have short body parts, loose joints, and misaligned bones of the fingers, toes, and hips. The muscles and reflexes are weak, and motor skills such as speech develop slowly. With medical care, trisomy 21 individuals live about fifty-five years. Early training can help affected individuals learn to care for themselves and to take part in normal activities. As a group, they tend to be cheerful.

The incidence of nondisjunction generally rises with the increasing age of the mother (Figure 12.14). Nondisjunction may occur in the father, although far less frequently. Trisomy 21 is one of the hundreds of conditions that can be detected easily through prenatal diagnosis (Section 12.8).

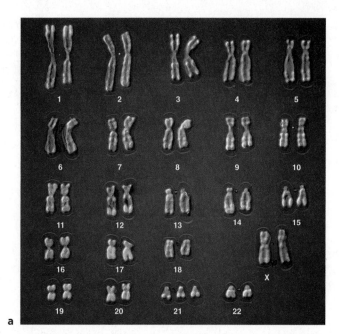

a

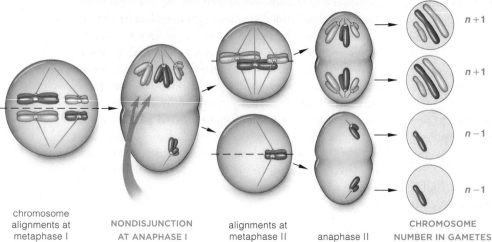

chromosome alignments at metaphase I

NONDISJUNCTION AT ANAPHASE I

alignments at metaphase II

anaphase II

CHROMOSOME NUMBER IN GAMETES

$n+1$

$n+1$

$n-1$

$n-1$

b

Figure 12.13 (**a**) A case of nondisjunction. This karyotype reveals the trisomic 21 condition of a human female. (**b**) One example of how nondisjunction arises. Of the two pairs of homologous chromosomes shown here, one fails to separate during anaphase I of meiosis. The chromosome number is altered in the gametes that form after meiosis.

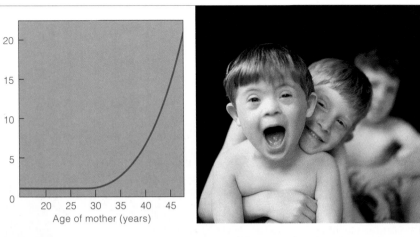

Figure 12.14 Relationship between the frequency of Down syndrome and mother's age at childbirth. The data are from a study of 1,119 affected children. The risk of having a trisomic 21 baby rises with the mother's age. About 80 percent of trisomic 21 individuals are born to women under thirty-five, but these women have the highest fertility rates, and they have more babies.

Figure 12.15
A 6-year-old with Turner's syndrome. Affected girls tend to be shorter than average, but daily hormone injections can help them reach normal height.

Change in the Sex Chromosome Number

Nondisjunction also causes alterations in the number of X and Y chromosomes, with a frequency of about 1 in 400 live births. Most often, such alterations lead to difficulties in learning and impaired motor skills such as a speech delay, but problems may be so subtle that the underlying cause is never diagnosed.

Female Sex Chromosome Abnormalities Individuals with Turner syndrome have an X chromosome and no corresponding X or Y chromosome (XO). The condition occurs about 75 percent of the time because of nondisjunction originating with the father. About 1 in 2,500 to 10,000 newborn girls are XO (Figure 12.15). At least 98 percent of XO embryos will spontaneously abort early in pregnancy, so there are fewer cases compared with other sex chromosome abnormalities.

XO individuals are not as disadvantaged as other aneuploids. They grow up well proportioned but short (with an average height of four feet, eight inches). Most do not have functional ovaries, so they do not make enough sex hormones to become sexually mature. The development of secondary sexual traits such as breasts is also affected.

A female may inherit three, four, or five X chromosomes. The resulting XXX syndrome occurs in about 1 of 1,000 births. Only one X chromosome is typically active in female cells, so having extra X chromosomes usually does not result in physical or medical problems.

Male Sex Chromosome Abnormalities About 1 out of every 500 males has an extra X chromosome (XXY). Most cases are an outcome of nondisjunction during meiosis. The resulting disorder, Klinefelter syndrome, develops at puberty. XXY males tend to be overweight,

tall, and within a normal range of intelligence. They make more estrogen and less testosterone than normal males, and this hormone imbalance has feminizing effects. Affected men tend to have small testes and prostate glands, low sperm counts, sparse facial and body hair, high-pitched voices, and enlarged breasts. Testosterone injections during puberty can reverse these feminized traits.

About 1 in 500 to 1,000 males has an extra Y chromosome (XYY). Adults tend to be taller than average and have mild mental impairment, but most are otherwise normal. XYY men were once thought to be predisposed to a life of crime. This misguided view was based on sampling error (too few cases in narrowly chosen groups such as prison inmates) and bias (the researchers who gathered the karyotypes also took the personal histories of the participants).

In 1976 a Danish geneticist reported results from his study of 4,139 tall males, all twenty-six years old, who had registered at their draft board. Besides their data from physical examinations and intelligence tests, the draft records offered clues to social and economic status, education, and any criminal convictions. Only twelve of the males studied were XYY, which meant that the "control group" had more than 4,000 males. The only findings? Mentally impaired, tall males who engage in criminal deeds are just more likely to get caught—irrespective of karyotype.

Take-Home Message

What are the effects of changes in the chromosome number?

■ Nondisjunction can change the number of autosomes or sex chromosomes in gametes. Such changes usually cause genetic disorders in offspring.

■ Sex chromosome abnormalities are usually associated with learning difficulties, speech delays, and motor skill impairment.

Human Genetic Analysis

- Charting genetic connections with pedigrees reveals inheritance patterns for certain alleles.

- Links to Sampling error 1.8, Meiosis 10.4

Some organisms, including pea plants and fruit flies, are ideal for genetic analysis. They have few chromosomes, and reproduce fast in small spaces under controlled conditions. It does not take long to track a trait through many generations.

Humans, however, are a different story. Unlike flies grown in laboratories, we humans live under variable conditions, in different places, and we live as long as the geneticists who study us. Most of us select our own mates and reproduce if and when we want to. Most human families are not large, which means that there are typically not enough offspring to clarify any inheritance patterns.

Thus, to minimize sampling error (Section 1.8), geneticists gather information from multiple generations. If a trait follows a simple Mendelian inheritance pattern, geneticists can predict the probability of its recurrence

Figure 12.16 An intriguing pattern of inheritance. Eight percent of the men in central Asia carry nearly identical Y chromosomes, which implies descent from a shared ancestor. If so, then 16 million males living between northeastern China and Afghanistan—close to 1 of every 200 men alive today—may be part of a lineage that started with the warrior and notorious womanizer Genghis Khan.

in future generations. Some inheritance patterns are clues to past events (Figure 12.16). Those who analyze pedigrees use their knowledge of probability and patterns of Mendelian inheritance. Such researchers have traced many genetic abnormalities and disorders to a dominant or recessive allele and often to its location on an autosome or a sex chromosome (Table 12.1).

Table 12.1 Examples of Human Genetic Disorders and Genetic Abnormalities

Disorder or Abnormality	Main Symptoms	Disorder or Abnormality	Main Symptoms
Autosomal Recessive Inheritance		**X-Linked Recessive Inheritance**	
Albinism	Absence of pigmentation	Androgen insensitivity syndrome	XY individual but having some female traits; sterility
Hereditary methemoglobinemia	Blue skin coloration		
Cystic fibrosis	Abnormal glandular secretions leading to tissue, organ damage	Red–green color blindness	Inability to distinguish among some or all shades of red and green
Ellis–van Creveld syndrome	Dwarfism, heart defects, polydactyly	Fragile X syndrome	Mental impairment
		Hemophilia	Impaired blood clotting ability
Fanconi anemia	Physical abnormalities, bone marrow failure	Muscular dystrophies	Progressive loss of muscle function
Galactosemia	Brain, liver, eye damage	X-linked anhidrotic dysplasia	Mosaic skin (patches with or without sweat glands); other effects
Phenylketonuria (PKU)	Mental impairment		
Sickle-cell anemia	Adverse pleiotropic effects on organs throughout body	**Changes in Chromosome Structure**	
		Chronic myelogenous leukemia (CML)	Overproduction of white blood cells in bone marrow; organ malfunctions
Autosomal Dominant Inheritance		Cri-du-chat syndrome	Mental impairment; abnormally shaped larynx
Achondroplasia	One form of dwarfism		
Camptodactyly	Rigid, bent fingers		
Familial hypercholesterolemia	High cholesterol levels in blood; eventually clogged arteries	**Changes in Chromosome Number**	
Huntington's disease	Nervous system degenerates progressively, irreversibly	Down syndrome	Mental impairment; heart defects
		Turner syndrome (XO)	Sterility; abnormal ovaries, abnormal sexual traits
Marfan syndrome	Abnormal or no connective tissue		
Polydactyly	Extra fingers, toes, or both	Klinefelter syndrome	Sterility; mild mental impairment
Progeria	Drastic premature aging	XXX syndrome	Minimal abnormalities
Neurofibromatosis	Tumors of nervous system, skin	XYY condition	Mild mental impairment or no effect

Figure 12.17
A pedigree for Huntington's disease, a progressive degeneration of the nervous system.

Researcher Nancy Wexler and her team constructed this extended family tree for nearly 10,000 Venezuelans. Their analysis of unaffected and affected individuals revealed that a dominant allele on human chromosome 4 is the culprit.

Wexler has a special interest in the disorder; it runs in her family.

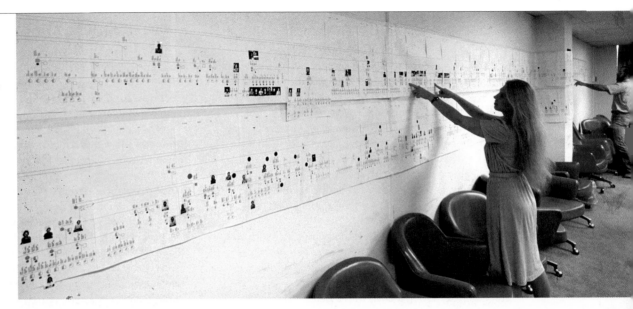

Inheritance patterns are often displayed as standardized charts of genetic connections called **pedigrees**. You already came across one pedigree in Figure 12.8. Figure 12.17 shows another.

What do we do with genetic information? The next section explores some of the options. When considering them, keep in mind some distinctions. First, a genetic abnormality is defined as a rare or uncommon version of a trait, such as when a person is born with six digits on each hand or foot instead of the usual five (Figure 12.18). Such abnormalities are not inherently life threatening, and how you view them is a matter of opinion. By contrast, a genetic disorder is an inherited condition that sooner or later causes mild to severe medical problems. A genetic disorder is characterized by a specific set of symptoms—a **syndrome**. By contrast, the term "disease" is usually reserved for an illness caused by infection or environmental factors.

One more point: Alleles that give rise to severe genetic disorders are generally rare in populations, because they put their bearers at risk. Why don't they disappear entirely? Rare mutations can reintroduce them. In heterozygotes, a normal allele may mask the effects of expression of a harmful recessive allele. Or, heterozygotes with a codominant allele may have an advantage in a particular environment. You will see an example of the latter case in Section 18.6.

Take-Home Message

How and why do we determine inheritance patterns in humans?

■ Geneticists track traits through a family tree using pedigrees.

■ Pedigrees reveal inheritance patterns for certain alleles. Such patterns may be used to determine the probability that future offspring will be affected by a genetic abnormality or disorder.

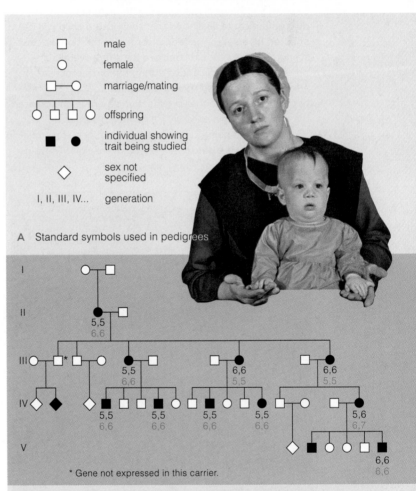

A Standard symbols used in pedigrees

□ male
○ female
□—○ marriage/mating
○-□-□-○ offspring
■ ● individual showing trait being studied
◇ sex not specified
I, II, III, IV... generation

* Gene not expressed in this carrier.

B A pedigree for *polydactyly*, which is characterized by extra fingers, toes, or both. The *black* numbers signify the number of fingers on each hand; the *blue* numbers signify the number of toes on each foot. Though it occurs on its own, polydactyly is also one of several symptoms of Ellis–van Creveld syndrome.

Figure 12.18 Animated Pedigrees.

12.8 Prospects in Human Genetics

- Genetic analyses can provide prospective parents with information about their future children.

With the first news of pregnancy, parents-to-be typically wonder if their baby will be normal. Quite naturally, they want their baby to be free of genetic disorders, and most babies are healthy. Many prospective parents have difficulty coming to terms with the possibility that a child of theirs might develop a severe genetic disorder, but sometimes that happens. What are their options?

Genetic Counseling Genetic counseling starts with diagnosis of parental genotypes, pedigrees, and genetic testing for known disorders. Using information gained from these tests, genetic counselors can predict a couple's probability of having a child with a genetic disorder.

Parents-to-be commonly ask genetic counselors to compare the risks associated with diagnostic procedures against the likelihood that their future child will be affected by a severe genetic disorder. At the time of counseling, they also should compare the small overall risk (3 percent) that complications during the birth process can affect any child. They should talk about how old they are, because the older either prospective parent is, the greater the risk of having a child with a genetic disorder.

As a case in point, suppose a first child or a close relative has a severe disorder. A genetic counselor will evalu-
ate the pedigrees of the parents, and the results of any genetic tests. Using this information, counselors can predict risks for disorders in future children. The same risk will apply to each pregnancy.

Prenatal Diagnosis Doctors and clinicians commonly use methods of prenatal diagnosis to determine the sex of embryos or fetuses and to screen for more than 100 known genetic problems. Prenatal means before birth. Embryo is a term that applies until eight weeks after fertilization, after which fetus is appropriate.

Suppose a forty-five-year-old woman becomes pregnant and worries about Down syndrome. Between fifteen and twenty weeks after conception, she might opt for amniocentesis (Figure 12.19). In this diagnostic procedure, a physician uses a syringe to withdraw a small sample of fluid from the amniotic cavity. The "cavity" is a fluid-filled sac, enclosed by a membrane—the amnion—that in turn encloses the fetus. The fetus normally sheds some cells into the fluid. Cells suspended in the fluid sample can be analyzed for many genetic disorders, including Down syndrome, cystic fibrosis, and sickle-cell anemia.

Chorionic villi sampling (CVS) is a diagnostic procedure similar to amniocentesis. A physician withdraws a few cells from the chorion, which is a membrane that surrounds the amnion and helps form the placenta. The placenta is an organ that keeps the blood of mother and embryo separate, while allowing substances to be exchanged between them. Unlike amniocentesis, CVS can be performed as early as eight weeks into pregnancy.

It is now possible to see a live, developing fetus with fetoscopy. The procedure uses an endoscope, a fiber-optic device, to directly visualize and photograph the fetus, umbilical cord, and placenta with high resolution (Figure 12.20a). Characteristic physical effects of certain genetic abnormalities or disorders can be diagnosed, and sometimes corrected, by fetoscopy.

There are risks to a fetus associated with all three procedures, including punctures or infections. If the amnion does not reseal itself quickly, too much fluid may leak out of the amniotic cavity. Amniocentesis increases the risk of miscarriage by 1 to 2 percent. CVS occasionally disrupts the placenta's development and thus causes underdeveloped or missing fingers and toes in 0.3 percent of newborns. Fetoscopy raises the miscarriage risk by 2 to 10 percent.

Preimplantation Diagnosis Preimplantation diagnosis is a procedure associated with *in vitro* fertilization. Sperm and eggs from prospective parents are mixed in a sterile culture medium. One or more eggs may become fertilized. Then, mitotic cell divisions can turn the fertilized egg into a ball of eight cells within forty-eight hours (Figure 12.20b).

According to one view, the tiny, free-floating ball is a pre-pregnancy stage. Like all of the unfertilized eggs that a woman's body discards monthly during her reproductive years, it has not attached to the uterus. All of its cells have

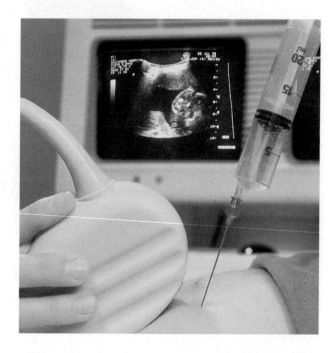

Figure 12.19 Animated Amniocentesis. A pregnant woman's doctor draws a sample of amniotic fluid into a syringe. The path of the needle is monitored by an ultrasound device. About 20 milliliters of fluid is withdrawn from the amniotic sac that holds the developing fetus. Amniotic fluid contains fetal cells and wastes that can be analyzed for genetic disorders.

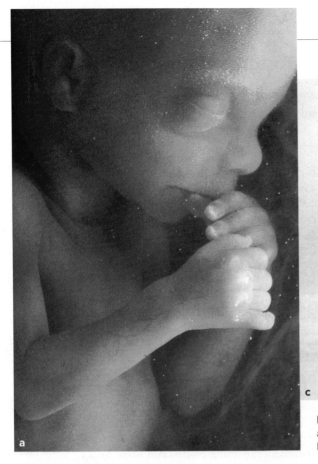

Figure 12.20 Stages of human development. (**a**) Fetoscopy reveals a fetus in high resolution. (**b**) Eight-celled and (**c**) multicelled stages of human development.

the same genes, but they are not yet committed to being specialized one way or another. Doctors can remove one of these undifferentiated cells and analyze its genes. If it has no detectable genetic defects, the ball is inserted into the mother's uterus. The withdrawn cell will not be missed, and the ball may go on to develop into an embryo. Many of the resulting "test-tube babies" are born in good health. Some couples who are at risk of passing on the alleles for cystic fibrosis, muscular dystrophy, or some other genetic disorder have opted for this procedure.

Regarding Abortion What happens after prenatal diagnosis reveals a severe problem? Some prospective parents opt for an induced abortion. An *abortion* is an induced expulsion of a pre-term embryo or fetus from the uterus. We can only say here that individuals must weigh their awareness of the severity of the genetic disorder against their ethical and religious beliefs. We return to methods of inducing abortion in Section 42.9, in the context of human embryonic development.

Phenotypic Treatments Surgery, prescription drugs, hormone replacement therapy, and often dietary controls can minimize and in some cases eliminate the symptoms of many genetic disorders. For instance, strict dietary controls work in cases of phenylketonuria, or PKU. Individuals affected by this genetic disorder are homozygous for a recessive allele on an autosome. They cannot make a functional form of an enzyme that catalyzes the conversion

of one amino acid (phenylalanine) to another (tyrosine). Because the conversion is blocked, phenylalanine accumulates and is diverted into other metabolic pathways. The outcome is an impairment of brain function.

Affected people who restrict phenylalanine intake can lead essentially normal lives. They must avoid diet soft drinks and other products that are sweetened with aspartame, a compound that contains phenylalanine.

Genetic Screening Genetic screening is the widespread, routine testing for alleles associated with genetic disorders. It provides information on reproductive risks, and helps families that are already affected by a genetic disorder. If a genetic disorder is detected early enough, phenotypic treatments may minimize the damage it causes in some cases.

Hospitals routinely screen newborns for certain genetic disorders. For example, most newborns in the United States are routinely tested for PKU. Affected infants receive early treatment, so we now see fewer individuals with symptoms of the disorder. Besides helping individuals, the information from genetic screening can help us estimate the prevalence and distribution of harmful alleles in populations.

There are social risks that must be considered. How would you feel if you were labeled as someone that carries a "bad" allele? Would the knowledge invite anxiety? If you become a parent even though you know you have a "bad" allele, how would you feel if your child ends up affected by a genetic disorder? There are no easy answers here.

Mutations that affect any of the steps in crucial metabolic pathways could impair brain chemistry, which in turn may result in NBDs. People with bipolar disorder or schizophrenia (such as John Nash, pictured at *right*) have altered gene expression, particularly in certain regions of the brain. Current research suggests that cells of people with these NBDs make too many or too few of the enzymes that carry out electron transfer phosphorylation. Remember, this stage of aero-

How would you vote?

Do you support legislation governing genetic testing that would identify individuals at risk for developing NBDs? See CengageNOW for details, then vote online.

bic respiration yields most of the body's ATP. In the brain, the disruption of electron transfer phosphorylation may alter cells in ways that boost creativity—but also invite illness.

Summary

Section 12.1 A human body cell has twenty-three pairs of homologous chromosomes. One is a pair of **sex chromosomes**. All of the others are pairs of **autosomes**. In both sexes, the two autosomes of each pair have the same length, shape, and centromere location, and they carry the same genes along their length.

Human females have identical sex chromosomes (XX) and males have nonidentical ones (XY). The *SRY* gene on the Y chromosome is the basis of male sex determination. Its expression causes an embryo to develop testes, which secrete testosterone. This hormone controls the development of male secondary sexual traits. An embryo with no Y chromosome (no *SRY* gene) develops into a female.

Karyotyping is a diagnostic tool that reveals missing or extra chromosomes, and some structural changes, in an individual's chromosomes. With this technique, a person's metaphase chromosomes are prepared for microscopy and then imaged. Images of the chromosomes are arranged by their defining features as a **karyotype**.

- *Use the interaction on CengageNOW to see how sex is determined in humans.*

- *Use the animation on CengageNOW to learn how to create a karyotype.*

Sections 12.2, 12.3 Certain dominant or recessive alleles on autosomes are associated with genetic abnormalities or genetic disorders.

- *Use the interaction on CengageNOW to investigate autosomal inheritance.*

Section 12.4 Certain dominant and recessive alleles on the X chromosome are inherited in Mendelian patterns. Mutated alleles on the X chromosome contribute to more than 300 known genetic disorders. Males cannot transmit a recessive X-linked allele to their sons; a female passes such alleles to male offspring.

- *Use the interaction on CengageNOW to investigate X-linked inheritance.*

Section 12.5 Rarely, a chromosome's structure becomes altered when part of it undergoes **duplication**, **deletion**, **inversion**, or **translocation**. Most alterations are harmful or lethal. Even so, many have accumulated in the chromosomes of all species over evolutionary time.

Section 12.6 The chromosome number of a cell can change permanently. Most often, such a change is an outcome of **nondisjunction**, which is the failure of one or more pairs of duplicated chromosomes to separate during meiosis. In **aneuploidy**, cells have too many or too few copies of a chromosome. In humans, the most common aneuploidy, trisomy 21, causes Down syndrome. Most other human autosomal aneuploids die before birth.

Polyploid individuals inherit three or more of each type of chromosome from their parents. About 70 percent of all flowering plants, and some insects, fishes, and other animals, are polyploid.

A change in the number of sex chromosomes usually results in impaired learning and motor skills. Problems can be so subtle that the underlying cause may not ever be diagnosed, as among XXY, XXX, and XYY children.

Sections 12.7, 12.8 A genetic abnormality is an uncommon version of a heritable trait that does not result in medical problems. A genetic disorder is a heritable condition that results a **syndrome** of mild or severe medical problems. Geneticists construct **pedigrees** to estimate the chance that a couple's offspring will inherit a genetic abnormality or disorder. Potential parents who may be at risk of transmitting a harmful allele to offspring have screening or treatment options.

- *Use the animations on CengageNOW to examine a human pedigree, and to explore amniocentesis.*

Self-Quiz
Answers in Appendix III

1. The _____ of chromosomes in a cell are compared to construct karyotypes.
   - a. length and shape
   - b. centromere location
   - c. gene sequence
   - d. both a and b

2. The _____ determines sex in humans.
   - a. X chromosome
   - b. *Dll* gene
   - c. *SRY* gene
   - d. both a and c

3. If one parent is heterozygous for a dominant autosomal allele and the other parent does not carry the allele, a child of theirs has a _____ chance of being heterozygous.
   - a. 25 percent
   - b. 50 percent
   - c. 75 percent
   - d. no chance; it will die

Data Analysis Exercise

A study in 1989 looked for a genetic relationship between mood disorders and intelligence. William Coryell and his colleagues found people with bipolar disorder, then tallied how many of their immediate family members had graduated from college. Some of the results are shown in Figure 12.21.

1. Which group of people, those with bipolar disorder or those without bipolar disorder, had the largest percentage of fathers with a college degree?

2. According to this data, if you had bipolar disorder, which one of your immediate relatives would be more likely to have a college degree? Which would be the least likely?

3. Are relatives of people with or without bipolar disorder more likely to graduate from college?

Proportion of College Graduates Among Relatives of People with Bipolar Disorder		
Relative	Nonbipolar (%)	Bipolar (%)
Father	16.1	27.3
Mother	10.5	18.2
Brother	23.9	38.7
Sister	16.1	31.8
Grandfather	7.6	13.0
Grandmother	5.1	7.1

Figure 12.21 Proportion of immediate relatives of nonbipolar and bipolar people that graduated from college.

4. Expansion mutations occur _____ within and between genes in human chromosomes.
   a. only rarely   c. not at all
   b. frequently   d. only in multiples of three

5. Name one X-linked recessive genetic disorder.

6. Men are about 16 times more likely to be affected by red–green color blindness than women. Why?

7. Is this statement true or false? A son can inherit an X-linked recessive allele from his father.

8. Color blindness is a case of _____ inheritance.
   a. autosomal dominant   c. X-linked dominant
   b. autosomal recessive   d. X-linked recessive

9. A(n) _____ can alter chromosome structure.
   a. deletion   c. inversion   e. all of the
   b. duplication   d. translocation   above

10. Nondisjunction may occur during _____ .
    a. mitosis   c. fertilization
    b. meiosis   d. both a and b

11. Is this statement true or false? Body cells may inherit three or more of each type of chromosome characteristic of the species, a condition called polyploidy.

12. The karyotype for Klinefelter syndrome is _____ .
    a. XO   c. XXY
    b. XXX   d. XYY

13. A recognized set of symptoms that characterize a specific disorder is a _____ .
    a. syndrome   b. disease   c. pedigree

14. Match the terms appropriately.
    ___ polyploidy   a. number and defining
    ___ deletion   features of an individual's
    ___ aneuploidy   metaphase chromosomes
    ___ translocation   b. segment of a chromosome
    ___ karyotype   moves to a nonhomologous
    ___ nondisjunction   chromosome
        during meiosis   c. extra sets of chromosomes
                            d. gametes with the wrong
                               chromosome number
                            e. a chromosome segment lost
                            f. one extra chromosome

■ *Visit CengageNOW for additional questions.*

Genetics Problems   *Answers in Appendix III*

1. Human females are XX and males are XY.
   a. Does a male inherit the X from his mother or father?
   b. With respect to X-linked alleles, how many different types of gametes can a male produce?
   c. If a female is homozygous for an X-linked allele, how many types of gametes can she produce with respect to that allele?
   d. If a female is heterozygous for an X-linked allele, how many types of gametes might she produce with respect to that allele?

2. In Section 11.4, you read about a mutation that causes a serious genetic disorder, Marfan syndrome. A mutated allele responsible for the disorder follows a pattern of autosomal dominant inheritance. What is the chance that any child will inherit it if one parent does not carry the allele and the other is heterozygous for it?

3. Somatic cells of individuals with Down syndrome usually have an extra chromosome 21; they contain forty-seven chromosomes.
   a. At which stages of meiosis I and II could a mistake alter the chromosome number?
   b. A few individuals with Down syndrome have forty-six chromosomes: two normal-appearing chromosomes 21, and a longer-than-normal chromosome 14. Speculate on how this chromosome abnormality may have arisen.

4. As you read earlier, Duchenne muscular dystrophy is a genetic disorder that arises through the expression of a recessive X-linked allele. Usually, symptoms start to appear in childhood. Gradual, progressive loss of muscle function leads to death, usually by age twenty or so. Unlike color blindness, the disorder is nearly always restricted to males. Suggest why.

5. In the human population, mutation of two genes on the X chromosome causes two types of X-linked hemophilia (A and B). In a few cases, a woman is heterozygous for both mutated alleles (one on each of the X chromosomes). All of her sons should have either hemophilia A or B.
   However, on very rare occasions, one of these women gives birth to a son who does not have hemophilia, and his one X chromosome does not have either mutated allele. Explain how such an X chromosome could arise.

13 DNA Structure and Function

Here, Kitty, Kitty, Kitty, Kitty, Kitty

By now, we have mentioned repeatedly that DNA holds heritable information. Has anybody actually demonstrated that it does? Well, yes. One jarring demonstration occurred in 1997, when Scottish geneticist Ian Wilmut made a genetic copy—a clone—of a fully grown sheep. His team removed the nucleus (and the DNA it contained) from an unfertilized sheep egg. They replaced it with the nucleus of a cell taken from the udder of a different sheep. The hybrid egg became an embryo, and then a lamb. The lamb, whom the researchers named Dolly, was genetically identical to the sheep that had donated the udder cell.

At first, Dolly looked and acted like a normal sheep. But five years later, she was as fat and arthritic as a twelve-year-old sheep. The following year, Dolly contracted a lung disease that is typical of geriatric sheep, and was euthanized.

Dolly's telomeres hinted that she had developed health problems because she was a clone. Telomeres are short, repeated DNA sequences at the ends of chromosomes. They become shorter and shorter as an animal ages. When Dolly was only two years old, her telomeres were as short as those of a six-year-old sheep—the exact age of the adult animal that had been her genetic donor.

Since Dolly was born, mice, rabbits, pigs, cattle, goats, mules, deer, horses, cats, dogs, and a wolf have been cloned, but cloning mammals is far from routine. Not very many clonings end successfully. It usually takes hundreds of attempts to produce one embryo, and most embryos that do form die before birth or shortly after. About 25 percent of the clones that survive have health problems. For example, cloned pigs tend to limp and have heart problems. One never did develop a tail or, even worse, an anus.

What causes the problems? Even though all cells of an individual *inherit* the same DNA, an adult cell *uses* only a fraction of it compared to an embryonic cell. To make a clone from an adult cell, researchers must reprogram its DNA to function like the DNA of an egg. As Dolly's story reminds us, we still have a lot to learn about doing that.

Why do geneticists keep at it? The potential benefits are enormous. Cells cloned from people with incurable diseases may be grown as replacement tissues or organs in laboratories. Endangered animals might be saved from extinction; extinct animals may be brought back. Livestock and pet animals are already being cloned commercially (Figure 13.1).

Perfecting the methods to make healthy animal clones brings us closer to the possibility of cloning humans, technically and also ethically. For example, if cloning a lost cat for a grieving pet owner is acceptable, why would it not be acceptable to clone a lost child for a grieving parent?

With this chapter, we turn to our understanding of how DNA functions as the foundation of inheritance.

See the video! Figure 13.1 Demonstration that DNA holds heritable information—cloning of an adult. Compare the markings on Tahini, a Bengal cat (*above*) with those of Tabouli and Baba Ganoush, two of her clones (*right*). Eye color changes as a Bengal cat matures; both clones later developed the same eye color as Tahini's.

Key Concepts

Discovery of DNA's function

The work of many scientists over more than a century led to the discovery that DNA is the molecule that stores hereditary information about traits. **Section 13.1**

Discovery of DNA's structure

A DNA molecule consists of two long chains of nucleotides coiled into a double helix. Four kinds of nucleotides make up the chains, which are held together along their length by hydrogen bonds. **Section 13.2**

How cells duplicate their DNA

Before a cell begins mitosis or meiosis, enzymes and other proteins replicate its chromosome(s). Newly forming DNA strands are monitored for errors. Uncorrected errors may become mutations. **Section 13.3**

Cloning animals

Knowledge about the structure and function of DNA is the basis of several methods of making clones, which are identical copies of organisms. **Section 13.4**

The Franklin footnote

Science proceeds as a joint effort. Many scientists contributed to the discovery of DNA's structure. **Section 13.5**

Links to Earlier Concepts

- This chapter builds on our earlier introduction to radioisotopes (Section 2.2), hydrogen bonding (2.4), condensation reactions (3.2), carbohydrates (3.3), proteins (3.5), and nucleic acids (3.7).

- Your knowledge of enzyme specificity (6.3) and the cell cycle (9.2) will help you understand how DNA replication works. We also see an example of the importance of checkpoint genes (9.5).

- What you know about mitosis, meiosis, and asexual reproduction (10.1) will help you understand cloning procedures. The cells used for therapeutic cloning are no more developed than those at the eight-cell stage of human development (12.8).

How would you vote? Some view sickly or deformed clones as unfortunate but acceptable casualties of animal cloning research that also yields medical advances for human patients. Should animal cloning be banned? See CengageNOW for details, then vote online.

■ Investigations that led to our understanding that DNA is the molecule of inheritance reveal how science advances.

■ Links to Radioisotopes 2.2, Proteins 3.5

Early and Puzzling Clues

About the time Gregor Mendel was born, a Swiss medical student, Johann Miescher, was ill with typhus. The typhus left Miescher partially deaf, so becoming a doctor was no longer an option for him. He switched to organic chemistry instead. By 1869, he was collecting white blood cells from pus-filled bandages and sperm from fish so he could study the composition of the nucleus. Such cells do not contain much cytoplasm, which made isolating the substances in their nucleus easy. Miescher found that nuclei contain an acidic substance composed mostly of nitrogen and phosphorus. Later, that substance would be called **deoxyribonucleic acid**, or **DNA**.

Sixty years later, a British medical officer, Frederick Griffith, was trying to make a vaccine for pneumonia. He isolated two strains (types) of *Streptococcus pneumoniae*, a bacteria that causes pneumonia. He named one strain *R*, because it grows in *R*ough colonies. He named the other strain *S*, because it grows in *S*mooth colonies. Griffith used both strains in a series of experiments that did not lead to the development of a vaccine, but did reveal a clue about inheritance (Figure 13.2).

First, he injected mice with live *R* cells. The mice did not develop pneumonia. *The R strain was harmless.*

Second, he injected other mice with live *S* cells. The mice died. Blood samples from them teemed with live *S* cells. *The S strain was pathogenic; it caused pneumonia.*

Third, he killed *S* cells by exposing them to high temperature. *Mice injected with dead S cells did not die.*

Fourth, he mixed live *R* cells with heat-killed *S* cells and injected the mixture into mice. The mice died, *and blood samples drawn from them teemed with live S cells!*

What happened in the fourth experiment? If heat-killed *S* cells in the mix were not really dead, then mice injected with them in the third experiment would have died. If the harmless *R* cells had changed into killer cells, then mice injected with *R* cells in experiment 1 would have died.

The simplest explanation was that heat had killed the *S* cells, but had not destroyed their hereditary material, including whatever part specified "infect mice." Somehow, that material had been transferred from the dead *S* cells into the live *R* cells, which put it to use.

The transformation was permanent and heritable. Even after hundreds of generations, the descendants of transformed *R* cells were infectious. What had caused the transformation? *Which substance encodes the information about traits that parents pass to offspring?*

In 1940, Oswald Avery and Maclyn McCarty set out to identify that substance, which they termed the "transforming principle." They used a process of elimination that tested each type of molecular component of *S* cells.

Avery and McCarty repeatedly froze and thawed *S* cells. Ice crystals that form during this process disrupt membranes, thus releasing cell contents. The researchers then filtered any intact cells from the resulting slush. At the end of this process, the researchers had a fluid that contained lipid, protein, and nucleic acid components of *S* cells.

The *S* cell extract could still transform *R* cells after it had been treated with lipid- and protein-destroying enzymes. Thus, the transforming principle could not be lipid or protein. Carbohydrates had been removed during the purification process, so Avery and McCarty realized that the substance they were seeking must be nucleic acid—RNA or DNA. The *S* cell extract could still transform *R* cells after treatment with RNA-degrading enzymes, but not after treatment with DNA-degrading enzymes. DNA had to be the transforming principle.

The result surprised Avery and McCarty, who, along with most other scientists, had assumed that proteins were the substance of heredity. After all, traits are diverse, and

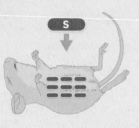

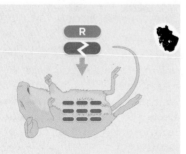

A Mice injected with live cells of harmless strain *R* do not die. Live *R* cells are in their blood.

B Mice injected with live cells of killer strain *S* die. Live *S* cells are in their blood.

C Mice injected with heat-killed *S* cells do not die. No live *S* cells are in their blood.

D Mice injected with live *R* cells plus heat-killed *S* cells die. Live *S* cells are in their blood.

Figure 13.2 Animated Fred Griffith's experiments, in which the hereditary material of harmful *Streptococcus pneumoniae* cells transformed cells of a harmless strain into killers.

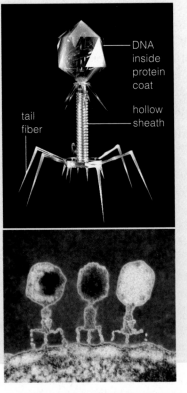

A   *Top*, model of a bacteriophage. *Bottom*, micrograph of three viruses injecting DNA into an *E. coli* cell.

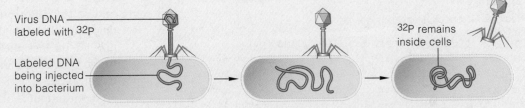

B   In one experiment, bacteria were infected with virus particles labeled with a radioisotope of sulfur (^{35}S). The sulfur had labeled only viral proteins. The viruses were dislodged from the bacteria by whirling the mixture in a kitchen blender. Most of the radioactive sulfur was detected in the viruses, not in the bacterial cells. The viruses had not injected protein into the bacteria.

C   In another experiment, bacteria were infected with virus particles labeled with a radioisotope of phosphorus (^{32}P). The phosphorus had labeled only viral DNA. When the viruses were dislodged from the bacteria, the radioactive phosphorus was detected mainly inside the bacterial cells. The viruses had injected DNA into the cells—evidence that DNA is the genetic material of this virus.

Figure 13.3 Animated The Hershey–Chase experiments. Alfred Hershey and Martha Chase tested whether the genetic material injected by bacteriophage into bacteria is DNA, protein, or both. The experiments were based on the knowledge that proteins contain more sulfur (S) than phosphorus (P), and DNA contains more phosphorus than sulfur.

proteins were thought to be the most diverse biological molecules. Other molecules just seemed too uniform. The two scientists were so skeptical that they published their results only after they had convinced themselves, by years of painstaking experimentation, that DNA was indeed the hereditary material. They were also careful to point out that they had not proven DNA was the *only* hereditary material.

Confirmation of DNA's Function

By 1950, researchers had discovered **bacteriophage**, a type of virus that infects bacteria (Figure 13.3*a*). Like all viruses, these infectious particles carry hereditary information about how to make new viruses. After a virus infects a cell, the cell starts making new virus particles. Bacteriophages inject genetic material into bacteria, but was that material DNA, protein, or both?

Alfred Hershey and Martha Chase decided to find out by exploiting the long-known properties of protein (high sulfur content) and DNA (high phosphorus content). They cultured bacteria in growth medium containing an isotope of sulfur, ^{35}S (Section 2.2). In this medium, the protein (but not the DNA) of bacteriophage that infected the bacteria became labeled with ^{35}S.

Hershey and Chase allowed the labeled viruses to infect bacteria. They knew from electron micrographs that phages attach to bacteria by their slender tails. They reasoned it would be easy to break this precarious attachment, so they poured the virus–bacteria mixture into a Waring blender and turned it on. (A Waring blender was one of the kitchen appliances that was at the time a common piece of laboratory equipment.)

After blending, the researchers separated the bacteria from the virus-containing fluid, and measured the ^{35}S content of each separately. The fluid contained most of the ^{35}S. Thus, the viruses had not injected protein into the bacteria (Figure 13.3*b*).

Hershey and Chase repeated the experiment using an isotope of phosphorus, ^{32}P, which labeled the DNA (but not the proteins) of the bacteriophage. This time, they found that the bacteria contained most of the ^{32}P. The viruses had injected DNA into the bacteria (Figure 13.3*c*).

Both of these experiments—and many others—supported the hypothesis that DNA, not protein, is the material of heredity common to all life on Earth.

The Discovery of DNA's Structure

- Watson and Crick's discovery of DNA's structure was based on almost fifty years of research by other scientists.
- Link to Carbohydrate rings 3.3

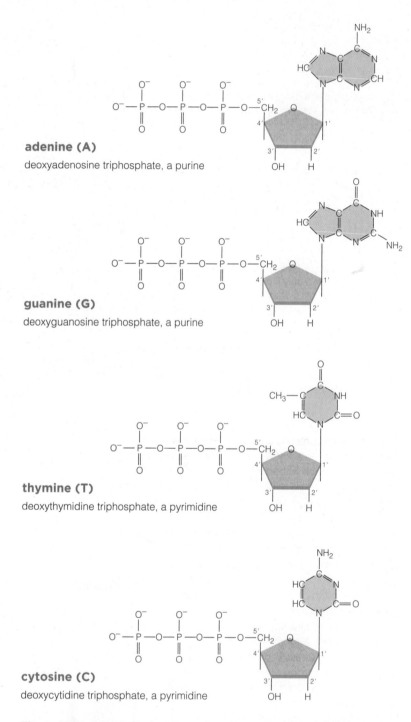

adenine (A)
deoxyadenosine triphosphate, a purine

guanine (G)
deoxyguanosine triphosphate, a purine

thymine (T)
deoxythymidine triphosphate, a pyrimidine

cytosine (C)
deoxycytidine triphosphate, a pyrimidine

Figure 13.4 Four kinds of nucleotides that are linked into strands of DNA. Each is nicknamed after its component base (in *blue*). Biochemist Phoebus Levene worked out the structure of these bases and how they are connected in DNA in the early 1900s. He worked with DNA for almost 40 years.

Numbering the carbons in the sugar rings (Section 3.7) allows us to keep track of the orientation of nucleotide chains, which is important in processes such as DNA replication. Compare Figure 13.6.

DNA's Building Blocks

Long before DNA's function was known, biochemists were investigating its composition. They had shown that DNA consists of only four kinds of nucleotide building blocks. A DNA **nucleotide** has a five-carbon sugar (deoxyribose), three phosphate groups, and one of four nitrogen-containing bases:

adenine	guanine	thymine	cytosine
A	G	T	C

Figure 13.4 shows the structures of these four nucleotides. Thymine and cytosine are called pyrimidines; their bases have single carbon rings. Adenine and guanine are purines; their bases have double carbon rings.

By 1952, biochemist Erwin Chargaff had made two important discoveries about the composition of DNA. First, the amounts of thymine and adenine in DNA are the same, as are the amounts of cytosine and guanine. Second, the proportion of adenine and guanine differs among species. We may show Chargaff's rules as:

$$A = T \quad \text{and} \quad G = C$$

The symmetrical proportions were an important clue to how nucleotides are arranged in DNA.

The first convincing evidence of that arrangement came from Rosalind Franklin, a researcher at King's College in London who specialized in x-ray crystallography. In this technique, x-rays are directed through a purified and crystallized substance. Atoms in the substance's molecules scatter the x-rays in a pattern that can be captured as an image. Researchers use the pattern to calculate the size, shape, and spacing between any repeating elements of the molecules—all of which are details of molecular structure.

Franklin made the first clear x-ray diffraction image of DNA in the form that occurs in cells. From the information in that image, she calculated that DNA is very long compared to its 2-nanometer diameter. She also identified a repeating pattern every 0.34 nanometers along its length, and another every 3.4 nanometers.

Franklin's image and data came to the attention of James Watson and Francis Crick, both at Cambridge University. Watson, an American biologist, and Crick, a British biophysicist, had been sharing their ideas about the structure of DNA. Biochemists Linus Pauling, Robert Corey, and Herman Branson had only recently described the alpha helix, a coiled pattern that occurs in many proteins (Section 3.5). Watson and Crick suspected that the DNA molecule was also a helix.

Watson and Crick spent many hours arguing about the size, shape, and bonding requirements of the four kinds of nucleotides that make up DNA. They pestered chemists to help them identify bonds they might have overlooked. They fiddled with cardboard cutouts, and made models from scraps of metal connected by suitably angled "bonds" of wire. Franklin's data provided them with the last piece of the puzzle. In 1953, Watson and Crick put together all of the clues that had been accumulating for the last fifty years and built the first accurate model of the DNA molecule (Figure 13.5).

Watson and Crick proposed that DNA's structure consists of two chains (or strands) of nucleotides, running in opposite directions and coiled into a double helix. Hydrogen bonds between the internally positioned bases hold the two strands together. Only two kinds of base pairings form: A to T, and G to C. Most scientists had assumed (incorrectly) that the bases had to be on the outside of the helix, because they would be more accessible to DNA-copying enzymes that way. You will see in Section 13.3 how those enzymes access the bases on the inside of the helix.

Patterns of Base Pairing

How do just two kinds of base pairings give rise to the stunning diversity of traits we see among living things? The answer is that the *order* in which one base pair follows the next—the DNA's **sequence**—is tremendously variable. For instance, a small piece of DNA from a petunia, a human, or any other organism might be:

one base pair [G A C T / C T G A] or [A T C G / T A G C] or [G G G C / C C C G]

Notice how the two strands of DNA match up; each base on one is suitably paired with a partner base on the other. This bonding pattern (A to T, G to C) is the same in all molecules of DNA. However, which base pair follows the next in line differs among species, and among individuals of the same species. Thus, *DNA, the molecule of inheritance in every cell, is the basis of life's unity. Variations in its base sequence from one individual or one species to the next is the basis of life's diversity.*

Take-Home Message

What is the structure of DNA?

■ A DNA molecule consists of two nucleotide chains (strands), running in opposite directions and coiled into a double helix.

■ Internally positioned nucleotide bases hydrogen-bond between the two strands. A always pairs with T, and G with C. The sequence of bases is the genetic information.

Figure 13.5 Animated Structure of DNA, as illustrated by a composite of different models.

Watson and Crick with their model

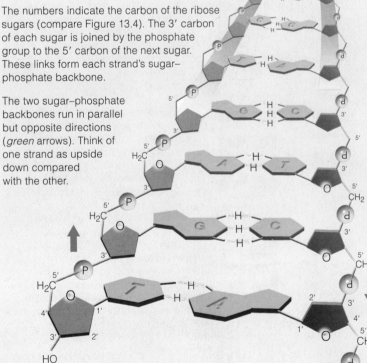

2-nanometer diameter

0.34 nanometer between each base pair

3.4-nanometer length of each full twist of the double helix

The numbers indicate the carbon of the ribose sugars (compare Figure 13.4). The 3' carbon of each sugar is joined by the phosphate group to the 5' carbon of the next sugar. These links form each strand's sugar–phosphate backbone.

The two sugar–phosphate backbones run in parallel but opposite directions (*green* arrows). Think of one strand as upside down compared with the other.

- A cell copies its DNA before mitosis or meiosis I.
- DNA repair mechanisms correct most replication errors.

- Links to Enzyme specificity 6.3, Cell cycle 9.2, Checkpoint genes 9.5

Remember, each cell copies its DNA before mitosis or meiosis I begins, so its descendant cells will inherit a complete set of chromosomes (Sections 9.2 and 10.1). **DNA polymerase** does the copying. This enzyme joins free nucleotides into a new strand of DNA. The process is driven by high-energy phosphate bonds in the nucleotides (Section 6.2). A free nucleotide has three phosphate groups, and DNA polymerase removes two of them when it attaches the nucleotide to a growing strand of DNA. That removal releases energy that the enzyme uses to attach the nucleotide to the strand.

How does an enzyme that assembles a single strand of DNA make a copy of a double-stranded molecule? Before replication begins, each chromosome is a single molecule of DNA—one double helix. During DNA replication, an enzyme called DNA helicase breaks the hydrogen bonds that hold the helix together, so the two DNA strands unwind. Both strands are then replicated independently. Each new DNA strand winds up with its "parent" strand into a new double helix. Thus, after replication, there are two double-stranded molecules of DNA (Figure 13.6). One strand of each molecule is old and the other is new; hence the name of the process, **semiconservative replication** (Figure 13.7).

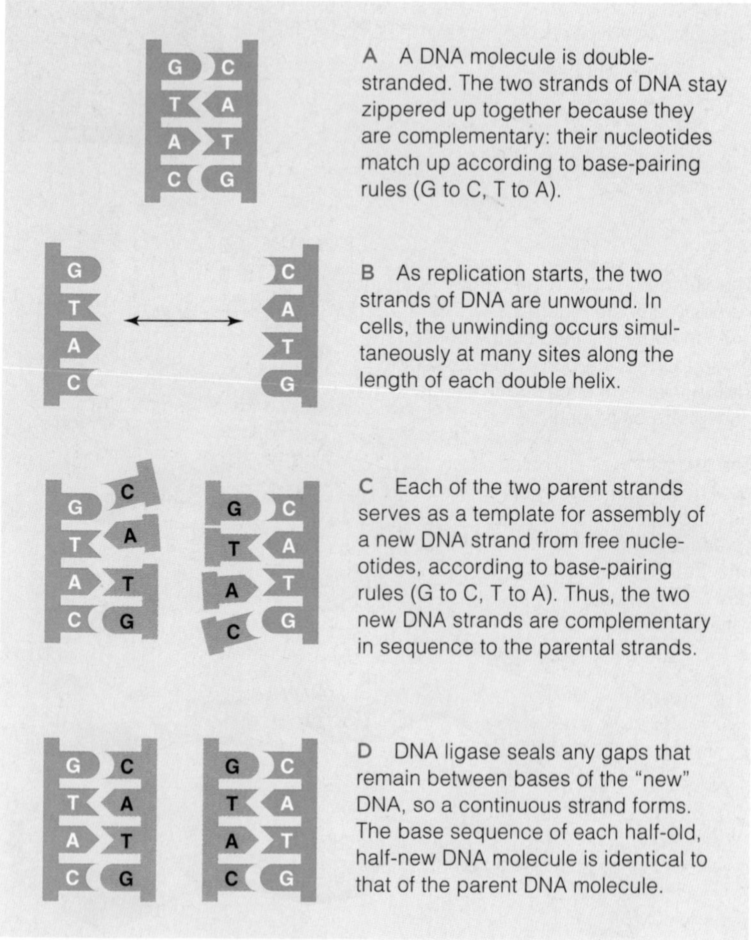

Numbering the carbons in nucleotides (Figure 13.5) allows us to keep track of the DNA strands in a double helix, because each strand has an unbonded 5′ carbon at one end and an unbonded 3′ carbon at the other:

DNA polymerase can attach free nucleotides only to a 3′ carbon. Thus, it can replicate only one strand of a DNA molecule continuously (Figure 13.8). Synthesis of the other strand occurs in segments, in the direction opposite that of unwinding. Another enzyme that participates in DNA replication, **DNA ligase**, joins those segments into a continuous strand of DNA.

There are only four kinds of nucleotides in DNA, but the order in which those nucleotides occur is very important. The nucleotide sequence is a cell's genetic information; descendant cells must get an exact copy of it, or inheritance will go awry. As a DNA polymerase moves along a strand of DNA, it uses the sequence of bases as a template, or guide, to assemble a new strand of DNA. The base sequence of the new strand is complementary to that of the template, because DNA polymerase follows base-pairing rules.

For example, the polymerase adds a T to the end of the new DNA strand when it reaches an A in the parent DNA sequence; it adds a G when it reaches a C; and so on. Because each new strand of DNA is complementary in sequence to the parent strand, both double-stranded molecules that result from DNA replication are duplicates of the parent molecule.

Checking for Mistakes

A DNA molecule is not always replicated with perfect fidelity. Sometimes the wrong base is added to a growing DNA strand; at other times, bases get lost, or extra ones are added. Either way, the new DNA strand will no longer match up perfectly with its parent strand.

Some of these errors occur after the DNA becomes damaged by exposure to radiation or toxic chemicals. DNA polymerases do not copy damaged DNA very

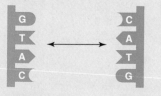

A A DNA molecule is double-stranded. The two strands of DNA stay zippered up together because they are complementary: their nucleotides match up according to base-pairing rules (G to C, T to A).

B As replication starts, the two strands of DNA are unwound. In cells, the unwinding occurs simultaneously at many sites along the length of each double helix.

C Each of the two parent strands serves as a template for assembly of a new DNA strand from free nucleotides, according to base-pairing rules (G to C, T to A). Thus, the two new DNA strands are complementary in sequence to the parental strands.

D DNA ligase seals any gaps that remain between bases of the "new" DNA, so a continuous strand forms. The base sequence of each half-old, half-new DNA molecule is identical to that of the parent DNA molecule.

Figure 13.6 Animated DNA replication. Each strand of a DNA double helix is copied; two double-stranded DNA molecules result.

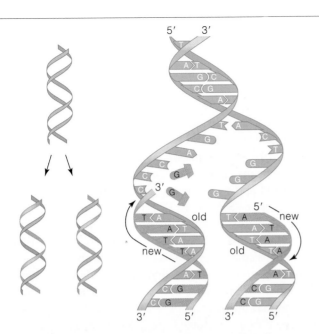

Figure 13.7 Semiconservative replication of DNA. Parent strands (*blue*) stay intact. A new strand (*purple*) is assembled on each parent in the direction shown by the arrows. The Y-shaped structure on the *right* is called a replication fork.

well. In most cases, **DNA repair mechanisms** fix DNA by enzymatically excising and replacing any damaged or mismatched bases before replication begins.

Most DNA replication errors occur simply because DNA polymerases catalyze a tremendous number of reactions very quickly—up to 1,000 bases per second. Mistakes are inevitable; some DNA polymerases make many of them. Luckily, most DNA polymerases proofread their own work. They correct any mismatches by immediately reversing the synthesis reaction to remove a mismatched nucleotide, and then resuming synthesis. If an error remains uncorrected, cellular controls may stop the cell cycle (Sections 9.2 and 9.5).

When proofreading and repair mechanisms fail, an error becomes a mutation—a permanent change in the DNA sequence. An individual or its offspring may not survive a mutation, because mutations can cause cancer in body cells. In cells that form eggs or sperm, they may lead to genetic disorders in offspring. However, not all mutations are dangerous. Some give rise to variations in traits that are the raw material of evolution.

Take-Home Message

How is DNA copied?

■ A cell replicates its DNA before mitosis or meiosis. Each strand of a DNA double helix serves as a template for synthesis of a new, complementary strand of DNA.

■ DNA repair mechanisms and proofreading maintain the integrity of a cell's genetic information. Unrepaired errors may be perpetuated as mutations.

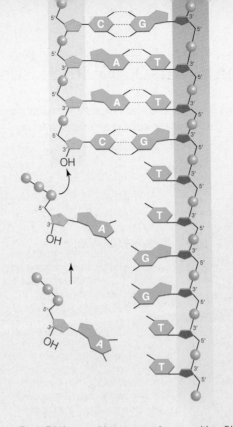

A Each DNA strand has two ends: one with a 5′ carbon, and one with a 3′ carbon. DNA polymerase can add nucleotides only at the 3′ carbon. In other words, DNA synthesis proceeds only in the 5′ to 3′ direction.

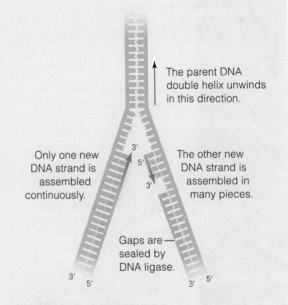

The parent DNA double helix unwinds in this direction.

Only one new DNA strand is assembled continuously.

The other new DNA strand is assembled in many pieces.

Gaps are sealed by DNA ligase.

B Because DNA synthesis proceeds only in the 5′ to 3′ direction, only one of the two new DNA strands can be assembled in a single piece.

The other new DNA strand forms in short segments, which are called Okazaki fragments after the two scientists who discovered them. DNA ligase joins the fragments into a continuous strand of DNA.

Figure 13.8 Discontinuous synthesis of DNA.
Figure It Out: What do the yellow balls represent?
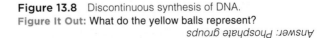
Answer: Phosphate groups

- Reproductive cloning is a reproductive intervention that results in an exact genetic copy of an adult individual.

- Link to Asexual reproduction 10.1

A A cow egg is held in place by suction through a hollow glass tube called a micropipette. The polar body (Section 10.5) and chromosomes are identified by a *purple* stain.

B A micropipette punctures the egg and sucks out the polar body and all of the chromosomes. All that remains inside the egg's plasma membrane is cytoplasm.

C A new micropipette prepares to enter the egg at the puncture site. The pipette contains a cell grown from the skin of a donor animal.

 skin cell

D The micropipette enters the egg and delivers the skin cell to a region between the cytoplasm and the plasma membrane.

E After the pipette is withdrawn, the donor's skin cell is visible next to the cytoplasm of the egg. The transfer is complete.

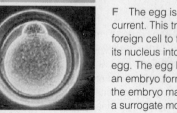

F The egg is exposed to an electric current. This treatment causes the foreign cell to fuse with and empty its nucleus into the cytoplasm of the egg. The egg begins to divide, and an embryo forms. After a few days, the embryo may be transplanted into a surrogate mother.

Figure 13.9 Animated Somatic cell nuclear transfer, using cattle cells. This series of micrographs was taken by scientists at Cyagra, a commercial company that specializes in cloning livestock.

The word "cloning" means making an identical copy of something. In biology, cloning can refer to a laboratory method by which researchers copy DNA fragments (we will discuss DNA cloning in Chapter 16). It can also refer to interventions in reproduction that result in an exact genetic copy of an organism.

Genetically identical organisms occur all the time in nature, and arise mainly by the process of asexual reproduction (Section 10.1). Embryo splitting, another natural process, results in identical twins. The first few divisions of a fertilized egg form a ball of cells that sometimes splits spontaneously. If both halves continue to develop independently, they become identical twins.

Embryo splitting has been a routine part of research and animal husbandry for decades. A ball of cells can be grown from a fertilized egg in a petri dish. If the ball is divided in two, each half will develop as a separate embryo. The embryos are implanted in surrogate mothers, which give birth to identical twins. Artificial twinning and any other technology that yields genetically identical individuals is called **reproductive cloning**.

Twins get their DNA from two parents, so they have a mixture of parental traits. Breeders that want no surprises may opt for a type of reproductive cloning that differs from embryo splitting. This process yields offspring with only one parent's traits; because it starts with nuclear DNA from an adult organism, it bypasses the genetic mixup of sexual reproduction (Section 10.5). All of the individuals that are produced by cloning an adult cell are genetically identical with the parent. However, the procedure presents more of a technical challenge than embryo splitting. A normal cell from an adult will not automatically start dividing as if it were a fertilized egg. It must first be tricked into rewinding its developmental clock.

All cells descended from a fertilized egg inherit the same DNA. As different cells in a developing embryo start using different subsets of their DNA, they differentiate, or become different in form and function. In animals, differentiation is usually a one-way path. Once a cell specializes, all of its descendant cells will be specialized the same way. By the time a liver cell, muscle cell, or other specialized cell forms, most of its DNA has been turned off, and is no longer used.

To clone an adult, scientists first transform one of its differentiated cells into an undifferentiated cell by turning its unused DNA back on. In **somatic cell nuclear transfer** (SCNT), a researcher removes the nucleus from an unfertilized egg, then inserts into the egg a nucleus from an adult animal cell (Figure 13.9). If all goes well, the egg's cytoplasm reprograms the transplanted DNA to direct the development of an embryo, which is then

Figure 13.10 Liz the cow and her clone. The clone was produced by somatic cell nuclear transfer, as in Figure 13.9.

implanted into a surrogate mother. The animal that is born to the surrogate is genetically identical with the donor of the nucleus (Figure 13.10). Dolly the sheep and the other animals described in the chapter introduction were produced using SCNT.

Adult cloning is now a common practice among people who breed prized livestock. Among other benefits, many more offspring can be produced in a given time frame by cloning than by traditional breeding methods, and offspring can be produced after a donor animal is castrated or even dead.

The controversial issue with adult cloning is not necessarily about livestock. The issue is that as the techniques become routine, cloning a human is no longer only within the realm of science fiction. Researchers are already using SCNT to produce human embryos for research, a practice called **therapeutic cloning**. The researchers harvest undifferentiated (stem) cells from the cloned human embryos. (We return to the topic of stem cells and their potential medical benefits in Chapter 32.) Reproductive cloning of humans is not the intent of such research, but somatic cell nuclear transfer would be the first step toward that end.

Take-Home Message

What is cloning?

■ Reproductive cloning technologies produce an exact copy of an individual—a clone.

■ Somatic cell nuclear transfer (SCNT) is a reproductive cloning method in which nuclear DNA of an adult donor is transferred to an enucleated egg. The hybrid cell develops into an embryo that is genetically identical to the donor individual.

■ Therapeutic cloning uses SCNT to produce human embryos for research purposes.

13.5 | Fame and Glory

■ In science, as in other professions, public recognition does not always include everyone who contributed to a discovery.

By the time she arrived at King's College, Rosalind Franklin was an expert x-ray crystallographer. She had solved the structure of coal, which is complex and unorganized (as are large biological molecules such as DNA), and she took a new mathematical approach to interpreting x-ray diffraction images. Like Pauling, she had built three-dimensional molecular models. Her assignment was to investigate DNA's structure. She did not know Maurice Wilkins was already doing the same thing just down the hall. Franklin had been told she would be the only one in the department working on the problem. When Wilkins proposed a collaboration with her, Franklin thought that Wilkins was oddly overinterested in her work and declined bluntly.

Wilkins and Franklin had been given identical samples of DNA, which had been carefully prepared by Rudolf Signer. Franklin's meticulous work with her sample yielded the first clear x-ray diffraction image of DNA as it occurs inside cells (Figure 13.11), and she gave a presentation on this work in 1952. DNA, she said, had two chains twisted into a double helix, with a backbone of phosphate groups on the outside, and bases arranged in an unknown way on the inside. She had calculated DNA's diameter, the distance between its chains and between its bases, the pitch (angle) of the helix, and the number of bases in each coil. Crick, with his crystallography background, would have recognized the significance of the work—if he had been there. Watson was in the audience but he was not a crystallographer, and he did not understand the implications of Franklin's x-ray diffraction image or her calculations.

Franklin started to write a research paper on her findings. Meanwhile, and perhaps without her knowledge, Watson reviewed Franklin's x-ray diffraction image with Wilkins, and Watson and Crick read a report containing Franklin's unpublished data. Crick, who had more experience with theoretical molecular modeling than Franklin, immediately understood what the image and the data meant. Watson and Crick used that information to build their model of DNA.

On April 25, 1953, Franklin's paper appeared third in a series of articles about the structure of DNA in the journal *Nature*. It supported with solid experimental evidence Watson and Crick's theoretical model, which appeared in the first article of the series.

Rosalind Franklin died at age 37, of ovarian cancer probably caused by extensive exposure to x-rays. Because the Nobel Prize is not given posthumously, she did not share in the 1962 honor that went to Watson, Crick, and Wilkins for the discovery of the structure of DNA.

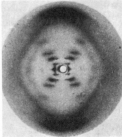

Figure 13.11 Rosalind Franklin and her famous x-ray diffraction image.

Human eggs are difficult to come by, so SCNT researchers are using adult human cells and enucleated cow eggs for therapeutic cloning. The nuclear DNA of the resulting hybrid eggs is human, and the cytoplasm is bovine. Remember, eukaryotic cytoplasm contains mitochondria, which have their own DNA and divide independently (Section 4.11). Thus, cells of embryos that develop from these hybrid eggs contain both human and cow DNA.

How would you vote?

Deformed or unhealthy clones, including Dolly (*right*) are unavoidable casualties of cloning research. Should cloning be banned? See CengageNOW for details, then vote online.

Summary

Section 13.1 Experiments with bacteria and **bacteriophage** offered solid evidence that **deoxyribonucleic acid (DNA)**, not protein, is hereditary material.

■ *Use the animation on CengageNOW to learn about experiments that revealed the function of DNA.*

Section 13.2 A DNA molecule consists of two strands of DNA coiled into a helix. Nucleotide monomers are joined to form each strand. A free **nucleotide** has a five-carbon sugar (deoxyribose), three phosphate groups, and one of four nitrogen-containing bases after which it is named: **adenine, thymine, guanine,** or **cytosine**.

Bases of the two DNA strands in a double helix pair in a consistent way: adenine with thymine (A–T), and guanine with cytosine (G–C). The order of the bases (the DNA **sequence**) varies among species and among individuals. The DNA of each species has unique sequences that set it apart from the DNA of all other species.

■ *Use the animation on CengageNOW to investigate the structure of DNA.*

Section 13.3 A cell replicates its DNA before mitosis or meiosis begins. By the process of **semiconservative replication**, one double-stranded molecule of DNA is copied, and two double-stranded DNA molecules identical to the parent are the result. One strand of each molecule is new, and the other is parental.

During the replication process, enzymes unwind the double helix at several sites along its length. **DNA polymerase** uses each strand as a template to assemble new, complementary strands of DNA from free nucleotides. DNA synthesis is discontinuous on one of the two strands of a DNA molecule. **DNA ligase** joins the segments into a continuous strand.

DNA repair mechanisms fix DNA damaged by chemicals or radiation. Proofreading by DNA polymerases corrects most base-pairing errors. Uncorrected errors can be perpetuated as mutations.

■ *Use the animation on CengageNOW to see how a DNA molecule is replicated.*

Section 13.4 Various **reproductive cloning** technologies produce genetically identical individuals (clones). In **somatic cell nuclear transfer** (SCNT), one cell from an adult is fused with an enucleated egg. The hybrid cell is treated with electric shocks or another stimulus that provokes the cell to divide and begin developing into a new individual. SCNT with human cells, which is called **therapeutic cloning**, produces embryos that are used for stem cell research.

■ *Use the animation on CengageNOW to observe the procedure used to create Dolly and other clones.*

Section 13.5 Sciences advances as a community effort. Ideally, individuals share their work and the recognition for achievement. As in all human endeavors, these ideals are not always achieved.

Self-Quiz   *Answers in Appendix III*

1. Bacteriophages are viruses that infect _____ .

2. Which is *not* a nucleotide base in DNA?
   a. adenine   c. uracil   e. cytosine
   b. guanine   d. thymine   f. All are in DNA.

3. What are the base-pairing rules for DNA?
   a. A–G, T–C   c. A–U, C–G
   b. A–C, T–G   d. A–T, G–C

4. One species' DNA differs from others in its _____ .
   a. sugars   c. base sequence
   b. phosphates   d. all of the above

5. When DNA replication begins, _____ .
   a. the two DNA strands unwind from each other
   b. the two DNA strands condense for base transfers
   c. two DNA molecules bond
   d. old strands move to find new strands

6. DNA replication requires _____ .
   a. template DNA   c. DNA polymerase
   b. free nucleotides   d. all of the above

7. DNA polymerase adds nucleotides to _____ (choose all that are correct).
   a. double-stranded DNA   c. double-stranded RNA
   b. single-stranded DNA   d. single-stranded RNA

8. Show the complementary strand of DNA that forms on this template DNA fragment during replication:

   5'–GGTTTCTTCAAGAGA–3'

9. _____ is an example of reproductive cloning.
   a. Somatic cell nuclear transfer (SCNT)
   b. Asexual reproduction
   c. Artificial embryo splitting
   d. a and c
   e. all of the above

Data Analysis Exercise

The graph in Figure 13.12 is reproduced from Alfred Hershey and Martha Chase's 1952 publication that showed DNA is the hereditary material of bacteriophage. The data are from the same two experiments described in Section 13.1, in which bacteriophage DNA and protein were labeled with radioactive tracers and allowed to infect bacteria. The virus–bacteria mixtures were whirled in a blender to dislodge the two, and the tracers were tracked inside and outside of the bacteria.

1. Before blending, what percentage of ^{35}S was outside the bacteria? What percentage was inside? What percentage of ^{32}P was outside the bacteria? What percentage was inside?

2. After 4 minutes in the blender, what percentage of ^{35}S was outside the bacteria? What percentage was inside? What percentage of ^{32}P was outside the bacteria? What percentage was inside?

3. How did the researchers know that the radioisotopes in the fluid came from outside the bacterial cells (extracellular) and not from bacteria that had broken apart?

4. The extracellular concentration of which isotope, ^{35}S or ^{32}P, increased the most with blending? DNA contains much more phosphorus than do proteins; proteins contain much more sulfur than do DNA. Do these results imply that the viruses inject DNA or protein into bacteria? Why?

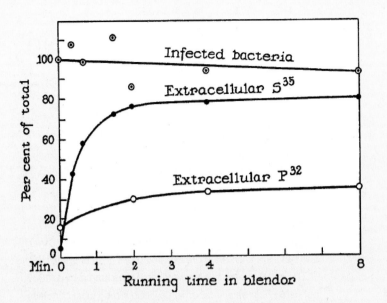

Figure 13.12 Detail of Alfred Hershey and Martha Chase's publication describing their experiments with bacteriophage. "Infected bacteria" refers to the percentage of bacteria that survived the blender.

From the Journal of General Physiology, *36(1), September 20, 1952: "Independent Functions of Viral Protein and Nucleic Acid in Growth of Bacteriophage."*

10. Match the terms appropriately.

____bacteriophage
____clone
____nucleotide
____purine
____DNA ligase
____DNA polymerase
____pyrimidine

a. nitrogen-containing base, sugar, phosphate groups
b. copy of an organism
c. nucleotide base with one carbon ring
d. injects DNA into bacteria
e. fills in gaps, seals breaks in a DNA strand
f. nucleotide base with two carbon rings
g. adds nucleotides to a growing DNA strand

■ *Visit CengageNOW for additional questions.*

Critical Thinking

1. Matthew Meselson and Franklin Stahl's experiments supported the semiconservative model of replication. These researchers obtained "heavy" DNA by growing *Escherichia coli* with ^{15}N, a radioactive isotope of nitrogen. They also prepared "light" DNA by growing *E. coli* in the presence of ^{14}N, the more common isotope. An available technique helped them identify which of the replicated molecules were heavy, light, or hybrid (one heavy strand and one light). Use different colored pencils to draw the heavy and light strands of DNA. Starting with a DNA molecule having two heavy strands, show the formation of daughter molecules after replication in a ^{14}N-containing medium. Show the four DNA molecules that would form if the daughter molecules were replicated a second time in the ^{14}N medium. Would the resulting DNA molecules be heavy, light, or mixed?

2. Mutations are permanent changes in a cell's DNA base sequence, the original source of genetic variation and the raw material of evolution. How can mutations accumulate, given that cells have repair systems that fix changes or breaks in DNA strands?

3. There may be millions of woolly mammoths frozen in the ice of Siberian glaciers. These huge elephant-like mammals have been extinct for about 10,000 years, but a team of privately funded Japanese scientists is planning to resurrect one of them by cloning DNA isolated from frozen remains. What are some of the pros and cons, both technical and ethical, of cloning an extinct animal?

4. Xeroderma pigmentosum is an autosomal recessive disorder characterized by rapid formation of skin sores (*right*) that can develop into cancers. Affected individuals must avoid all forms of radiation—including sunlight and fluorescent lights. They have no mechanism for dealing with the damage that ultraviolet (UV) light can inflict on skin cells because they lack a DNA repair mechanism that corrects thymine dimers. When the nitrogen-containing bases in DNA absorb UV light, a covalent bond can form between two thymine bases in the same strand of DNA (*left*). The resulting thymine dimer makes a kink in the DNA strand. Propose what consequences might occur because of a thymine dimer during DNA replication.

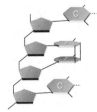

From DNA to Protein

Ricin and Your Ribosomes

Ricin is a highly toxic protein. It is present in all tissues of the castor-oil plant (*Ricinus communis*), which is the source of the castor oil that is an ingredient in many plastics, cosmetics, paints, textiles, and adhesives. The oil—and the ricin—is most concentrated in castor-oil beans (Figure 14.1), but the ricin is discarded when the oil is extracted.

Injected, a dose of ricin as small as a few grains of salt can kill an adult. Inhaled or ingested, it takes more. Only plutonium and botulism toxin are more deadly. There is no antidote.

The lethal effects of ricin were known as long ago as 1888, but using ricin as a weapon is now banned by most countries under the Geneva Protocol. It takes no special skills or equipment to manufacture the toxin from easily obtained raw materials, so controlling its production is impossible. Thus, ricin appears periodically in the news.

For example, at the height of the Cold War, Georgi Markov, a Bulgarian writer, had defected to England and was working as a journalist for the BBC. As he made his way to a bus stop on a London street, an assassin used the tip of a modified umbrella to jam a small, ricin-laced ball into Markov's leg. Markov died in agony three days later.

In 2003, police acted on an intelligence tip and stormed a London apartment, where they found laboratory glassware

and castor-oil beans. Traces of ricin were found in a United States Senate mailroom and State Department building, and also in an envelope addressed to the White House. In 2005, the FBI arrested a man who had castor-oil beans and an assault rifle stashed in his Florida home. Jars of banana baby food laced with ground castor-oil beans also made the news in 2005. In 2006, police found pipe bombs and a baby food jar full of ricin in a Tennessee man's shed. In 2008, castor beans, firearms, and several vials of ricin were found in a Las Vegas motel room after its occupant was hospitalized for ricin exposure.

Ricin is toxic because it inactivates ribosomes, the organelles upon which amino acids are assembled into proteins in all cells. Proteins are critical to all life processes. Cells that cannot make them die very quickly. Someone who inhales ricin typically dies from low blood pressure and respiratory failure within a few days of exposure.

This chapter details how the information encoded by a gene becomes converted to a gene product—an RNA or a protein. Though it is extremely unlikely that your ribosomes will ever encounter ricin, protein synthesis is nevertheless worth appreciating for how it keeps you—and all other organisms—alive.

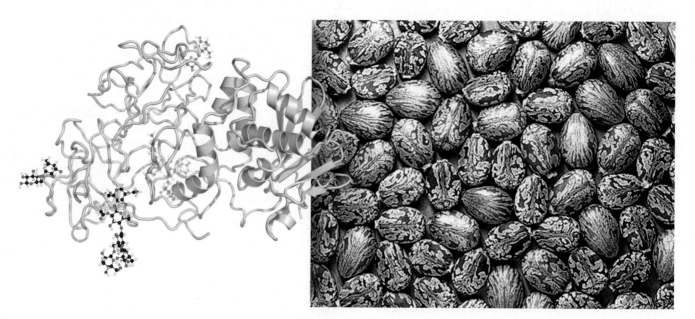

See the video! Figure 14.1 *Left*, model of ricin. One of its polypeptide chains (*green*) helps ricin penetrate a living cell. The other chain (*tan*) destroys the cell's capacity for protein synthesis. Ricin is a glycoprotein; sugars attached to the protein are shown. *Right*, seeds of the castor-oil plant, source of ribosome-busting ricin.

Key Concepts

DNA to RNA to protein

Proteins consist of polypeptide chains. The chains are sequences of amino acids that correspond to sequences of nucleotide bases in DNA called genes. The path leading from genes to proteins has two steps: transcription and translation. **Section 14.1**

DNA to RNA: transcription

During transcription, one strand of a DNA double helix is a template for assembling a single, complementary strand of RNA (a transcript). Each transcript is an RNA copy of a gene. **Section 14.2**

RNA

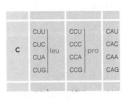

Messenger RNA (mRNA) carries DNA's protein-building instructions. Its nucleotide sequence is read three bases at a time. Sixty-four mRNA base triplets—codons—represent the genetic code. Two other types of RNA interact with mRNA during translation of that code. **Section 14.3**

RNA to protein: translation

Translation is an energy-intensive process by which a sequence of codons in mRNA is converted to a sequence of amino acids in a polypeptide chain. **Section 14.4**

Mutations

Small-scale, permanent changes in the nucleotide sequence of DNA may result from replication errors, the activity of transposable elements, or exposure to environmental hazards. Such mutation can change a gene's product. **Section 14.5**

Links to Earlier Concepts

- This chapter builds on your understanding of enzymatic reactions (Section 3.2) and energy in metabolism (6.2). You will see how information coded in nucleic acids (3.7) becomes translated into proteins (3.5, 3.6).

- You will use what you know about genes (11.1) and base pairing (13.2) to understand transcription, which has many features in common with DNA replication (13.3).

- A review of peptide bond formation (3.5) will be helpful as you learn about translation.

- The last section of this chapter revisits mutations (1.4): their molecular basis, how environmental factors (2.3, 6.3, 7.1) cause them, and some of their consequences (9.5, 12.2).

- Transcription converts information in a gene to RNA; translation converts information in an mRNA to protein.

- Links to Enzymatic reactions 3.2, Proteins 3.5, Nucleotides 3.7, Genes 11.1, DNA replication 13.3

The Nature of Genetic Information

A cell's DNA contains all of its genetic information, but how does the cell convert that information into structural and functional components? Let's start with the nature of the information itself.

DNA is like a book, an encyclopedia that contains all of the instructions for building a new individual. You already know the alphabet used to write the book: the four letters: A, T, G, and C, for the four nucleotide bases adenine, thymine, guanine, and cytosine. Each strand of DNA consists of a chain of those four kinds

of nucleotides. The linear order, or sequence, of the four bases in the strand is the genetic information. That information occurs in subsets called genes, which are Mendel's "units of inheritance" that you read about in Chapter 11.

Converting a Gene to an RNA

Converting the information encoded by a gene into a product starts with RNA synthesis, or **transcription**. By this process, enzymes use the nucleotide sequence of a gene as a template to synthesize a strand of RNA (ribonucleic acid):

$$DNA \xrightarrow{transcription} RNA$$

Except for the double-stranded RNA that is the genetic material of some types of viruses, RNA usually occurs in single-stranded form. A strand of RNA is structurally similar to a single strand of DNA. For example, both are chains of four kinds of nucleotides. Like a DNA nucleotide, an RNA nucleotide has three phosphate groups, a ribose sugar, and one of four bases. However, DNA and RNA nucleotides are slightly different. Three of the bases (adenine, cytosine, and guanine) are the same in DNA and RNA nucleotides, but the fourth base in RNA is uracil, not thymine, and the ribose sugar differs in RNA (Figure 14.2).

Despite these small differences in structure, DNA and RNA have very different functions (Figure 14.3). DNA's only role is to store a cell's heritable information. By contrast, a cell transcribes several kinds of RNAs, each of which has a different function. MicroRNAs are important in gene control, which is the subject of the next chapter. Three types of RNA have roles in protein synthesis. **Ribosomal RNA (rRNA)** is the main component of ribosomes, structures upon which polypeptide chains are built (Sections 4.4 and 4.6). **Transfer RNA (tRNA)** delivers amino acids to ribosomes, one by one, in the order specified by a **messenger RNA (mRNA)**.

Converting mRNA to Protein

mRNA is the only kind of RNA that carries a protein-building message. That message is encoded within the sequence of the mRNA itself by sets of three nucleotide bases, "genetic words" that follow one another along the length of the mRNA. Like the words of a sentence, a series of genetic words can form a meaningful parcel of information—in this case, the sequence of amino acids of a protein.

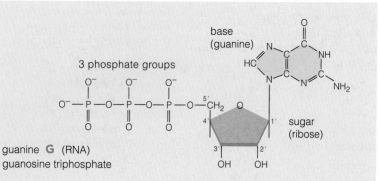

A Guanine, one of the four nucleotides in RNA. The others (adenine, uracil, and cytosine) differ only in their component bases. Three of the four bases in RNA nucleotides are identical to the bases in DNA nucleotides.

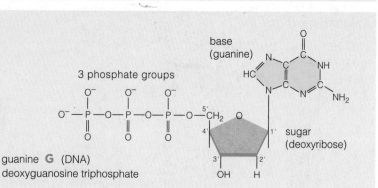

B Compare the DNA nucleotide guanine. The only structural difference between the RNA and DNA versions of guanine (or adenine, or cytosine) is the functional group on the 2' carbon of the sugar.

Figure 14.2 Ribonucleotides and nucleotides compared.

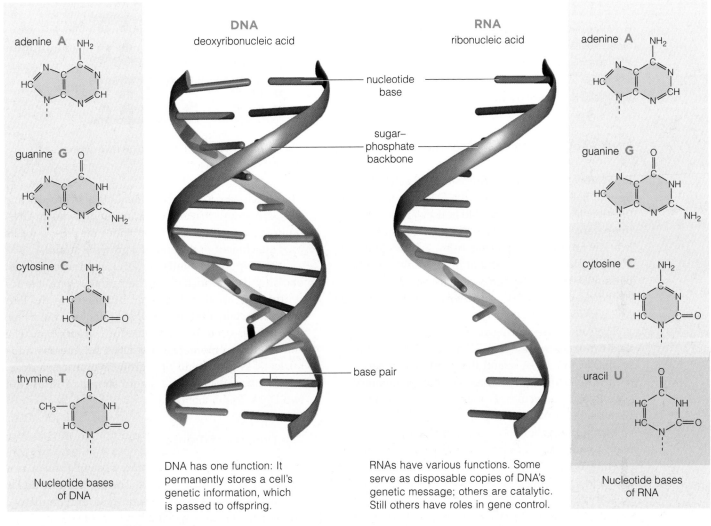

adenine **A**

guanine **G**

cytosine **C**

thymine **T**

Nucleotide bases
of DNA

DNA
deoxyribonucleic acid

nucleotide
base

sugar–
phosphate
backbone

base pair

RNA
ribonucleic acid

adenine **A**

guanine **G**

cytosine **C**

uracil **U**

Nucleotide bases
of RNA

DNA has one function: It
permanently stores a cell's
genetic information, which
is passed to offspring.

RNAs have various functions. Some
serve as disposable copies of DNA's
genetic message; others are catalytic.
Still others have roles in gene control.

Figure 14.3 DNA and RNA compared.

By the process of **translation**, the protein-building information in an mRNA is decoded (translated) into a sequence of amino acids. The result is a polypeptide chain that twists and folds into a protein:

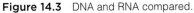

mRNA $\xrightarrow{\textit{translation}}$ PROTEIN

Sections 14.3 and 14.4 describe how rRNA and tRNA interact to translate the sequence of base triplets in an mRNA into the sequence of amino acids in a protein.

The processes of transcription and translation are part of **gene expression**, a multistep process by which genetic information encoded by a gene is converted into a structural or functional part of a cell or body:

DNA $\xrightarrow{\textit{transcription}}$ mRNA $\xrightarrow{\textit{translation}}$ PROTEIN

A cell's DNA sequence contains all of the information it needs to make the molecules of life. Each gene encodes an RNA, and different types of RNAs interact to assemble proteins from amino acids (Section 3.5). Proteins—enzymes—can assemble lipids and complex carbohydrates from simple building blocks (Section 3.2), replicate DNA (Section 13.3), and make RNA, as you will see in the next section.

Take-Home Message

What is the nature of genetic information carried by DNA?

■ The nucleotide sequence of a gene encodes instructions for building an RNA or protein product.

■ A cell transcribes the nucleotide sequence of a gene into RNA.

■ Although RNA is structurally similar to a single strand of DNA, the two types of molecules differ functionally.

■ A messenger RNA (mRNA) carries a protein-building code in its nucleotide sequence. rRNAs and tRNAs interact to translate that sequence into a protein.

14.2 | Transcription: DNA to RNA

- RNA polymerase links RNA nucleotides into a chain, in the order dictated by the base sequence of a gene.
- A new RNA strand is complementary in sequence to the DNA strand from which it was transcribed.

- Links to Base pairing 13.2, DNA replication 13.3

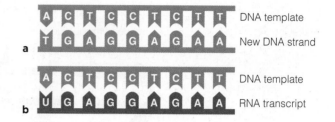

Figure 14.4 Base pairing during (**a**) DNA synthesis and (**b**) transcription.

DNA Replication and Transcription Compared

Remember that DNA replication begins with one DNA double helix and ends with two DNA double helices (Section 13.3). The two double helices are identical to the parent molecule because the process of DNA replication follows base-pairing rules. A nucleotide can be added to a growing strand of DNA only if it base-pairs with the corresponding nucleotide of the parent strand: G pairs with C, and A pairs with T (Section 13.2 and Figure 14.4a).

The same base-pairing rules also govern RNA synthesis in transcription. An RNA strand is structurally so similar to a DNA strand that the two can base-pair if their nucleotide sequences are complementary. In such hybrid molecules, G pairs with C; A pairs with U—uracil (Figure 14.4b).

During transcription, a strand of DNA acts as a template upon which a strand of RNA—a transcript—is assembled from RNA nucleotides. A nucleotide can be added to a growing strand of RNA only if it is complementary to the corresponding nucleotide of the parent strand of DNA: G pairs with C, and A pairs with U.

Thus, each new RNA is complementary in sequence to the DNA strand that served as its template. As in DNA replication, each nucleotide provides the energy for its own attachment to the end of a growing strand.

Transcription is similar to DNA replication in that one strand of a nucleic acid serves as a template for synthesis of another. However, in contrast with DNA replication, only part of one DNA strand, not the whole molecule, is used as a template for transcription. The enzyme **RNA polymerase**, not DNA polymerase, adds nucleotides to the end of a growing transcript. Also, transcription results in a single strand of RNA, not two DNA double helices.

The Process of Transcription

Transcription begins with a chromosome, which is a double helix molecule of DNA. The process gets under way when an RNA polymerase and several regulatory proteins attach to a specific binding site in the DNA

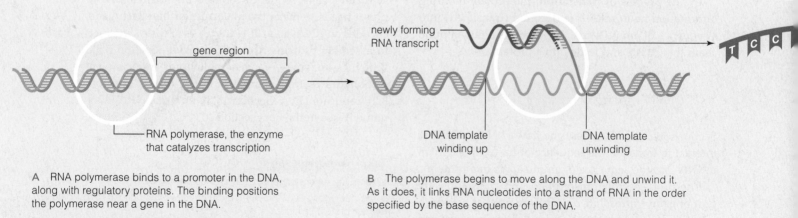

A RNA polymerase binds to a promoter in the DNA, along with regulatory proteins. The binding positions the polymerase near a gene in the DNA.

In most cases, the nucleotide sequence of the gene occurs on only one of the two strands of DNA. Only the complementary strand will be translated into RNA.

B The polymerase begins to move along the DNA and unwind it. As it does, it links RNA nucleotides into a strand of RNA in the order specified by the base sequence of the DNA.

The DNA double helix winds up again after the polymerase passes. The structure of the "opened" DNA molecule at the transcription site is called a transcription bubble, after its appearance.

Figure 14.5 Animated Transcription.

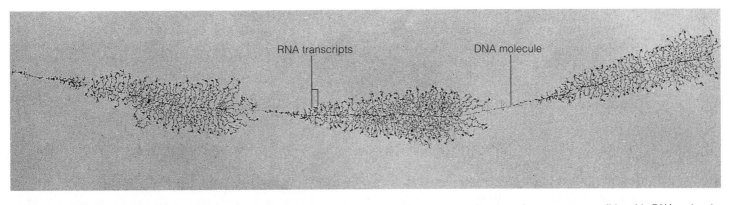

RNA transcripts   DNA molecule

Figure 14.6 Typically, many RNA polymerases simultaneously transcribe the same gene, producing a conglomerate structure often called a "Christmas tree" after its shape. Here, three genes next to one another on the same chromosome are being transcribed.

Figure It Out: Are the polymerases transcribing this DNA molecule moving from left to right or from right to left? *Answer: Left to right*

called a **promoter** (Figure 14.5a). The binding positions the polymerase at a transcription start site close to a gene. The polymerase starts moving along the DNA, in the 5′ to 3′ direction over the gene (Figure 14.5b). As it moves, the polymerase unwinds the double helix just a bit so it can "read" the base sequence of the noncoding DNA strand. The polymerase joins free RNA nucleotides into a chain, in the order dictated by that DNA sequence. As in DNA replication, the synthesis is directional: An RNA polymerase adds nucleotides only to the 3′ end of a growing strand of RNA.

When the polymerase reaches the end of the gene, the DNA and the new RNA strand are released. RNA polymerase follows base-pairing rules, so the new RNA

strand is complementary in base sequence to the DNA strand from which it was transcribed (Figure 14.5c,d). It is an RNA copy of a gene.

Typically, many polymerases transcribe a particular gene region at the same time, so many new RNA strands can be produced very quickly (Figure 14.6).

Take-Home Message

How is RNA assembled?

■ In transcription, RNA polymerase uses the nucleotide sequence of a gene region in a chromosome as a template to assemble a strand of RNA.

■ The new strand of RNA is a copy of the gene from which it was transcribed.

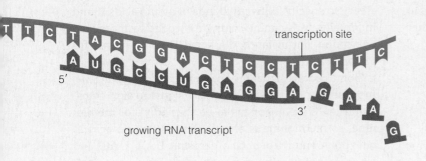

transcription site

growing RNA transcript

C What happened in the gene region? RNA polymerase catalyzed the covalent bonding of many nucleotides to one another to form an RNA strand. The base sequence of the new RNA strand is complementary to the base sequence of its DNA template—a copy of the gene.

D *Top*, at the end of the gene region, the last stretch of the new transcript unwinds and detaches from the DNA template. *Bottom*, a ball-and-stick model of a strand of RNA.

RNA and the Genetic Code

■ Base triplets in an mRNA are words in a protein-building message. Two other classes of RNA—rRNA and tRNA—translate those words into a polypeptide chain.

Post-Transcriptional Modifications

In eukaryotes, transcription takes place in the nucleus, where new RNA is modified before being shipped to the cytoplasm. Just as a dressmaker may snip off loose threads or add bows to a dress before it leaves the shop, so do eukaryotic cells tailor their RNA before it leaves the nucleus.

For example, most eukaryotic genes contain **introns**, nucleotide sequences that are removed from a new RNA. Introns intervene between **exons**, sequences that stay in the RNA (Figure 14.7). Introns are transcribed along with the exons, but are removed before the RNA leaves the nucleus. Either all exons remain in the mature RNA, or some are removed and the rest are spliced in various combinations. By such **alternative splicing**, one gene can encode different proteins.

New transcripts that will become mRNAs are further tailored after splicing. A modified guanine "cap" gets attached to the 5' end of each. Later, the cap will help the mRNA bind to a ribosome. A tail of 50 to 300 adenines is also added to the 3' end of a new mRNA; hence the name, poly-A tail.

mRNA—The Messenger

DNA stores heritable information about proteins, but making those proteins requires mRNA, tRNA, and rRNA. The three types of RNA interact to translate DNA's information into a protein.

An mRNA is a disposable copy of a gene; its job is to carry DNA's protein-building information to the other

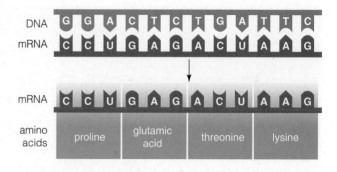

Figure 14.8 Example of the correspondence between DNA and proteins. A DNA strand is transcribed into mRNA, and the codons of the mRNA specify a chain of amino acids.

two types of RNA for translation. Like sentences, the genetic message carried by an mRNA can be understood by those who know the language. Each mRNA is a linear sequence of genetic "words," all spelled with an alphabet of just four nucleotides. Each "word" is three nucleotides long, and each is a code—a **codon**—for a particular amino acid. One codon follows the next along the length of an mRNA. Thus, the order of codons in an mRNA determines the order of amino acids in the polypeptide chain that will be translated from it (Figure 14.8).

With four different nucleotides possible in each of three positions, there are a total of sixty-four (or 4^3) mRNA codons. Collectively, the codons constitute the **genetic code** (Figure 14.9). Which of the four nucleotides is in the first, second, and third position of a triplet determines which amino acid the codon specifies. For instance, the codon AUG (adenine–uracil–guanine) encodes the amino acid methionine, and UGG encodes tryptophan. There are many more codons than are necessary to specify all twenty kinds of amino acids found in proteins. Most amino acids are encoded by more than one codon. For instance, GAA and GAG both code for glutamic acid.

Some codons signal the beginning and end of a gene. In most species, the first AUG is a signal to start translation. AUG also happens to be the codon for methionine, so methionine is always the first amino acid in new polypeptides of such organisms. UAA, UAG, and UGA do not specify an amino acid. They are signals that stop translation—stop codons. A stop codon marks the end of a coding sequence in an mRNA.

The genetic code is highly conserved, which means that many organisms use the same code and probably always have. Prokaryotes and some protists have a few codons that vary, as do mitochondria and chloroplasts. The variation was a clue that led to the theory of how organelles evolved, which we discuss in Section 20.4.

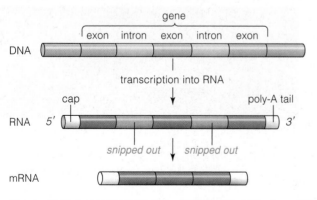

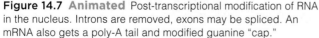

Figure 14.7 **Animated** Post-transcriptional modification of RNA in the nucleus. Introns are removed, exons may be spliced. An mRNA also gets a poly-A tail and modified guanine "cap."

first base ▼	second base U	C	A	G	third base ▼
U	UUU } phe UUC UUA } leu UUG	UCU } UCC } ser UCA UCG	UAU } tyr UAC UAA **STOP** UAG **STOP**	UGU } cys UGC UGA **STOP** UGG trp	U C A G
C	CUU } CUC } leu CUA CUG	CCU } CCC } pro CCA CCG	CAU } his CAC CAA } gln CAG	CGU } CGC } arg CGA CGG	U C A G
A	AUU } AUC } ile AUA AUG met	ACU } ACC } thr ACA ACG	AAU } asn AAC AAA } lys AAG	AGU } ser AGC AGA } arg AGG	U C A G
G	GUU } GUC } val GUA GUG	GCU } GCC } ala GCA GCG	GAU } asp GAC GAA } glu GAG	GGU } GGC } gly GGA GGG	U C A G

Figure 14.9 Animated The sixty four codons of the genetic code. The *left* column lists a codon's first base. The *top* row lists the second base. The *right* column lists the third. Appendix V shows the amino acids. **Figure It Out: Which codons specify the amino acid lysine (lys)?** Answer: AAA and AAG

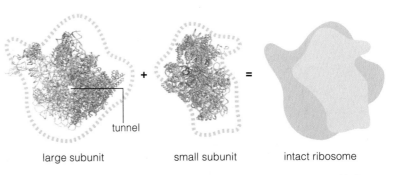

large subunit   small subunit   intact ribosome

Figure 14.10 The ribosome consists of a large and a small subunit. Notice the tunnel through the interior of the large subunit. rRNA components of the ribosome (*tan*) catalyze assembly of polypeptide chains, which thread through this tunnel as they form. We show an mRNA (*red*) attached to the small subunit.

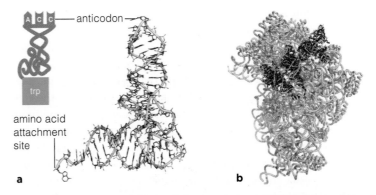

Figure 14.11 tRNA. (**a**) Models of the tRNA that carries the amino acid tryptophan. Each tRNA's anticodon is complementary to an mRNA codon. Each also carries the amino acid specified by that codon. (**b**) During translation, tRNAs dock at an intact ribosome. Here, three tRNAs (*brown*) are docked at the small ribosomal subunit (the large subunit is not shown, for clarity). The anticodons of the tRNAs line up with complementary codons in an mRNA (*red*).

rRNA and tRNA—The Translators

A ribosome has a large and a small subunit. Each consists of proteins and rRNA (Figure 14.10). rRNA is one of the few examples of RNA with enzymatic activity: The rRNA of a ribosome, not the protein, catalyzes the formation of a peptide bond between amino acids.

As you will see in the next section, two ribosomal subunits converge as an intact ribosome on an mRNA during translation. tRNAs bring amino acids to this complex. A tRNA has two attachment sites: One is an **anticodon**, a triplet of nucleotides that base-pairs with an mRNA codon (Figure 14.11). The other binds to a free amino acid—the one specified by the codon.

Some tRNAs can base-pair with more than one type of codon. For example, the codons AUU, AUC, and AUA all specify isoleucine; a tRNA that carries isoleucine can base-pair with all of them.

As you will see in the next section, tRNAs deliver amino acids, one after the next, to a ribosome–mRNA complex during translation. The order of codons in the mRNA is the order in which tRNAs deliver their amino acid cargoes to the ribosome. As the amino acids are delivered, the ribosome joins them via peptide bonds into a new polypeptide chain (Section 3.5). Thus, the order of codons in an mRNA—DNA's protein-building message—is translated into a protein.

Take-Home Message

What are the functions of mRNA, tRNA, and rRNA?

■ Nucleotide bases in mRNA are "read" in sets of three during protein synthesis. Most of these base triplets (codons) code for amino acids. The genetic code comprises all sixty-four codons.

■ A tRNA has an anticodon complementary to an mRNA codon, and it has a binding site for the amino acid specified by that codon. tRNAs deliver amino acids to ribosomes.

■ Ribosomes, which consist of two subunits of rRNA and proteins, link amino acids into polypeptide chains.

14.4 | Translation: RNA to Protein

- Translation converts the information carried by an mRNA into a new polypeptide chain.
- The order of the codons in the mRNA determines the order of the amino acids in the polypeptide chain.

- Links to Peptide bonds 3.5, Energy in metabolism 6.2

Translation, the second part of protein synthesis, occurs in the cytoplasm of all cells. It has three stages: initiation, elongation, and termination.

The initiation stage begins when a small ribosomal subunit binds to an mRNA. Next, the anticodon of a special initiator tRNA base-pairs with the first AUG codon of the mRNA. Then, a large ribosomal subunit joins the small subunit. The cluster is now called an initiation complex (Figure 14.12a,b).

In the elongation stage, the ribosome assembles a polypeptide chain as it moves along the mRNA, threading the strand between its two subunits. The initiator tRNA carries the amino acid methionine, so the first amino acid of the new polypeptide chain is methionine. Other tRNAs bring successive amino acids to the complex as their anticodons base-pair with the codons in the mRNA, one after the next. The ribosome joins each amino acid to the end of the growing polypeptide chain by way of a peptide bond (Figure 14.12c–e and Section 3.5).

Termination occurs when the ribosome encounters a stop codon in the mRNA. Proteins called release factors recognize this codon and bind to the ribosome. The binding triggers enzyme activity that detaches the mRNA and the polypeptide chain from the ribosome (Figure 14.12f).

In cells that are making a lot of protein, new initiation complexes may form on an mRNA before other ribosomes finish translating it. Many ribosomes may simultaneously translate the same mRNA, in which case they are called polysomes (*left*). Transcription and translation both occur in the cytoplasm of prokaryotes, and these processes are closely linked in time and in space. Translation begins before transcription is done, so in these cells, a transcription Christmas tree (Figure 14.6) often appears decorated with polysome "balls."

Translation is a biosynthetic process that requires a lot of energy to run (Section 6.2). That energy is provided mainly in the form of phosphate-group transfers from the RNA nucleotide GTP (Figure 14.2a). GTP caps eukaryotic mRNAs, and its hydrolysis also fuels formation of the initiation complex, binding of tRNA to the ribosome, movement of the ribosome along the mRNA, formation of peptide bonds, and release of the ribosomal subunits from mRNA during termination. ATP is used to attach amino acids to free tRNAs.

polysome

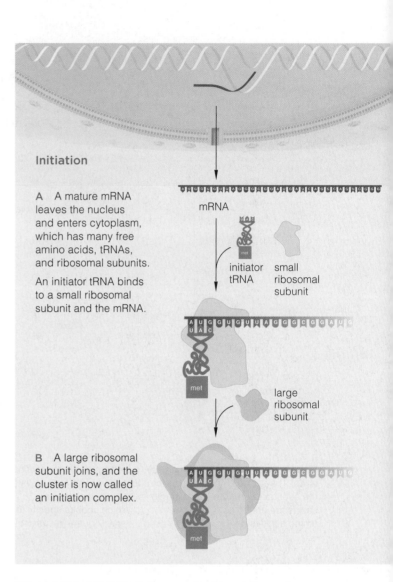

Initiation

A A mature mRNA leaves the nucleus and enters cytoplasm, which has many free amino acids, tRNAs, and ribosomal subunits.

An initiator tRNA binds to a small ribosomal subunit and the mRNA.

mRNA

initiator tRNA   small ribosomal subunit

large ribosomal subunit

B A large ribosomal subunit joins, and the cluster is now called an initiation complex.

Figure 14.12 Animated An example of translation as it occurs in eukaryotic cells.

(**a,b**) In initiation, an mRNA, an intact ribosome, and an initiator tRNA form an initiation complex.

(**c–e**) In elongation, the new polypeptide chain grows as the ribosome catalyzes the formation of peptide bonds between amino acids delivered by tRNAs.

(**f**) In termination, the mRNA and the new polypeptide chain are released, and the ribosome disassembles.

Take-Home Message

How is mRNA translated into protein?

- Translation is an energy-requiring process that begins as an mRNA joins with an initiator tRNA and two ribosomal subunits.
- Amino acids are delivered to the complex by tRNAs in the order dictated by successive mRNA codons. As they arrive, the ribosome joins each to the end of the polypeptide chain.
- Translation ends when the ribosome encounters a stop codon in the mRNA.

Elongation

C An initiator tRNA carries the amino acid methionine, so the first amino acid of the new polypeptide chain will be methionine. A second tRNA binds the second codon of the mRNA (here, that codon is GUG, so the tRNA that binds carries the amino acid valine).

A peptide bond forms between the first two amino acids (here, methionine and valine).

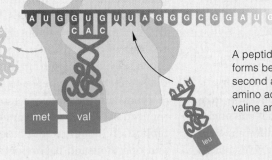

D The first tRNA is released and the ribosome moves to the next codon in the mRNA. A third tRNA binds to the third codon of the mRNA (here, that codon is UUA, so the tRNA carries the amino acid leucine).

A peptide bond forms between the second and third amino acids (here, valine and leucine).

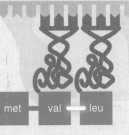

E The second tRNA is released and the ribosome moves to the next codon. A fourth tRNA binds the fourth mRNA codon (here, that codon is GGG, so the tRNA carries the amino acid glycine).

A peptide bond forms between the third and fourth amino acids (here, leucine and glycine).

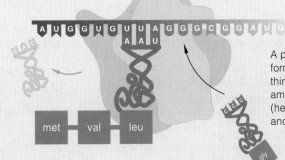

Termination

F Steps **d** and **e** are repeated over and over until the ribosome encounters a stop codon in the mRNA. The mRNA transcript and the new polypeptide chain are released from the ribosome. The two ribosomal subunits separate from each other. Translation is now complete. Either the chain will join the pool of proteins in the cytoplasm or it will enter rough ER of the endomembrane system (Section 4.9).

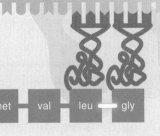

Mutated Genes and Their Protein Products

■ If the nucleotide sequence of a gene changes, it may result in an altered gene product, with harmful effects.

■ Links to Mutation 1.4, Electrons 2.3, Protein structure 3.6, Free radicals 6.3, Radiant energy 7.1, Cancer 9.5, Huntington's disease 12.2, DNA replication 13.3

We have repeatedly mentioned mutations in reference to the harm they can cause, and also as the raw material of evolution. Mutations are small-scale changes in the nucleotide sequence of a cell's DNA. One or more nucleotides may be substituted for another or lost, or extra ones inserted. Such changes can alter the genetic instructions encoded in the DNA, and the result may be an altered gene product. Remember, more than one codon can specify the same amino acid, so cells have a margin of safety. For example, a mutation that changes a UCU to a UCC in an mRNA may not have further effects, because both codons specify serine. However, many mutations have negative consequences.

Common Mutations

A nucleotide mispaired during DNA replication may end up as a **base-pair substitution**, in which one nucleotide and its partner are replaced by a different base pair. A substitution may result in an amino acid change or a premature stop codon in a gene's protein product. Sickle-cell anemia is caused by a base-pair substitution in the hemoglobin beta chain gene (Figure 14.13b).

A **deletion** mutation, in which one or more bases is lost, is smaller than a chromosomal deletion (Section 12.5), but either can cause the reading frame of mRNA codons to shift. The shift garbles the genetic message (Figure 14.13c). Frameshifts are also caused by **insertion** mutations, in which extra bases are inserted into DNA. The expansion mutation that causes Huntington's disease (Section 12.2) is a type of insertion.

What Causes Mutations?

Insertion mutations are often caused by the activity of **transposable elements**, which are segments of DNA that can insert themselves anywhere in a chromosome (Figure 14.14). Transposable elements can be hundreds or thousands of base pairs long. When one interrupts a gene sequence, it becomes a major insertion that changes the gene's product. Transposable elements occur in the DNA of all species; about 45 percent of human DNA consists of them or their remnants. Certain kinds can move spontaneously from one place to another within the same chromosome, or to a different chromosome.

Many mutations occur spontaneously during DNA replication. That is not surprising, given the fast pace of replication (about twenty bases per second in humans, and one thousand bases per second in bacteria). DNA polymerases make mistakes at predictable rates, but most types fix errors as they occur (Section 13.3). Errors that remain uncorrected are mutations.

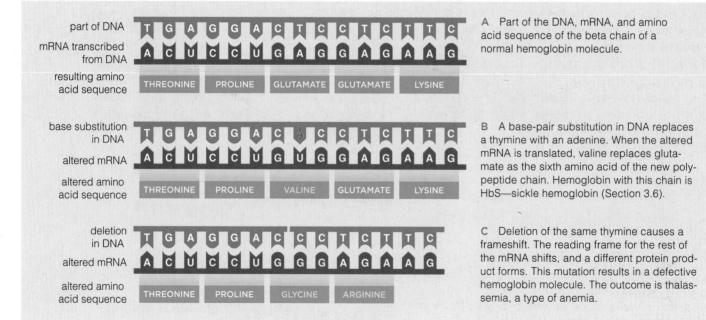

part of DNA	T G A G G A C T C C T C T T C	A Part of the DNA, mRNA, and amino acid sequence of the beta chain of a normal hemoglobin molecule.
mRNA transcribed from DNA	A C U C C U G A G G A G A A G	
resulting amino acid sequence	THREONINE PROLINE GLUTAMATE GLUTAMATE LYSINE	

base substitution in DNA	T G A G G A C A C C T C T T C	B A base-pair substitution in DNA replaces a thymine with an adenine. When the altered mRNA is translated, valine replaces glutamate as the sixth amino acid of the new polypeptide chain. Hemoglobin with this chain is HbS—sickle hemoglobin (Section 3.6).
altered mRNA	A C U C C U G U G G A G A A G	
altered amino acid sequence	THREONINE PROLINE VALINE GLUTAMATE LYSINE	

deletion in DNA	T G A G G A C C C T C T T C	C Deletion of the same thymine causes a frameshift. The reading frame for the rest of the mRNA shifts, and a different protein product forms. This mutation results in a defective hemoglobin molecule. The outcome is thalassemia, a type of anemia.
altered mRNA	A C U C C U G G G A G A A G	
altered amino acid sequence	THREONINE PROLINE GLYCINE ARGININE	

Figure 14.13 Animated Examples of mutation.

Figure 14.14 Barbara McClintock discovered transposable elements, which slip into and out of different locations in DNA. The curiously nonuniform coloration of individual kernels in Indian corn (*Zea mays*) sent her on the road to the discovery. She won a Nobel Prize for her research in 1983.

Several genes govern the formation and deposition of pigments in corn kernels, which are a type of seed. Interactions among these genes and their products result in yellow, white, red, orange, blue, or purple kernels. McClintock realized that unstable mutations in the genes cause streaks or spots of color in individual kernels.

The same pigment genes occur in all cells of a kernel, but those near a transposable element are inactive. Transposable elements move while a kernel's tissues are forming, so they can end up in different locations in the DNA of different cell lineages. Streaks and spots on the kernels are evidence of transposable element movement that inactivated and reactivated different pigment genes in different cell lineages.

Harmful environmental agents can cause mutations. For example, some forms of energy such as x-rays can ionize atoms by knocking electrons right out of them. Such ionizing radiation can break chromosomes into pieces, some of which may get lost during DNA replication (Figure 14.15*a*). Ionizing radiation also damages DNA indirectly when it penetrates living tissue, because it leaves a trail of destructive free radicals. Free radicals, remember, damage DNA (Section 6.3). That is why doctors and dentists use the lowest possible doses of x-rays on their patients.

Nonionizing radiation boosts electrons to a higher energy level, but not enough to knock them out of an atom. DNA absorbs one kind, ultraviolet (UV) light. Exposure to UV light can cause two adjacent thymine bases to bond covalently to one another. This bond, a thymine dimer, kinks the DNA (Figure 14.15*b*). During replication, the kinked part may be copied incorrectly, so a mutation is introduced into the DNA. Mutations that cause certain kinds of cancers begin with thymine dimers. They are the reason that exposing unprotected skin to sunlight increases the risk of skin cancer.

Some natural or synthetic chemicals can also cause mutations. For instance, certain chemicals in cigarette smoke transfer small hydrocarbon groups to the bases in DNA. The altered bases mispair during replication, or stop replication entirely.

The Proof Is in the Protein

A mutation that occurs in a somatic cell of a sexually reproducing individual is not passed to the individual's offspring, so its effects do not endure. A mutation that arises in a germ cell or a gamete, however, may enter the evolutionary arena. It may also do so when passed on to offspring by asexual reproduction. Either way, an inherited mutation may affect an individual's capacity to function in its prevailing environment. The effects of uncountable mutations in millions of species have had spectacular evolutionary consequences—and that is a topic of later chapters.

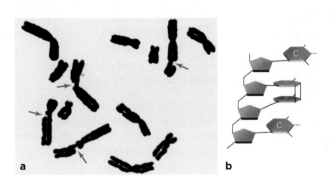

Figure 14.15 Two types of DNA damage that can lead to mutations. (**a**) Chromosomes from a human cell after exposure to gamma rays (ionizing radiation). The broken pieces (*red arrows*) may get lost during DNA replication. The extent of damage in an exposed cell typically depends on how much radiation it absorbed. (**b**) A thymine dimer.

Take-Home Message

What is a mutation?

■ A mutation is a permanent small-scale change in the nucleotide sequence of DNA. A base-pair substitution, insertion, or deletion may alter a gene product.

■ Most mutations arise during DNA replication as a result of unrepaired DNA polymerase errors. Some mutations occur after exposure to harmful radiation or chemicals.

■ An inherited mutation may have positive or negative effects on an individual's capacity to function in its environment.

Ricin and Your Ribosomes

One of ricin's two polypeptide chains binds to a receptor on animal cell membranes that triggers endocytosis. The other chain is an enzyme; it removes a specific adenine base from one of the rRNA chains in the large ribosomal subunit. Once that happens, the ribosome stops working. A single molecule of ricin can inactivate about 1,500 ribosomes per minute. Protein synthesis grinds to a halt as ricin inactivates the rest of the cell's ribosomes.

How would you vote?

Terrorists may try to poison food or water supplies with ricin. Do you want to be vaccinated for ricin exposure? See CengageNOW for details, then vote online.

Summary

Section 14.1 The process of **gene expression** includes two steps, **transcription** and **translation** (Figure 14.16). It requires the participation of **messenger RNA (mRNA)**, **transfer RNA (tRNA)**, and **ribosomal RNA (rRNA)**.

Section 14.2 In eukaryotic cells, transcription occurs in the nucleus, and translation occurs in the cytoplasm. Both processes occur in the cytoplasm of prokaryotic cells.

In transcription, **RNA polymerase** binds to a **promoter** in the DNA near a gene, then assembles a strand of RNA by linking RNA nucleotides in the order dictated by the base sequence of the DNA.

■ *Use the animation on CengageNOW to explore transcription.*

Section 14.3 The RNA of eukaryotes is modified before it leaves the nucleus. **Introns** are removed. Some **exons** may be removed also, and the remaining ones spliced in different combinations (**alternative splicing**). A cap and a poly-A tail are also added to a new mRNA.

mRNA carries DNA's protein-building information. Its genetic message is written in **codons**, sets of three nucleotides. Sixty-four codons, most of which specify amino acids, constitute the **genetic code**. Variations occur among prokaryotes, organelles, and single-celled eukaryotes.

Each tRNA has an **anticodon** that can base-pair with a codon, and it binds to the kind of amino acid specified by the codon. Catalytic rRNA and proteins make up the two subunits of ribosomes.

■ *Use the interaction on CengageNOW to learn about transcript processing and the genetic code.*

Section 14.4 Genetic information carried by an mRNA directs the synthesis of a polypeptide chain during translation. First, an mRNA, an initiator tRNA, and two ribosomal subunits converge. The intact ribosome then catalyzes formation of a peptide bond between successive amino acids, which are are delivered by tRNAs in the order specified by the codons in the mRNA. Translation ends when the polymerase encounters a stop codon.

■ *Use the animation on CengageNOW to see the translation of an mRNA transcript.*

Section 14.5 Insertions, deletions, and **base-pair substitutions** may change a gene's product. These mutations may arise by replication error, **transposable element** activity, or exposure to environmental hazards.

■ *Use the animation on CengageNOW to investigate the effects of mutation.*

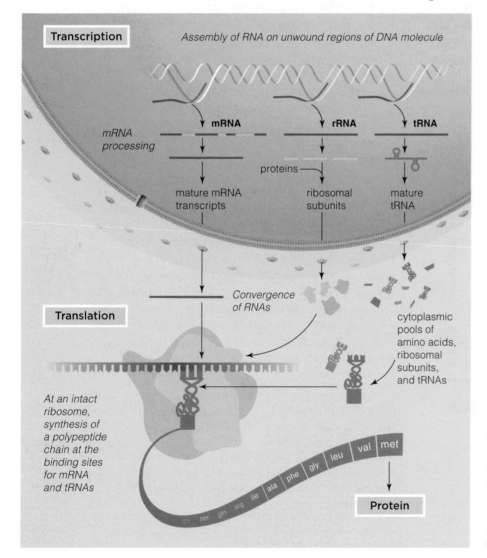

Transcription — Assembly of RNA on unwound regions of DNA molecule

mRNA processing

mRNA   rRNA   tRNA

proteins

mature mRNA transcripts   ribosomal subunits   mature tRNA

Convergence of RNAs

Translation

cytoplasmic pools of amino acids, ribosomal subunits, and tRNAs

At an intact ribosome, synthesis of a polypeptide chain at the binding sites for mRNA and tRNAs

gly ser gln arg ile ala phe gly leu val met

Protein

Figure 14.16 Animated Summary of protein synthesis as it occurs in eukaryotic cells.

Data Analysis Exercise

About one out of 3,500 people carry a mutation that affects the product of the NF1 gene, which is a tumor suppressor (Section 9.5). People who are heterozygous for one of these mutations have neurofibromatosis, an autosomal dominant genetic disorder (Section 12.2). Among other problems, soft, fibrous tumors (neurofibromas) form in the skin and nervous system. The homozygous condition may be lethal.

Most mutations associated with neurofibromatosis result in defective splicing of the gene's 60 exons. Each neurofibroma typically arises from a new mutation that disrupts the individual's one functional allele. In a 1997 study, Eduard Serra and his colleagues tested several tumors from an individual with the disorder for such mutations (Figure 14.17).

1. Which tumors are missing marker D17S250? Is this sequence inside or outside of the *NF1* gene?

2. In four of these six tumors, the entire large arm of chromosome 17 was deleted. Which four?

3. People affected by neurofibromatosis are 200 to 500 times more likely to develop malignant tumors than unaffected people. Why do you think that is the case?

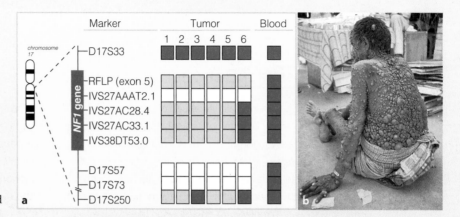

Figure 14.17 Neurofibromatosis. (**a**) Genetic analysis of six tumors from a single individual affected by neurofibromatosis. Each tumor was checked for the presence of nine nucleotide sequences (markers) in or near the *NF1* gene.

For each tumor (1–6), *green* boxes indicate that the marker is present; *yellow* boxes indicate the marker is missing; *white* boxes indicate inconclusive results. Blood was also tested as a control.

(**b**) An individual affected by neurofibromatosis.

Self-Quiz   *Answers in Appendix III*

1. A chromosome contains many genes that are transcribed into different _____ .
   a. proteins   c. RNAs
   b. polypeptides   d. a and b

2. A binding site for RNA polymerase is a _____ .

3. Energy that drives transcription is provided by _____ .

4. An RNA molecule is typically _____ -stranded.

5. RNAs form by _____ ; proteins form by _____ .
   a. replication; translation   c. translation; transcription
   b. transcription; translation   d. replication; transcription

6. _____ remain in mRNA.
   a. Introns   b. Exons

7. How many codons constitute the genetic code?

8. Most codons specify a(n) _____ .
   a. protein   c. amino acid
   b. polypeptide   d. mRNA

9. Anticodons pair with _____ .
   a. mRNA codons   c. RNA anticodons
   b. DNA codons   d. amino acids

10. Energy that drives translation is provided by _____ .
   a. ATP   c. UTP
   b. GTP   d. a and b are correct

11. Using Figure 14.9, translate this nucleotide sequence into an amino acid sequence, starting at the first base:

5′—GGUUUCUUCAAGAGA—3′

12. Name one cause of mutations.

13. Match each term with the most suitable description.
   ___ genetic message   a. protein-coding mRNA
   ___ sequence   b. gets around
   ___ polysome   c. read as base triplets
   ___ exon   d. linear order of bases
   ___ genetic code   e. occurs only in groups
   ___ intron   f. set of 64 codons
   ___ transposable   g. removed before translation
       element

■ *Visit CengageNOW for additional questions.*

Critical Thinking

1. Each position of a codon can be occupied by one of four (4) nucleotides. If codons were two (2) nucleotides long, they could encode a maximum of $4^2 = 16$ amino acids. What is the minimum number of nucleotides per codon necessary to specify all 20 biological amino acids?

2. Cigarette smoke contains at least fifty-five different chemicals identified as carcinogenic (cancer-causing) by the International Agency for Research on Cancer (IARC). When these carcinogens enter the bloodstream, enzymes convert them to a series of chemical intermediates that are easier to excrete. Some of the intermediates bind irreversibly to DNA. Propose a mechanism by which such binding causes cancer.

3. Termination of prokaryotic DNA transcription often depends on the structure of a newly forming RNA. Transcription stops where the mRNA folds back on itself, forming a hairpin-looped structure such as the one at *right*. How do you think this structure stops transcription?

...CCCAC
```
  C
U   C
G—C
A—U
C—G
C—G
G—C
C—G
C—G
G—C
C—G
  A   A
```
...AUUUUU...

15 | Controls Over Genes

IMPACTS, ISSUES | Between You and Eternity

You are in college, your whole life ahead of you. Your risk of developing cancer is as remote as old age, an abstract statistic that is easy to forget. "There is a moment when everything changes—when the width of two fingers can suddenly be the total distance between you and eternity." Robin Shoulla wrote those words after being diagnosed with breast cancer. She was seventeen. At an age when most young women are thinking about school, parties, and potential careers, Robin was dealing with radical mastectomy—the removal of a breast, all lymph nodes under the arm, and skeletal muscles in the chest wall under the breast. She was pleading with her oncologist not to use her jugular vein for chemotherapy and wondering if she would survive to see the next year (Figure 15.1).

Robin's ordeal became part of a statistic—one of more than 200,000 new cases of breast cancer diagnosed in the United States each year. About 5,700 of those cases occur in women and men under thirty-four years of age.

Mutations in some genes predispose individuals to develop certain kinds of cancer. Tumor suppressor genes are named because tumors are more likely to occur when these genes mutate. Two examples are *BRCA1* and *BRCA2*. A mutated version of one or both of these genes is often found in breast and ovarian cancer cells. If a *BRCA* gene mutates in one of three especially dangerous ways, a woman has an 80 percent chance of developing breast cancer before the age of seventy.

Tumor suppressors are part of a system of stringent controls over gene expression that keeps the cells of multicelled organisms functioning normally. Such controls govern when and how fast specific genes are transcribed and translated. You will be considering the impact of gene controls in chapters throughout the book—and in some chapters of your life.

Robin Shoulla survived. Although radical mastectomy is rarely performed today (a modified procedure is less disfiguring), it is the only option when cancer cells invade muscles under the breast. It was Robin's only option. She may never know which mutation caused her cancer. Now, sixteen years later, she has what she calls a normal life—career, husband, children. Her goal as a cancer survivor: "To grow very old with gray hair and spreading hips, smiling."

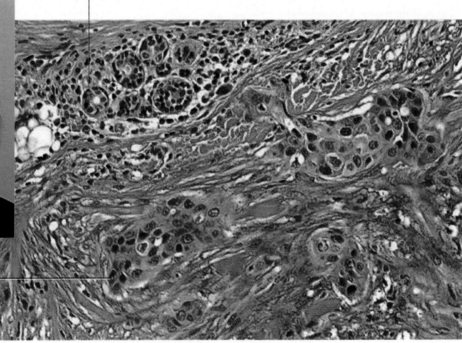

normal cells in organized clusters

irregular clusters of malignant cells

See the video! Figure 15.1 A case of breast cancer. *Right*, this light micrograph revealed irregular clusters of carcinoma cells that infiltrated milk ducts in breast tissue. *Above*, Robin Shoulla. Diagnostic tests revealed abnormal cells such as these in her body.

Key Concepts

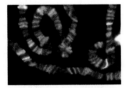

Overview of controls over gene expression

A variety of molecules and processes alter gene expression in response to changing conditions both inside and outside the cell. Selective gene expression also results in cell differentiation, by which different cell lineages become specialized. **Section 15.1**

Examples from eukaryotes

The orderly, localized expression of certain genes in embryos gives rise to the body plan of complex multicelled organisms. In female mammals, most of the genes on one of the two X chromosomes are inactivated in every cell. **Section 15.2**

Fruit fly development

Drosophila research revealed how a complex body plan emerges. All cells in a developing embryo inherit the same genes, but they use different subsets of those genes. **Section 15.3**

Examples from prokaryotes

Prokaryotic gene controls govern responses to short-term changes in nutrient availability and other aspects of the environment. The main gene controls bring about fast adjustments in the rate of transcription. **Section 15.4**

Links to Earlier Concepts

■ A review of what you know about metabolic controls (Section 6.4) will be helpful as we revisit the concept of gene expression (14.1) in more detail. You may wish to review alleles (11.1), autosomal inheritance (12.2), and mutation (14.5).

■ You will be applying what you know about the organization of chromosomal DNA (9.1, 9.2), and sex determination and X-linked inheritance in humans (12.1, 12.4), as we delve into controls over transcription (14.2), post-transcriptional processing (14.3), translation (14.4), and other processes that affect gene expression.

■ You will revisit carbohydrates (3.3) and fermentation (8.5) as you learn about gene control in prokaryotes.

How would you vote? Some women at high risk of developing breast cancer opt for preventive surgical removal of their breasts before cancer develops. Many of those women never would have developed cancer. Should surgery be restricted to cancer treatment? See CengageNOW for details, then vote online.

- Gene controls govern the kinds and amounts of substances that are present in a cell at any given interval.

- Links to Histones 9.1, Gene expression 14.1, Transcription 14.2, Post-transcriptional modification 14.3, Translation 14.4

Which Genes Get Tapped?

All of the cells in your body are descended from the same fertilized egg, so they all contain the same DNA with the same genes. Some of the genes are transcribed by all cells; such genes affect structural features and metabolic pathways common to all cells.

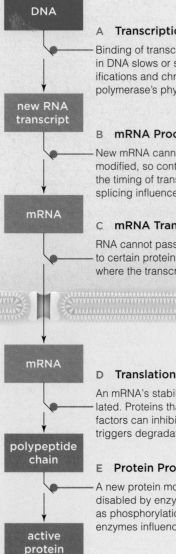

NUCLEUS

DNA

A Transcription

Binding of transcription factors to special sequences in DNA slows or speeds transcription. Chemical modifications and chromosome duplications affect RNA polymerase's physical access to genes.

new RNA transcript

B mRNA Processing

New mRNA cannot leave the nucleus before being modified, so controls over mRNA processing affect the timing of transcription. Controls over alternative splicing influence the final form of the protein.

mRNA

C mRNA Transport

RNA cannot pass through a nuclear pore unless bound to certain proteins. Transport protein binding affects where the transcript will be delivered in the cell.

CYTOPLASM

mRNA

D Translation

An mRNA's stability influences how long it is translated. Proteins that attach to ribosomes or initiation factors can inhibit translation. Double-stranded RNA triggers degradation of complementary mRNA.

polypeptide chain

E Protein Processing

A new protein molecule may become activated or disabled by enzyme-mediated modifications, such as phosphorylation or cleavage. Controls over these enzymes influence many other cell activities.

active protein

Figure 15.2 Animated Points of control over eukaryotic gene expression.

enhancer

Figure 15.3 Hypothetical part of a chromosome that contains a gene. Molecules that affect the rate of transcription of the gene bind at promoter (*yellow*) and enhancer (*green*) sequences.

In other ways, however, nearly all of your body cells are specialized. **Differentiation**, the process by which cells become specialized, occurs as different cell lineages begin to express different subsets of their genes. Which genes a cell uses determines the molecules it will produce, which in turn determines what kind of cell it will be.

For example, most of your body cells express the genes that encode the enzymes of glycolysis, but only immature red blood cells use the genes that code for globin chains. Only your liver cells express genes for enzymes that neutralize certain toxins.

A cell rarely uses more than 10 percent of its genes at once. Which genes are expressed at any given time depends on many factors, such as conditions in the cytoplasm and extracellular fluid, and the type of cell. The factors affect controls governing all steps of gene expression, starting with transcription and ending with delivery of an RNA or protein product to its final destination. Such controls consist of processes that start, enhance, slow, or stop gene expression.

Control of Transcription Many controls affect whether and how fast certain genes are transcribed into RNA (Figure 15.2*a*). Those that prevent an RNA polymerase from attaching to a promoter near a gene also prevent transcription of the gene. Controls that help RNA polymerase bind to DNA also speed up transcription.

Some types of proteins affect the rate of transcription by binding to special nucleotide sequences in the DNA. For example, an **activator** speeds up transcription when it binds to a promoter. Activators also bind to DNA sequences called **enhancers**. An enhancer is not necessarily close to the gene it affects, and may even be on a different chromosome (Figure 15.3). As another example, a **repressor** slows or stops transcription when it binds to certain sites in DNA.

Regulatory proteins such as activators and repressors are called **transcription factors**. Whether and how fast a gene is transcribed depends on which transcription factors are bound to the DNA.

Interactions between DNA and the histone proteins it wraps around also affect transcription. RNA polymerase can only attach to DNA that is unwound from histones (Section 9.1). Attachment of methyl groups

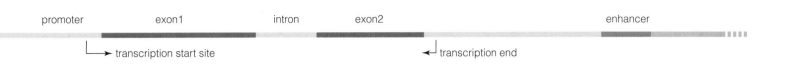

promoter   exon1   intron   exon2   enhancer

└─→ transcription start site   ↵ transcription end

(—CH₃) causes DNA to wind tightly around histones; thus, methylation of DNA prevents its transcription.

The number of copies of a gene also affects how fast its product is made. For example, in some cells, DNA is copied repeatedly with no cytoplasmic division between replications. The result is a cell full of polytene chromosomes, each of which consists of hundreds or thousands of side-by-side copies of the same DNA molecule. All of the DNA strands carry the same genes. Translation of one gene, which occurs simultaneously on all of the identical DNA strands, produces a lot of mRNA, which is translated quickly into a lot of protein. Polytene chromosomes are common in the saliva gland cells of some insect larva and immature amphibian eggs (Figure 15.4).

mRNA Processing As you know, before eukaryotic mRNAs leave the nucleus, they are modified—spliced, capped, and finished with a poly-A tail (Section 14.3). Controls over these modifications can affect the form of a protein product and when it will appear in the cell (Figure 15.2b). For example, controls that determine which exons are spliced out of an mRNA affect which form of a protein will be translated from it.

mRNA Transport mRNA transport is another point of control (Figure 15.2c). For example, in eukaryotes, transcription occurs in the nucleus, and translation in the cytoplasm. A new RNA can pass through pores of the nuclear envelope only after it has been processed appropriately. Controls that delay the processing also delay an mRNA's appearance in the cytoplasm, and thereby delay its translation.

Controls also govern mRNA localization. A short base sequence near an mRNA's poly-A tail is like a zip code. Certain proteins that attach to the zip code drag the mRNA along cytoskeletal elements and deliver it to a particular organelle or area of the cytoplasm. Other proteins that attach to the zip code region prevent the mRNA from being translated before it reaches its destination. mRNA localization allows cells to grow or move in specific directions. It is also crucial for proper embryonic development.

Translational Control Most controls over eukaryotic gene expression affect translation (Figure 15.2d). Many govern the production or function of the various mole-

cules that carry out translation. Others affect mRNA stability: The longer an mRNA lasts, the more protein can be made from it. Enzymes begin to disassemble a new mRNA as soon as it arrives in the cytoplasm. The fast turnover allows cells to adjust their protein synthesis quickly in response to changing needs. How long an mRNA persists depends on its base sequence, the length of its poly-A tail, and which proteins are attached to it.

As a different example, microRNAs inhibit translation of other RNA. Part of a microRNA folds back on itself and forms a small double-stranded region. By a process called RNA interference, any double-stranded RNA (including a microRNA) is cut up into small bits that are taken up by special enzyme complexes. These complexes destroy every mRNA in a cell that can base-pair with the bits. So, expression of a microRNA complementary in sequence to a gene inhibits expression of that gene.

Post-Translational Modification
Many newly-synthesized polypeptide chains must get modified before they become functional (Figure 15.2e). For example, some enzymes become active only after they have been phosphorylated (another enzyme has attached a phosphate group to them). Such post-translational modifications can inhibit, activate, or stabilize many molecules, including the enzymes that participate in transcription and translation.

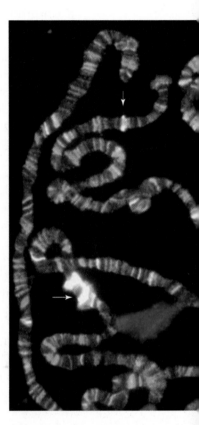

Figure 15.4 *Drosophila* polytene chromosomes. *Drosophila* larvae eat continuously, so they use a lot of saliva. In their salivary gland cells, giant polytene chromosomes form by repeated DNA replication.

Each of these chromosomes consists of hundreds or thousands of copies of the same DNA strand, aligned side by side. Transcription is visible as puffs, where the DNA has loosened (*arrows*).

Take-Home Message

What is gene expression control?

■ Most cells of multicelled organisms differentiate when they start expressing a unique subset of their genes. Which genes a cell expresses depends on the type of organism, its stage of development, and environmental conditions.

■ Various control processes regulate all steps between gene and gene product.

15.2 | A Few Outcomes of Eukaryotic Gene Controls

■ Many traits are evidence of selective gene expression.

■ Links to Chromosome number 9.2, Alleles 11.1, Sex chromosomes 12.1, X-linked inheritance 12.4, Mutation 14.5

X Chromosome Inactivation

Remember, in humans and other mammals, a female's cells each contain two X chromosomes, one inherited from her mother, the other one from her father (Section 12.1). One X chromosome is always tightly condensed, even during interphase (Figure 15.5a). We call the condensed X chromosomes "Barr bodies," after Murray Barr, who discovered them. RNA polymerase cannot access most of the genes on the condensed chromosome. **X chromosome inactivation** ensures that only one of the two X chromosomes in a female's cells is active.

X chromosome inactivation occurs when an embryo is a ball of about 200 cells. In humans and many other mammals, it occurs independently in every cell of a female embryo. The maternal X chromosome may get inactivated in one cell, and the paternal or maternal X chromosome may get inactivated in a cell next to it. Once the selection is made in a cell, all of that cell's descendants make the same selection as they continue dividing and forming tissues.

As a result of the X chromosome inactivation, an adult female mammal is a "mosaic" for the expression of X-linked genes. She has patches of tissue in which

genes of the maternal X chromosome are expressed, and patches in which genes of the paternal X chromosome are expressed.

The homologous X chromosomes of most females have at least some alleles that are not identical. Thus, most females have variations in traits among patches of tissue. A female's mosaic tissues are visible if she is heterozygous for certain X chromosome mutations.

For example, incontinentia pigmenti is an X-linked disorder that affects the skin, teeth, nails, and hair. In heterozygous human females, mosaic tissues show up as lighter and darker patches of skin. The darker skin consists of cells in which the active X chromosome has the mutated allele; the lighter skin consists of cells in which the active X chromosome has the normal allele (Figure 15.5c).

Mosaic tissues are visible in other female mammals as well. For example, a gene on the X chromosomes of cats influences fur color. The expression of an allele (O) results in orange fur, and expression of another allele (o) results in black fur. Heterozygous cats (Oo) have patches of orange and black fur. Orange patches grow from skin cells in which the active X chromosome carries the O allele; black patches grow from skin cells in which the active X chromosome carries the o allele (Figure 15.6).

According to the theory of **dosage compensation**, X chromosome inactivation equalizes expression of X

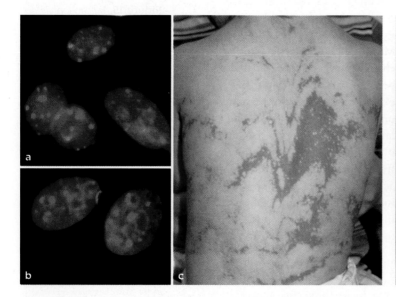

Figure 15.5 X chromosome inactivation. (**a**) Barr bodies (*red*) in the nucleus of four XX cells. (**b**) Compare the nucleus of two XY cells. (**c**) Mosaic tissues show up in human females who are heterozygous for mutations that cause incontinentia pigmenti. In darker patches of this girl's skin, the X chromosome with the mutation is active. In lighter skin, the X chromosome with the normal allele is active.

Figure 15.6 Animated Why is this cat "calico"? When she was an embryo, one or the other X chromosome was inactivated in each of her cells. The descendants of the cells formed mosaic patches of tissue. Orange or black fur results from expression of different alleles on the active X chromosome. (White patches are the outcome of a different gene, the product of which blocks synthesis of all pigment.)

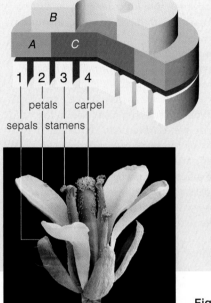

A The pattern in which the floral identity genes *A*, *B*, and *C* are expressed affects differentiation of cells growing in whorls in the plant's tips. Their gene products guide expression of other genes in cells of each whorl; a flower results.

B Mutations in *Arabidopsis* floral identity genes result in mutant flowers. *Top left*, *right*, some mutations lead to flowers with no petals. *Bottom left*, B gene mutations lead to flowers with sepals instead of petals. *Bottom right*, C gene mutations lead to flowers with petals instead of stamens and carpels. Compare the normal flower in (**a**).

Figure 15.7 Animated Control of flower formation, revealed by mutations in *Arabidopsis thaliana*.

chromosome genes between the sexes. The body cells of male mammals (XY) have one set of X chromosome genes. The body cells of female mammals (XX) have two sets, but only one is expressed. Normal development of female embryos depends on this control.

How does just one of two X chromosomes get inactivated? An X chromosome gene called *XIST* does the trick. This gene is transcribed on only one of the two X chromosomes. The gene's product, a large RNA, sticks to the chromosome that expresses the gene. The RNA coats the chromosome and causes it to condense into a Barr body. Thus, transcription of the *XIST* gene keeps the chromosome from transcribing other genes. The other chromosome does not express *XIST*, so it does not get coated with RNA; its genes remain available for transcription. It is still unknown how the cell chooses which chromosome will express *XIST*.

Flower Formation

When it is time for a plant to flower, populations of cells that would otherwise give rise to leaves instead differentiate into floral parts—sepals, petals, stamens, and carpels. How does the switch happen? Studies of mutations in the common wall cress plant, *Arabidopsis thaliana*, support the **ABC model**. This model explains how the specialized parts of a flower develop. Three sets of master genes—*A*, *B*, and *C*—guide the process. **Master genes** encode products that affect expression of many other genes. The expression of a master gene initiates cascades of expression of other genes, with the outcome being the completion of an intricate task such as the formation of a flower.

The master genes that control flower formation are switched on by environmental cues such as daylength, as you will see in Section 31.5. At the tip of a floral shoot (a modified stem), cells form whorls of tissue, one over the other like layers of an onion. Cells in each whorl give rise to different tissues depending on which of their *ABC* genes is activated. In the outer whorl, only the *A* genes are switched on, and their products trigger events that cause sepals to form. Cells in the next whorl express both *A* and *B* genes; they give rise to petals. Cells farther in express *B* and *C* genes; they give rise to male floral structures called stamens. The cells of the innermost whorl express only the *C* genes; they give rise to female floral structures called carpels (Figure 15.7*a*). Studies of the phenotypic effects of *ABC* gene mutations support this model (Figure 15.7*b*).

Take-Home Message

What are some examples of gene expression control?

■ Most genes on one X chromosome in female mammals (XX) are inactivated, which balances gene expression with males (XY).

■ Gene control also guides flower formation. ABC master genes are expressed differently in tissues of floral shoots.

■ Research with fruit flies yielded the insight that body plans are a result of patterns of gene expression in embryos.

For about a hundred years, *Drosophila melanogaster* has been the subject of choice for many research experiments. Why? It costs almost nothing to feed this fruit fly, which is only about 3 millimeters long (*right*) and can live in bottles. *D. melanogaster* also reproduces fast and has a short life cycle. In addition, experimenting on insects that are considered nuisance pests presents few ethical dilemmas.

fruit fly, actual size

Many important discoveries about how gene expression guides development have come from *Drosophila* research. The discoveries help us understand similar processes in humans and other organisms, and provide clues to our shared evolutionary history.

Discovery of Homeotic Genes We now know of 13,767 genes on *Drosophila's* four chromosomes. As in most other eukaryotic species, some are **homeotic genes**: master genes that control formation of specific body parts (eyes, legs, segments, and so on) during the development of embryos. All homeotic genes encode transcription factors with a homeodomain, a region of about sixty amino acids that can bind to a promoter or some other sequence in DNA.

Localized expression of homeotic genes in tissues of a developing embryo gives rise to details of the adult body plan. The process begins long before body parts develop, as various master genes are expressed in local areas of the early embryo. The products of these master genes are transcription factors. They form in concentration gradients that span the entire embryo. Depending on where they are located within the gradients, embryonic cells begin to transcribe different homeotic genes. Products of these homeotic genes form in specific areas of the embryo.

—homeodomain

—DNA

The different products cause cells to differentiate into tissues that form specific structures such as wings or a head.

Researchers discovered homeotic genes by analyzing the DNA of mutant fruit flies that had body parts growing in the wrong places. As an example, the homeotic gene *antennapedia* is transcribed in embryonic tissues that give rise to a thorax, complete with legs. Normally, it is never transcribed in cells of any other tissue. Figure 15.8*b* shows what happens after a mutation causes *antennapedia* to be transcribed in embryonic tissue that gives rise to the head.

Over 100 homeotic genes have been identified. They control development by the same mechanisms in all eukaryotes, and many are interchangeable between species. Thus, we can expect that they evolved in the most ancient eukaryotic cells. Homeodomains often differ among species only in conservative substitutions—one amino acid has replaced another with similar chemical properties.

Knockout Experiments By controlling the expression of genes in *Drosophila* one at a time, researchers have made other important discoveries about how embryos of many organisms develop. In **knockout experiments**, researchers inactivate a gene by introducing a mutation into it. Then they observe how an organism that carries the mutation differs from normal individuals. The differences are clues to the function of the missing gene product.

Researchers name homeotic genes based on what happens in their absence. For instance, flies that have had their *eyeless* gene knocked out develop with no eyes. *Dunce* is required for learning and memory. *Wingless*, *wrinkled*, and *minibrain* genes are self-explanatory. *Tinman* is necessary for development of a heart. Flies with a mutated *groucho* gene have too many bristles above their eyes. One gene was named *toll*, after what a German researcher exclaimed upon seeing the disastrous effects of its mutation (*toll* is German for "cool!"). Figure 15.8 shows some mutant flies.

Humans, squids, mice, and many other animals have a homologue of the *eyeless* gene called *PAX6*. In humans, mutations in *PAX6* cause eye disorders such as aniridia—underdeveloped or missing irises. Altered expression of

Figure 15.8 Homeotic gene experiments.
(**a**) Normal fly head. (**b**) Transcription of the *antennapedia* gene in the embryonic tissues of the thorax causes legs to form on the body. A mutation that causes *antennapedia* to be transcribed in the embryonic tissues of the head causes legs to form there too. The *antennapedia* homeodomain is modeled *above*, in *green*.

(**c**) Eyes form wherever the *eyeless* gene is expressed in fly embryos—here, on a wing.

(**d**) *Facing page*, more *Drosophila* mutations that yielded clues to homeotic gene function.

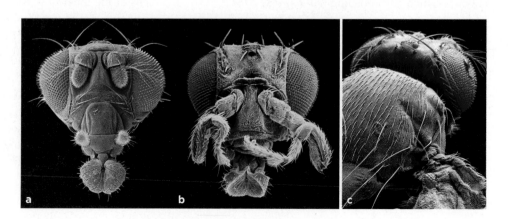

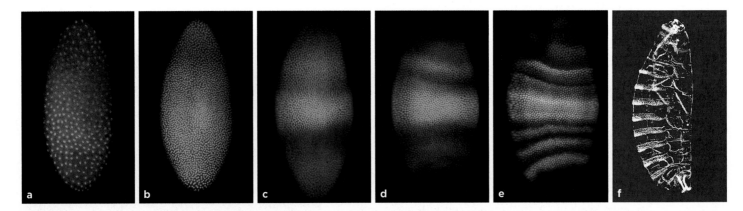

Figure 15.9 How gene expression control makes a fly, as illuminated by segmentation. The expression of different master genes is shown by different colors in fluorescence microscopy images of whole *Drosophila* embryos at successive stages of development. The bright dots are individual cell nuclei.

(**a,b**) The master gene *even-skipped* is expressed (in *red*) only where two maternal gene products (*blue* and *green*) overlap.

(**c–e**) The products of several master genes, including the two shown here in *green* and *blue*, confine the expression of *even-skipped* (*red*) to seven stripes. (**f**) One day later, seven segments develop that correspond to the position of the stripes.

the *eyeless* gene causes eyes to form not only on a fruit fly's head but also on its wings and legs (Figure 15.8c). *PAX6* works the same way in frogs—it causes eyes to form wherever it is expressed in tadpoles.

Researchers also discovered that *PAX6* is one of the homeotic genes that works across different species. If *PAX6* from a human, mouse, or squid is inserted into an *eyeless* mutant fly, it has the same effect as the *eyeless* gene: An eye forms wherever it is expressed. Such studies are evidence of a shared ancestor among these evolutionarily distant animals.

Filling In Details of Body Plans As an embryo develops, its differentiating cells form tissues, organs, and body parts. Some cells that alternately migrate and stick to other cells develop into nerves, blood vessels, and other structures that weave through the tissues. Events like these fill in the body's details, and all are driven by cascades of master gene expression. **Pattern formation** is the process by which a complex body forms from local processes

in an embryo. Patterning begins as maternal mRNAs are delivered to opposite ends of an unfertilized egg as it forms. The localized maternal mRNAs get translated right after the egg is fertilized, and their protein products diffuse away in gradients that span the entire embryo. Cells of the developing embryo begin to translate different master genes, depending on where they fall within those gradients. The products of those genes also form in overlapping gradients. Cells of the embryo translate still other master genes depending on where they fall within the gradients, and so on.

Such regional gene expression during development results in a three-dimensional map that consists of overlapping concentration gradients of master gene products. Which master genes are active at any given time changes, and so does the map. Some master gene products cause undifferentiated cells to differentiate, and specialized tissues are the outcome. The formation of body segments in a fruit fly embryo is an example of how pattern formation works (Figure 15.9). Section 43.4 returns to this topic.

d *Left to right*, normal fly; yellow miniature; curly wings; vestigial wings; and ultrabithorax—a double thorax mutant.

- Prokaryotes control gene expression mainly by adjusting the rate of transcription.

- Links to Carbohydrates 3.3, Controls over metabolism 6.4, Lactate fermentation 8.5, Autosomal inheritance patterns 12.2

Prokaryotes do not undergo development and become multicelled organisms, so these cells do not use master genes. However, they do use gene controls. By adjusting gene expression, they can respond to environmental conditions. For example, when a certain nutrient becomes available, a prokaryotic cell will begin transcribing genes whose products allow the cell to use that nutrient. When the nutrient is not available, transcription of those genes stops. Thus, the cell does not waste energy and resources producing gene products that are not needed at a particular moment.

Prokaryotes control their gene expression mainly by adjusting the rate of transcription. Genes that are used together often occur together on the chromosome, one after the other. All of them are transcribed together into a single RNA strand, so their transcription is controllable in one step.

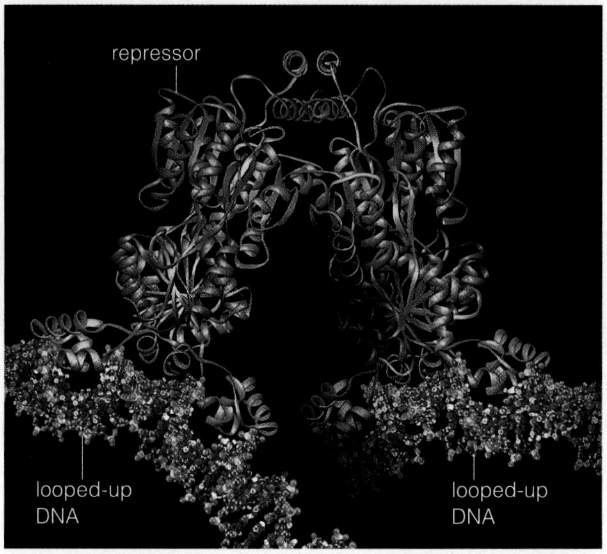

Figure 15.10 Model of the lactose operon repressor, shown here bound to operators. Binding twists the bacterial chromosome into a loop, which in turn prevents RNA polymerase from binding to the lac operon promoter.

The Lactose Operon

Escherichia coli lives in the gut of mammals, where it dines on nutrients traveling past. Its carbohydrate of choice is glucose, but it can make use of other sugars, such as the lactose in milk. *E. coli* cells can harvest the glucose subunit of lactose molecules with a set of three enzymes. However, unless there is lactose in the gut, *E. coli* cells keep the three genes for those enzymes turned off.

There is one promoter for all three genes. Flanking the promoter are two **operators**—DNA regions that are binding sites for a repressor. (Repressors, remember, stop transcription.) A promoter and one or more operators that together control the transcription of multiple genes are collectively called an **operon**.

When lactose is not present, lactose (lac) operon repressors bind *E. coli* DNA, and lactose-metabolizing genes stay switched off. One repressor molecule binds to both operators, and twists the DNA region with the promoter into a loop (Figure 15.10). RNA polymerase cannot bind to the twisted promoter, so it cannot transcribe the operon genes.

When lactose *is* in the gut, some of it is converted to another sugar, allolactose. Allolactose binds to the repressor and changes its shape. The altered repressor can no longer bind to the operators. The looped DNA unwinds, and the promoter is now free for RNA polymerase to begin transcription (Figure 15.11).

Note that *E. coli* cells use extra enzymes to metabolize lactose compared with glucose, so it is more efficient for them to use glucose. Accordingly, when both sugars are present, *E. coli* cells will use up all of the glucose before they switch to lactose metabolism.

How do *E. coli* shut off lactose metabolism in the presence of lactose? They have an additional level of control. Transcription of the lac operon genes occurs very slowly unless an activator binds to the promoter along with RNA polymerase. The activator consists of a protein with a bound nucleotide called cAMP (cyclic adenosine monophosphate). When glucose is plentiful, synthesis of cAMP is blocked, and the activator does not form. When glucose is scarce, cAMP is made. The activator forms and binds to the lac operon promoter. Lac operon genes are transcribed quickly, and lactose-metabolizing enzymes are produced at top speed.

Lactose Intolerance

Like infants of other mammals, human infants drink milk. Cells in the lining of the small intestine secrete lactase, an enzyme that cleaves the lactose in milk into its subunit monosaccharides. In most people, lactase production starts to decline at the age of five.

After that, it becomes more difficult to digest lactose in food—a condition called lactose intolerance.

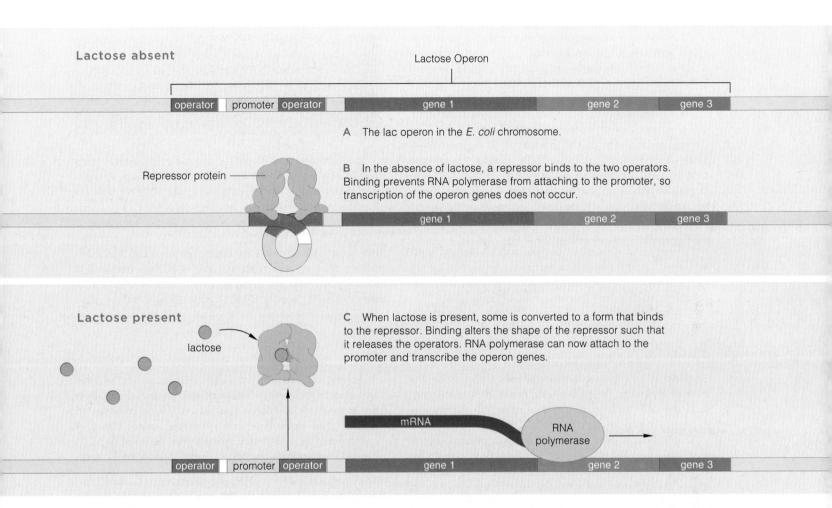

A   The lac operon in the *E. coli* chromosome.

B   In the absence of lactose, a repressor binds to the two operators. Binding prevents RNA polymerase from attaching to the promoter, so transcription of the operon genes does not occur.

Repressor protein

Lactose absent

Lactose Operon

operator | promoter | operator | gene 1 | gene 2 | gene 3

Lactose present

lactose

C   When lactose is present, some is converted to a form that binds to the repressor. Binding alters the shape of the repressor such that it releases the operators. RNA polymerase can now attach to the promoter and transcribe the operon genes.

mRNA

RNA polymerase

operator | promoter | operator | gene 1 | gene 2 | gene 3

Figure 15.11 **Animated** Example of gene control in prokaryotes: the lactose operon on a bacterial chromosome. The operon consists of a promoter flanked by two operators, and three genes for lactose-metabolizing enzymes.

Figure It Out: What portion of the operon binds RNA polymerase when lactose is present?

Answer: The promoter

Lactose is not absorbed directly by the intestine. Thus, any that is not broken down in the small intestine ends up in the large intestine, which hosts *E. coli* and a variety of other prokaryotes. These resident organisms respond to the abundant sugar supply by switching on their lac operons. Carbon dioxide, methane, hydrogen, and other gaseous products of their various fermentation reactions accumulate quickly in the large intestine, distending its wall and causing pain. The other products of their metabolism (undigested carbohydrates) disrupt the solute–water balance inside the large intestine, and diarrhea results.

Not everybody is lactose intolerant. Many people carry a mutation in one of the genes responsible for the programmed lactase shutdown. The mutation is autosomal dominant (Section 12.2), so even heterozygotes make enough lactase to continue drinking milk without problems into adulthood.

Take-Home Message

Do prokaryotes have gene expression controls?

■ In prokaryotes, the main gene expression controls regulate transcription in response to shifts in nutrient availability and other outside conditions.

BRCA proteins promote transcription of genes that encode some of the DNA repair enzymes (Section 13.3). Any mutations that alter this function also alter a cell's capacity to repair damaged DNA. Other mutations are likely to accumulate, and that sets the stage for cancer (Section 14.5).

BRCA proteins also bind to receptors for the hormones estrogen and progesterone, which are abundant in breast and ovarian tissues. Binding regulates the transcription of growth factor genes. Among other things, growth factors (Section 9.5) stimulate cells to divide during normal, cyclic renewals of breast and ovarian tissues. When a mutation results in a BRCA protein that cannot bind to hormone receptors, the growth factors are overproduced. Cell division goes out of control, and tissue growth becomes disorganized. In other words, cancer develops.

Two groups of researchers, one at the Dana-Farber Cancer Institute at Harvard, the other at the University of Milan, recently found that XIST RNA localization is abnormal in breast cancer cells. In those cells, both X chromosomes are active.

How would you vote?

Some women at high risk of developing breast cancer opt for preventive breast removal. Many of them never would have developed cancer. Should the surgery be restricted to cancer treatment? See CengageNOW for details, then vote online.

It makes sense that having two active X chromosomes would have something to do with abnormal gene expression in breast and ovarian cancer cells, but why unmutated XIST RNA does not localize properly in such cells remains a mystery.

Mutations in the *BRCA1* gene may be part of the answer. A mutated *BRCA1* or *BRCA2* gene is often found in breast and ovarian cancer cells. The Harvard researchers found that the BRCA1 protein physically associates with XIST RNA. They were able to restore proper XIST RNA localization—and proper X chromosome inactivation—by rescuing BRCA1 function in breast cancer cells.

Summary

Section 15.1 Which genes a cell uses depends on the type of organism, the type of cell, factors inside and outside the cell, and, in complex multicelled species, the organism's stage of development.

Controls over gene expression are part of homeostasis in all organisms. They also drive development in multicelled eukaryotes. All cells of an embryo share the same genes. As different cell lineages use different subsets of genes during development, they become specialized, a process called **differentiation**. Specialized cells form tissues and organs in the adult.

Different molecules and processes govern every step between transcription of a gene and delivery of the gene's product to its final destination. Most controls operate at transcription; **transcription factors** such as **activators** and **repressors** influence transcription by binding to promoters, **enhancers**, or other sequences in DNA.

■ *Use the animation on CengageNOW to review the points of control for gene expression.*

Section 15.2 In female mammals, most genes on one of the two X chromosomes are permanently inaccessible. This **X chromosome inactivation** balances gene expression between the sexes (**dosage compensation**).

In plants, three sets of **master genes** guide cell differentiation in the whorls of a floral shoot (**ABC model**).

■ *Use the animation on CengageNOW to see how controls over gene expression affect eukaryotic development.*

Section 15.3 **Knockout experiments** involving **homeotic genes** in fruit flies (*Drosophila melanogaster*) revealed local controls over gene expression that govern the embryonic development of all complex, multicelled bodies, a process called **pattern formation**. Various master genes are expressed locally in different parts of an embryo as

it develops. Their products diffuse through the embryo and affect expression of other master genes, which affect the expression of others, and so on. These cascades of master gene products form a dynamic spatial map of overlapping gradients that spans the entire embryo body. Cells differentiate according to their location on the map.

Section 15.4 Most prokaryotic gene controls adjust transcription rates in response to environmental conditions, especially nutrient availability. The lactose **operon** governs expression of three genes active in lactose metabolism. Two **operators** that flank the promoter are binding sites for a repressor that blocks transcription.

■ *Use the animation on CengageNOW to explore the structure and function of the lactose operon.*

Self-Quiz   *Answers in Appendix III*

1. The expression of a given gene depends on _____ .
   a. the type of organism   c. the type of cell
   b. environmental conditions   d. all of the above

2. Gene expression in cells of multicelled eukaryotes changes in response to _____ .
   a. conditions outside the cell   c. operation of operons
   b. master gene products   d. a and b

3. Binding of _____ to _____ in DNA can increase the rate of transcription of specific genes.
   a. activators; promoters   c. repressors; operators
   b. activators; enhancers   d. both a and b

4. Proteins that influence gene expression by binding to DNA are called _____ .

5. Polytene chromosomes form in cells that _____ .
   a. have a lot of chromosomes   c. are polyploid
   b. are making a lot of protein   d. b and c are correct

Data Analysis Exercise

Investigating a correlation between specific cancer-causing mutations and risk of mortality in humans is challenging, in part because each cancer patient is given the best treatment available at the time. There are no "untreated control" cancer patients, and the idea of what treatments are the best changes quickly as new drugs become available and new discoveries are made.

Figure 15.12 shows a study in which 442 women who had been diagnosed with breast cancer were checked for BRCA mutations, and their treatments and progress were followed over several years. All of the women in the study had at least two affected close relatives, so their risk of developing breast cancer due to an inherited factor was estimated to be greater than that of the general population.

1. According to this study, what is a woman's risk of dying of cancer if two of her close relatives have breast cancer?

2. What is her risk of dying of cancer if she carries a mutated BRCA1 gene?

3. Is a BRCA1 or BRCA2 mutation more dangerous in breast cancer cases?

4. What other data would you have to see in order to make a conclusion about the effectiveness of preventive surgeries?

BRCA Mutations in Women Diagnosed With Breast Cancer				
	BRCA1	BRCA2	No BRCA Mutation	Total
Total number of patients	89	35	318	442
Avg. age at diagnosis	43.9	46.2	50.4	
Preventive mastectomy	6	3	14	23
Preventive oophorectomy	38	7	22	67
Number of deaths	16	1	21	38
Percent died	18.0	2.8	6.9	8.6

Figure 15.12 Results from a 2007 study investigating BRCA mutations in women diagnosed with breast cancer. All women in the study had a family history of breast cancer.

Some of the women underwent preventive mastectomy (removal of the noncancerous breast) during their course of treatment. Others had preventive oophorectomy (surgical removal of the ovaries) to prevent the possibility of getting ovarian cancer.

6. Controls over eukaryotic gene expression guide _____ .
a. natural selection   c. development
b. nutrient availability   d. all of the above

7. Incontinentia pigmenti is a rare example of _____ .
a. autosomal dominant X-linked inheritance
b. uneven pigmentation in humans

8. By the ABC model, _____ .
a. *Antecedents* trigger *Behavior* that has *Consequences*
b. three master gene sets (A,B,C) control flower formation
c. gene A affects gene B, which affects gene C
d. both b and c

9. During X chromosome inactivation, _____ .
a. female cells shut down   c. pigments form
b. RNA coats chromosomes   d. both a and b

10. A cell with a Barr body is _____ .
a. prokaryotic   c. from a female mammal
b. a sex cell   d. infected by Barr virus

11. Homeotic gene products _____ .
a. flank a bacterial operon
b. map out the overall body plan in embryos
c. control the formation of specific body parts

12. Knockout experiments _____ genes.
a. delete   c. express
b. inactivate   d. either a or b

13. Gene expression in prokaryotic cells changes in response to _____ .
a. activators; promoters   c. repressors; operators
b. activators; enhancers   d. both a and c

14. A promoter and a set of operators that control access to two or more prokaryotic genes is a(n) _____ .

15. Match the terms with the most suitable description.
___ ABC genes   a. a big RNA is its product
___ XIST gene   b. binding site for repressor
___ operator   c. cells become specialized
___ Barr body   d. —CH$_3$ additions to DNA
___ differentiation   e. inactivated X chromosome
___ methylation   f. guide flower development

■ *Visit CengageNOW for additional questions.*

Critical Thinking

1. Unlike most rodents, guinea pigs are well developed at the time of birth. Within a few days, they can eat grass, vegetables, and other plant material.

Suppose a breeder decides to separate baby guinea pigs from their mothers three weeks after they were born. He wants to raise the males and the females in different cages. However, he has trouble identifying the sex of young guinea pigs. Suggest how a quick look through a microscope can help him identify the females.

2. Calico cats are almost always female. Male calico cats are rare, and usually they are sterile. Why?

3. Geraldo isolated an *E. coli* strain in which a mutation has hampered the capacity of the cAMP activator to bind the promoter of the lactose operon. How will this mutation affect transcription of the lactose operon when the *E. coli* cells are exposed to the following conditions?

a. Lactose and glucose are both available.

b. Lactose is available but glucose is not.

c. Both lactose and glucose are absent.

16 Studying and Manipulating Genomes

IMPACTS, ISSUES Golden Rice or Frankenfood?

Vitamin A is necessary for good vision, growth, and immunity. A small child can get enough of it just by eating a carrot every few days, yet each year about 140 million children under the age of six suffer from serious health problems due to vitamin A deficiency. These children do not grow as they should, and they succumb easily to infection. As many as 500,000 of them go blind because of vitamin A deficiency, and half of them die within a year of losing their sight.

It is no coincidence that populations with the highest incidence of vitamin A deficiency also are the poorest. Most people in such populations tend to eat few animal products, vegetables, or fruits—all foods that are rich sources of vitamin A. Correcting and preventing vitamin A deficiency can be as simple as supplementing the diet with these foods, but changes in dietary habits are often limited by cultural traditions and poverty. Political and economic issues hamper long-term vitamin supplementation programs.

Geneticists Ingo Potrykus and Peter Beyer wanted to help such people by improving the nutritional value of rice. Why rice? Rice is the dietary staple for 3 billion people in impover-ished countries around the world. Economies, traditions, and cuisines are based on growing and eating rice. So, growing and eating rice that happens to contain enough vitamin A to prevent disease would be compatible with prevailing methods of agriculture and traditional dietary preferences.

The body can easily convert beta-carotene, an orange photosynthetic pigment, into vitamin A. However, getting rice grains to make beta-carotene is beyond the scope of conventional methods of plant breeding. For example, corn seeds (kernels) make and store beta-carotene, but even the best gardener cannot induce rice plants to breed with corn plants.

Potrykus and Beyer genetically modified rice plants to make beta-carotene in their seeds—in the grains of Golden Rice (Figure 16.1). Like many other genetically modified organisms (GMOs), Golden Rice is transgenic, which means it carries genes from a different species. GMOs are made in laboratories, not on farms, but they are an extension of breeding practices used for many thousands of years to coax new plants and new breeds of animals from wild species.

No one wants children to suffer or die. However, many people are opposed to any GMO. Some worry that our ability to tinker with genetics has surpassed our ability to evaluate its impact. Should we be more cautious? Two people created a way to keep millions of children from dying. How much of a risk should we as a society take to help those children?

At this time, geneticists hold molecular keys to the kingdom of inheritance. As you will see, what they are unlocking is already having an impact on life in the biosphere.

See the video! Figure 16.1 Golden Rice, a miracle of modern science. (**a**) Vitamin A deficiency is common in Southeast Asia and other regions where people subsist mainly on rice.

(**b**) Rice plants with artificially inserted genes make and store the orange pigment beta-carotene in their seeds, or rice grains. The grains of this Golden Rice may help prevent vitamin A deficiency in developing countries. Compare unmodified rice grains in (**c**).

Key Concepts

DNA cloning

Researchers routinely make recombinant DNA by cutting and pasting together DNA from different species. Plasmids and other vectors can carry foreign DNA into host cells. **Section 16.1**

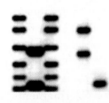

Needles in haystacks

Researchers manipulate targeted genes by isolating and making many copies of particular DNA fragments. **Section 16.2**

Deciphering DNA fragments

Sequencing reveals the linear order of nucleotides in a fragment of DNA. A DNA fingerprint is an individual's unique array of DNA sequences. **Sections 16.3, 16.4**

Mapping and analyzing whole genomes

Genomics is the study of genomes. Efforts to sequence and compare different genomes offer insights about our own genes. **Section 16.5**

Using the new technologies

Genetic engineering, the directed modification of an organism's genes, is now used in research, and it is being tested in medical applications. It continues to raise ethical questions. **Sections 16.6–16.10**

Links to Earlier Concepts

- This chapter builds on earlier explanations of DNA's structure (Section 13.2), and the molecules and processes that bring about DNA replication (13.3).

- We revisit mRNA (14.3) and bacteriophage (13.1) in the context of DNA cloning.

- Gene expression (14.1) and knockout experiments (15.3) are important concepts in genetic engineering, particularly as they apply to research on human genetic disorders (12.7).

- You may want to refer back to triglycerides (3.4) and lignin (4.12).

- You will also see more examples of how researchers use light-emitting molecules (6.5) as tracers (2.2).

16.1 | Cloning DNA

- Researchers cut up DNA from different sources, then paste the resulting fragments together.
- Cloning vectors can carry foreign DNA into host cells.
- Links to Bacteriophage 13.1, Base pairing 13.2, DNA ligase 13.3, mRNA 14.3, Introns 14.3

Cut and Paste

In the 1950s, excitement over the discovery of DNA's structure gave way to frustration: No one could determine the order of nucleotides in a molecule of DNA. Identifying a single base among thousands or millions of others turned out to be a huge technical challenge.

A seemingly unrelated discovery offered a solution. Some types of bacteria resist infection by bacteriophage, which are viruses that inject their DNA into bacterial cells. Werner Arber, Hamilton Smith, and their coworkers discovered that special enzymes inside the bacteria chop up any injected bacteriophage DNA before it has a chance to integrate into the bacterial chromosome.

The enzymes restrict bacteriophage growth; hence their name, restriction enzymes. A **restriction enzyme** will cut DNA wherever a specific nucleotide sequence occurs. For example, the enzyme *Eco*RI (named after the organism from which it was isolated, *E. coli*) cuts DNA only at the sequence GAATTC (Figure 16.2*a*).

Other restriction enzymes cut different sequences. Many of them leave single-stranded tails, or "sticky ends," on DNA fragments (Figure 16.2*b*). Researchers realized that matching sticky ends base-pair together, regardless of the origin of the DNA (Figure 16.2*c*).

The enzyme DNA ligase speeds formation of covalent bonds between matching sticky ends in a mixture of DNA fragments (Figure 16.2*d*). Thus, using appropriate restriction enzymes and DNA ligase, research-

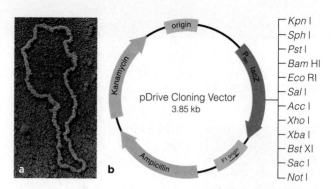

Figure 16.3 Cloning vectors. (**a**) Micrograph of a plasmid. (**b**) A commercial plasmid cloning vector. Restriction enzyme recognition sequences are indicated on the *right* by the name of the enzyme that cuts them. Researchers insert foreign DNA into the vector at these sites.

Bacterial genes help researchers identify host cells that take up a vector with inserted DNA. This vector carries two antibiotic resistance genes (*purple*) and the lactose operon (*red*).

ers can cut and paste DNA from different organisms. The result, a hybrid molecule composed of DNA from two or more organisms, is **recombinant DNA**.

Making recombinant DNA is the first step in **DNA cloning**, a set of laboratory methods that uses living cells to make many copies of specific DNA fragments.

For example, researchers often insert specific DNA fragments into **plasmids**, small circles of bacterial DNA that are independent of the chromosome (Figure 16.3*a*). Before a bacterium divides, it copies its chromosome and any plasmids, so both descendant cells get one of each. If a plasmid carries a fragment of foreign DNA, that fragment gets copied and distributed to descendant cells along with the plasmid.

Thus, plasmids can be **cloning vectors**, molecules that carry foreign DNA into host cells (Figure 16.3*b*). A host cell that takes up a cloning vector can be grown in the laboratory to yield a huge population of geneti-

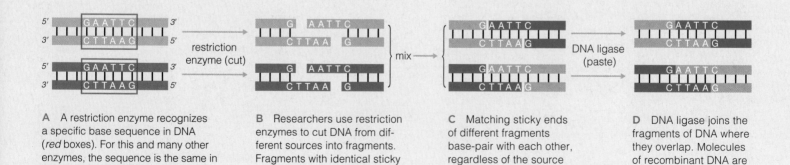

A A restriction enzyme recognizes a specific base sequence in DNA (*red* boxes). For this and many other enzymes, the sequence is the same in the 5′ to 3′ direction on both strands.

B Researchers use restriction enzymes to cut DNA from different sources into fragments. Fragments with identical sticky ends are mixed together.

C Matching sticky ends of different fragments base-pair with each other, regardless of the source of the DNA.

D DNA ligase joins the fragments of DNA where they overlap. Molecules of recombinant DNA are the result.

Figure 16.2 Making recombinant DNA.

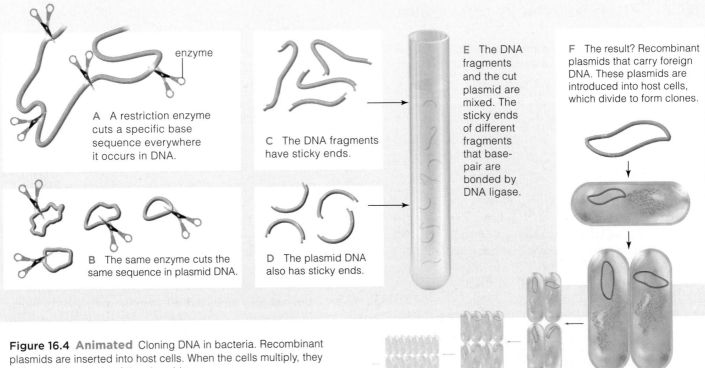

A A restriction enzyme cuts a specific base sequence everywhere it occurs in DNA.

B The same enzyme cuts the same sequence in plasmid DNA.

C The DNA fragments have sticky ends.

D The plasmid DNA also has sticky ends.

E The DNA fragments and the cut plasmid are mixed. The sticky ends of different fragments that base-pair are bonded by DNA ligase.

F The result? Recombinant plasmids that carry foreign DNA. These plasmids are introduced into host cells, which divide to form clones.

Figure 16.4 Animated Cloning DNA in bacteria. Recombinant plasmids are inserted into host cells. When the cells multiply, they make multiple copies of the plasmids.

cally identical cells called **clones**. Each clone contains a copy of the vector and the foreign DNA it carries (Figure 16.4). Researchers then harvest the DNA from the clones, and use it for experiments.

cDNA Cloning

Eukaryotic DNA, remember, contains introns. Unless you happen to be a eukaryotic cell, it is not easy to figure out which parts of the DNA encode gene products, and which do not. Researchers who study eukaryotic genes and their expression work with mRNA, because introns have already been snipped out (Section 14.3).

Messenger RNA cannot be cloned directly, because restriction enzymes and DNA ligase cut and paste only double-stranded DNA. However, mRNA can be used as a template to make double-stranded DNA in a test tube. **Reverse transcriptase**, a replication enzyme from certain types of viruses, transcribes mRNA into DNA. This enzyme uses the mRNA as a template to assemble a strand of complementary DNA, or **cDNA**:

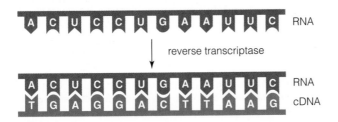

DNA polymerase added to the mixture strips the RNA from the hybrid molecule as it copies the cDNA into a second strand of DNA. The outcome is a double-stranded DNA copy of the original mRNA:

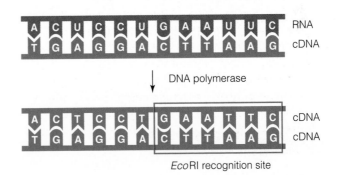

Like any other DNA, double-stranded cDNA may be cut with restriction enzymes, and the fragments can be pasted into a cloning vector using DNA ligase.

Take-Home Message

What is DNA cloning?

■ DNA cloning uses living cells to make identical copies of a particular fragment of DNA. Restriction enzymes cut DNA into fragments, then DNA ligase seals the fragments into cloning vectors. Recombinant DNA molecules result.

■ A cloning vector that holds foreign DNA can enter a living cell. The host cell can divide and give rise to huge populations of genetically identical cells (clones), each of which contains a copy of the foreign DNA.

- DNA libraries and the polymerase chain reaction (PCR) help researchers isolate particular DNA fragments.

- Links to Tracers 2.2, Base pairing 13.2

A Individual bacterial cells from a DNA library are spread over the surface of a solid growth medium. The cells divide repeatedly and form colonies—clusters of millions of genetically identical daughter cells.

B A piece of special paper pressed onto the surface of the growth medium will bind some cells from each colony.

C The paper is soaked in a solution that ruptures the cells and releases their DNA. The DNA clings to the paper in spots mirroring the distribution of colonies.

D A probe is added to the liquid bathing the paper. The probe hybridizes with (sticks to) only the spots of DNA that contain complementary base sequences.

E The bound probe makes a spot. Here, one radioactive spot darkens x-ray film. The position of the spot on the film is compared to the positions of all the original bacterial colonies. Cells from the colony that made the spot are cultured, and the DNA they contain is harvested.

The entire set of genetic material—the **genome**—of an organism typically comprises hundreds or thousands of genes. To study or manipulate one of those genes, researchers must first separate it from all of the others.

Researchers can isolate a gene by cutting an organism's DNA into pieces, and cloning all the pieces. The result is a genomic library, a set of clones that collectively host all of the DNA in a genome. Researchers can also harvest mRNA, make cDNA copies of it, and then clone the cDNA to make a cDNA library. A cDNA library represents only those genes being expressed at the time the mRNA was harvested.

Genomic and cDNA libraries are **DNA libraries**—sets of cells that host various cloned DNA fragments. In such libraries, a cell that hosts a particular gene of interest may be mixed up with thousands or millions of others that do not. All the cells look the same, so researchers get tricky to find that one clone among all of the others—the needle in the haystack.

Using a probe is one trick. A **probe** is a fragment of DNA labeled with a tracer (Section 2.2). Researchers design probes to match a targeted DNA sequence. For example, they might synthesize an oligomer (a short chain of nucleotides) based on a known DNA sequence, then attach a radioactive phosphate group to it.

The nucleotide sequence of a probe is complementary to that of the targeted gene, so the probe can base-pair with the gene. Base pairing between DNA (or DNA and RNA) from more than one source is called **nucleic acid hybridization**. A probe mixed with DNA from a library hybridizes with (sticks to) the targeted gene (Figure 16.5). Researchers pinpoint a clone that hosts the gene by detecting the label on the probe. The clone is cultured so a huge population of genetically identical cells forms. The DNA can then be extracted in bulk from the cells.

Big-Time Amplification: PCR

Researchers can isolate and mass-produce a particular DNA fragment without cloning. They do so with the *Polymerase Chain Reaction*, or **PCR**. This hot-and-cold cycled reaction uses a heat-tolerant DNA polymerase to copy a fragment of DNA by the billions.

PCR can transform one needle in a haystack—that one-in-a-million DNA fragment—into a huge stack of

Figure 16.5 Animated Nucleic acid hybridization. In this example, a radioactive probe helps identify a bacterial colony that contains a targeted sequence of DNA.

needles with a little hay in it (Figure 16.6). The starting material for PCR is a sample of DNA with at least one molecule of a target sequence. It might be DNA from a mixture of 10 million different clones, one sperm, a hair left at a crime scene, or a mummy. Essentially any sample that has DNA in it can be used for PCR.

First, the starting material is mixed with DNA polymerase, nucleotides, and primers. **Primers** are oligomers that base-pair with DNA at a certain sequence—here, on either end of the DNA to be amplified.

Researchers expose the reaction mixture to repeated cycles of high and low temperature. High temperature disrupts the hydrogen bonds that hold the two strands of a DNA double helix together (Section 13.2). During a high temperature cycle, every molecule of double-stranded DNA unwinds and becomes single-stranded. During a low temperature cycle, single DNA strands hybridize with complementary partners, and double-stranded DNA forms again.

Most DNA polymerases are destroyed by the high temperatures required to separate DNA strands. The kind that is used in PCR reactions, *Taq* polymerase, is from *Thermus aquaticus*. This bacterial species lives in superheated springs (Chapter 20 introduction), so its polymerase is necessarily heat-tolerant. Like other DNA polymerases, the *Taq* polymerase recognizes hybridized primers as places to start DNA synthesis. During a low temperature cycle, it starts synthesizing DNA where primers have hybridized with template. Synthesis proceeds along the template strand until the temperature rises and the DNA separates into single strands. The newly synthesized DNA is a copy of the target.

When the mixture cools, primers rehybridize, and DNA synthesis begins again. With each temperature cycle, the number of copies of target DNA can double. After about thirty PCR cycles, the number of template molecules may be amplified by about a billionfold.

A DNA template (*purple*) is mixed with primers (*red*), free nucleotides, and heat-tolerant *Taq* DNA polymerase.

B When the mixture is heated, DNA strands separate. When it is cooled, some primers hydrogen-bond to the template DNA.

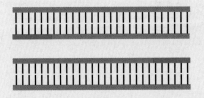

C *Taq* polymerase uses the primers to initiate synthesis, and complementary strands of DNA form. The first round of PCR is now complete.

D The mixture is heated again, and all of the DNA separates into single strands. When the mixture is cooled, some of the primers hydrogen-bond to the DNA.

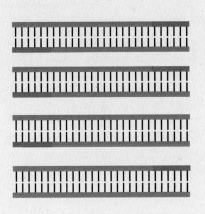

E *Taq* polymerase uses the primers to initiate DNA synthesis, and complementary strands of DNA form. The second round of PCR is complete.

Each round can double the number of DNA molecules. After 30 rounds, the mixture contains huge numbers of DNA fragments, all copies of the template DNA.

Take-Home Message

How do researchers study one gene in the context of many?

■ Researchers make DNA libraries or use PCR to isolate one gene from the many other genes in a genome.

■ Probes are used to identify one clone that hosts a DNA fragment of interest among many other clones in a DNA library.

■ PCR, the polymerase chain reaction, rapidly increases the number of molecules of a particular DNA fragment.

Figure 16.6 Animated Two rounds of PCR. Thirty cycles of this polymerase chain reaction may increase the number of starting DNA template molecules a billionfold.

16.3 DNA Sequencing

- DNA sequencing reveals the order of nucleotide bases in a fragment of DNA.

- Links to Tracers 2.2, Nucleotides 13.2, DNA replication 13.3

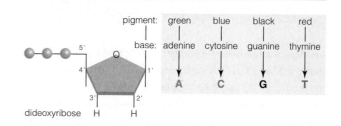

Figure 16.7 Structure of a dideoxynucleotide. Each of the four bases is labeled with a different colored pigment.

The order of the nucleotide bases in a DNA fragment is determined with **DNA sequencing**. The most commonly used method of sequencing DNA is similar to DNA replication, in that the DNA fragment is used as a template for DNA synthesis.

Researchers mix a DNA template with nucleotides, DNA polymerase, and a primer that hybridizes to the DNA. Starting at the primer, the polymerase joins free nucleotides into a new strand of DNA, in the order dictated by the sequence of the template.

DNA polymerase joins a nucleotide to a DNA strand only at the hydroxyl group on the strand's 3′ carbon (Section 13.3). The reaction mixture includes four kinds of *dideoxy*nucleotides, which have no hydroxyl group on their 3′ carbon (Figure 16.7). During the sequencing reaction, a polymerase randomly adds either a regular nucleotide or a dideoxynucleotide to the end of a growing DNA strand. If it adds a dideoxynucleotide, the 3′ carbon of the strand will not have a hydroxyl group, so synthesis of the strand ends there (Figure 16.8a,b).

After about 10 minutes, there are millions of DNA fragments of all different lengths; most are incomplete copies of the template DNA. All of the copies end with one of the four dideoxynucleotides (Figure 16.8c). For example, there will be many ten base-pair-long copies of the template in the mixture. If the tenth base in the template was adenine, every one of those fragments will end with a dideoxyadenine.

The fragments are then separated by **electrophoresis**. With this technique, an electric field pulls all the DNA fragments through a semisolid gel. DNA fragments of different sizes move through the gel at different rates. The shorter the fragment, the faster it moves, because shorter fragments slip through the tangled molecules of the gel faster than longer fragments do.

All fragments of the same length move through the gel at the same speed, so they gather into bands. All of the fragments in a given band have the same dideoxynucleotide at their ends. Each of the four types of dideoxynucleotides (A, C, G, or T) carries a different colored pigment label, and those tracers now impart distinct colors to the bands (Figure 16.8d). Each color designates one of the four dideoxynucleotides, so the order of colored bands in the gel represents the DNA sequence (Figure 16.8e).

A The fragment of DNA to be sequenced is mixed with a primer, DNA polymerase, and nucleotides. The mixture also includes the four dideoxynucleotides labeled with four different colored pigments.

B The polymerase uses the DNA as a template to synthesize new strands again and again. Synthesis of each new strand stops when a dideoxynucleotide is added.

C At the end of the reaction, there are many truncated copies of the DNA template in the mixture.

D An electrophoresis gel separates the fragments into bands according to length. All fragments in each band end with the same dideoxynucleotide; thus, each band is the color of that dideoxynucleotide.

E A computer detects and records the color of each band on the gel. The order of colors of the bands represents the sequence of the template DNA.

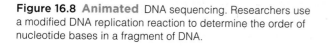

Figure 16.8 Animated DNA sequencing. Researchers use a modified DNA replication reaction to determine the order of nucleotide bases in a fragment of DNA.

Take-Home Message

How is the order of nucleotides in DNA determined?

- With DNA sequencing, a strand of DNA is partially replicated. Electrophoresis is used to separate the resulting fragments of DNA, which are tagged with tracers, by length.

16.4 | DNA Fingerprinting

■ One individual can be distinguished from all others on the basis of DNA fingerprints.

Each human has a unique set of fingerprints. In addition, like members of other sexually reproducing species, each also has a **DNA fingerprint**—a unique array of DNA sequences. More than 99 percent of the DNA in all humans is the same, but the other fraction of 1 percent is unique to each individual. Some of these unique sequences are sprinkled throughout the human genome as **short tandem repeats**—many copies of the same 2- to 10-base-pair sequences, positioned one after the next along the length of a chromosome.

For example, one person's DNA might contain fifteen repeats of the bases TTTTC in a certain location. Another person's DNA might have TTTTC repeated two times in the same location. One person might have ten repeats of CGG; another might have fifty. Such repetitive sequences slip spontaneously into DNA during replication, and their numbers grow or shrink over generations. The mutation rate is relatively high around tandem repeat regions.

DNA fingerprinting reveals differences in the tandem repeats among individuals. With this technique, PCR is used to copy a region of a chromosome known to have tandem repeats of 4 or 5 nucleotides. The size of the copied DNA fragment differs among most individuals, because the number of tandem repeats in that region also differs.

Thus, the genetic differences between individuals can be detected by electrophoresis. As in DNA sequencing, the fragments form bands according to length as they migrate through a gel. Several regions of chromosomal DNA are typically tested. The resulting banding patterns on the electrophoresis gel constitute an individual's DNA fingerprint—which, for all practical purposes, is unique. Unless two people are identical twins, the chances that they have identical tandem repeats in even three regions of DNA is 1 in 1,000,000,000,000,000,000—or one in a quintillion—which is far more than the number of people that live on Earth.

A few drops of blood, semen, or cells from a hair follicle at a crime scene or on a suspect's clothing yield enough DNA to amplify with PCR for DNA fingerprinting (Figure 16.9). DNA fingerprints have been established as accurate and unambiguous, and are often used as evidence in court. For example, DNA fingerprints are now routinely submitted as evidence in paternity disputes. The technique is being widely used not only to convict the guilty, but also to exonerate the innocent: As of this writing, DNA fingerprinting evidence has helped release more than 160 innocent people from prison.

DNA fingerprint analysis has many applications. For instance, DNA fingerprinting was used to identify the remains of many individuals who died in the World Trade Center on September 11, 2001. It confirmed that human bones exhumed from a shallow pit in Siberia belonged to five individuals of the Russian imperial family, all shot to death in secrecy in 1918.

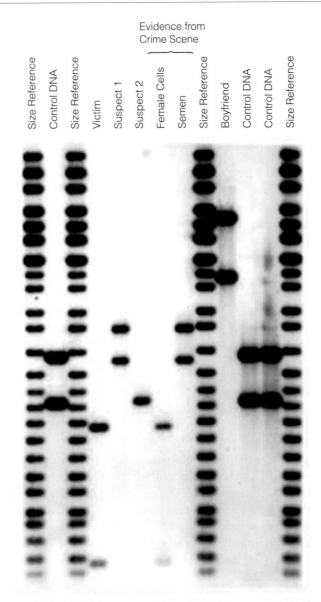

Figure 16.9 DNA fingerprinting in an investigation of sexual assault. A single short tandem repeat region was amplified from evidence found at the crime scene—the perpetrator's semen and the victim's cells. The two samples were compared with the same tandem repeat region amplified from DNA of the victim, her boyfriend, and two suspects (1 and 2).

The photo shows an x-ray film image of an electrophoresis gel from a forensics laboratory. The dark bands represent different sized DNA fragments that had been labeled with a radioactive tracer. Note the three samples of control DNA (to confirm that the assay was working correctly), and the four size reference samples. **Figure It Out:** Which suspect is guilty?

Answer: Suspect 1

Researchers also use DNA fingerprinting to study population dispersal in humans and other animals. Because only a tiny amount of DNA is necessary, such studies are not necessarily limited to living populations. Short tandem repeats on the Y chromosome are also used to determine genetic relationships among male relatives and descendants, and to trace an individual's ethnic heritage.

Studying Genomes

■ Comparing the sequence of our genome with that of other species is giving us insights into how the human body works.

■ Links to Triglycerides 3.4, Discovery of DNA structure 13.2, Knockout experiments 15.3

The Human Genome Project

Around 1986, people were arguing about sequencing the human genome. Many insisted that deciphering it would have enormous payoffs for medicine and pure research. Others said sequencing would divert funds from more urgent work that also had a better chance of success. At that time, sequencing 3 billion bases seemed like a daunting task: It would take at least 6 million sequencing reactions, all done by hand. Given the techniques available, the work would have taken more than fifty years to complete.

But techniques kept getting better, so more bases could be sequenced in less time. Automated (robotic) DNA sequencing and PCR had just been invented. Both of these techniques were still cumbersome and expensive, but many researchers sensed their potential. Waiting for faster technologies seemed the most efficient way to sequence 3 billion bases, but how fast did they need to be before the project could begin?

In 1987, private companies started to sequence the human genome. Walter Gilbert, one of the early inventors of DNA sequencing, started one that intended to sequence and patent the human genome. This development provoked widespread outrage, but it also spurred commitments in the public sector. In 1988, the National Institutes of Health (NIH) effectively annexed the entire project by hiring James Watson (of DNA structure fame) to head the official Human Genome Project, and providing $200 million per year to fund it. A consortium formed between the NIH and international institutions that were sequencing different parts of the genome. Watson set aside 3 percent of the funding for studies of ethical and social issues arising from the research. He resigned later over a patent disagreement.

Amid ongoing squabbles over patent issues, Celera Genomics formed in 1998 (Figure 16.10). With Craig Venter at its helm, the company intended to commercialize genetic information. Celera started to sequence the genome using new, faster techniques, because the first to have the complete sequence had a legal basis for patenting it. The competition motivated the public consortium to move its efforts into high gear.

Then, in 2000, United States President Bill Clinton and British Prime Minister Tony Blair jointly declared that the sequence of the human genome could not be patented. Celera kept on sequencing anyway. Celera and the public consortium separately published about 90 percent of the sequence in 2001.

By 2003, fifty years after the discovery of the structure of DNA, the sequence of the human genome was officially completed. At this writing, about 99 percent of its coding regions—28,976 genes—have been identified. Researchers have not discovered what all of the genes encode, only where they are in the genome.

What do we do with this vast amount of data? The next step is to find out just what the sequence means.

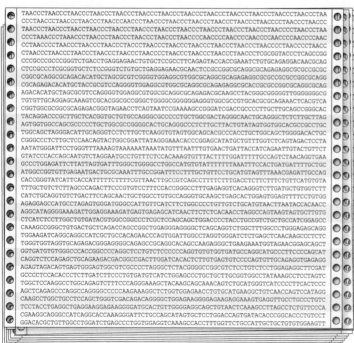

Figure 16.10 Some of the bases of the human genome—and a few of the supercomputers used to sequence it—at Venter's Celera Genomics in Maryland.

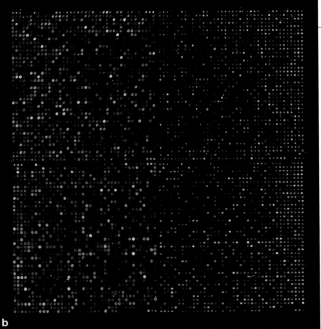

Figure 16.11 (**a**) A DNA chip. (**b**) DNA chips are often used in gene expression research. Here, RNA from yeast cells carrying out fermentation was used to make cDNA, which was labeled with a green tracer; RNA from the same type of cells carrying out aerobic respiration was used to make cDNA labeled with a red tracer.

The probes were dropped onto a DNA chip with a 19-millimeter (3/4-inch) array of the complete yeast genome—around 6,000 genes. *Green* spots indicate the genes active during fermentation; *red* spots, genes active during aerobic respiration. *Yellow* spots are a combination of red and green; they indicate genes active in both pathways.

Genomics

Investigations into the genomes of humans and other species have converged into the new research field of **genomics**. Structural genomics focuses on determining the three-dimensional structure of proteins encoded by a genome. Comparative genomics compares genomes of different species; similarities and differences reflect evolutionary relationships.

The human genome sequence is a massive collection of seemingly cryptic data. Currently, the only way we are able to decipher it is by comparing it to genomes of other organisms, the premise being that all organisms are descended from shared ancestors, so all genomes are related to some extent. We see evidence of such genetic relationships just by comparing the sequence data. For example, the human and mouse sequences are about 78 percent identical; the human and banana sequences are about 50 percent identical.

Intriguing as these percentages might be, gene-by-gene comparisons offer more practical benefits. We have learned about the function of many human genes by studying their counterpart genes in other species. For example, researchers studying a human gene might disable the same gene in mice. The effects of the gene's absence on mice are clues to its function in humans.

These types of knockout experiments are revealing the function of many human genes. For example, researchers comparing the human and mouse genomes discovered a human version of the mouse gene *APOA5*. This gene encodes a protein that carries lipids in the blood. Mice with an *APOA5* knockout have four times the normal level of triglycerides in their blood. The researchers then looked for—and found—a correlation between *APOA5* mutations and high triglyceride levels in humans. High triglycerides are a risk factor for coronary artery disease.

DNA Chips

Genomics researchers often use **DNA chips**, which are microscopic arrays (microarrays) of DNA samples that have been stamped in separate spots on small glass plates (Figure 16.11*a*). Typically, one microarray comprises hundreds or thousands of DNA fragments that collectively represent an entire genome.

Using DNA chips, researchers can compare the patterns of gene expression among cells—perhaps different types of cells from one individual, or the same cells at different times or under different conditions (Figure 16.11*b*). Which genes are expressed, at which times, in which cells, is information useful in research and other applications. For example, DNA chips have been used to determine which genes are deregulated in cancer cells. Soon, they may be used to quickly screen people for genetic predisposition to disease, to identify pathogens, and in forensic investigations.

Take-Home Message

What do we do with DNA sequence information?

■ Analysis of the human genome sequence is yielding new information about human genes and how they work. The information has practical applications in medicine, research, and other fields.

16.6 | Genetic Engineering

■ The most common genetically modified organisms are bacteria and yeast.

■ Links to Bioluminescence 6.5, Gene expression 14.1

Traditional cross-breeding methods can alter genomes, but only if individuals with the desired traits will interbreed. Genetic engineering takes gene-swapping to an entirely new level. **Genetic engineering** is a laboratory process by which deliberate changes are introduced into an individual's genome. A gene from one species may be transferred to another to produce a **transgenic** organism, or a gene may be altered and reinserted into an individual of the same species. Both methods result in **genetically modified organisms** (GMOs).

The most common GMOs are bacteria and yeast. These cells have the metabolic machinery to make complex organic molecules, and they are easily modified.

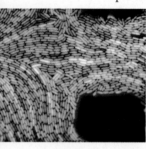

genetically modified bacteria expressing a jellyfish gene emit green light

Genetically modifying bacteria or yeast has practical applications. For example, the *E. coli* on the *left* have been modified to produce a fluorescent protein from jellyfish. The cells are genetically identical, so the visible variation in fluorescence among them reveals differences in gene expression. These GMOs may help us discover why some individual bacteria become dangerously resistant to antibiotics.

Some genetically modified bacteria and yeast are "factories" for medically important proteins. Diabetics were among the first beneficiaries of such organisms. Insulin for their injections was once extracted from animals, but it provoked an allergic reaction in some people. Human insulin, which does not provoke allergic reactions, has been produced by transgenic *E. coli* since 1982. Slight modifications of the gene have also yielded fast-acting and slow-release human insulin.

Engineered microorganisms also produce proteins used in food manufacturing. For example, cheese is traditionally made with an extract of calf stomachs, which contain the enzyme chymotrypsin. Most cheese manufacturers now use chymotrypsin made by genetically engineered bacteria. Other examples are GMO-made enzymes that improve the taste and clarity of beer and fruit juice, slow bread staling, or modify fats.

Take-Home Message

What is genetic engineering?

■ Genetic engineering is the directed alteration of an individual's genome, and it results in a genetically modified organism (GMO).

■ A transgenic organism carries a gene from a different species. Transgenic bacteria and yeast are used in research, medicine, and industry.

16.7 | Designer Plants

■ Genetically engineered crop plants are widespread in the United States.

■ Links to Lignin 4.12, Luciferase 6.5

Agrobacterium tumefaciens is a species of bacteria that infects many plants, including peas, beans, potatoes, and other important crops. Its plasmid contains genes that cause tumors to form on infected plants; hence the name Ti plasmid (for *Tumor-inducing*). Researchers use the Ti plasmid as a vector to transfer foreign or modified genes into plants. They remove the tumor-inducing genes from the plasmid, then insert desired genes into it. Whole plants can be grown from plant cells that take up the modified plasmid (Figure 16.12).

Modified *A. tumefaciens* bacteria are used to deliver genes into some food crop plants, including soybeans, squash, and potatoes. Researchers also transfer genes into plants by way of electric or chemical shocks, or by blasting them with DNA-coated pellets.

Genetically Engineered Plants

As crop production expands to keep pace with human population growth, it places unavoidable pressure on ecosystems everywhere. Irrigation leaves mineral and salt residues in soils. Tilled soil erodes, taking topsoil with it. Runoff clogs rivers, and fertilizer in it causes algae to grow so fast that fish suffocate. Pesticides can harm humans, other animals, and beneficial insects.

Pressured to produce more food at lower cost and with less damage to the environment, many farmers have begun to rely on genetically modified crop plants. Some of these modified plants carry genes that impart resistance to devastating plant diseases. Others offer improved yields, such as a strain of transgenic wheat that has double the yield of unmodified wheat.

GMO crops such as *Bt* corn and soy help farmers use smaller amounts of toxic pesticides. Organic farmers often spray their crops with spores of *Bt* (*Bacillus thuringiensis*), a bacterial species that makes a protein toxic only to insect larvae. Researchers transferred the gene encoding the *Bt* protein into plants. The engineered plants produce the *Bt* protein, but otherwise they are identical with unmodified plants. Insect larvae die shortly after eating their first (and only) GMO meal. Farmers can use much less pesticide on crops that make their own (Figure 16.13*a*).

Transgenic crop plants are also being developed for regions that are affected by severe droughts, such as Africa. Genes that confer drought tolerance and insect resistance are being transferred into crop plants such as corn, beans, sugarcane, cassava, cowpeas, banana,

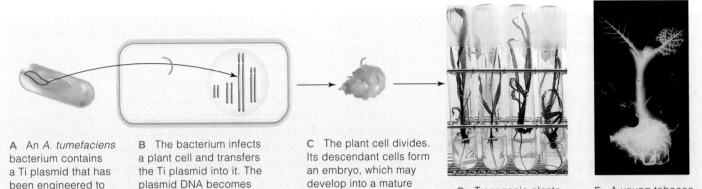

A An *A. tumefaciens* bacterium contains a Ti plasmid that has been engineered to carry a foreign gene.

B The bacterium infects a plant cell and transfers the Ti plasmid into it. The plasmid DNA becomes integrated into one of the plant cell's chromosomes.

C The plant cell divides. Its descendant cells form an embryo, which may develop into a mature plant that can express the foreign gene.

D Transgenic plants

E A young tobacco plant visibly expressing a foreign gene.

Figure 16.12 Animated (**a–d**) Ti plasmid transfer of an *Agrobacterium tumefaciens* gene to a plant cell. (**e**) Transgenic plant expressing a gene for the firefly enzyme luciferase (Section 6.5).

Figure 16.13 Examples of genetically modified plants. (**a**) Some GMO crops help farmers use less insecticide. *Top*, the *Bt* gene conferred insect resistance to the genetically modified plants that produced this corn. *Bottom*, unmodified corn is more vulnerable to pests.

(**b**) Lignin strengthens the secondary cell walls of many kinds of woody plants. Before paper can be made from wood, the lignin must be extracted from wood pulp. Paper products and clean-burning fuels such as ethanol may be easier to manufacture from the wood of trees engineered to produce less lignin.

Shown are a control plant (*left*) and three aspen seedlings (*right*) in which expression of a control gene in the lignin synthesis pathway was suppressed. The modified plants made normal lignin, but not as much of it.

and wheat. Such crops may help people that rely on agriculture for food and income in drought-stricken, impoverished regions of the world.

The USDA Animal and Plant Health Inspection Service (APHIS) regulates the introduction of GMOs into the environment. At this writing, the agency has deregulated seventy-three genetically modified plants, which means the plants are approved for unregulated use in the United States. Hundreds more are pending such deregulation.

The most widely planted GMO crops include corn, sorghum, cotton, soy, canola, and alfalfa engineered for resistance to glyphosate, an herbicide. Rather than tilling the soil to control weeds, farmers can spray their fields with glyphosate, which kills the weeds but not the engineered crops. However, weeds are becom-

ing resistant to glyphosate, so spraying it no longer kills the weeds in glyphosphate-resistant crop fields. The engineered gene is also appearing in wild plants and in nonengineered crops, which means that transgenes can—and do—escape into the environment.

Controversy raised by such GMO use invites you to read the research and form your own opinions. The alternative is to be swayed by media hype (the term "Frankenfood," for instance), or by reports from possibly biased sources (such as herbicide manufacturers).

Take-Home Message

Are there genetically engineered plants?

■ Plants with modified or foreign genes are now common farm crops.

■ Genetically engineered animals are invaluable in medical research and in other applications.

■ Link to Knockouts 15.3

Of Mice and Men

Traditional cross-breeding has produced animals so unusual that transgenic animals may seem mundane by comparison (Figure 16.14a). Cross-breeding is also a form of genetic manipulation, but transgenics that carry genes from other genera would probably never occur in nature (Figure 16.14b,c).

The first transgenic animals—mice—were produced in 1982. Researchers inserted a gene that codes for a rat growth hormone into a plasmid, then injected the recombinant plasmid into fertilized mouse eggs. The eggs were implanted in surrogate mother mice. A third of the mice that were born to the surrogates grew much larger than their littermates (Figure 16.15). The larger mice were transgenic: The rat gene had integrated into their chromosomes, and was being expressed.

Today, transgenic mice are commonplace, and they are invaluable in research into human genes. For example, researchers have discovered the function of many human genes by inactivating their counterparts in mice (Section 16.5).

Genetically modified animals are also used as models of many human diseases. For example, researchers inactivated the molecules involved in the control of glucose metabolism, one by one, in mice. Studying the effects of the knockouts has resulted in much of our current understanding of how diabetes works in humans. Genetically modified animals such as these mice are allowing researchers to study human diseases (and their potential cures) without experimenting on humans.

Genetically engineered animals also make proteins that have medical and industrial applications. Various transgenic goats produce proteins used to treat cystic fibrosis, heart attacks, blood clotting disorders, and even nerve gas exposure. Milk from goats transgenic for lysozyme, an antibacterial protein in human milk, may protect infants and children in developing countries from acute diarrheal disease. Goats transgenic for a spider silk gene produce the protein in their milk. Once researchers figure out how to spin it like spiders do, the silk may be used to manufacture fashionable fabrics, bulletproof vests, sports equipment, and biodegradable medical supplies.

Rabbits make human interleukin-2, a protein that triggers divisions of immune cells. Genetic engineering has also given us dairy goats with heart-healthy milk, low-fat pigs, pigs with environmentally friendly low-phosphate manure, extra-large sheep, and cows that are resistant to mad cow disease.

Tinkering with the genes of animals raises ethical dilemmas. For example, many people view transgenic animal research as unconscionable. Many others see it as simply an extension of thousands of years of acceptable animal husbandry practices: The techniques have changed, but not the intent. We humans still have a vested interest in improving our livestock.

Knockout Cells and Organ Factories

Millions of people suffer with organs or tissues that are damaged beyond repair. In any given year, more than 80,000 of them are on waiting lists for an organ

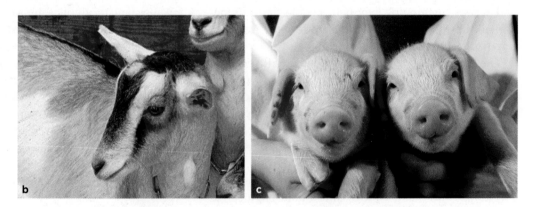

Figure 16.14 Genetically modified animals. (**a**) Featherless chicken developed by cross-breeding methods in Israel. Such chickens survive in deserts where cooling systems are not an option. (**b**) Mira, a goat transgenic for human antithrombin III (a protein that inhibits blood clotting). (**c**) The pig on the *left* is transgenic for a yellow fluorescent protein; its nontransgenic littermate is on the *right*.

Figure 16.15 Expression of a rat growth hormone gene in a transgenic mouse. These two mice are littermates. The mouse on the *left* weighed 29 grams (1 ounce); the one on the *right*, 44 grams (1.5 ounces).

transplant in the United States alone. Human donors are in such short supply that illegal organ trafficking is now a common problem.

Pigs are a potential source of organs for transplantation, because pig and human organs are about the same in both size and function. However, the human immune system battles anything it recognizes as nonself. It rejects a pig organ at once, because it recognizes a foreign glycoprotein on the plasma membrane of pig cells. Within a few hours, blood coagulates inside the organ's vessels and dooms the transplant. Drugs can suppress the immune response, but they also render organ recipients particularly vulnerable to infection.

Researchers have produced genetically engineered pigs that lack the offending glycoprotein on their cells. The human immune system may not reject tissues or organs transplanted from these pigs.

Transferring an organ from one species into another is called **xenotransplantation**. Critics of xenotransplantation are concerned that, among other things, pig-to-human transplants would invite pig viruses to cross the species barrier and infect humans, perhaps catastrophically. Their concerns are not unfounded. Evidence suggests that some of the worst pandemics arose because animal viruses adapted to new hosts—humans.

16.9 | Safety Issues

- The first transfer of foreign DNA into bacteria ignited an ongoing debate about potential dangers of transgenic organisms that may enter the environment.

When James Watson and Francis Crick presented their model of the DNA double helix in 1953, they ignited a global blaze of optimism about genetic research. The very book of life seemed to be open for scrutiny. In reality, no one could read it. Scientific breakthroughs are not very often accompanied by the simultaneous discovery of the tools to study them. New techniques would have to be invented before that book would become readable.

Twenty years later, Paul Berg and his coworkers discovered how to make recombinant organisms by fusing DNA from two species of bacteria. By isolating DNA in manageable subsets, researchers now had the tools to be able to study its sequence in detail. They began to clone and analyze DNA from many different organisms. The technique of genetic engineering was born, and suddenly everyone was worried about it.

Researchers knew that DNA itself was not toxic, but they could not predict with certainty what would happen every time they fused genetic material from different organisms. Would they accidentally make a superpathogen? Could they make a new, dangerous form of life by fusing DNA of two normally harmless organisms? What if that new form escaped from the laboratory and transformed other organisms?

In a remarkably quick and responsible display of self-regulation, scientists reached a consensus on new safety guidelines for DNA research. Adopted at once by the NIH, these guidelines included precautions for laboratory procedures. They covered the design and use of host organisms that could survive only under the narrow range of conditions inside the laboratory. Researchers stopped using DNA from pathogenic or toxic organisms for recombination experiments until proper containment facilities were developed.

Now, all genetic engineering research is done under these laboratory guidelines. Releasing and importing genetically modified organisms is carefully regulated by the USDA. Such regulations are our best effort to minimize any risk involved in the research or as a result of it, but they are not a guarantee.

Take-Home Message

Why do we genetically engineer animals?

- Animals that would be impossible to produce by traditional breeding methods are being created by genetic engineering. Such animals are used in research, medicine, and industry.

Take-Home Message

Is genetic engineering safe?

- Rigorous safety guidelines for DNA research have been in place for decades in the United States and other countries. Researchers are expected to comply with these stringent standards.

16.10 | Modified Humans?

■ We as a society continue to work our way through the ethical implications of applying new DNA technologies.

■ The manipulation of individual genomes continues even as we are weighing the risks and benefits of this research.

■ Link to Human genetic disorders 12.7

Getting Better We know of more than 15,000 serious genetic disorders. Collectively, they cause 20 to 30 percent of infant deaths each year, and account for half of all mentally impaired patients and a fourth of all hospital admissions. They also contribute to many age-related disorders, including cancer, Parkinson's disease, and diabetes.

Drugs and other treatments can minimize the symptoms of some genetic diseases, but gene therapy is the only cure. **Gene therapy** is the transfer of recombinant DNA into an individual's body cells, with the intent to correct a genetic defect or treat a disease. The transfer, which occurs by way of viral vectors or lipid clusters, inserts an unmutated gene into an individual's chromosomes.

Human gene therapy is a compelling reason to embrace genetic engineering research. It is now being tested as a treatment for cystic fibrosis, hemophilia A, several types of cancer, inherited retinal disease, and inherited immune disorders, among other diseases. The results are encouraging.

For example, little Rhys Evans (*left*) was born with a severe immune disorder, SCID-X1. SCID-X1 stems from mutations in the *IL2RG* gene, which encodes a receptor for an immune signaling molecule. Children affected by this disorder can survive only in germ-free isolation tents, because they cannot fight infections.

In 1998, a viral vector was used to insert unmutated copies of *IL2RG* into cells taken from the bone marrow of eleven boys with SCID-X1. Each child's modified cells were infused back into his bone marrow. Months later, ten of the boys left their isolation tents for good. Their immune systems had been repaired by the gene therapy. Since then, gene therapy has freed many other SCID-X1 patients from life in an isolation tent. Rhys is one of them.

Getting Worse Manipulating a gene within the context of a living individual is unpredictable even when we know its sequence and where it is within the genome. No one, for example, can predict where a virus-injected gene will insert into chromosomes. Its insertion might disrupt other genes. If it interrupts a gene that is part of the controls over cell division, then cancer might be the outcome.

For example, three boys from the 1998 SCID-X1 clinical trial have since developed leukemia, and one of them died. The researchers had wrongly predicted that cancer related to the gene therapy would be rare. Research now implicates the very gene targeted for repair, especially when combined with the viral vector that delivered it.

Other unanticipated problems sometimes occur with gene therapy. Jesse Gelsinger had a rare genetic deficiency of ornithine transcarbamylase. This liver enzyme helps the body rid itself of ammonia, a toxic by-product of protein breakdown. Jesse's health was fairly stable while he was on a low-protein diet, but he had to take a lot of medication. In 1999, Jesse volunteered to be in a clinical trial of a gene therapy. He had a severe allergic reaction to the viral vector, and four days after receiving the treatment, his organs shut down and he died. He was 18.

Our understanding of how the human genome works clearly lags behind our ability to modify it.

Getting Perfect The idea of using human gene therapy to cure genetic disorders seems like a socially acceptable goal to most people. However, go one step further. Would it also be acceptable to modify genes of an individual who is within a normal range in order to minimize or enhance a particular trait? Researchers have already produced mice that have enhanced memory, bigger muscles, or improved learning abilities. Why not people?

The idea of selecting the most desired human traits is called eugenic engineering. Yet who decides which forms of traits are most desirable? Realistically, cures for many severe but rare genetic disorders will not be found, because the financial payback will not even cover the cost of the research. Eugenics, however, might just turn a profit. How much would potential parents pay to be sure that their child will be tall or blue-eyed? Would it be okay to engineer "superhumans" with breathtaking strength or intelligence? How about a treatment that can help you lose that extra weight, and keep it off permanently? The gray area between interesting and abhorrent can be very different depending on who you ask.

In a survey conducted in the United States, more than 40 percent of those interviewed said it would be fine to use gene therapy to make smarter and cuter babies. In one poll of British parents, 18 percent would be willing to use genetic enhancement to keep their child from being aggressive, and 10 percent would use it to keep a child from growing up to be homosexual.

Getting There Some people are adamant that we must never alter the DNA of anything. The concern is that gene therapy puts us on a slippery slope that may result in irreversible damage to ourselves and to nature. We as a society may not have the wisdom to know how to stop once we set foot on that slope. One is reminded of our peculiar human tendency to leap before we look.

And yet, something about the human experience allows us to dream of such things as wings of our own making, a capacity that carried us to the frontiers of space. In this brave new world, the questions before you are these: What do we stand to lose if serious risks are not taken? And, do we have the right to impose the consequences of taking such risks on those who would choose not to take them?

Beta-carotene is an orange photosynthetic pigment (Section 7.1) that is remodeled by cells of the small intestine into vitamin A. Potrykus and Beyer transferred two genes in the beta-carotene synthesis pathway into rice plants—one gene from corn and one from bacteria. All three genes were under the control of a promoter that works only in seeds. The transgenic rice plants began to make beta-carotene in their seeds—in the grains of Golden Rice. One cup of Golden Rice has enough beta-carotene to satisfy a child's daily recommended amount of vitamin A. The rice was ready in 2005, but is still not available for human consumption. The biosafety experiments required by regulatory agencies

How would you vote?

Should food distributors be required to identify food products made from genetically modified organisms? See CengageNOW for details, then vote online.

are far too expensive for a humanitarian agency in the public sector. Most of the transgenic organisms used for food today were carried through the deregulation process by private companies.

Summary

Section 16.1 **Recombinant DNA** consists of the fused DNA of different organisms. In **DNA cloning**, **restriction enzymes** cut DNA into pieces, then DNA ligase splices the pieces into **plasmids** or other **cloning vectors**. The resulting hybrid molecules are inserted into host cells such as bacteria. When a host cell divides, it forms huge populations of genetically identical descendant cells, or **clones**. Each clone has a copy of the foreign DNA.

RNA cannot be cloned directly. **Reverse transcriptase**, a viral enzyme, is used to convert single-stranded RNA into **cDNA** for cloning.

■ *Use the animation on CengageNOW to survey the tools of researchers who make recombinant DNA.*

Section 16.2 A **DNA library** is a collection of cells that host different fragments of DNA, often representing an organism's entire **genome**. Researchers can use **probes** to identify cells that host a specific fragment of DNA. Base-pairing between nucleic acids from different sources is called **nucleic acid hybridization**.

The polymerase chain reaction (**PCR**) uses **primers** and a heat-resistant DNA polymerase to rapidly increase the number of molecules of a DNA fragment.

■ *Use the interaction on CengageNOW to learn how researchers isolate and copy genes.*

Section 16.3 **DNA sequencing** can reveal the order of nucleotide bases in a fragment of DNA. DNA polymerase is used to partially replicate a DNA template. The reaction produces a mixture of DNA fragments of different lengths. **Electrophoresis** separates the fragments into bands.

■ *Use the animation on CengageNOW to investigate the technique of DNA sequencing.*

Section 16.4 **Short tandem repeats** are multiple copies of a short DNA sequence that follow one another along a chromosome. The number and distribution of short tandem repeats, unique in each individual, is revealed by electrophoresis as a **DNA fingerprint**.

■ *Use the animation on CengageNOW to observe the process of DNA fingerprinting.*

Section 16.5 The genomes of several organisms have been sequenced. **Genomics**, or the study of genomes, is providing insights into the function of the human genome. **DNA chips** are used to study gene expression.

Sections 16.6–16.8 Recombinant DNA technology and genome analysis are the basis of **genetic engineering**: directed modification of an organism's genetic makeup. Genes from one species are inserted into an individual of a different species to make a **transgenic** organism, or a gene is modified and reinserted into an individual of the same species. The result of either process is a **genetically modified organism** (GMO). GMO animals provide a source of organs for **xenotransplantation**.

■ *Use the animation on CengageNOW to see how the Ti plasmid is used to genetically engineer plants.*

Section 16.9 Rigorous safety procedures minimize potential risks to researchers in genetic engineering labs. Although these and other strict government regulations limit the release of genetically modified organisms into the environment, such laws are not guarantees against accidental releases or unforeseen environmental effects.

Section 16.10 With **gene therapy**, a gene is transferred into body cells to correct a genetic defect or treat a disease.

Self-Quiz   *Answers in Appendix III*

1. Researchers can cut DNA molecules at specific sites by using _____ .
   a. DNA polymerase   c. restriction enzymes
   b. DNA probes   d. reverse transcriptase

2. Fill in the blank: A _____ is a small circle of bacterial DNA that contains only a few genes and is separate from the bacterial chromosome.

3. By reverse transcription, _____ is assembled on a(n) _____ template.
   a. mRNA; DNA   c. DNA; ribosome
   b. cDNA; mRNA   d. protein; mRNA

4. For each species, all _____ in the complete set of chromosomes is the _____ .
   a. genomes; phenotype   c. mRNA; start of cDNA
   b. DNA; genome   d. cDNA; start of mRNA

Data Analysis Exercise

Autism is a neurobiological disorder with a range of symptoms that include impaired social interactions, stereotyped patterns of behavior such as hand-flapping or rocking, and, occasionally, greatly enhanced intellectual abilities.

Autism may have a genetic basis. Some autistic people have a mutation in neuroligin 3, a type of cell adhesion protein (Section 5.2) that connects brain cells. One mutation changes amino acid 451 from arginine to cysteine.

Mouse and human neuroligin 3 are very similar. In 2007, Katsuhiko Tabuchi and his colleagues genetically modified mice to carry the same arginine-to-cysteine substitution in their neuroligin 3. The mutation caused an increase in transmission of some types of signals between brain cells. Mice with the mutation had impaired social behavior, and, unexpectedly, enhanced spatial learning ability (Figure 16.16).

1. Did the modified or the unmodified mice learn the location of the platform faster in the first test?

2. Which mice learned faster the second time around?

3. Which mice showed the greatest improvement in memory between the first and the second test?

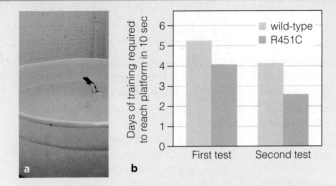

Figure 16.16 Enhanced spatial learning ability in mice with a mutation in neuroligin 3 (R451C), compared with unmodified (wild-type) mice. The mice were tested in a water maze, in which a platform is submerged a few millimeters below the surface of a deep pool of warm water (**a**). The platform, which is not visible to swimming mice, was moved for the second test.

Mice do not particularly enjoy swimming, so they locate a hidden platform as fast as they can. When tested again, they can remember its location by checking visual cues around the edge of the pool. How quickly they remember the platform's location is a measure of their spatial learning ability (**b**).

---

5. A set of cells that host various DNA fragments collectively representing an organism's entire set of genetic information is a _____ .

6. PCR can be used _____ .
   a. to increase the number of specific DNA fragments
   b. to make DNA fingerprints
   c. in a DNA sequencing reaction
   d. a and b are correct

7. Fragments of DNA can be separated by electrophoresis according to _____ .
   a. sequence   b. length   c. species

8. DNA sequencing relies on _____ .
   a. standard and labeled nucleotides
   b. primers and DNA polymerase
   c. electrophoresis
   d. all of the above

9. Which of the following can be used to carry foreign DNA into host cells? Choose all correct answers.
   a. RNA   e. lipid clusters
   b. viruses   f. blasts of pellets
   c. PCR   g. xenotransplantation
   d. plasmids   h. DNA microarrays

10. _____ can be used to correct a genetic defect.

11. Match the terms with the most suitable description.
   ___DNA fingerprint   a. carries a foreign gene
   ___Ti plasmid   b. slows bacteriophage growth
   ___nucleic acid   c. a person's unique collection
      hybridization   of short tandem repeats
   ___eugenic   d. base pairing of DNA or
      engineering   DNA and RNA from
   ___transgenic   different sources
   ___GMO   e. selecting "desirable" traits
   ___restriction enzyme f. genetically modified
                          g. used in some gene transfers

■ *Visit CengageNOW for additional questions.*

Critical Thinking

1. The *FOXP2* gene encodes a transcription factor associated with vocal learning in mice, bats, birds, and humans. Mutations in *FOXP2* result in altered vocalizations in mice, and severe language disorders in humans. The chimpanzee, gorilla, and rhesus *FOXP2* proteins are identical; the human version differs in two of 715 amino acids. The change of two amino acids may have contributed to the development of language in humans. Would it be okay to transfer the human *FOXP2* gene into a nonhuman primate? What do you think might happen if the transgenic animal learned to speak?

2. Animal viruses can mutate so that they infect humans, occasionally with disastrous results. In 1918, an influenza pandemic that apparently originated with a strain of avian flu killed 50 million people. Researchers isolated samples of that virus, the influenza A(H1N1) strain, from bodies of infected people preserved in Alaskan permafrost since 1918. From the samples, the researchers reconstructed the DNA sequence of the viral genome, then reconstructed the virus. Being 39,000 times more infectious than modern influenza strains, the reconstructed A(H1N1) virus proved to be 100 percent lethal in mice.

Understanding how the A(H1N1) strain works can help us defend ourselves against other strains that may be like it. For example, researchers are using the reconstructed virus to discover which of its mutations made it so infectious and deadly in humans. Their work is urgent. A deadly new strain of avian influenza in Asia shares some mutations with the A(H1N1) strain. Even now, researchers are working to test the effectiveness of antiviral drugs and vaccines on the reconstructed virus, and to develop new ones.

Critics of the A(H1N1) reconstruction are concerned. If the virus escapes the containment facilities (even though it has not done so yet), it might cause another pandemic. Worse, terrorists could use the published DNA sequence and methods to make the virus for horrific purposes. Do you think this research makes us more or less safe?

Appendix I. Classification System

This revised classification scheme is a composite of several that microbiologists, botanists, and zoologists use. The major groupings are agreed upon, more or less. However, there is not always agreement on what to name a particular grouping or where it might fit within the overall hierarchy. There are several reasons why full consensus is not possible at this time.

First, the fossil record varies in its completeness and quality. Therefore, the phylogenetic relationship of one group to other groups is sometimes open to interpretation. Today, comparative studies at the molecular level are firming up the picture, but the work is still under way. Also, molecular comparisons do not always provide definitive answers to questions about phylogeny. Comparisons based on one set of genes may conflict with those comparing a different part of the genome. Or comparisons with one member of a group may conflict with comparisons based on other group members.

Second, ever since the time of Linnaeus, systems of classification have been based on the perceived morphological similarities and differences among organisms. Although some original interpretations are now open to question, we are so used to thinking about organisms in certain ways that reclassification often proceeds slowly.

A few examples: Traditionally, birds and reptiles were grouped in separate classes (Reptilia and Aves); yet there are compelling arguments for grouping the lizards and snakes in one group and the crocodilians, dinosaurs, and birds in another. Many biologists still favor a six-kingdom system of classification (archaea, bacteria, protists, plants, fungi, and animals). Others advocate a switch to the more recently proposed three-domain system (archaea, bacteria, and eukarya).

Third, researchers in microbiology, mycology, botany, zoology, and other fields of inquiry inherited a wealth of literature, based on classification systems that have been developed over time in each field of inquiry. Many are reluctant to give up established terminology that offers access to the past.

For example, botanists and microbiologists often use *division*, and zoologists *phylum*, for taxa that are equivalent in hierarchies of classification.

Why bother with classification frameworks if we know they only imperfectly reflect the evolutionary history of life? We do so for the same reasons that a writer might break up a history of civilization into several volumes, each with a number of chapters. Both are efforts to impart structure to an enormous body of knowledge and to facilitate retrieval of information from it. More importantly, to the extent that modern classification schemes accurately reflect evolutionary relationships, they provide the basis for comparative biological studies, which link all fields of biology.

Bear in mind that we include this appendix for your reference purposes only. Besides being open to revision, it is not meant to be complete. Names shown in "quotes" are polyphyletic or paraphyletic groups that are undergoing revision. For example, "reptiles" comprise at least three and possibly more lineages.

The most recently discovered species, as from the mid-ocean province, are not listed. Many existing and extinct species of the more obscure phyla are also not represented. Our strategy is to focus primarily on the organisms mentioned in the text or familiar to most students. We delve more deeply into flowering plants than into bryophytes, and into chordates than annelids.

PROKARYOTES AND EUKARYOTES COMPARED

As a general frame of reference, note that almost all bacteria and archaea are microscopic in size. Their DNA is concentrated in a nucleoid (a region of cytoplasm), not in a membrane-bound nucleus. All are single cells or simple associations of cells. They reproduce by prokaryotic fission or budding; they transfer genes by bacterial conjugation.

Table A lists representative types of autotrophic and heterotrophic prokaryotes. The authoritative reference, *Bergey's Manual of Systematic Bacteriology*, has called this a time of taxonomic transition. It references groups mainly by numerical taxonomy (Section 19.1) rather than by phylogeny. Our classification system does reflect evidence of evolutionary relationships for at least some bacterial groups.

The first life forms were prokaryotic. Similarities between Bacteria and Archaea have more ancient origins relative to the traits of eukaryotes.

Unlike the prokaryotes, all eukaryotic cells start out life with a DNA-enclosing nucleus and other membrane-bound organelles. Their chromosomes have many histones and other proteins attached. They include spectacularly diverse single-celled and multicelled species, which can reproduce by way of meiosis, mitosis, or both.

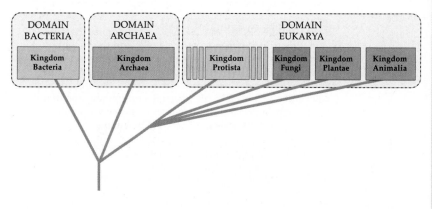

DOMAIN OF BACTERIA

KINGDOM BACTERIA   The largest, and most diverse group of prokaryotic cells. Includes photosynthetic autotrophs, chemosynthetic autotrophs, and heterotrophs. All prokaryotic pathogens of vertebrates are bacteria.

PHYLUM AQIFACAE   Most ancient branch of the bacterial tree. Gram-negative, mostly aerobic chemoautotrophs, mainly of volcanic hot springs. *Aquifex.*

PHYLUM DEINOCOCCUS-THERMUS   Gram-positive, heat-loving chemoautotrophs. *Deinococcus* is the most radiation resistant organism known. *Thermus* occurs in hot springs and near hydrothermal vents.

PHYLUM CHLOROFLEXI   Green nonsulfur bacteria. Gram-negative bacteria of hot springs, freshwater lakes, and marine habitats. Act as nonoxygen-producing photoautotrophs or aerobic chemoheterotrophs. *Chloroflexus.*

PHYLUM ACTINOBACTERIA   Gram-positive, mostly aerobic heterotrophs in soil, freshwater and marine habitats, and on mammalian skin. *Propionibacterium, Actinomyces, Streptomyces.*

PHYLUM CYANOBACTERIA   Gram-negative, oxygen-releasing photoautotrophs mainly in aquatic habitats. They have chlorophyll *a* and photosystem I. Includes many nitrogen-fixing genera. *Anabaena, Nostoc, Oscillatoria.*

PHYLUM CHLOROBIUM   Green sulfur bacteria. Gram-negative nonoxygen-producing photosynthesizers, mainly in freshwater sediments. *Chlorobium.*

PHYLUM FIRMICUTES   Gram-positive walled cells and the cell wall-less mycoplasmas. All are heterotrophs. Some survive in soil, hot springs, lakes, or oceans. Others live on or in animals. *Bacillus, Clostridium, Heliobacterium, Lactobacillus, Listeria, Mycobacterium, Mycoplasma, Streptococcus.*

PHYLUM CHLAMYDIAE   Gram-negative intracellular parasites of birds and mammals. *Chlamydia.*

PHYLUM SPIROCHETES   Free-living, parasitic, and mutualistic gram-negative spring-shaped bacteria. *Borelia, Pillotina, Spirillum, Treponema.*

PHYLUM PROTEOBACTERIA   The largest bacterial group. Includes photoautotrophs, chemoautotrophs, and heterotrophs; free-living, parasitic, and colonial groups. All are gram-negative.

Class Alphaproteobacteria. *Agrobacterium, Azospirillum, Nitrobacter, Rickettsia, Rhizobium.*

Class Betaproteobacteria. *Neisseria.*

Class Gammaproteobacteria. *Chromatium, Escherichia, Haemopilius, Pseudomonas, Salmonella, Shigella, Thiomargarita, Vibrio, Yersinia.*

Class Deltaproteobacteria. *Azotobacter, Myxococcus.*

Class Epsilonproteobacteria. *Campylobacter, Helicobacter.*

DOMAIN OF ARCHAEA

KINGDOM ARCHAEA   Prokaryotes that are evolutionarily between eukaryotic cells and the bacteria. Most are anaerobes. None are photosynthetic. Originally discovered in extreme habitats, they are now known to be widely dispersed. Compared with bacteria, the archaea have a distinctive cell wall structure and unique membrane lipids, ribosomes, and RNA sequences. Some are symbiotic with animals, but none are known to be animal pathogens.

PHYLUM EURYARCHAEOTA   Largest archean group. Includes extreme thermophiles, halophiles, and methanogens. Others are abundant in the upper waters of the ocean and other more moderate habitats. *Methanocaldococcus, Nanoarchaeum.*

PHYLUM CRENARCHAEOTA   Includes extreme theromophiles, as well as species that survive in Antarctic waters, and in more moderate habitats. *Sulfolobus, Ignicoccus.*

PHYLUM KORARCHAEOTA   Known only from DNA isolated from hydrothermal pools. As of this writing, none have been cultured and no species have been named.

DOMAIN OF EUKARYOTES

KINGDOM "PROTISTA"   A collection of single-celled and multicelled lineages, which does not constitute a monophyletic group. Some biologists consider the groups listed below to be kingdoms in their own right.

PARABASALIA   Parabasalids. Flagellated, single-celled anaerobic heterotrophs with a cytoskeletal "backbone" that runs the length of the cell. There are no mitochondria, but a hydrogenosome serves a similar function. *Trichomonas, Trichonympha.*

DIPLOMONADIDA   Diplomonads. Flagellated, anaerobic single-celled heterotrophs that do not have mitochondria or Golgi bodies and do not form a bipolar spindle at mitosis. May be one of the most ancient lineages. *Giardia.*

EUGLENOZOA   Euglenoids and kinetoplastids. Free-living and parasitic flagellates. All with one or more mitochondria. Some photosynthetic euglenoids with chloroplasts, others heterotrophic. *Euglena, Trypanosoma, Leishmania.*

RHIZARIA   Formaminiferans and radiolarians. Free-living, heterotrophic amoeboid cells that are enclosed in shells. Most live in ocean waters or sediments. *Pterocorys, Stylosphaera.*

ALVEOLATA   Single cells having a unique array of membrane-bound sacs (alveoli) just beneath the plasma membrane.

Ciliata. Ciliated protozoans. Heterotrophic protists with many cilia. *Paramecium, Didinium.*

Dinoflagellates. Diverse heterotrophic and photosynthetic flagellated cells that deposit cellulose in their alveoli. *Gonyaulax, Gymnodinium, Karenia, Noctiluca.*

Apicomplexans. Single-celled parasites of animals. A unique microtubular device is used to attach to and penetrate a host cell. *Plasmodium.*

STRAMENOPHILA   Stramenophiles. Single-celled and multicelled forms; flagella with tinsel-like filaments.

Oomycotes. Water molds. Heterotrophs. Decomposers, some parasites. *Saprolegnia, Phytophthora, Plasmopara.*

Chrysophytes. Golden algae, yellow-green algae, diatoms, coccolithophores. Photosynthetic. *Emiliania, Mischococcus.*

Phaeophytes. Brown algae. Photosynthetic; nearly all live in temperate marine waters. All are multicellular. *Macrocystis, Laminaria, Sargassum, Postelsia.*

RHODOPHYTA   Red algae. Mostly photosynthetic, some parasitic. Nearly all marine, some in freshwater habitats. Most multicellular. *Porphyra, Antithamion.*

CHLOROPHYTA   Green algae. Mostly photosynthetic, some parasitic. Most freshwater, some marine or terrestrial. Single-celled, colonial, and multicellular forms. Some biologists place the chlorophytes and charophytes with the land plants in a kingdom called the Viridiplantae. *Acetabularia, Chlamydomonas, Chlorella, Codium, Udotea, Ulva, Volvox.*

CHAROPHYTA   Photosynthetic. Closest living relatives of plants. Include both single-celled and multicelled forms. Desmids, stoneworts. *Micrasterias, Chara, Spirogyra.*

AMOEBOZOA   True amoebas and slime molds. Heterotrophs that spend all or part of the life cycle as a single cell that uses pseudopods to capture food. *Amoeba, Entoamoeba* (amoebas), *Dictyostelium* (cellular slime mold), *Physarum* (plasmodial slime mold).

KINGDOM FUNGI

Nearly all multicelled eukaryotic species with chitin-containing cell walls. Heterotrophs, mostly saprobic decomposers, some parasites. Nutrition based upon extracellular digestion of organic matter and absorption of nutrients by individual cells. Multicelled species form absorptive mycelia and reproductive structures that produce asexual spores (and sometimes sexual spores).

PHYLUM CHYTRIDIOMYCOTA   Chytrids. Primarily aquatic; saprobic decomposers or parasites that produce flagellated spores. *Chytridium*.

PHYLUM ZYGOMYCOTA   Zygomycetes. Producers of zygospores (zygotes inside thick wall) by way of sexual reproduction. Bread molds, related forms. *Rhizopus, Philobolus*.

PHYLUM ASCOMYCOTA   Ascomycetes. Sac fungi. Sac-shaped cells form sexual spores (ascospores). Most yeasts and molds, morels, truffles. *Saccharomyces, Morchella, Neurospora, Claviceps, Candida, Aspergillus, Penicillium*.

PHYLUM BASIDIOMYCOTA   Basidiomycetes. Club fungi. Most diverse group. Produce basidiospores inside club-shaped structures. Mushrooms, shelf fungi, stinkhorns. *Agaricus, Amanita, Craterellus, Gymnophilus, Puccinia, Ustilago*.

"IMPERFECT FUNGI"   Sexual spores absent or undetected. The group has no formal taxonomic status. If better understood, a given species might be grouped with sac fungi or club fungi. *Arthobotrys, Histoplasma, Microsporum, Verticillium*.

"LICHENS"   Mutualistic interactions between fungal species and a cyanobacterium, green alga, or both. *Lobaria, Usnea*.

KINGDOM PLANTAE

Most photosynthetic with chlorophylls *a* and *b*. Some parasitic. Nearly all live on land. Sexual reproduction predominates.

BRYOPHYTES (NONVASCULAR PLANTS)

Small flattened haploid gametophyte dominates the life cycle; sporophyte remains attached to it. Sperm are flagellated; require water to swim to eggs for fertilization.

PHYLUM HEPATOPHYTA   Liverworts. *Marchantia*.

PHYLUM ANTHOCEROPHYTA   Hornworts.

PHYLUM BRYOPHYTA   Mosses. *Polytrichum, Sphagnum*.

SEEDLESS VASCULAR PLANTS

Diploid sporophyte dominates, free-living gametophytes, flagellated sperm require water for fertilization.

PHYLUM LYCOPHYTA   Lycophytes, club mosses. Small single-veined leaves, branching rhizomes. *Lycopodium, Selaginella*.

PHYLUM MONILOPHYTA

Subphylum Psilophyta. Whisk ferns. No obvious roots or leaves on sporophyte, very reduced. *Psilotum*.

Subphylum Sphenophyta. Horsetails. Reduced scalelike leaves. Some stems photosynthetic, others spore-producing. *Calamites* (extinct), *Equisetum*.

Subphylum Pterophyta. Ferns. Large leaves, usually with sori. Largest group of seedless vascular plants (12,000 species), mainly tropical, temperate habitats. *Pteris, Trichomanes, Cyathea* (tree ferns), *Polystichum*.

SEED-BEARING VASCULAR PLANTS

PHYLUM CYCADOPHYTA   Cycads. Group of gymnosperms (vascular, bear "naked" seeds). Tropical, subtropical. Compound leaves, simple cones on male and female plants. Plants usually palm-like. Motile sperm. *Zamia, Cycas*.

PHYLUM GINKGOPHYTA   Ginkgo (maidenhair tree). Type of gymnosperm. Motile sperm. Seeds with fleshy layer. *Ginkgo*.

PHYLUM GNETOPHYTA   Gnetophytes. Only gymnosperms with vessels in xylem and double fertilization (but endosperm does not form). *Ephedra, Welwitchia, Gnetum*.

PHYLUM CONIFEROPHYTA   Conifers. Most common and familiar gymnosperms. Generally cone-bearing species with needle-like or scale-like leaves. Includes pines (*Pinus*), redwoods (*Sequoia*), yews (*Taxus*).

PHYLUM ANTHOPHYTA   Angiosperms (the flowering plants). Largest, most diverse group of vascular seed-bearing plants. Only organisms that produce flowers, fruits. Some families from several representative orders are listed:

BASAL FAMILIES

Family Amborellaceae. *Amborella*.
Family Nymphaeaceae. Water lilies.
Family Illiciaceae. Star anise.

MAGNOLIIDS

Family Magnoliaceae. Magnolias.
Family Lauraceae. Cinnamon, sassafras, avocados.
Family Piperaceae. Black pepper, white pepper.

EUDICOTS

Family Papaveraceae. Poppies.
Family Cactaceae. Cacti.
Family Euphorbiaceae. Spurges, poinsettia.
Family Salicaceae. Willows, poplars.
Family Fabaceae. Peas, beans, lupines, mesquite.
Family Rosaceae. Roses, apples, almonds, strawberries.
Family Moraceae. Figs, mulberries.
Family Cucurbitaceae. Squashes, melons, cucumbers.
Family Fagaceae. Oaks, chestnuts, beeches.
Family Brassicaceae. Mustards, cabbages, radishes.
Family Malvaceae. Mallows, okra, cotton, hibiscus, cocoa.
Family Sapindaceae. Soapberry, litchi, maples.
Family Ericaceae. Heaths, blueberries, azaleas.
Family Rubiaceae. Coffee.
Family Lamiaceae. Mints.
Family Solanaceae. Potatoes, eggplant, petunias.
Family Apiaceae. Parsleys, carrots, poison hemlock.
Family Asteraceae. Composites. Chrysanthemums, sunflowers, lettuces, dandelions.

MONOCOTS

Family Araceae. Anthuriums, calla lily, philodendrons.
Family Liliaceae. Lilies, tulips.
Family Alliaceae. Onions, garlic.
Family Iridaceae. Irises, gladioli, crocuses.
Family Orchidaceae. Orchids.
Family Arecaceae. Date palms, coconut palms.
Family Bromeliaceae. Bromeliads, pineapples.
Family Cyperaceae. Sedges.
Family Poaceae. Grasses, bamboos, corn, wheat, sugarcane.
Family Zingiberaceae. Gingers.

KINGDOM ANIMALIA

Multicelled heterotrophs, nearly all with tissues and organs, and organ systems, that are motile during part of the life cycle. Sexual reproduction occurs in most, but some also reproduce asexually. Embryos develop through a series of stages.

PHYLUM PORIFERA   Sponges. No symmetry, tissues.

PHYLUM PLACOZOA   Marine. Simplest known animal. Two cell layers, no mouth, no organs. *Trichoplax*.

PHYLUM CNIDARIA   Radial symmetry, tissues, nematocysts.
Class Hydrozoa. Hydrozoans. *Hydra, Obelia, Physalia, Prya*.
Class Scyphozoa. Jellyfishes. *Aurelia*.
Class Anthozoa. Sea anemones, corals. *Telesto*.

PHYLUM PLATYHELMINTHES   Flatworms. Bilateral, cephalized; simplest animals with organ systems. Saclike gut.

Class Turbellaria. Triclads (planarians), polyclads. *Dugesia.*
Class Trematoda. Flukes. *Clonorchis, Schistosoma.*
Class Cestoda. Tapeworms. *Diphyllobothrium, Taenia.*

PHYLUM ROTIFERA   Rotifers. *Asplancha, Philodina.*

PHYLUM MOLLUSCA   Mollusks.

Class Polyplacophora. Chitons. *Cryptochiton, Tonicella.*

Class Gastropoda. Snails, sea slugs, land slugs. *Aplysia, Ariolimax, Cypraea, Haliotis, Helix, Liguus, Limax, Littorina.*

Class Bivalvia. Clams, mussels, scallops, cockles, oysters, shipworms. *Ensis, Chlamys, Mytelus, Patinopectin.*

Class Cephalopoda. Squids, octopuses, cuttlefish, nautiluses. *Dosidiscus, Loligo, Nautilus, Octopus, Sepia.*

PHYLUM ANNELIDA   Segmented worms.

Class Polychaeta. Mostly marine worms. *Eunice, Neanthes.*

Class Oligochaeta. Mostly freshwater and terrestrial worms, many marine. *Lumbricus* (earthworms), *Tubifex.*

Class Hirudinea. Leeches. *Hirudo, Placobdella.*

PHYLUM NEMATODA   Roundworms. *Ascaris, Caenorhabditis elegans, Necator* (hookworms), *Trichinella.*

PHYLUM ARTHROPODA

Subphylum Chelicerata. Chelicerates. Horseshoe crabs, spiders, scorpions, ticks, mites.

Subphylum Crustacea. Shrimps, crayfishes, lobsters, crabs, barnacles, copepods, isopods (sowbugs).

Subphylum Myriapoda. Centipedes, millipedes.

Subphylum Hexapoda. Insects and sprintails.

PHYLUM ECHINODERMATA   Echinoderms.

Class Asteroidea. Sea stars. *Asterias.*
Class Ophiuroidea. Brittle stars.
Class Echinoidea. Sea urchins, heart urchins, sand dollars.
Class Holothuroidea. Sea cucumbers.
Class Crinoidea. Feather stars, sea lilies.
Class Concentricycloidea. Sea daisies.

PHYLUM CHORDATA   Chordates.

Subphylum Urochordata. Tunicates, related forms.
Subphylum Cephalochordata. Lancelets.

CRANIATES

Class Myxini. Hagfishes.

VERTEBRATES (SUBGROUP OF CRANIATES)

Class Cephalaspidomorphi. Lampreys.

Class Chondrichthyes. Cartilaginous fishes (sharks, rays, skates, chimaeras).

Class "Osteichthyes." Bony fishes. Not monophyletic (sturgeons, paddlefish, herrings, carps, cods, trout, seahorses, tunas, lungfishes, and coelocanths).

TETRAPODS (SUBGROUP OF VERTEBRATES)

Class Amphibia. Amphibians. Require water to reproduce.
Order Caudata. Salamanders and newts.
Order Anura. Frogs, toads.
Order Apoda. Apodans (caecilians).

AMNIOTES (SUBGROUP OF TETRAPODS)

Class "Reptilia." Skin with scales, embryo protected and nutritionally supported by extraembryonic membranes.

Subclass Anapsida. Turtles, tortoises.
Subclass Lepidosaura. *Sphenodon*, lizards, snakes.
Subclass Archosaura. Crocodiles, alligators.

Class Aves. Birds. In some classifications birds are grouped in the archosaurs.

Order Struthioniformes. Ostriches.
Order Sphenisciformes. Penguins.
Order Procellariiformes. Albatrosses, petrels.
Order Ciconiiformes. Herons, bitterns, storks, flamingoes.
Order Anseriformes. Swans, geese, ducks.
Order Falconiformes. Eagles, hawks, vultures, falcons.
Order Galliformes. Ptarmigan, turkeys, domestic fowl.
Order Columbiformes. Pigeons, doves.
Order Strigiformes. Owls.
Order Apodiformes. Swifts, hummingbirds.
Order Passeriformes. Sparrows, jays, finches, crows, robins, starlings, wrens.
Order Piciformes. Woodpeckers, toucans.
Order Psittaciformes. Parrots, cockatoos, macaws.

Class Mammalia. Skin with hair; young nourished by milk-secreting mammary glands of adult.

Subclass Prototheria. Egg-laying mammals (monotremes; duckbilled platypus, spiny anteaters).

Subclass Metatheria. Pouched mammals or marsupials (opossums, kangaroos, wombats, Tasmanian devils).

Subclass Eutheria. Placental mammals.

Order Edentata. Anteaters, tree sloths, armadillos.
Order Insectivora. Tree shrews, moles, hedgehogs.
Order Chiroptera. Bats.
Order Scandentia. Insectivorous tree shrews.
Order Primates.

Suborder Strepsirhini (prosimians). Lemurs, lorises.
Suborder Haplorhini (tarsioids and anthropoids).

Infraorder Tarsiiformes. Tarsiers.
Infraorder Platyrrhini (New World monkeys).

Family Cebidae. Spider monkeys, howler monkeys, capuchin.

Infraorder Catarrhini (Old World monkeys and hominoids).

Superfamily Cercopithecoidea. Baboons, macaques, langurs.

Superfamily Hominoidea. Apes and humans.

Family Hylobatidae. Gibbon.

Family "Pongidae." Chimpanzees, gorillas, orangutans.

Family Hominidae. Existing and extinct human species (*Homo*) and humanlike species, including the australopiths.

Order Lagomorpha. Rabbits, hares, pikas.

Order Rodentia. Most gnawing animals (squirrels, rats, mice, guinea pigs, porcupines, beavers, etc.).

Order Carnivora. Carnivores (wolves, cats, bears, etc.).

Order Pinnipedia. Seals, walruses, sea lions.

Order Proboscidea. Elephants, mammoths (extinct).

Order Sirenia. Sea cows (manatees, dugongs).

Order Perissodactyla. Odd-toed ungulates (horses, tapirs, rhinos).

Order Tubulidentata. African aardvarks.

Order Artiodactyla. Even-toed ungulates (camels, deer, bison, sheep, goats, antelopes, giraffes, etc.).

Order Cetacea. Whales, porpoises.

Appendix II. Annotations to A Journal Article

This journal article reports on the movements of a female wolf during the summer of 2002 in northwestern Canada. It also reports on a scientific process of inquiry, observation and interpretation to learn where, how and why the wolf traveled as she did. In some ways, this article reflects the story of "how to do science" told in section 1.5 of this textbook. These notes are intended to help you read and understand how scientists work and how they report on their work.

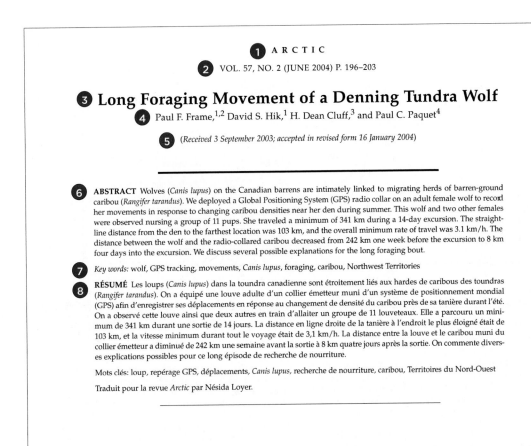

(1) ARCTIC

(2) VOL. 57, NO. 2 (JUNE 2004) P. 196–203

(3) Long Foraging Movement of a Denning Tundra Wolf

(4) Paul F. Frame,[1,2] David S. Hik,[1] H. Dean Cluff,[3] and Paul C. Paquet[4]

(5) (Received 3 September 2003; accepted in revised form 16 January 2004)

(6) ABSTRACT Wolves (*Canis lupus*) on the Canadian barrens are intimately linked to migrating herds of barren-ground caribou (*Rangifer tarandus*). We deployed a Global Positioning System (GPS) radio collar on an adult female wolf to record her movements in response to changing caribou densities near her den during summer. This wolf and two other females were observed nursing a group of 11 pups. She traveled a minimum of 341 km during a 14-day excursion. The straight-line distance from the den to the farthest location was 103 km, and the overall minimum rate of travel was 3.1 km/h. The distance between the wolf and the radio-collared caribou decreased from 242 km one week before the excursion to 8 km four days into the excursion. We discuss several possible explanations for the long foraging bout.

(7) *Key words:* wolf, GPS tracking, movements, *Canis lupus*, foraging, caribou, Northwest Territories

(8) **RÉSUMÉ** Les loups (*Canis lupus*) dans la toundra canadienne sont étroitement liés aux hardes de caribous des toundras (*Rangifer tarandus*). On a équipé une louve adulte d'un collier émetteur muni d'un système de positionnement mondial (GPS) afin d'enregistrer ses déplacements en réponse au changement de densité du caribou près de sa tanière durant l'été. On a observé cette louve ainsi que deux autres en train d'allaiter un groupe de 11 louveteaux. Elle a parcouru un minimum de 341 km durant une sortie de 14 jours. La distance en ligne droite de la tanière à l'endroit le plus éloigné était de 103 km, et la vitesse minimum durant tout le voyage était de 3,1 km/h. La distance entre la louve et le caribou muni du collier émetteur a diminué de 242 km une semaine avant la sortie à 8 km quatre jours après la sortie. On commente diverses explications possibles pour ce long épisode de recherche de nourriture.

Mots clés: loup, repérage GPS, déplacements, *Canis lupus*, recherche de nourriture, caribou, Territoires du Nord-Ouest.

Traduit pour la revue *Arctic* par Nésida Loyer.

(9) Introduction

Wolves (*Canis lupus*) that den on the central barrens of mainland Canada follow the seasonal movements of their main prey, migratory barren-ground caribou (*Rangifer tarandus*) (Kuyt, 1962; Kelsall, 1968; Walton et al., 2001). However, most wolves do not den near caribou calving grounds, but select sites farther south, closer to the tree line (Heard and Williams, 1992). Most caribou migrate beyond primary wolf denning areas by mid-June and do not return until mid-to-late July (Heard et al., 1996; Gunn et al., 2001). Conse-quently, caribou density near dens is low for part of the summer.

During this period of spatial separation from the main caribou herds, wolves must either search near (10) the homesite for scarce caribou or alternative prey (or both), travel to where prey are abundant, or use a combination of these strategies.

Walton et al. (2001) postulated that the travel of (11) tundra wolves outside their normal summer ranges is a response to low caribou availability rather than a pre-dispersal exploration like that observed in territorial wolves (Fritts and Mech, 1981; Messier, 1985). The authors postulated this because most such travel was directed toward caribou calving grounds. We report details of such a long-distance excursion by a breeding female tundra wolf wearing a GPS radio collar. We discuss the relationship of the excursion to movements of satellite-collared caribou (Gunn et al., 2001), supporting the hypothesis that tundra wolves make directional, rapid, long-distance movements in response to seasonal prey availability.

[1] Department of Biological Sciences, University of Alberta, Edmonton, Alberta T6G 2E9, Canada
[2] Corresponding author: pframe@ualberta.ca
[3] Department of Resources, Wildlife, and Economic Development, North Slave Region, Government of the Northwest Territories, P.O. Box 2668, 3803 Bretzlaff Dr., Yellowknife, Northwest Territories X1A 2P9, Canada; Dean_Cluff@gov.nt.ca
[4] Faculty of Environmental Design, University of Calgary, Calgary, Alberta T2N 1N4, Canada; current address: P.O. Box 150, Meacham, Saskatchewan S0K 2V0, Canada

© The Arctic Institute of North America

196

---

1 Title of the journal, which reports on science taking place in Arctic regions.

2 Volume number, issue number and date of the journal, and page numbers of the article.

3 Title of the article: a concise but specific description of the subject of study—one episode of long-range travel by a wolf hunting for food on the Arctic tundra.

4 Authors of the article: scientists working at the institutions listed in the footnotes below. Note #2 indicates that P. F. Frame is the *corresponding author*—the person to contact with questions or comments. His email address is provided.

5 Date on which a draft of the article was received by the journal editor, followed by date one which a revised draft was accepted for publication. Between these dates, the article was reviewed and critiqued by other scientists, a process called peer review. The authors revised the article to make it clearer, according to those reviews.

6 ABSTRACT: A brief description of the study containing all basic elements of this report. First sentence summarizes the *background* material. Second sentence encapsulates the *methods* used. The rest of the paragraph sums up the *results*. Authors introduce the main *subject* of the study—a female wolf (#388) with pups in a den—and refer to later *discussion* of possible explanations for her behavior.

7 Key words are listed to help researchers using computer databases. Searching the databases using these key words will yield a list of studies related to this one.

8 RÉSUMÉ: The French translation of the abstract and key words. Many researchers in this field are French Canadian. Some journals provide such translations in French or in other languages.

9 INTRODUCTION: Gives the background for this wolf study. This paragraph tells of known or suspected wolf behavior that is important for this study. Note that (a) major species mentioned are always accompanied by scientific names, and (b) statements of fact or *postulations* (claims or assumptions about what is likely to be true) are followed by references to studies that established those facts or supported the postulations.

10 This paragraph focuses directly on the wolf behaviors that were studied here.

11 This paragraph starts with a statement of the *hypothesis* being tested, one that originated in other studies and is supported by this one. The hypothesis is restated more succinctly in the last sentence of this paragraph. This is the *inquiry* part of the scientific process—asking questions and suggesting possible answers.

12 This map shows the study area and depicts wolf and caribou locations and movements during one summer. Some of this information is explained below.

13 STUDY AREA: This section sets the stage for the study, locating it precisely with latitude and longitude coordinates and describing the area (illustrated by the map in Figure 1).

14 Here begins the story of how prey (caribou) and predators (wolves) interact on the tundra. Authors describe movements of these nomadic animals throughout the year.

15 We focus on the denning season (summer) and learn how wolves locate their dens and travel according to the movements of caribou herds.

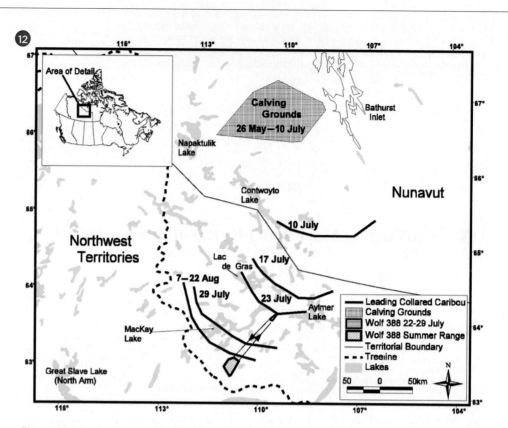

Figure 1. Map showing the movements of satellite radio-collared caribou with respect to female wolf 388's summer range and long foraging movement, in summer 2002.

13 Study Area

Our study took place in the northern boreal forest–low Arctic tundra transition zone (63° 30′ N, 110° 00′ W; Figure 1; Timoney et al., 1992). Permafrost in the area changes from discontinuous to continuous (Harris, 1986). Patches of spruce (*Picea mariana, P. glauca*) occur in the southern portion and give way to open tundra to the northeast. Eskers, kames, and other glacial deposits are scattered throughout the study area. Standing water and exposed bedrock are characteristic of the area.

14 *Details of the Caribou-Wolf System*

The Bathurst caribou herd uses this study area. Most caribou cows have begun migrating by late April, reaching calving grounds by June (Gunn et al., 2001;

Figure 1). Calving peaks by 15 June (Gunn et al., 2001), and calves begin to travel with the herd by one week of age (Kelsall, 1968). The movement patterns of bulls are less known, but bulls frequent areas near calving grounds by mid-June (Heard et al., 1996; Gunn et al., 2001). In summer, Bathurst caribou cows generally travel south from their calving grounds and then, parallel to the tree line, to the northwest. The rut usually takes place at the tree line in October (Gunn et al., 2001). The winter range of the Bathurst herd varies among years, ranging through the taiga and along the tree line from south of Great Bear Lake to southeast of Great Slave Lake. Some caribou spend the winter on the tundra (Gunn et al., 2001; Thorpe et al., 2001).

In winter, wolves that prey on Bathurst caribou do **15** not behave territorially. Instead, they follow the herd throughout its winter range (Walton et al., 2001; Musiani, 2003). However, during denning (May–

Table 1. Daily distances from wolf 388 and the den to the nearest radio-collared caribou during a long excursion in summer 2002.

Date (2002)	Mean distance from caribou to wolf (km)	Daily distance from closest caribou to den
12 July	242	241
13 July	210	209
14 July	200	199
15 July	186	180
16 July	163	162
17 July	151	148
18 July	144	137
19 July[1]	126	124
20 July	103	130
21 July	73	130
22 July	40	110
23 July[2]	9	104
29 July[3]	16	43
30 July	32	43
31 July	28	44
1 August	29	46
2 August[4]	54	52
3 August	53	53
4 August	74	74
5 August	75	75
6 August	74	75
7 August	72	75
8 August	76	75
9 August	79	79

[1] Excursion starts.
[2] Wolf closest to collared caribou.
[3] Previous five days' caribou locations not available.
[4] Excursion ends.

August, parturition late May to mid-June), wolf movements are limited by the need to return food to the den. To maximize access to migrating caribou, many wolves select den sites closer to the tree line than to caribou calving grounds (Heard and Williams, 1992). Because of caribou movement patterns, tundra denning wolves are separated from the main caribou herds by several hundred kilometers at some time during summer (Williams, 1990:19; Figure 1; Table 1).

 Muskoxen do not occur in the study area (Fournier and Gunn, 1998), and there are few moose there (H.D. Cluff, pers. obs.). Therefore, alternative prey for wolves includes waterfowl, other ground-nesting birds, their eggs, rodents, and hares (Kuyt, 1972; Williams, 1990:16; H.D. Cluff and P.F. Frame, unpubl. data). During 56 hours of den observations, we saw no ground squirrels or hares, only birds. It appears that the abundance of alternative prey was relatively low in 2002.

Methods

Wolf Monitoring

We captured female wolf 388 near her den on 22 June 2002, using a helicopter net-gun (Walton et al., 2001). She was fitted with a releasable GPS radio collar (Merrill et al., 1998) programmed to acquire locations at 30-

minute intervals. The collar was electronically released (e.g., Mech and Gese, 1992) on 20 August 2002. From 27 June to 3 July 2002, we observed 388's den with a 78 mm spotting scope at a distance of 390 m.

Caribou Monitoring

In spring of 2002, ten female caribou were captured by helicopter net-gun and fitted with satellite radio collars, bringing the total number of collared Bathurst cows to 19. Eight of these spent the summer of 2002 south of Queen Maud Gulf, well east of normal Bathurst caribou range. Therefore, we used 11 caribou for this analysis. The collars provided one location per day during our study, except for five days from 24 to 28 July. Locations of satellite collars were obtained from Service Argos, Inc. (Landover, Maryland).

Data Analysis

Location data were analyzed by ArcView GIS software (Environmental Systems Research Institute Inc., Redlands, California). We calculated the average distance from the nearest collared caribou to the wolf and the den for each day of the study.

Wolf foraging bouts were calculated from the time 388 exited a buffer zone (500 m radius around the den) until she re-entered it. We considered her to be traveling when two consecutive locations were spatially separated by more than 100 m. Minimum distance traveled was the sum of distances between each location and the next during the excursion.

We compared pre- and post-excursion data using Analysis of Variance (ANOVA; Zar, 1999). We first tested for homogeneity of variances with Levene's test (Brown and Forsythe, 1974). No transformations of these data were required.

Results

Wolf Monitoring

Pre-Excursion Period: Wolf 388 was lactating when captured on 22 June. We observed her and two other females nursing a group of 11 pups between 27 June and 3 July. During our observations, the pack consisted of at least four adults (3 females and 1 male) and 11 pups. On 30 June, three pups were moved to a location 310 m from the other eight and cared for by an uncollared female. The male was not seen at the den after the evening of 30 June.

Before the excursion, telemetry indicated 18 foraging bouts. The mean distance traveled during these bouts was 25.29 km (± 4.5 SE, range 3.1–82.5 km). Mean greatest distance from the den on foraging

16 Other variables are considered—prey other than caribou and their relative abundance in 2002.

17 METHODS: There is no one scientific method. Procedures for each and every study must be explained carefully.

18 Authors explain when and how they tracked caribou and wolves, including tools used and the exact procedures followed.

19 This important subsection explains what data were calculated (average distance ...) and how, including the software used and where it came from. (The calculations are listed in Table 1.) Note that the behavior measured (traveling) is carefully defined.

20 RESULTS: The heart of the report and the *observation* part of the scientific process. This section is organized parallel to the Methods section.

21 This subsection is broken down by periods of observation. Pre-excursion period covers the time between 388's capture and the start of her long-distance travel. The investigators used visual observations as well as telemetry (measurements taken using the global positioning system (GPS)) to gather data. They looked at how 388 cared for her pups, interacted with other adults, and moved about the den area.

22 The key in the lower right-hand corner of the map shows areas (shaded) within which the wolves and caribou moved, and the dotted trail of 388 during her excursion. From the results depicted on this map, the investigators tried to determine when and where 388 might have encountered caribou and how their locations affected her traveling behavior.

23 The wolf's excursion (her long trip away from the den area) is the focus of this study. These paragraphs present detailed measurements of daily movements during her two-week trip—how far she traveled, how far she was from collared caribou, her time spent traveling and resting, and her rate of speed. Authors use the phrase "minimum distance traveled" to acknowledge they couldn't track every step but were measuring samples of her movements. They knew that she went at least as far as they measured. This shows how scientists try to be exact when reporting results. Results of this study are depicted graphically in the map in Figure 2.

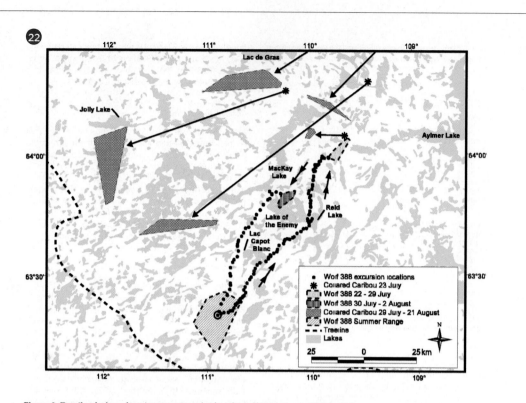

Figure 2. Details of a long foraging movement by female wolf 388 between 19 July and 2 August 2002. Also shown are locations and movements of three satellite radio-collared caribou from 23 July to 21 August 2002. On 23 July, the wolf was 8 km from a collared caribou. The farthest point from the den (103 km distant) was recorded on 27 July. Arrows indicate direction of travel.

bouts was 7.1 km (± 0.9 SE, range 1.7–17.0 km). The average duration of foraging bouts for the period was 20.9 h (± 4.5 SE, range 1–71 h).

The average daily distance between the wolf and the nearest collared caribou decreased from 242 km on 12 July, one week before the excursion period, to 126 km on 19 July, the day the excursion began (Table 1).

23 **Excursion Period:** On 19 July at 2203, after spending 14 h at the den, 388 began moving to the northeast and did not return for 336 h (14 d; Figure 2). Whether she traveled alone or with other wolves is unknown. During the excursion, 476 (71%) of 672 possible locations were recorded. The wolf crossed the southeast end of Lac Capot Blanc on a small land bridge, where she paused for 4.5 h after traveling for 19.5 h (37.5 km). Following this rest, she traveled for 9 h (26.3 km) onto a peninsula in Reid Lake, where she spent 2 h before backtracking and stopping for 8 h just off the peninsula. Her next period of travel lasted 16.5 h (32.7 km), terminating in a pause of 9.5 h just 3.8 km from a concentration of locations at the far end of her excursion, where we presume she encountered caribou. The mean duration of these three movement periods was 15.7 h (± 2.5 SE), and that of the pauses, 7.3 h (± 1.5). The wolf required 72.5 h (3.0 d) to travel a minimum of 95 km from her den to this area near caribou (Figure 2). She remained there (35.5 km2) for 151.5 h (6.3 d) and then moved south to Lake of the Enemy, where she stayed (31.9 km^2) for 74 h (3.1 d) before returning to her den. Her greatest distance from the den, 103 km, was recorded 174.5 h (7.3 d) after the excursion began, at 0433 on 27 July. She was 8 km from a collared caribou on 23 July, four days after the excursion began (Table 1).

The return trip began at 0403 on 2 August, 318 h (13.2 d) after leaving the den. She followed a relatively direct path for 18 h back to the den, a distance of 75 km.

The minimum distance traveled during the excursion was 339 km. The estimated overall minimum travel rate was 3.1 km/h, 2.6 km/h away from the den and 4.2 km/h on the return trip.

(24) Post-Excursion Period: We saw three pups when recovering the collar on 20 August, but others may have been hiding in vegetation.

Telemetry recorded 13 foraging bouts in the post-excursion period. The mean distance traveled during these bouts was 18.3 km (+ 2.7 SE, range 1.2–47.7 km), and mean greatest distance from the den was 7.1 km (+ 0.7 SE, range 1.1–11.0 km). The mean duration of these post-excursion foraging bouts was 10.9 h (+ 2.4 SE, range 1–33 h).

When 388 reached her den on 2 August, the distance to the nearest collared caribou was 54 km. On 9 August, one week after she returned, the distance was 79 km (Table 1).

Pre- and Post-Excursion Comparison

(25) We found no differences in the mean distance of foraging bouts before and after the excursion period (F = 1.5, df = 1, 29, p = 0.24). Likewise, the mean greatest distance from the den was similar pre- and post-excursion (F = 0.004, df = 1, 29, p = 0.95). However, the mean duration of 388's foraging bouts decreased by 10.0 h after her long excursion (F = 3.1, df = 1, 29, p = 0.09).

(26) *Caribou Monitoring*

Summer Movements: On 10 July, 5 of 11 collared caribou were dispersed over a distance of 10 km, 140 km south of their calving grounds (Figure 1). On the same day, three caribou were still on the calving grounds, two were between the calving grounds and the leaders, and one was missing. One week later (17 July), the leading radio-collared cows were 100 km farther south (Figure 1). Two were within 5 km of each other in front of the rest, who were more dispersed. All radio-collared cows had left the calving grounds by this time. On 23 July, the leading radio-collared caribou had moved 35 km farther south, and all of them were more widely dispersed. The two cows closest to the leader were 26 km and 33 km away, with 37 km between them. On the next location (29 July), the most southerly caribou were 60 km

farther south. All of the caribou were now in the areas where they remained for the duration of the study (Figure 2).

A Minimum Convex Polygon (Mohr and Stumpf, 1966) around all caribou locations acquired during the study encompassed 85 119 km².

Relative to the Wolf Den: (27) The distance from the nearest collared caribou to the den decreased from 241 km one week before the excursion to 124 km the day it began. The nearest a collared caribou came to the den was 43 km away, on 29 and 30 July. During the study, four collared caribou were located within 100 km of the den. Each of these four was closest to the wolf on at least one day during the period reported.

(28) Discussion

Prey Abundance

Caribou are the single most important prey of tundra (29) wolves (Clark, 1971; Kuyt, 1972; Stephenson and James, 1982; Williams, 1990). Caribou range over vast areas, and for part of the summer, they are scarce or absent in wolf home ranges (Heard et al., 1996). Both the long distance between radio-collared caribou and the den the week before the excursion and the increased time spent foraging by wolf 388 indicate that caribou availability near the den was low. Observations of the pups' being left alone for up to 18 h, presumably while adults were searching for food, provide additional support for low caribou availability locally. Mean foraging bout duration decreased by 10.0 h after the excursion, when collared caribou were closer to the den, suggesting an increase in caribou availability nearby.

Foraging Excursion

One aspect of central place foraging theory (CPFT) (30) deals with the optimality of returning different-sized food loads from varying distances to dependents at a central place (i.e., the den) (Orians and Pearson, 1979). Carlson (1985) tested CPFT and found that the predator usually consumed prey captured far from the central place, while feeding prey captured nearby to dependants. Wolf 388 spent 7.2 days in one area near caribou before moving to a location 23 km back towards the den, where she spent an additional 3.1 days, likely hunting caribou. She began her return trip from this closer location, traveling directly to the den. While away, she may have made one or more successful kills and spent time meeting her own energetic needs before returning to the den. Alternatively, it may have taken several attempts to make a kill,

24 Post-excursion measurements of 388's movements were made to compare with those of the pre-excursion period. In order to compare, scientists often use *means*, or averages, of a series of measurements—mean distances, mean duration, etc.

25 In the comparison, authors used statistical calculations (F and df) to determine that the differences between pre- and post-excursion measurements were *statistically insignificant*, or close enough to be considered essentially the same or similar.

26 As with wolf 388, the investigators measured the movements of caribou during the study period. The areas within which the caribou moved are shown in Figure 2 by shaded polygons mentioned in the second paragraph of this subsection.

27 This subsection summarizes how distances separating predators and prey varied during the study period.

28 DISCUSSION: This section is the *interpretation* part of the scientific process.

29 This subsection reviews observations from other studies and suggests that this study fits with patterns of those observations.

30 Authors discuss a prevailing *theory* (CBFT) which might explain why a wolf would travel far to meet her own energy needs while taking food caught closer to the den back to her pups. The results of this study seem to fit that pattern.

31 Here our authors note other possible explanations for wolves' excursions presented by other investigators, but this study does not seem to support those ideas.

32 Authors discuss possible reasons for why 388 traveled directly to where caribou were located. They take what they learned from earlier studies and apply it to this case, suggesting that the lay of the land played a role. Note that their description paints a clear picture of the landscape.

33 Authors suggest that 388 may have learned in traveling during previous summers where the caribou were. The last two sentences suggest ideas for future studies.

34 Or maybe 388 followed the scent of the caribou. Authors acknowledge difficulties of proving this, but they suggest another area where future studies might be done.

35 Authors suggest that results of this study support previous studies about how fast wolves travel to and from the den. In the last sentence, they speculate on how these observed patterns would fit into the theory of evolution.

36 Authors also speculate on the fate of 388's pups while she was traveling. This leads to . . .

which she then fed on before beginning her return trip. We do not know if she returned food to the pups, but such behavior would be supported by CPFT.

(31) Other workers have reported wolves' making long round trips and referred to them as "extraterritorial" or "pre-dispersal" forays (Fritts and Mech, 1981; Messier, 1985; Ballard et al., 1997; Merrill and Mech, 2000). These movements are most often made by young wolves (1–3 years old), in areas where annual territories are maintained and prey are relatively sedentary (Fritts and Mech, 1981; Messier, 1985). The long excursion of 388 differs in that tundra wolves do not maintain annual territories (Walton et al., 2001), and the main prey migrate over vast areas (Gunn et al., 2001).

Another difference between 388's excursion and those reported earlier is that she is a mature, breeding female. No study of territorial wolves has reported reproductive adults making extraterritorial movements in summer (Fritts and Mech, 1981; Messier, 1985; Ballard et al., 1997; Merrill and Mech, 2001). However, Walton et al. (2001) also report that breeding female tundra wolves made excursions.

Direction of Movement

(32) Possible explanations for the relatively direct route 388 took to the caribou include landscape influence and experience. Considering the timing of 388's trip and the locations of caribou, had the wolf moved northwest, she might have missed the caribou entirely, or the encounter might have been delayed.

A reasonable possibility is that the land directed 388's route. The barrens are crisscrossed with trails worn into the tundra over centuries by hundreds of thousands of caribou and other animals (Kelsall, 1968; Thorpe et al., 2001). At river crossings, lakes, or narrow peninsulas, trails converge and funnel towards and away from caribou calving grounds and summer range. Wolves use trails for travel (Paquet et al., 1996; Mech and Boitani, 2003; P. Frame, pers. observation). Thus, the landscape may direct an animal's movements and lead it to where cues, such as the odor of caribou on the wind or scent marks of other wolves, may lead it to caribou.

(33) Another possibility is that 388 knew where to find caribou in summer. Sexually immature tundra wolves sometimes follow caribou to calving grounds (D. Heard, unpubl. data). Possibly, 388 had made such journeys in previous years and killed caribou. If this were the case, then in times of local prey scarcity she might travel to areas where she had hunted successfully before. Continued monitoring of tundra wolves may answer questions about how their food needs are met in times of low caribou abundance near dens.

Caribou often form large groups while moving **(34)** south to the tree line (Kelsall, 1968). After a large aggregation of caribou moves through an area, its scent can linger for weeks (Thorpe et al., 2001:104). It is conceivable that 388 detected caribou scent on the wind, which was blowing from the northeast on 19–21 July (Environment Canada, 2003), at the same time her excursion began. Many factors, such as odor strength and wind direction and strength, make systematic study of scent detection in wolves difficult under field conditions (Harrington and Asa, 2003). However, humans are able to smell odors such as forest fires or oil refineries more than 100 km away. The olfactory capabilities of dogs, which are similar to wolves, are thought to be 100 to 1 million times that of humans (Harrington and Asa, 2003). Therefore, it is reasonable to think that under the right wind conditions, the scent of many caribou traveling together could be detected by wolves from great distances, thus triggering a long foraging bout.

Rate of Travel

Mech (1994) reported the rate of travel of Arctic **(35)** wolves on barren ground was 8.7 km/h during regular travel and 10.0 km/h when returning to the den, a difference of 1.3 km/h. These rates are based on direct observation and exclude periods when wolves moved slowly or not at all. Our calculated travel rates are assumed to include periods of slow movement or no movement. However, the pattern we report is similar to that reported by Mech (1994), in that homeward travel was faster than regular travel by 1.6 km/h. The faster rate on return may be explained by the need to return food to the den. Pup survival can increase with the number of adults in a pack available to deliver food to pups (Harrington et al., 1983). Therefore, an increased rate of travel on homeward trips could improve a wolf's reproductive fitness by getting food to pups more quickly.

Fate of 388's Pups

Wolf 388 was caring for pups during den observa- **(36)** tions. The pups were estimated to be six weeks old, and were seen ranging as far as 800 m from the den. They received some regurgitated food from two of the females, but were unattended for long periods. The excursion started 16 days after our observations, and it is improbable that the pups could have traveled the distance that 388 moved. If the pups died, this would have removed parental responsibility, allowing the long movement.

Our observations and the locations of radio-collared caribou indicate that prey became scarce in

Foraging Movement of A Tundra Wolf **201**

Appendix II

the area of the den as summer progressed. Wolf 388 may have abandoned her pups to seek food for herself. However, she returned to the den after the excursion, where she was seen near pups. In fact, she foraged in a similar pattern before and after the excursion, suggesting that she again was providing for pups after her return to the den.

 A more likely possibility is that one or both of the other lactating females cared for the pups during 388's absence. The three females at this den were not seen with the pups at the same time. However, two weeks earlier, at a different den, we observed three females cooperatively caring for a group of six pups. At that den, the three lactating females were observed providing food for each other and trading places while nursing pups. Such a situation at the den of 388 could have created conditions that allowed one or more of the lactating females to range far from the den for a period, returning to her parental duties afterwards. However, the pups would have been weaned by eight weeks of age (Packard et al., 1992), so nonlactating adults could also have cared for them, as often happens in wolf packs (Packard et al., 1992; Mech et al., 1999).

Cooperative rearing of multiple litters by a pack could create opportunities for long-distance foraging movements by some reproductive wolves during summer periods of local food scarcity. We have recorded multiple lactating females at one or more tundra wolf dens per year since 1997. This reproductive strategy may be an adaptation to temporally and spatially unpredictable food resources. All of these possibilities require further study, but emphasize both the adaptability of wolves living on the barrens and their dependence on caribou.

Long-range wolf movement in response to caribou availability has been suggested by other researchers (Kuyt, 1972; Walton et al., 2001) and traditional ecological knowledge (Thorpe et al., 2001). Our report demonstrates the rapid and extreme response of wolves to caribou distribution and movements in summer. Increased human activity on the tundra (mining, road building, pipelines, ecotourism) may influence caribou movement patterns and change the interactions between wolves and caribou in the region. Continued monitoring of both species will help us to assess whether the association is being affected adversely by anthropogenic change.

Acknowledgements

This research was supported by the Department of Resources, Wildlife, and Economic Development, Government of the Northwest Territories; the Department of Biological Sciences at the University of Alberta; the Natural Sciences and Engineering Research Council of Canada; the Department of Indian and Northern Affairs Canada; the Canadian Circumpolar Institute; and DeBeers Canada, Ltd. Lorna Ruechel assisted with den observations. A. Gunn provided caribou location data. We thank Dave Mech for the use of GPS collars. M. Nelson, A. Gunn, and three anonymous reviewers made helpful comments on earlier drafts of the manuscript. This work was done under Wildlife Research Permit – WL002948 issued by the Government of the Northwest Territories, Department of Resources, Wildlife, and Economic Development.

References

BALLARD, W.B., AYRES, L.A., KRAUSMAN, P.R., REED, D.J., and FANCY, S.G. 1997. Ecology of wolves in relation to a migratory caribou herd in northwest Alaska. Wildlife Monographs 135. 47 p.

BROWN, M.B., and FORSYTHE, A.B. 1974. Robust tests for the equality of variances. Journal of the American Statistical Association 69:364–367.

CARLSON, A. 1985. Central place foraging in the red-backed shrike (Lanius collurio L.): Allocation of prey between forager and sedentary consumer. Animal Behaviour 33:664–666.

CLARK, K.R.F. 1971. Food habits and behavior of the tundra wolf on central Baffin Island. Ph.D. Thesis, University of Toronto, Ontario, Canada.

ENVIRONMENT CANADA. 2003. National climate data information archive. Available online: http://www.climate.weatheroffice.ec.gc.ca/Welcome_e.html

FOURNIER, B., and GUNN, A. 1998. Musk ox numbers and distribution in the NWT, 1997. File Report No. 121. Yellowknife: Department of Resources, Wildlife, and Economic Development, Government of the Northwest Territories. 55 p.

FRITTS, S.H., and MECH, L.D. 1981. Dynamics, movements, and feeding ecology of a newly protected wolf population in northwestern Minnesota. Wildlife Monographs 80. 79 p.

GUNN, A., DRAGON, J., and BOULANGER, J. 2001. Seasonal movements of satellite-collared caribou from the Bathurst herd. Final Report to the West Kitikmeot Slave Study Society, Yellowknife, NWT. 80 p. Available online: http://www.wkss.nt.ca/HTML/08_ProjectsReports/PDF/Seasonal MovementsFinal.pdf

HARRINGTON, F.H., and ASA, C.S. 2003. Wolf communication. In: Mech, L.D., and Boitani, L., eds. Wolves: Behavior, ecology, and conservation. Chicago: University of Chicago Press. 66–103.

HARRINGTON, F.H., MECH, L.D., and FRITTS, S.H. 1983. Pack size and wolf pup survival: Their relationship under varying ecological conditions. Behavioral Ecology and Sociobiology 13:19–26.

HARRIS, S.A. 1986. Permafrost distribution, zonation and stability along the eastern ranges of the cordillera of North America. Arctic 39(1):29–38.

HEARD, D.C., and WILLIAMS, T.M. 1992. Distribution of wolf dens on migratory caribou ranges in the Northwest

---

37 Discussion of cooperative rearing of pups and, in turn, to speculation on how this study and what is known about cooperative rearing might fit into the animal's strategies for survival of the species. Again, the authors approach the broader theory of evolution and how it might explain some of their results.

38 And again, they suggest that this study points to several areas where further study will shed some light.

39 In conclusion, the authors suggest that their study supports the hypothesis being tested here. And they touch on the implications of increased human activity on the tundra predicted by their results.

40 ACKNOWLEDGEMENTS: Authors note the support of institutions, companies and individuals. They thank their reviewers ad list permits under which their research was carried on.

41 REFERENCES: List of all studies cited in the report. This may seem tedious, but is a vitally important part of scientific reporting. It is a record of the sources of information on which this study is based. It provides readers with a wealth of resources for further reading on this topic. Much of it will form the foundation of future scientific studies like this one.

Territories, Canada. Canadian Journal of Zoology 70:1504–1510.

HEARD, D.C., WILLIAMS, T.M., and MELTON, D.A. 1996. The relationship between food intake and predation risk in migratory caribou and implication to caribou and wolf population dynamics. Rangifer Special Issue No. 2:37–44.

KELSALL, J.P. 1968. The migratory barren-ground caribou of Canada. Canadian Wildlife Service Monograph Series 3. Ottawa: Queen's Printer. 340 p.

KUYT, E. 1962. Movements of young wolves in the Northwest Territories of Canada. Journal of Mammalogy 43:270–271.

———. 1972. Food habits and ecology of wolves on barren-ground caribou range in the Northwest Territories. Canadian Wildlife Service Report Series 21. Ottawa: Information Canada. 36 p.

MECH, L.D. 1994. Regular and homeward travel speeds of Arctic wolves. Journal of Mammalogy 75:741–742.

MECH, L.D., and BOITANI, L. 2003. Wolf social ecology. In: Mech, L.D., and Boitani, L., eds. Wolves: Behavior, ecology, and conservation. Chicago: University of Chicago Press. 1–34.

MECH, L.D., and GESE, E.M. 1992. Field testing the Wildlink capture collar on wolves. Wildlife Society Bulletin 20:249–256.

MECH, L.D., WOLFE, P., and PACKARD, J.M. 1999. Regurgitative food transfer among wild wolves. Canadian Journal of Zoology 77:1192–1195.

MERRILL, S.B., and MECH, L.D. 2000. Details of extensive movements by Minnesota wolves (Canis lupus). American Midland Naturalist 144:428–433.

MERRILL, S.B., ADAMS, L.G., NELSON, M.E., and MECH, L.D. 1998. Testing releasable GPS radiocollars on wolves and white-tailed deer. Wildlife Society Bulletin 26:830–835.

MESSIER, F. 1985. Solitary living and extraterritorial movements of wolves in relation to social status and prey abundance. Canadian Journal of Zoology 63:239–245.

MOHR, C.O., and STUMPF, W.A. 1966. Comparison of methods for calculating areas of animal activity. Journal of Wildlife Management 30:293–304.

MUSIANI, M. 2003. Conservation biology and management of wolves and wolf-human conflicts in western North America. Ph.D. Thesis, University of Calgary, Calgary, Alberta, Canada.

ORIANS, G.H., and PEARSON, N.E. 1979. On the theory of central place foraging. In: Mitchell, R.D., and Stairs, G.F., eds. Analysis of ecological systems. Columbus: Ohio State University Press. 154–177.

PACKARD, J.M., MECH, L.D., and REAM, R.R. 1992. Weaning in an arctic wolf pack: Behavioral mechanisms. Canadian Journal of Zoology 70:1269–1275.

PAQUET, P.C., WIERZCHOWSKI, J., and CALLAGHAN, C. 1996. Summary report on the effects of human activity on gray wolves in the Bow River Valley, Banff National Park, Alberta. In: Green, J., Pacas, C., Bayley, S., and Cornwell, L., eds. A cumulative effects assessment and futures outlook for the Banff Bow Valley. Prepared for the Banff Bow Valley Study. Ottawa: Department of Canadian Heritage.

STEPHENSON, R.O., and JAMES, D. 1982. Wolf movements and food habits in northwest Alaska. In: Harrington, F.H., and Paquet, P.C., eds. Wolves of the world. New Jersey: Noyes Publications. 223–237.

THORPE, N., EYEGETOK, S., HAKONGAK, N., and QITIR-MIUT ELDERS. 2001. The Tuktu and Nogak Project: A caribou chronicle. Final Report to the West Kitikmeot/Slave Study Society, Ikaluktuuttiak, NWT. 160 p.

TIMONEY, K.P., LA ROI, G.H., ZOLTAI, S.C., and ROBINSON, A.L. 1992. The high subarctic forest-tundra of northwestern Canada: Position, width, and vegetation gradients in relation to climate. Arctic 45(1):1–9.

WALTON, L.R., CLUFF, H.D., PAQUET, P.C., and RAMSAY, M.A. 2001. Movement patterns of barren-ground wolves in the central Canadian Arctic. Journal of Mammalogy 82:867–876.

WILLIAMS, T.M. 1990. Summer diet and behavior of wolves denning on barren-ground caribou range in the Northwest Territories, Canada. M.Sc. Thesis, University of Alberta, Edmonton, Alberta, Canada.

ZAR, J.H. 1999. Biostatistical analysis. 4th ed. New Jersey: Prentice Hall. 663 p.

Appendix III. Answers to Self-Quizzes and Genetics Problems

Italicized numbers refer to relevant section numbers

CHAPTER 1

1. Atoms — *1.1*
2. cell — *1.1*
3. Animals — *1.3*
4. energy, nutrients — *1.2*
5. Homeostasis — *1.2*
6. Domains — *1.3*
7. d — *1.2*
8. d — *1.2*
9. Reproduction — *1.2*
10. observable — *1.5*
11. Mutations — *1.4*
12. adaptive — *1.4*
13. b — *1.6*
14. c — *1.1*
    e — *1.4*
    d — *1.6*
    f — *1.6*
    a — *1.6*
    b — *1.3*

CHAPTER 2

1. Tracer — *2.2*
2. b — *2.3*
3. compound — *2.3*
4. electronegativity — *2.3*
5. polar covalent — *2.4*
6. atomic number — *2.1*
7. e — *2.5*
8. hydrophobic — *2.5*
9. d — *2.6*
10. solute — *2.5*
11. acid — *2.6*
12. hydrogen ions (H$^+$) or hydroxyl ions (OH$^-$) — *2.6*
13. buffer system — *2.6*
12. c — *2.5*
    b — *2.1*
    d — *2.1*
    a — *2.5*

CHAPTER 3

1. four — *3.1*
2. carbohydrate — *3.3*
3. f — *3.3, 3.7*
4. double covalent bonds — *3.4*
5. False — *3.4*
6. fatty acid tails — *3.4*
7. e — *3.4*
8. d — *3.3, 3.5*
9. d — *3.6*
10. d — *3.7*
11. c — *3.4*
12. c — *3.5*
    e — *3.7*
    b — *3.4*
    d — *3.7*
    a — *3.3*
    f — *3.4*

CHAPTER 4

1. cell — *4.2*
2. False (all protists are eukaryotes) — *4.6*
3. phospholipids — *4.2*
4. c — *4.2*
5. eukaryotic — *4.6*
6. lipds, proteins — *4.9*
7. nucleus — *4.8*
8. cell wall — *4.12*
9. False (cell walls enclose the plasma membrane of many cells) — *4.12*
10. lysosomes — *4.9*
11. c — *4.11*
    f — *4.11*
    a — *4.2*
    e — *4.9*
    d — *4.9*
    b — *4.9*

CHAPTER 5

1. c — *5.1*
2. c — *5.1*
3. fluid mosaic — *5.1*
4. a — *5.2*
5. a — *5.2*
6. adhesion — *5.2*
7. more, less — *5.3*
8. oxygen (CO$_2$, water, etc.) — *5.3*
9. b — *5.4*
10. a — *5.6*
11. hydrostatic pressure (or turgor) — *5.6*
12. e — *5.5*
13. d, b, e, a — *5.5*
14. d — *5.5*
    g — *5.4*
    a — *5.2*
    e — *5.4*
    c — *5.1*
    b — *5.3*
    f — *5.2*

CHAPTER 6

1. c — *6.1*
2. d — *6.1*
3. b, c — *6.1*
4. d — *6.2*
5. b — *6.2*
6. c, d — *6.2, 6.4*
7. d — *6.3*
8. e — *6.4*
9. c — *6.4*
10. c — *6.3*
11. a — *6.3*
12. c — *6.2*
    g — *6.3*
    d — *6.1*
    b — *6.2*
    f — *6.4*
    a — *6.3*
    e — *6.1*

CHAPTER 7

1. carbon dioxide, light (or sunlight) — *7.1, 7.8*
2. b — *7.1*
3. a — *7.3*
4. b — *7.3*
5. c — *7.3*
6. d — *7.4*
7. c — *7.3*
8. b — *7.6*
9. e — *7.6*
10. PGA; oxaloacelate — *7.7*
11. oxygen gas (O$_2$) — *7.8*
12. The cat, bird, and caterpillar are heterotrophs. The weed is an autotroph. — *7.8*
13. c — *7.6*
    a — *7.6*
    b — *7.4*
    d — *7.4*

CHAPTER 8

1. False (plants make ATP by aerobic respiration too) — *8.1*
2. d — *8.2*
3. a — *8.1*
4. c — *8.2*
5. b — *8.1*
6. e — *8.3*
7. b — *8.3*
8. c — *8.4*
9. c — *8.5*
10. b — *8.5*
11. d — *8.7*
12. b — *8.1*
    c — *8.5*
    a — *8.3*
    d — *8.4*

CHAPTER 9

1. d — *9.1*
2. b — *9.1*
3. c — *9.1*
4. d — *9.2*
5. a — *9.2*
6. c — *9.2*
7. a — *9.2*
8. See Figure 9.6 — *9.3*
9. b — *9.2*
10. a — *9.5*
11. kinase, growth factor, epidermal growth factor, tumor suppressor are all mentioned in this chapter — *9.5*
12. d — *9.3*
    b — *9.3*
    c — *9.3*
    a — *9.3*

CHAPTER 10

1. c — *10.1*
2. d — *10.1*
3. alleles — *10.1*
4. d — *10.1*
5. d — *10.2*
6. b — *10.2*
7. d — *10.2*
8. d — *10.3*
9. meiosis gives rise to nonparental combinations of alleles — *10.1, 10.4*
10. sister chromatids have separated — *10.3*
11. d — *10.2*
    a — *10.1*
    c — *10.3*
    b — *10.2*

CHAPTER 11

1. a — *11.1*
2. continuous variation — *11.7*
3. b — *11.1*
4. a — *11.1*
5. b — *11.1*
6. c — *11.2*
7. a — *11.4*
8. b — *11.2*
9. d — *11.3*
10. c — *11.5*
11. a — *11.5*
12. b — *11.3*
    d — *11.2*
    a — *11.1*
    c — *11.1*

CHAPTER 12

1. d — *12.1*
2. c — *12.1*
3. b — *12.2*
4. b — *12.2*
5. the three mentioned in text are hemophilia, red–green color blindness, DMD (Duchenne muscular dystrophy) — *12.4*
6. Genes for red and green light photoreceptors are located on the X chromosome — *12.4*
7. False — *12.4*
8. d — *12.4*
9. e — *12.5*
10. d — *12.6*
11. True — *12.6*
12. c — *12.6*
13. a — *12.7*
14. c — *12.6*
    e — *12.5*
    f — *12.6*
    b — *12.5*
    a — *12.1*
    d — *12.6*

CHAPTER 13

1. bacteria — *13.1*
2. c — *13.2*
3. d — *13.2*
4. c — *13.2*
5. a — *13.3*
6. d — *13.3*
7. b — *13.3*
8. 3'-CCAAAGAAGTTCTCT-5' — *13.3*
9. d — *13.4*
10. d — *13.1*
    b — *13.4*
    a — *13.2*
    f — *13.2*
    e — *13.3*
    g — *13.3*
    c — *13.2*

CHAPTER 14

1. c — *14.1*
2. promoter — *14.2*
3. high-energy phosphate bonds of free nucleotides — *14.1*
4. single — *14.1*
5. b — *14.1*
6. b — *14.3*
7. 64 — *14.3*
8. c — *14.3*
9. a — *14.3*
10. d — *14.4*
11. gly-phe-phe-lys-arg — *14.3*
12. replication error, ionizing or nonionizing radiation, transposable elements, and toxic chemicals are mentioned in the text — *14.5*
13. c — *14.3*
    d — *14.1*
    e — *14.4*
    a — *14.3*
    f — *14.3*
    g — *14.3*
    b — *14.5*

CHAPTER 15

1. d — *15.1*
2. b — *15.1*
3. d — *15.1*
4. transcription factors — *15.1*
5. b — *15.1*
6. c — *15.1*
7. a — *15.2*
8. b — *15.2*
9. b — *15.2*
10. c — *15.2*
11. c — *15.3*
12. d — *15.3*
13. c — *15.4*
14. operon — *15.4*
15. f — *15.2*
    a — *15.2*
    b — *15.4*
    e — *15.2*
    c — *15.1*
    d — *15.1*

CHAPTER 16

1. c — *16.1*
2. plasmid — *16.1*
3. b — *16.1*
4. b — *16.2*
5. DNA library — *16.2*
6. d — *16.2*
7. b — *16.3*
8. d — *16.3*
9. b — *16.10*
    d — *16.1*
    e — *16.10*
    f — *16.7*
10. Gene therapy — *16.10*
11. c — *16.4*
    g — *16.7*
    d — *16.2*
    e — *16.10*
    a — *16.6*
    f — *16.6*
    b — *16.1*

CHAPTER 11: GENETIC PROBLEMS

1. a. *AB*
    b. *AB, aB*
    c. *Ab, ab*
    d. *AB, Ab, aB, ab*

2. a. All offspring will be *AaBB*.
    b. 1/4 *AABB*   (25% each genotype)
        1/4 *AABb*
        1/4 *AaBB*
        1/4 *AaBb*
    c. 1/4 *AaBb*   (25% each genotype)
        1/4 *Aabb*
        1/4 *aaBb*
        1/4 *aabb*
    d. 1/16 *AABB* (6.25% of genotype)
        1/8 *AaBB*   (12.5%)
        1/16 *aaBB* (6.25%)
        1/8 *AABb*   (12.5%)
        1/4 *AaBb*   (25%)
        1/8 *aaBb*   (12.5%)
        1/16 *AAbb* (6.25%)
        1/8 *Aabb*   (12.5%)
        1/16 *aabb* (6.25%)

3. a. *ABC*
    b. *ABC, aBC*
    c. *ABC, aBC, ABc, aBc*
    d. *ABC*
        aBC
        AbC
        abC
        ABc
        aBc
        Abc
        abc

4. a. Both parents are heterozygotes (*Aa*). Their children may be albino (*aa*) or unaffected (*AA* or *Aa*).
    b. All are homozygous recessive (*aa*).
    c. Homozygous recessive (*aa*) father, and heterozygous (*Aa*) mother. The albino child is *aa*, the unaffected children *Aa*.

5. A mating of two M^L cats yields 1/4 *MM*, 1/2 $M^L M$, and 1/4 $M^L M^L$. Because $M^L M^L$ is lethal, the probability that any one kitten among the survivors will be heterozygous is 2/3.

6. Possible outcomes of an experimental cross between F_1 rose plants heterozygous for height (*Aa*):

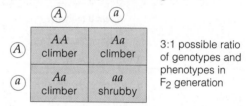

3:1 possible ratio of genotypes and phenotypes in F_2 generation

Possible outcomes of a testcross between an F_1 rose plant heterozygous for height and a shrubby rose plant:

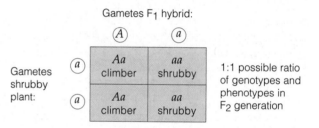

1:1 possible ratio of genotypes and phenotypes in F_2 generation

7. Yellow is recessive. Because F_1 plants have a green phenotype and must be heterozygous, green must be dominant over the recessive yellow.

8. A mating between a mouse from a true-breeding, white-furred strain and a mouse from a true-breeding,

brown-furred strain would provide you with the most direct evidence. Because true-breeding strains of organisms typically are homozygous for a trait being studied, all F_1 offspring from this mating should be heterozygous. Record the phenotype of each F_1 mouse, then let them mate with one another. Assuming only one gene locus is involved, these are possible outcomes for the F_1 offspring:

a. All F_1 mice are brown, and their F_2 offspring segregate: 3 brown : 1 white. *Conclusion*: Brown is dominant to white.

b. All F_1 mice are white, and their F_2 offspring segregate: 3 white : 1 brown. *Conclusion*: White is dominant to brown.

c. All F_1 mice are tan, and the F_2 offspring segregate: 1 brown : 2 tan : 1 white. *Conclusion*: The alleles at this locus show incomplete dominance.

CHAPTER 12: GENETIC PROBLEMS

1. a. Human males (XY) inherit their X chromosome from their mother.

b. A male can produce two kinds of gametes. Half carry an X chromosome and half carry a Y chromosome. All the gametes that carry the X chromosome carry the same X-linked allele.

c. A female homozygous for an X-linked allele produces only one kind of gamete.

d. Fifty percent of the gametes of a female who is heterozygous for an X-linked allele carry one of the two alleles at that locus; the other fifty percent carry its partner allele for that locus.

2. Because Marfan syndrome is a case of autosomal dominant inheritance and because one parent bears the allele, the probability that any child of theirs will inherit the mutant allele is 50 percent.

3. a. Nondisjunction might occur during anaphase I or anaphase II of meiosis.

b. As a result of translocation, chromosome 21 may get attached to the end of chromosome 14. The new

9. The data reveal that these genes do not assort independently because the observed ratio is very far from the 9:3:3:1 ratio expected with independent assortment. Instead, the results can be explained if the genes are located close to each other on the same chromosome, which is called linkage.

10. a. <u>1/2</u> red <u>1/2</u> pink _____ white
    b. _____ red <u>All</u> pink _____ white
    c. <u>1/4</u> red <u>1/2</u> pink <u>1/4</u> white
    d. _____ red <u>1/2</u> pink <u>1/2</u> white

11. Because both parents are heterozygotes (Hb^AHb^S), the following are the probabilities for each child:

a. 1/4 Hb^SHb^S

b. 1/4 Hb^AHb^A

c. 1/2 Hb^AHb^S

individual's chromosome number would still be 46, but its somatic cells would have the translocated chromosome 21 in addition to two normal chromosomes 21.

4. A daughter could develop this muscular dystrophy only if she inherited two X-linked recessive alleles— one from each parent. Males who carry the allele are unlikely to father children because they develop the disorder and die early in life.

5. In the mother, a crossover between the two genes at meiosis generates an X chromosome that carries neither mutant allele.

6. The phenotype appeared in every generation shown in the diagram, so this must be a pattern of autosomal dominant inheritance.

7. There is no scientific answer to this question, which simply invites you to reflect on the difference between a scientific and a subjective interpretation of this individual's condition.

Appendix IV. Periodic Table of the Elements

Group

Atomic number → **11**
Symbol → **Na**
Atomic mass → **22.99**

Atomic masses are based on carbon-12. Numbers in parentheses are mass numbers of most stable or best known isotopes of radioactive elements.

Noble Gases (18)

Period	IA(1)	IIA(2)	IIIB(3)	IVB(4)	VB(5)	VIB(6)	VIIB(7)	(8)	VIII (9)	(10)	IB(11)	IIB(12)	IIIA(13)	IVA(14)	VA(15)	VIA(16)	VIIA(17)	(18)
1	1 H 1.008																	2 He 4.003
2	3 Li 6.941	4 Be 9.012											5 B 10.81	6 C 12.01	7 N 14.01	8 O 16.00	9 F 19.00	10 Ne 20.18
3	11 Na 22.99	12 Mg 24.31											13 Al 26.98	14 Si 28.09	15 P 30.97	16 S 32.06	17 Cl 35.45	18 Ar 39.95
4	19 K 39.10	20 Ca 40.08	21 Sc 44.96	22 Ti 47.90	23 V 50.94	24 Cr 52.00	25 Mn 54.94	26 Fe 55.85	27 Co 58.93	28 Ni 58.7	29 Cu 63.55	30 Zn 65.38	31 Ga 69.72	32 Ge 72.59	33 As 74.92	34 Se 78.96	35 Br 79.90	36 Kr 83.80
5	37 Rb 85.47	38 Sr 87.62	39 Y 88.91	40 Zr 91.22	41 Nb 92.91	42 Mo 95.94	43 Tc 98.91	44 Ru 101.1	45 Rh 102.9	46 Pd 106.4	47 Ag 107.9	48 Cd 112.4	49 In 114.8	50 Sn 118.7	51 Sb 121.8	52 Te 127.6	53 I 126.9	54 Xe 131.3
6	55 Cs 132.9	56 Ba 137.3	57* La 138.9	72 Hf 178.5	73 Ta 180.9	74 W 183.9	75 Re 186.2	76 Os 190.2	77 Ir 192.2	78 Pt 195.1	79 Au 197.0	80 Hg 200.6	81 Tl 204.4	82 Pb 207.2	83 Bi 209.0	84 Po (210)	85 At (210)	86 Rn (222)
7	87 Fr (223)	88 Ra 226.0	89** Ac (227)	104 Unq (261)	105 Unp (262)	106 Unh (263)	107 Uns (262)	108 Uno (265)	109 Une (266)									

— Transition Elements —

Inner Transition Elements

Lanthanide Series 6 *

58 Ce 140.1	59 Pr 140.9	60 Nd 144.2	61 Pm (145)	62 Sm 150.4	63 Eu 152.0	64 Gd 157.3	65 Tb 158.9	66 Dy 162.5	67 Ho 164.9	68 Er 167.3	69 Tm 168.9	70 Yb 173.0	71 Lu 175.0

Actinide Series 7 **

90 Th 232.0	91 Pa 231.0	92 U 238.0	93 Np 237.0	94 Pu (244)	95 Am (243)	96 Cm (247)	97 Bk (247)	98 Cf (251)	99 Es (252)	100 Fm (257)	101 Md (258)	102 No (259)	103 Lr (260)

Appendix V. Molecular Models

A molecule's structure can be depicted by different kinds of molecular models. Such models allow us to visualize different characteristics of the same structure.

Structural models show how atoms in a molecule connect to one another:

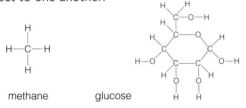

methane   glucose

In such models, each line indicates one covalent bond: Double bonds are shown as two lines; triple bonds as three lines. Some atoms or bonds may be implied but not shown. For example, carbon ring structures such as those of glucose and other sugars are often represented as polygons. If no atom is shown at the corner of a polygon, a carbon atom is implied. Hydrogen atoms bonded to one of the atoms in the carbon backbone of a molecule may also be omitted:

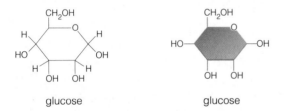

glucose   glucose

Ball-and-stick models show the relative sizes of the atoms and their positions in three dimensions:

methane   glucose

All types of covalent bonds (single, double, or triple) are shown as one stick. Typically, the elements in such models are coded in standardized colors:

carbon   hydrogen   oxygen   nitrogen

Space-filling models show the outer boundaries of the atoms in three dimensions:

methane   glucose

A model of a large molecule can be quite complex if all the atoms are shown. This space-filling model of hemoglobin is an example:

To reduce visual complexity, other types of models omit individual atoms. Surface models of large molecules can show features such as an active site crevice (Figure 5.7). In this surface model of hemoglobin, you can see two heme groups (red) nestled in pockets of the protein:

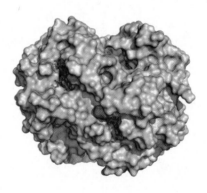

Large molecules such as proteins are often shown as ribbon models. Such models highlight secondary structure such as coils or sheets. In this ribbon model of hemoglobin, you can see the four coiled polypeptide chains, each of which folds around a heme group:

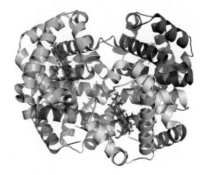

Such structural details are clues to how a molecule functions. Hemoglobin is the main oxygen carrier in vertebrate blood. Oxygen binds at the hemes, so one hemoglobin molecule can hold four molecules of oxygen.

The Amino Acids

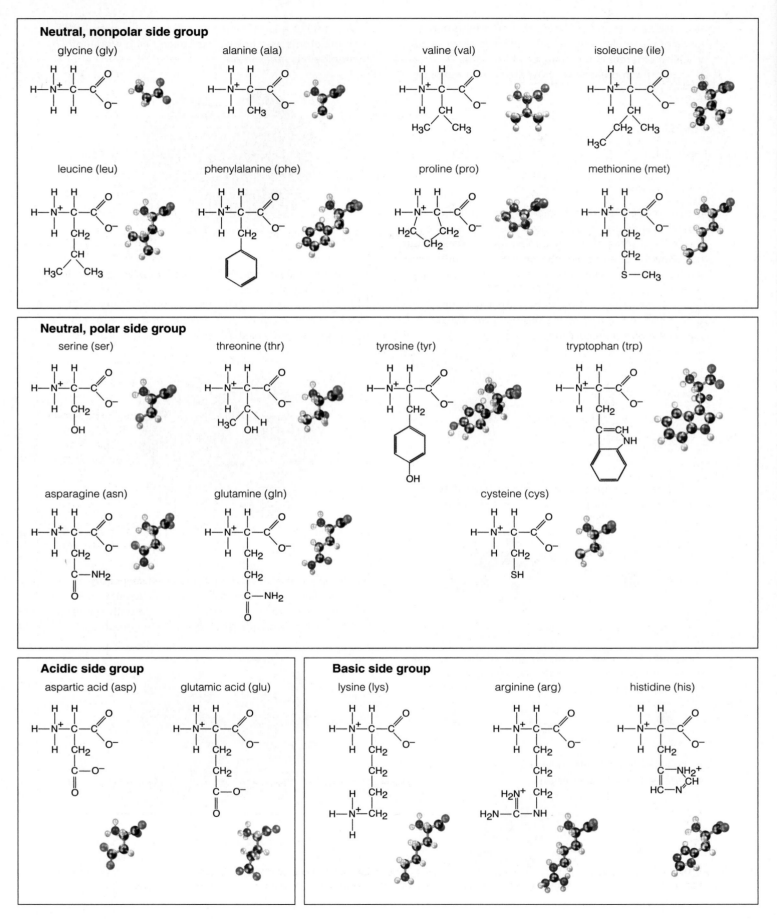

Neutral, nonpolar side group

glycine (gly) alanine (ala) valine (val) isoleucine (ile)

leucine (leu) phenylalanine (phe) proline (pro) methionine (met)

Neutral, polar side group

serine (ser) threonine (thr) tyrosine (tyr) tryptophan (trp)

asparagine (asn) glutamine (gln) cysteine (cys)

Acidic side group

aspartic acid (asp) glutamic acid (glu)

Basic side group

lysine (lys) arginine (arg) histidine (his)

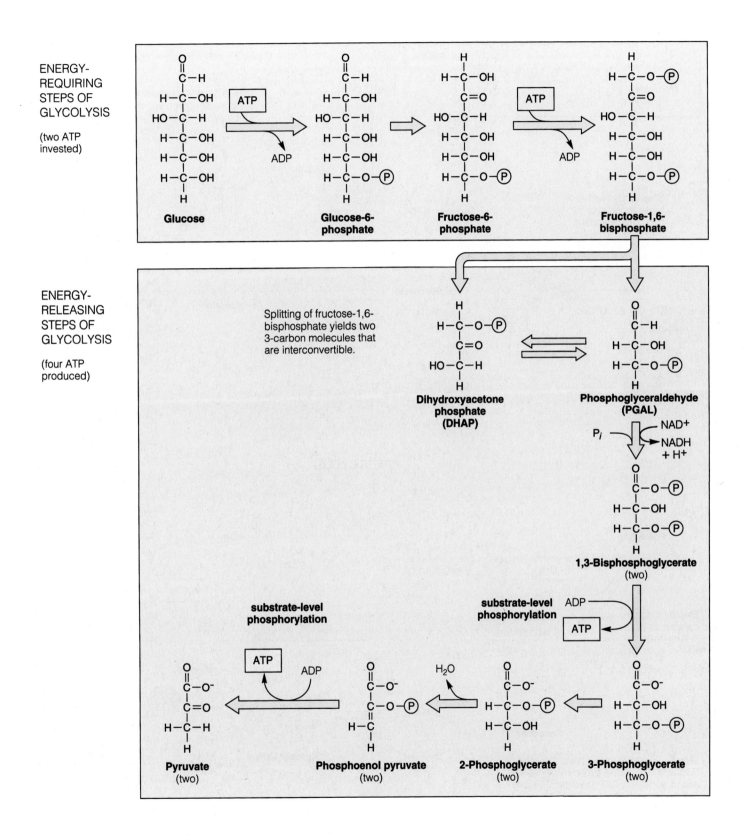

Figure A Glycolysis, ending with two 3-carbon pyruvate molecules for each 6-carbon glucose molecule entering the reactions. The *net* energy yield is two ATP molecules (two invested, four produced).

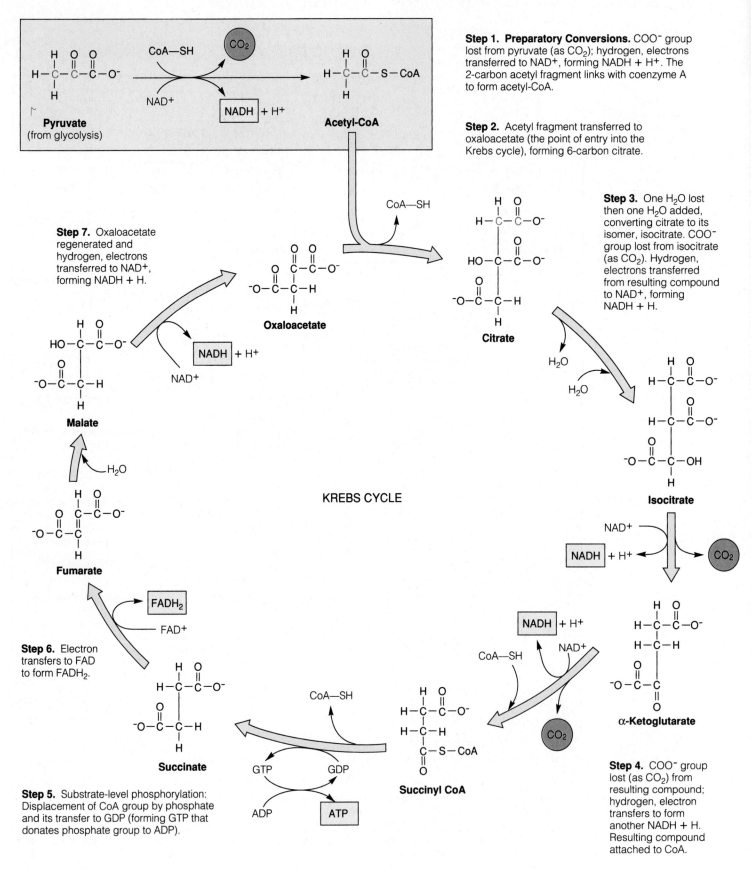

Step 1. Preparatory Conversions. COO^- group lost from pyruvate (as CO_2); hydrogen, electrons transferred to NAD^+, forming $NADH + H^+$. The 2-carbon acetyl fragment links with coenzyme A to form acetyl-CoA.

Step 2. Acetyl fragment transferred to oxaloacetate (the point of entry into the Krebs cycle), forming 6-carbon citrate.

Step 3. One H_2O lost then one H_2O added, converting citrate to its isomer, isocitrate. COO^- group lost from isocitrate (as CO_2). Hydrogen, electrons transferred from resulting compound to NAD^+, forming $NADH + H$.

Step 7. Oxaloacetate regenerated and hydrogen, electrons transferred to NAD^+, forming $NADH + H$.

Step 6. Electron transfers to FAD to form $FADH_2$.

Step 5. Substrate-level phosphorylation: Displacement of CoA group by phosphate and its transfer to GDP (forming GTP that donates phosphate group to ADP).

Step 4. COO^- group lost (as CO_2) from resulting compound; hydrogen, electron transfers to form another $NADH + H$. Resulting compound attached to CoA.

KREBS CYCLE

Figure B Krebs cycle, also known as the citric acid cycle. *Red* identifies carbon atoms entering the cyclic pathway (by way of acetyl-CoA) and leaving (by way of carbon dioxide). These cyclic reactions run twice for each glucose molecule that has been degraded to two pyruvate molecules.

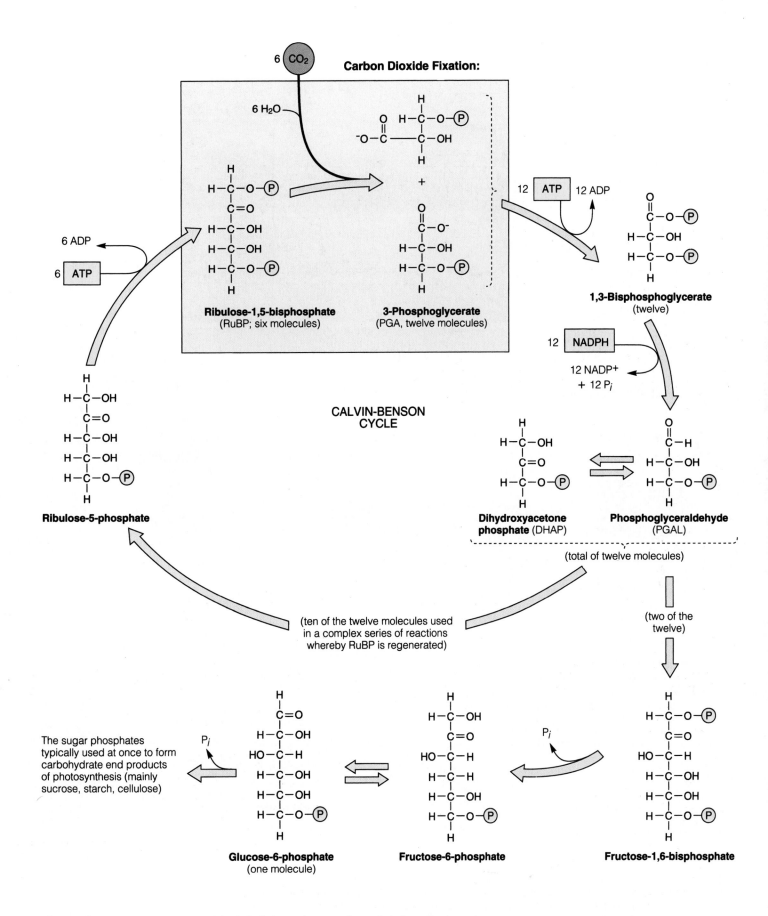

Figure C Calvin–Benson cycle of the light-independent reactions of photosynthesis.

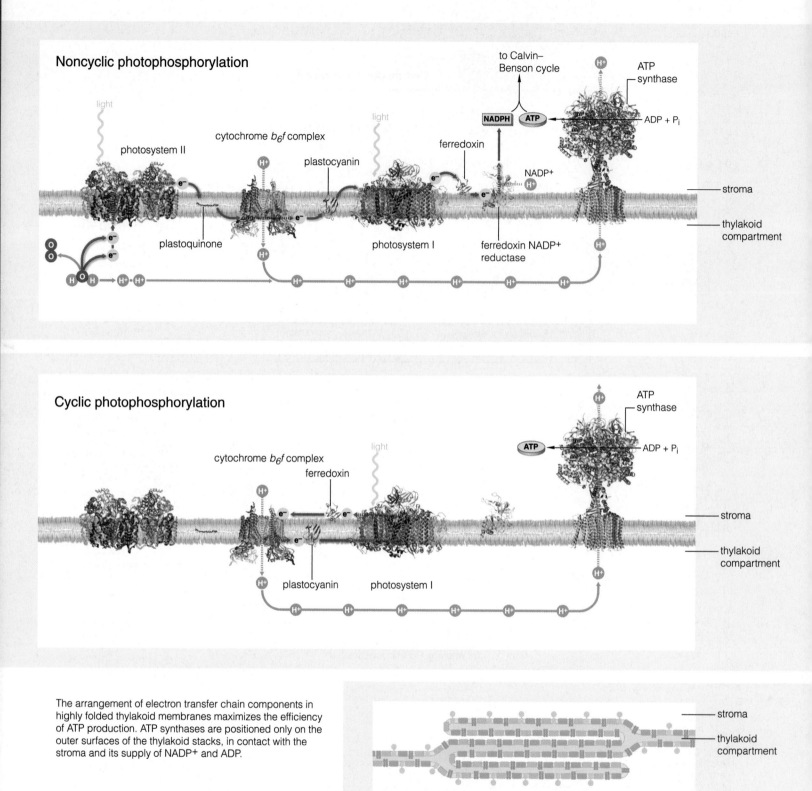

Noncyclic photophosphorylation

to Calvin–Benson cycle

ATP synthase

light

NADPH ATP

ADP + P$_i$

cytochrome b_6f complex

photosystem II

light

ferredoxin

plastocyanin

e$^-$

stroma

NADP$^+$

H$^+$

e$^-$

thylakoid compartment

plastoquinone

e$^-$
e$^-$

photosystem I

ferredoxin NADP$^+$ reductase

H$^+$

H O H

H$^+$ H$^+$

H$^+$ H$^+$ H$^+$ H$^+$ H$^+$ H$^+$

Cyclic photophosphorylation

ATP synthase

ATP

ADP + P$_i$

cytochrome b_6f complex

ferredoxin

light

e$^-$ e$^-$

stroma

e$^-$

thylakoid compartment

H$^+$

plastocyanin

photosystem I

H$^+$

H$^+$ H$^+$ H$^+$ H$^+$ H$^+$ H$^+$

The arrangement of electron transfer chain components in highly folded thylakoid membranes maximizes the efficiency of ATP production. ATP synthases are positioned only on the outer surfaces of the thylakoid stacks, in contact with the stroma and its supply of NADP$^+$ and ADP.

stroma

thylakoid compartment

ATP synthase cytochrome b_6f complex
photosystem I photosystem II

Figure D Electron transfer in the light-dependent reactions of photosynthesis. Members of the electron transfer chains are densely packed in thylakoid membranes; electrons are transferred directly from one molecule to the next. For clarity, we show the components of the chains widely spaced.

Appendix VII. A Plain English Map of the Human Chromosomes

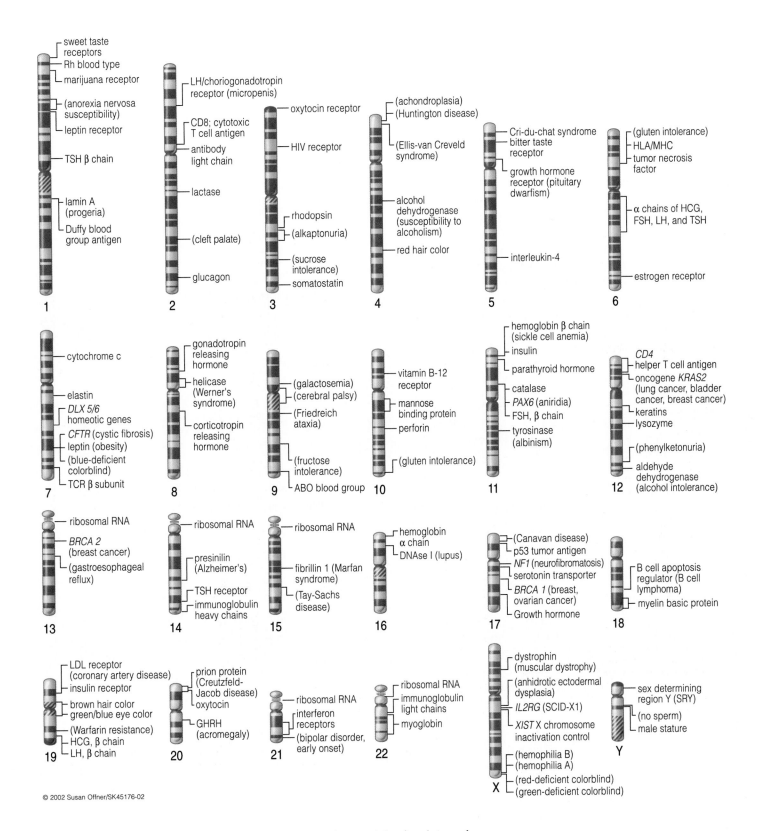

Haploid set of human chromosomes. The banding patterns characteristic of each type of chromosome appear after staining with a reagent called Giemsa. The locations of some of the 20,065 known genes (as of November, 2005) are indicated. Also shown are locations that, when mutated, cause some of the genetic diseases discussed in the text.

Length

1 kilometer (km) = 0.62 miles (mi)
1 meter (m) = 39.37 inches (in)
1 centimeter (cm) = 0.39 inches

To convert	multiply by	to obtain
inches	2.25	centimeters
feet	30.48	centimeters
centimeters	0.39	inches
millimeters	0.039	inches

Area

1 square kilometer = 0.386 square miles
1 square meter = 1.196 square yards
1 square centimeter = 0.155 square inches

Volume

1 cubic meter = 35.31 cubic feet
1 liter = 1.06 quarts
1 milliliter = 0.034 fluid ounces = 1/5 teaspoon

To convert	multiply by	to obtain
quarts	0.95	liters
fluid ounces	28.41	milliliters
liters	1.06	quarts
milliliters	0.03	fluid ounces

Weight

1 metric ton (mt) = 2,205 pounds (lb) = 1.1 tons (t)
1 kilogram (kg) = 2.205 pounds (lb)
1 gram (g) = 0.035 ounces (oz)

To convert	multiply by	to obtain
pounds	0.454	kilograms
pounds	454	grams
ounces	28.35	grams
kilograms	2.205	pounds
grams	0.035	ounces

Temperature

Celcius (°C) to Fahrenheit (°F) :
$$°F = 1.8 (°C) + 32$$

Fahrenheit (°F) to Celsius:
$$°C = \frac{(°F - 32)}{1.8}$$

	°C	°F
Water boils	100	212
Human body temperature	37	98.6
Water freezes	0	32

Appendix X. A Comparative View of Mitosis in Plant and Animal Cells

For step-by-step description of the stages of mitosis, refer to Figure 9.6.

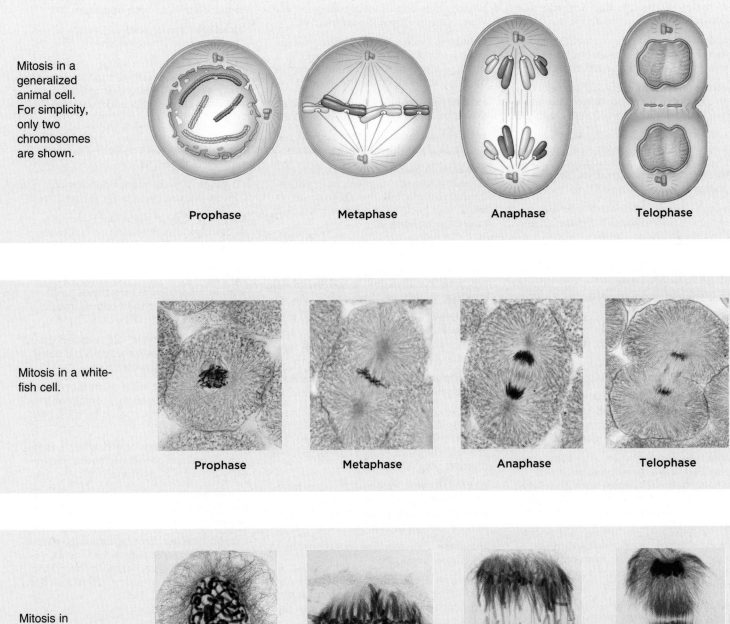

Mitosis in a generalized animal cell. For simplicity, only two chromosomes are shown.

Prophase **Metaphase** **Anaphase** **Telophase**

Mitosis in a white-fish cell.

Prophase **Metaphase** **Anaphase** **Telophase**

Mitosis in a lily cell.

Prophase **Metaphase** **Anaphase** **Telophase**

Glossary of Biological Terms

ABC model Model of the genetic basis of flower formation; products of three master genes (*A*, *B*, *C*) control the development of sepals, petals, and stamens and carpels. **233**

acid Any substance that releases hydrogen ions in water. **30**

activation energy Minimum amount of energy required to start a reaction; enzymes lower it in metabolic reactions. **96**

activator A regulatory protein that increases the rate of transcription when it binds to a promoter or enhancer. **230**

active site Chemically stable crevice in an enzyme where substrates bind and a reaction can be catalyzed repeatedly. **98**

active transport Mechanism by which a solute is moved across a cell membrane against its concentration gradient, through a transport protein. Requires energy input, as from ATP. **85**

adaptive trait A heritable trait that enhances an individuals fitness; an evolutionary adaptation. **10**

adenine (A) A type of nitrogen-containing base in nucleotides; also, a nucleotide with an adenine base. Base-pairs with thymine in DNA and uracil in RNA. **206**

adhesion protein In multicelled species, a membrane protein that helps cells stick to each other or to extracellular matrix. **80**

aerobic Oxygen-requiring. **124**

aerobic respiration Metabolic pathway that breaks down carbohydrates to produce ATP by using oxygen. Typical yield: 36 ATP per molecule of glucose. **124**

alcoholic fermentation Anaerobic pathway that breaks down glucose, forms ethanol and ATP. Begins with glycolysis; end reactions regenerate NAD⁺ so glycolysis continues. Net yield: 2 ATP per glucose. **132**

allele One of two or more forms of a gene; alleles arise by mutation and encode slightly different versions of the same gene product. **156**

allosteric A region of an enzyme other than the active site that can bind regulatory molecules. **100**

alternative splicing mRNA processing event in which some exons are removed or joined in various combinations. By this process, one gene can specify two or more slightly different proteins. **220**

amino acid A small organic compound with a carboxylic acid group, an amino group, and a characteristic side group (R); monomer of polypeptide chains. **44**

anaerobic Occurring in the absence of oxygen. **124**

anaphase Stage of mitosis in which sister chromatids separate and move to opposite spindle poles. **146**

aneuploidy A chromosome abnormality in which there are too many or too few copies of a particular chromosome; e.g., having three copies of chromosome 21, which causes Down syndrome. **194**

animal A multicelled heterotroph with unwalled cells. It develops through a series of embryonic stages and is motile during part or all of the life cycle. **9**

anticodon Set of three nucleotides in a tRNA; base-pairs with mRNA codon. **221**

antioxidant Substance that neutralizes free radicals or other strong oxidizers. **99**

archaean A member of the prokaryotic domain Archaea. Members have some unique features but also share some traits with bacteria and other traits with eukaryotic species. **8**

asexual reproduction Any reproductive mode by which offspring arise from one parent and inherit that parent's genes only; e.g., prokaryotic fission, transverse fission, budding, vegetative propagation. **156**

atom Particle that is a fundamental building block of matter; consists of varying numbers of electrons, protons, and neutrons. **4, 22**

atomic number The number of protons in the nucleus of atoms of a given element. **22**

ATP Adenosine triphosphate. Nucleotide that consists of an adenine base, the five-carbon sugar ribose, and three phosphate groups. The main energy carrier between reaction sites in cells. **48, 97**

ATP/ADP cycle How a cell regenerates its ATP supply. ADP forms when ATP loses a phosphate group, then ATP forms as ADP gains a phosphate group. **97**

autosome Any chromosome other than a sex chromosome. **186**

autotroph Organism that makes its own food using carbon from inorganic molecules such as CO_2, and energy from light or chemical reactions. **118**

bacteria Members of the prokaryotic domain Bacteria; the most diverse and most ancient prokaryotic lineage. **8**

bacteriophage Type of virus that infects bacteria. **205**

base A substance that accepts hydrogen ions as it dissolves in water. **30**

base-pair substitution Type of mutation; a single base-pair change. **224**

bell curve Curve that typically results when range of variation for a continuous trait is plotted against frequency in the population. **181**

biofilm Community of different types of microorganisms living within a shared mass of slime. **61**

bioluminescence Light emitted as a result of reactions in a living organism. **102**

biosphere All regions of Earth's waters, crust, and air where organisms live. **5**

bipolar spindle In a eukaryotic cell, a dynamically assembled and dissasembled array of microtubules that moves chromosomes during mitosis or meiosis. **145**

buffer system Set of chemicals that can keep the pH of a solution stable by alternately donating and accepting ions that contribute to pH. **31**

C3 plant Type of plant that uses only the Calvin–Benson cycle to fix carbon. **116**

C4 plant Type of plant that minimizes photorespiration by fixing carbon twice, using a C4 pathway in addition to the Calvin–Benson cycle. **116**

calcium pump Active transport protein; pumps calcium ions across a cell membrane against their concentration gradient. **85**

Calvin–Benson cycle Light-independent reactions of photosynthesis; cyclic pathway that forms glucose from CO_2. **115**

CAM plant Type of C4 plant that conserves water by opening stomata only at night, when it fixes carbon by a C4 pathway. **117**

cancer Disease that occurs when a malignant neoplasm physically and metabolically disrupts body tissues. **151**

carbohydrate Organic molecule that consists primarily of carbon, hydrogen, and oxygen atoms in a 1:2:1 ratio. **40**

carbon fixation Process by which carbon from an inorganic source such as CO_2 is incorporated into an organic compound. Occurs in the light-independent reactions of photosynthesis. **115**

cDNA DNA synthesized from RNA by the enzyme reverse transcriptase. **243**

cell Smallest unit with the properties of life—the capacity for metabolism, growth, homeostasis, and reproduction. **4, 56**

cell cortex Mesh of microfilaments that reinforces the plasma membrane. **72**

cell cycle A series of events from the time a cell forms until it reproduces. In eukaryotes, a cycle consists of interphase, mitosis, and cytoplasmic division. **144**

cell differentiation *See* differentiation.

cell junction Structure that connects a cell to another cell or to extracellular matrix; e.g., gap junction, adhering junction, tight junction. **71**

cell plate After nuclear division in a plant cell, a disk-shaped structure that forms a cross-wall between the two new nuclei. **149**

cell theory All organisms consist of one or more cells; the cell is the smallest unit of life; each new cell arises from another cell; and a cell passes hereditary material to its offspring. **55**

cell wall In many cells (not animal cells), a semirigid permeable structure around the plasma membrane. **60**

central vacuole A fluid-filled organelle in many plant cells. **69**

centriole A barrel-shaped structure that has a role in microtubule formation in cilia, flagella, and eukaryotic spindles. **73**

centromere Constricted region in a eukaryotic chromosome where sister chromatids are attached. **143**

charge An electrical property. Opposite charges attract; like charges repel. **22**

chemical bond An attractive force that arises between two atoms when their electrons interact. **25**

chemoautotroph Organism that makes its own food using carbon from inorganic sources such as carbon dioxide, and energy from chemical reactions. **118**

chlorophyll *a* Main photosynthetic pigment in plants, algae, and cyanobacteria. **109**

chloroplast Organelle of photosynthesis in plants and some protists. Two outer membranes enclose a semifluid stroma. A third membrane forms a compartment that functions in ATP and NADPH formation; sugars form in the stroma. **69, 111**

chromatin All of the DNA molecules and associated proteins in a nucleus. **65**

chromosome A complete molecule of DNA and its attached proteins; carries part or all of an organism's genes. Linear in eukaryotic cells; circular in prokaryotes. **65**

chromosome number The sum of all chromosomes in a cell of a given type; e.g., it is 46 in human body cells. **144**

cilium, plural **cilia** Short movable structure that projects from the plasma membrane of certain eukaryotic cells. **73**

clone A genetically identical copy of DNA, a cell, or an organism. **156, 243**

cloning vector A DNA molecule that can accept foreign DNA, be transferred to a host cell, and get replicated in it. **242**

codominance Nonidentical alleles that are both fully expressed in heterozygotes; neither is dominant or recessive. **176**

codon In mRNA, a nucleotide base triplet that codes for an amino acid or stop signal during translation. *See* genetic code. **220**

coenzyme An organic cofactor. **99**

cofactor A metal ion or a coenzyme that associates with an enzyme and is necessary for its function; e.g., NAD^+. **99**

cohesion Tendency of molecules to stick together under tension; a property of liquid water. **29**

community All populations of all species in a habitat. **5**

compound Type of molecule that has atoms of more than one element. **25**

concentration The number of molecules or ions per unit volume of fluid. **82**

concentration gradient Difference in concentration between adjoining regions of fluid. **82**

condensation Chemical reaction in which two molecules become covalently bonded as a larger molecule; water often forms as a by-product. **39**

consumer Heterotroph that gets energy and carbon by feeding on tissues, wastes, or remains of other organisms. **6**

continuous variation In a population, a range of small differences in a trait; result of polygenic inheritance. **180**

contractile ring A thin band of actin and myosin filaments that wraps around the midsection of an animal cell undergoing cytoplasmic division. It contracts and pinches the cytoplasm in two. **148**

control group In experiments, a group that is the same as an experimental group except for one variable; used as a standard of comparison. **13**

cotransporter Transport protein that can move two or more substances across a membrane; e.g. sodium-potassium pump. **85**

covalent bond Chemical bond in which two atoms share a pair of electrons. **26**

critical thinking Mental process of judging information before accepting it. **11**

crossing over Process in which homologous chromosomes exchange corresponding segments during prophase I of meiosis. Puts nonparental combinations of alleles in gametes. **160**

cuticle Of plants, a cover of waxes and cutin on the outer wall of epidermal cells. Of annelids, a thin, flexible secreted layer. Of arthropods, a lightweight exoskeleton hardened with chitin. **70**

cytokinesis Cytoplasmic division. **148**

cytoplasm The semifluid matrix between a cell's plasma membrane and its nucleus or nucleoid. **56**

cytosine (C) A type of nitrogen-containing base in nucleotides; also, a nucleotide with a cytosine base. Base-pairs with guanine in DNA and RNA. **206**

cytoskeleton Dynamic framework of protein filaments that structurally support, organize, and move eukaryotic cells and their internal structures. Prokaryotic cells have similar protein filaments. **72**

deletion Loss of a part of a chromosome; also, a mutation in which one or a few base pairs are lost. **192, 224**

denature To unravel the shape of a protein or other lare biological molecule, as by high temperature or pH. **46**

deoxyribonucleic acid *See* DNA.

development The process that transforms a zygote into an adult with specialized tissues and, usually, organs. **7**

differentiation The process by which cells become specialized; occurs as different cell lineages begin to express different subsets of their genes. **230**

diffusion Net movement of molecules or ions from a region where they are more concentrated to a region where they are less concentrated. **82**

dihybrid experiment An experiment in which individuals with different alleles at two loci are crossed or self-fertilized; e.g., *AaBb* × *AaBb*. The ratio of phenotypes in the resulting offspring offers information about dominance relationships between the alleles. **174**

diploid Having two of each type of chromosome characteristic of the species (2*n*). **145**

DNA Deoxyribonucleic acid. Double-stranded nucleic acid twisted into a helix; hereditary material for all living organisms and many viruses. Information in its base sequence is the basis of an organism's form and function. **7, 48**

DNA chip Microscopic array of DNA fragments that collectively represent a genome; used to study gene expression. **249**

DNA cloning A set of procedures that uses living cells such as bacteria to make many identical copies of a DNA fragment. **242**

DNA fingerprint An individual's unique array of short tandem repeats. **247**

DNA library A collection of cells that hosts different fragments of foreign DNA, often representing an organism's entire genome. **244**

DNA ligase Enzyme that seals breaks in double-stranded DNA. **208**

DNA polymerase DNA replication enzyme; assembles a new strand of DNA from free nucleotides based on the sequence of a DNA template. **208**

DNA repair mechanism One of several processes by which enzymes repair broken or mismatched DNA strands. **209**

DNA replication Process by which a cell duplicates its DNA before it divides. **144**

DNA sequencing Method of determining the order of nucleotides in DNA. **246**

dominant With regard to an allele, having the ability to mask the effects of a recessive allele paired with it. **171**

dosage compensation Theory that X chromosome inactivation equalizes gene expression between males and females. **232**

duplication Base sequence in DNA that is repeated two or more times. **192**

ecosystem Community interacting with its environment through a one-way flow of energy and cycling of materials. **5**

egg Mature female gamete, or ovum. **162**

electron Negatively charged subatomic particle that occupies orbitals around the atomic nucleus. **22**

electron transfer chain Array of enzymes and other molecules in a cell membrane that accept and give up electrons in sequence, thus releasing the energy of the electrons in small, usable increments. **101**

electron transfer phosphorylation Third stage of aerobic respiration; electron flow through electron transfer chains in inner mitochondrial membrane sets up an H+ gradient that drives ATP formation. **130**

electronegativity A measure of an atom's ability to pull electrons away from other atoms. **25**

electrophoresis Technique of separating DNA fragments by size. **246**

element A substance that consists only of atoms with the same number of protons. **22**

emergent property A property of a system that does not appear in any of its component parts; e.g., cells (which are alive) are composed of many molecules (which are not alive). **5**

endergonic Type of reaction in which reactants have less free energy than products; requires a net energy input to proceed. **96**

endocytosis Process by which a cell takes in a substance by engulfing it in a vesicle formed from a bit of plasma membrane. **86**

endomembrane system Series of interacting organelles between the nucleus and plasma membrane; produces lipids and proteins for secretion or insertion into cell membranes. Includes endoplasmic reticulum, Golgi bodies, vesicles. **66**

endoplasmic reticulum (ER) Membranous organelle, a continuous system of sacs and tubes that is an extension of the nuclear envelope. Rough ER is studded with ribosomes; smooth ER is not. **66**

energy A capacity to do work. **6, 94**

enhancer Binding site in DNA for proteins that enhance the rate of transcription. **230**

enzyme Protein or RNA that catalyzes (speeds) a reaction without being changed by it. **80, 98**

epistasis Interacting products of two or more gene pairs influence a trait. **177**

eukaryote Organism whose cells characteristically start out life with a nucleus and other membrane-enclosed organelles; a protist, plant, fungus, or animal. **8**

eukaryotic cell Type of cell that starts life with a nucleus. **56**

eukaryotic flagella *See* flagellum.

evaporation Transition of a liquid to a gas; requires energy input. **29**

evolution Change in a line of descent. **10**

exergonic Type of reaction in which products have less free energy than reactants; ends with net release of energy. **96**

exocytosis Fusion of a cytoplasmic vesicle with the plasma membrane; as it becomes part of the membrane, its contents are released to extracellular fluid. **86**

exon Nucleotide sequence that is not spliced out of RNA during processing. **220**

experiment A test designed to support or falsify a prediction. Involves experimental and control groups. **13**

experimental group In experiments, a group of objects or individuals that display or are exposed to a variable under investigation. Experimental results for this group are compared with results for a control group. **13**

extracellular matrix (ECM) Complex mixture of fibrous proteins and polysaccharides secreted by cells; supports and anchors cells, separates tissues, and has functions in cell signaling; e.g., basement membrane, bone. **70**

fat Lipid with one, two, or three fatty acid tails attached to a glycerol. **42**

fatty acid Simple organic compound with a carboxyl group and a backbone of four to thirty-six carbon atoms; component of many lipids. Backbone of saturated types has single bonds only; that of unsaturated types has one or more double bonds. **42**

feedback inhibition Mechanism by which a change that results from some activity decreases or stops the activity. **100**

fermentation An anaerobic metabolic pathway by which cells harvest energy from organic molecules. *See* alcoholic fermentation and lactate fermentation. **124**

fertilization Fusion of a sperm nucleus and an egg nucleus, the result being a single-celled zygote. **162**

first law of thermodynamics Energy cannot be created or destroyed. **94**

flagellum, plural flagella Long, slender cellular structure used for motility. Eukaryotic flagella whip from side to side; prokaryotic flagella rotate like a propeller. **60, 73**

fluid mosaic model A cell membrane has a mixed composition (mosaic) of lipids and proteins, the interactions and motions of which impart fluidity to it. **78**

free energy The amount of energy that is available (free) to do work. **96**

functional group An atom or a group of atoms covalently bonded to carbon; imparts certain chemical properties to an organic compound. **38**

fungus, plural fungi Type of eukaryotic heterotroph; can be multicelled or single-celled; cell walls contain chitin; obtains nutrients by extracellular digestion and absorption. **8**

gamete Mature, haploid reproductive cell; e.g., an egg or sperm. **156**

gametophyte A haploid, multicelled body in which gametes form during the life cycle of plants and some algae. **162**

gene Heritable unit of information in DNA; occupies a particular location (locus) on a chromosome. **156, 171**

gene expression Process by which the information contained in a gene becomes converted to a structural or functional part of a cell. **171, 217**

gene therapy The transfer of a normal or modified gene into an individual with the goal of treating a genetic disorder. **254**

genetic code Set of sixty-four mRNA codons, each of which specifies an amino acid or stop signal in translation. **220**

genetic engineering Process by which deliberate changes are introduced into an individual's chromosome(s). **250**

genetically modified organism (GMO) An organism whose genome has been deliberately modified; e.g., a transgenic organism. **250**

genome An organism's complete set of genetic material. **244**

genomics The study of genomes. The structural branch investigates the three-dimensional structure of proteins encoded by a genome; comparative branch compares genomes of different species. **249**

genotype The particular alleles carried by an individual. **171**

genus, plural **genera** A group of species that share a unique set of traits. **8**

germ cell Animal cell that can undergo meiosis and give rise to gametes. **156**

glycolysis First stage of aerobic respiration and fermentation; glucose or another sugar molecule is broken down to two pyruvate for a net yield of 2 ATP. **124**

Golgi body Organelle of endomembrane system; enzymes inside its much-folded membrane modify polypeptide chains and lipids; the products are sorted and packaged into vesicles. **67**

growth factor Checkpoint gene product that stimulates cell division. **150**

guanine (G) A type of nitrogen-containing base in nucleotides; also, a nucleotide with a guanine base. Base-pairs with cytosine in DNA and RNA. **206**

haploid Having one of each type of chromosome characteristic of the species (n); e.g., a human gamete is haploid. **157**

heterotroph Organism that obtains carbon from organic compounds assembled by other organisms. **118**

heterozygous Having two different alleles at a gene locus; e.g., *Aa*. **171**

histone Type of protein that structurally organizes eukaryotic chromosomes. Part of nucleosomes. **143**

homeostasis The collection of processes by which the conditions in a multicelled organism's internal environment are kept within tolerable ranges. **7**

homeotic gene Type of master gene; its expression controls formation of specific body parts during development. **234**

homologous chromosome One of a pair of chromosomes in body cells of diploid organisms; except for the nonidentical sex chromosomes, members of a pair have the same length, shape, and genes. **156**

homozygous Having identical alleles at a gene locus; e.g., *AA*. **171**

homozygous dominant Having a pair of dominant alleles at a locus on homologous chromosomes; e.g., *AA*. **171**

homozygous recessive Having a pair of recessive alleles at a locus on homologous chromosomes; e.g., *aa*. **171**

hybrid Heterozygote. Individual with two different alleles at a gene locus. **171**

hydrogen bond Attraction that forms between a covalently bonded hydrogen atom and an electronegative atom taking part in a separate covalent bond. **27**

hydrolysis A type of cleavage reaction in which an enzyme breaks a bond by attaching a hydroxyl group to one atom and a hydrogen atom to the other. The hydrogen atom and the hydroxyl group are derived from a water molecule. **39**

hydrophilic Describes a substance that dissolves easily in water; e.g., a salt. **28**

hydrophobic Describes a substance that resists dissolving in water; e.g., an oil. **28**

hypertonic Describes a fluid with a high solute concentration relative to another fluid. **88**

hypothesis, scientific Testable explanation of a natural phenomenon. **12**

hypotonic Describes a fluid with a low solute concentration relative to another fluid. **88**

incomplete dominance Condition in which one allele is not fully dominant over another, so the heterozygous phenotype is somewhere between the two homozygous phenotypes. **176**

independent assortment Theory that alleles of one gene become distributed into gametes independently of alleles of all other genes during meiosis. **174**

induced-fit model Explanation of how some enzymes work; an active site bends or squeezes a substrate, which brings on the transition state. **98**

inheritance Transmission of DNA from parents to offspring. **7**

insertion A mutation in which extra base pairs become inserted into DNA. **224**

intermediate filament Cytoskeletal element that mechanically strengthens cell and tissue structures. **72**

interphase In a eukaryotic cell cycle, the interval between mitotic divisions when a cell grows in mass, roughly doubles the number of its cytoplasmic components, and replicates its DNA. **144**

intron Nucleotide sequence that intervenes between exons; excised during RNA processing. **220**

inversion Structural rearrangement of a chromosome in which part of it becomes oriented in the reverse direction. **192**

ion Atom that carries a charge because of an unequal number of protons and electrons. **25**

ionic bond Type of chemical bond; strong mutual attraction between ions of opposite charge. **26**

isotonic Describes a fluid with the same solute concentration relative to another fluid. **88**

isotopes Forms of an element that differ in the number of neutrons their atoms carry. **22**

karyotype Image of an individual's complement of chromosomes arranged by size, length, shape, and centromere location. **187**

knockout experiment An experiment in which an organism is genetically engineered so one of its genes does not function. **234**

Krebs cycle The second stage of aerobic respiration; breaks down two pyruvate to CO_2 and H_2O for a net yield of two ATP and many reduced coenzymes. **128**

lactate fermentation Anaerobic pathway that breaks down glucose, forms ATP and lactate. Begins with glycolysis; regenerates NAD^+ so glycolysis continues. Net yield: 2 ATP per glucose. **133**

light-dependent reactions First stage of photosynthesis; one of two metabolic pathways (cyclic or noncyclic) in which light energy is converted to the chemical energy of ATP. NADPH and O_2 also form in the noncyclic pathway. **111**

light-independent reactions Second stage of photosynthesis; metabolic pathway in which the enzyme rubisco fixes carbon, and glucose forms. Runs on ATP and NADPH produced in the light-dependent reactions. *See also* Calvin–Benson cycle. **111**

lignin Organic compound that strengthens cell walls of vascular plants; reinforces stems and thus helps plant stand upright. **70**

linkage group All genes on a chromosome; tend to stay together during meiosis but may be separated by crossovers. **178**

lipid Fatty, oily, or waxy organic compound; often has one or more fatty acid components. **42**

lipid bilayer Structural foundation of cell membranes; mainly phospholipids arranged tail-to-tail in two layers. **57**

locus, plural **loci** The location of a gene on a chromosome. **171**

lysosome Enzyme-filled vesicle; functions in intracellular digestion. **67**

mass number Total number of protons and neutrons in the nucleus of an element's atoms. **22**

master gene Gene encoding a product that affects the expression of many other genes; cascades of master gene expression often result in the completion of a complex task such as flower formation. **233**

meiosis Nuclear division process that halves the chromosome number, to the haploid (*n*) number. Basis of sexual reproduction. **142, 156**

messenger RNA (mRNA) Type of RNA that carries a protein-building message; intermediary between DNA and protein synthesis. **216**

metabolic pathway Series of enzyme-mediated reactions by which cells build, remodel, or break down organic molecules; e.g., photosynthesis. **100**

metabolism All the enzyme-mediated chemical reactions by which cells acquire and use energy as they build, remodel, and break down organic molecules. **39**

metaphase Stage of mitosis during which the cell's chromosomes align midway between poles of the spindle. **146**

microfilament Cytoskeletal element that helps strengthen or change the shape of a cell. Fiber of actin subunits. **72**

microtubule Cytoskeletal element involved in the movement of a cell or its components; hollow filament of tubulin subunits. **72**

mitochondrion Double-membraned organelle of ATP formation; site of second and third stages of aerobic respiration in eukaryotes. **68**

mitosis Nuclear division mechanism that maintains the chromosome number. Basis of body growth, tissue repair and replacement in multicelled eukaryotes, as well as asexual reproduction in some plants, animals, fungi, and protists. **142**

mixture Two or more types of molecules intermingled in proportions that vary. **25**

model Analogous system used to test an object or event that cannot itself be tested directly. **12**

molecule Group of two or more atoms joined by chemical bonds. **4, 25**

monohybrid experiment An experiment in which individuals with different alleles at one locus are crossed or self-fertilized; e.g., *Aa* × *Aa*. The phenotype ratio of the resulting offspring offers information about dominance relationships between the alleles. **172**

monomer A small molecule that is a repeating subunit in a polymer; e.g., glucose is a monomer of starch. **39**

motor protein Type of protein that, when energized by ATP hydrolysis, interacts with cytoskeletal elements to move cell parts or the whole cell; e.g., myosin. **72**

multiple allele system Three or more alleles persist in a population. **176**

mutation Permanent, small-scale change in DNA. Primary source of new alleles and, thus, of life's diversity. **10, 171**

natural selection A process of evolution in which individuals of a population who vary in the details of heritable traits survive and reproduce with differing success. **10**

nature Everything in the universe except what humans have manufactured. **4**

neoplasm Tumor; abnormal mass of cells that lost control over their cell cycle. **150**

neutron Uncharged subatomic particle in the atomic nucleus. **22**

nondisjunction Failure of sister chromatids or homologous chromosomes to separate during meiosis or mitosis. Resulting cells get too many or too few chromosomes. **194**

nonpolar Having an even distribution of charge. Two atoms share electrons equally in a nonpolar covalent bond. **27**

nuclear envelope A double membrane that constitutes the outer boundary of the nucleus. **64**

nucleic acid Single- or double-stranded chain of nucleotides joined by sugar–phosphate bonds; e.g., DNA, RNA. **48**

nucleic acid hybridization Base-pairing between DNA or RNA from different sources. **244**

nucleoid Of a prokaryotic cell, region of cytoplasm where the DNA is concentrated. **56**

nucleolus In a nucleus, a dense, irregularly shaped region where ribosomal subunits are assembled. **65**

nucleoplasm Of a nucleus, the viscous fluid enclosed by the nuclear envelope. **65**

nucleosome Smallest unit of structural organization in eukaryotic chromosomes; a length of DNA wound twice around a spool of histone proteins. **143**

nucleotide Organic compound with a five-carbon sugar, a nitrogen-containing base, and at least one phosphate group. Monomer of nucleic acids. **48, 206**

nucleus In eukaryotic cells only, organelle with an outer envelope of two pore-studded lipid bilayers; separates the cell's DNA from its cytoplasm. **22, 56**

nutrient An element or type of molecule with an essential role in an individual's survival or growth. **6**

operator Part of an operon; a DNA binding site for a repressor. **236**

operon Group of genes together with a promoter–operator DNA sequence that controls their transcription. **236**

organelle Structure that carries out a specialized metabolic function inside a cell; e.g., a nucleus in eukaryotes. **62**

organic Molecule that consists primarily of carbon and hydrogen atoms; many types have functional groups. **36**

organism An individual that consists of one or more cells. **4**

osmosis Diffusion of water in response to a concentration gradient. **88**

osmotic pressure Amount of hydrostatic pressure that prevents osmosis into cytoplasm or other hypertonic fluid. **89**

oxidation–reduction reaction Reaction in which one molecule accepts electrons (it becomes reduced) from another molecule (which becomes oxidized). **101**

passive transport Mechanism by which a concentration gradient drives the movement of a solute across a cell membrane through a transport protein; no energy input is required. **84**

pattern formation The process by which a complex body forms from local processes during embryonic development. **235**

PCR Polymerase chain reaction. Method that rapidly generates many copies of a specific DNA fragment. **244**

pedigree Chart showing the pattern of inheritance of a gene in a family. **197**

periodic table of the elements Tabular arrangement of the known atomic elements by atomic number. **22**

peroxisome Enzyme-filled vesicle that breaks down amino acids, fatty acids, and toxic substances. **67**

pH A measure of the number of hydrogen ions in a solution. pH 7 is neutral. **30**

phagocytosis "Cell eating," an endocytic pathway by which a cell engulfs particles such as microbes or cellular debris. **86**

phenotype An individual's observable traits. **171**

phospholipid A lipid with a phosphate group in its hydrophilic head, and two nonpolar fatty acid tails; main constituent of cell membranes. **43**

phosphorylation Transfer of a phosphate group to a recipient molecule. **97**

photoautotroph Photosynthetic autotroph; e.g., nearly all plants, most algae, and a few bacteria. **118**

photolysis Reaction in which light energy breaks down a molecule. Photolysis of water molecules during noncyclic photosynthesis releases electrons and hydrogen ions used in the reactions, and molecular oxygen. **112**

photophosphorylation Any light-driven phosphorylation reaction. **114**

photorespiration Reaction in which rubisco attaches oxygen instead of carbon dioxide to ribulose bisphosphate; occurs in C4 plants when stomata close and oxygen levels rise. Produces no ATP. **116**

photosynthesis The metabolic pathway by which photoautotrophs capture light energy and use it to make sugars from CO_2 and water. **6, 108**

photosystem In photosynthetic cells, a cluster of pigments and proteins that, as a unit, converts light energy to chemical energy in photosynthesis. **111**

pigment An organic molecule that absorbs light of certain wavelengths. Reflected light imparts a characteristic color. **108**

pilus, plural **pili** A protein filament that projects from the surface of some bacterial cells. **60**

plant A multicelled photoautotroph, typically with well-developed roots and shoots. Primary producer on land. **9**

plasma membrane Outer cell membrane; encloses the cytoplasm. **56**

plasmid A small, circular DNA molecule in bacteria, replicated independently of the chromosome. **242**

plastid In plants and algae, an organelle that functions in photosynthesis or storage; e.g., chloroplast, amyloplast. **69**

pleiotropy The effect of a single gene on multiple traits. **177**

polar Having an uneven distribution of charge. Two atoms share electrons unequally in a polar covalent bond. **27**

polarity Any separation of charge into distinct positive and negative regions. **27**

polymer Large molecule of multiple linked monomers. **39**

polypeptide Chain of amino acids linked by peptide bonds. **44**

polyploid Having three or more of each type of chromosome characteristic of the species. **194**

population A group of individuals of the same species in a specified area. **5**

prediction A statement, based on a hypothesis, about a condition that should exist if the hypothesis is not wrong; often called the "if–then process." **12**

primary wall The first thin, pliable wall of young plant cells. **70**

primer Short, single strand of DNA designed to hybridize with a template; DNA polymerases initiate synthesis at primers during PCR or sequencing. **245**

probability Chance that a particular outcome of an event will occur; depends on the total number of outcomes possible. **173**

probe Short fragment of DNA labeled with a tracer; designed to hybridize with a nucleotide sequence of interest. **244**

producer Autotroph; an organism that makes its own food using carbon from inorganic molecules such as CO_2. Most are photosynthetic. **6**

product A molecule remaining at the end of a reaction. **96**

prokaryote Single-celled organism in which the DNA is not contained in a nucleus; a bacterium or archaean. **56**

prokaryotic cell *See* Prokaryote.

promoter In DNA, a nucleotide sequence to which RNA polymerase binds. **219**

prophase Stage of mitosis and meiosis in which chromosomes condense and become attached to a newly forming spindle. **146**

protein Organic compound that consists of one or more polypeptide chains. **44**

protists Informal name for eukaryotes that are not plants, fungi, or animals. **8**

proton Positively charged subatomic particle in the nucleus of all atoms. The number of protons (the atomic number) defines the element. **22**

pseudopod A dynamic lobe of membrane-enclosed cytoplasm; functions in motility and phagocytosis by amoebas, amoeboid cells, and phagocytic white blood cells. **73**

Punnett square A diagram used to predict the outcome of a testcross. **173**

pyruvate Three-carbon end product of glycolysis. **124**

radioactive decay Process by which atoms of a radioisotope spontaneously emit energy and subatomic particles when their nucleus disintegrates. **23**

radioisotope Isotope with an unstable nucleus; decays into predictable daughter elements at a predictable rate. **23**

reactant Molecule that enters a reaction. **96**

reaction Process of chemical change. **96**

receptor A molecule or structure that can respond to a form of stimulation such as light energy, or to binding of a signaling molecule such as a hormone. **6**

receptor protein Plasma membrane protein that binds to a particular substance outside of the cell. **80**

recessive With regard to an allele, having effects that are masked by a dominant allele on the homologous chromosome. **171**

recognition protein Plasma membrane protein that identifies a cell as belonging to *self* (one's own body tissue). **80**

recombinant DNA A DNA molecule that contains genetic material from more than one organism. **242**

repressor Transcription factor that blocks transcription by binding to a (eukaryotic) promoter or (prokaryotic) operator. **230**

reproduction An asexual or sexual process by which a parent cell or organism produces offspring. **7**

reproductive cloning Technology that produces genetically identical individuals; e.g., artificial twinning, SCNT. **210**

restriction enzyme Type of enzyme that cuts specific base sequences in DNA. **242**

reverse transcriptase A viral enzyme that catalyzes the assembly of nucleotides into DNA, using RNA as a template. **243**

ribosomal RNA (rRNA) A type of RNA that becomes part of ribosomes; some catalyze formation of peptide bonds. **216**

ribosome Site of protein synthesis. An intact ribosome has two subunits, each composed of rRNA and proteins. **56**

RNA Ribonucleic acid. Type of nucleic acid, typically single-stranded; important in transcription, translation, and gene control; some are catalytic. *See also* ribosomal RNA, transfer RNA, messenger RNA. **48**

RNA polymerase Enzyme that catalyzes transcription of DNA into RNA. **218**

rubisco Ribulose bisphosphate carboxylase, or RuBP. Carbon-fixing enzyme of light-independent photosynthesis reactions. **115**

salt Compound that dissolves easily in water and releases ions other than H^+ and OH^-. **31**

sampling error Difference between results derived from testing an entire group of events or individuals, and results derived from testing a subset of the group. **16**

science Systematic study of nature. **11**

scientific theory Hypothesis that has not been disproven after many years of rigorous testing, and is useful for making predictions about other phenomena. **12**

second law of thermodynamics Energy tends to disperse spontaneously. **94**

secondary wall Lignin-reinforced wall inside the primary wall of a plant cell. **70**

segregation Theory that the two members of each pair of genes on homologous chromosomes separate during meiosis. **173**

selective permeability Membrane property that allows some substances, but not others, to cross. **82**

semiconservative replication Describes the process of DNA replication, by which one strand of each copy of a DNA molecule is new, and the other is a strand of the original DNA. **208**

sequence The order of nucleotides in a strand of DNA or RNA. **207**

sex chromosome Member of a pair of chromosomes that differs between males and females. **186**

sexual reproduction Production of genetically variable offspring by gamete formation and fertilization. **156**

shell model Model of electron distribution in an atom; orbitals are shown as nested circles, electrons as dots. **24**

short tandem repeat Stretch of DNA that consists of many copies of a short sequence; basis of DNA fingerprinting. **247**

sister chromatid One of two attached members of a duplicated eukaryotic chromosome. **142**

solute A dissolved substance. **28**

solvent Substance, typically a liquid, that can dissolve other substances; e.g., water. **28**

somatic cell nuclear transfer (SCNT) Method of reproductive cloning in which genetic material is transferred from an adult somatic cell into an unfertilized, enucleated egg. **210**

species A type of organism. Of sexually reproducing species, one or more groups of individuals that potentially can interbreed, produce fertile offspring, and do not interbreed with other groups. **8**

sperm Mature male gamete. **162**

spindle *See* bipolar spindle.

sporophyte Diploid, spore-producing body of a plant or multicelled alga. **162**

steroid A type of lipid with four carbon rings and no fatty acid tails. **43**

stoma, plural **stomata** Gap that opens between two guard cells; lets water vapor and gases diffuse across the epidermis of a leaf or primary stem. **116**

stroma The semifluid matrix between the thylakoid membrane and the two outer membranes of a chloroplast; site of light-independent photosynthesis reactions. **111**

substrate A reactant molecule that is specifically acted upon by an enzyme. **98**

substrate-level phosphorylation The direct transfer of a phosphate group from a substrate to ADP; forms ATP. **126**

surface-to-volume ratio A relationship in which the volume of an object increases

with the cube of the diameter, but the surface area increases with the square. **56**

syndrome The set of symptoms that characterize a medical condition. **197**

telophase Stage of mitosis during which chromosomes arrive at the spindle poles and decondense, and new nuclei form. **146**

temperature Measure of molecular motion. **29**

testcross Method of determining genotype; a cross between an individual of unknown genotype and a homozygous recessive individual. Offspring phenotypes are analyzed. **172**

therapeutic cloning Producing human embryos by SCNT. **211**

thylakoid membrane A chloroplast's inner membrane system, often folded as flattened sacs, that forms a continuous compartment in the stroma. In the first stage of photosynthesis, pigments and enzymes in the membrane function in the formation of ATP and NADPH. **111**

thymine (T) A type of nitrogen-containing base in nucleotides; also, a nucleotide with a thymine base. Base-pairs with adenine; does not occur in RNA. **206**

tracer A molecule with a detectable label attached; researchers can track it after delivering it into a cell or other system. **23**

trait A physical, biochemical, or behavioral characteristic of an individual. **7**

transcription Process by which an RNA is assembled from nucleotides using a gene region in DNA as a template. First step in protein synthesis. **216**

transcription factor Regulatory protein that influences transcription; e.g., activator, repressor. **230**

transfer RNA (tRNA) Type of RNA that delivers amino acids to a ribosome during translation. Its anticodon pairs with an mRNA codon. **216**

transgenic Refers to an organism that has been genetically engineered to carry a gene from a different species. **250**

transition state In a chemical reaction, the point at which reactant bonds are at their breaking point. **98**

translation At ribosomes, information encoded in an mRNA guides synthesis

of a polypeptide chain from amino acids. Second stage of protein synthesis. **217**

translocation Attachment of a piece of a broken chromosome to another chromosome. Also, the movement of organic compounds through phloem. **192**

transport protein Membrane protein that passively or actively assists specific ions or molecules into or out of a cell. **80**

transposable element Small segment of DNA that can spontaneously move to a new location in the chromosomal DNA of a cell. **224**

triglyceride A lipid with three fatty acid tails attached to a glycerol backbone. **42**

tumor Abnormal mass of cells. Benign tumor cells stay in their home tissue; malignant ones invade other places in the body and start new tumors. *See also* neoplasm. **150**

turgor Hydrostatic pressure. Pressure that a fluid exerts against a wall, membrane, or some other structure that contains it. **88**

uracil (u) A type of nitrogen-containing base in nucleotides; also, a nucleotide with a uracil base. Base-pairs with adenine; occurs in RNA, not in DNA. **216**

vacuole A fluid-filled organelle that isolates or disposes of waste, debris, or toxic materials. **67**

variable In experiments, a characteristic or event that differs among individuals and that may change over time. **13**

vesicle Small, membrane-enclosed, saclike organelle; different kinds store, transport, or degrade their contents. **67**

wavelength Distance between crests of two successive waves of radiant energy. **108**

wax Water-repellent lipid with long fatty acid tails bonded to long-chain alcohols or carbon rings. **43**

X chromosome inactivation Shutdown of one of the two X chromosomes in the cells of female mammals. *See also* Dosage compensation. **232**

xenotransplantation Transplant of an organ from one species into another. **253**

zygote Cell formed by fusion of gametes; first cell of a new individual. **157**

Art Credits and Acknowledgments

This page constitutes an extension of the book copyright page. We have made every effort to trace the ownership of all copyrighted material and secure permission from copyright holders. In the event of any question arising as to the use of any material, we will be pleased to make the necessary corrections in future printings. Thanks are due to the following authors, publishers, and agents for permission to use the material indicated.

TABLE OF CONTENTS **Page vi** top, © Raymond Gehman/ Corbis. **Page vii** from left, ArchiMeDes; R. Calentine/ Visuals Unlimited; © Kenneth Bart. **Page viii** from left, Hemoglobin models: PDB ID: 1GZX; Paoli, M., Liddington, R., Tame, J., Wilkinson, A., Dodson, G.; Crystal structure of T state hemoglobin with oxygen bound at all four haems. *J.Mol.Biol.*, v256, pp. 775–792, 1996; Larry West/ FPG / Getty Images; © Professors P. Motta and T Naguro/SPL/ Photo Researchers, Inc. **Page ix** from left, Dr. Pascal Madaule, France; Image courtesy of Carl Zeiss MicroImaging, Thornwood, NY; Moravian Museum, Brno; © Russ Schleipman/ Corbis. **Page x** from left, A C. Barrington Brown © 1968 J. D. Watson; P. J. Maughan. **Page xi** from left, © Jürgen Berger, Max-Planck-Institut for Developmental Biology, Türbingen; © Visuals Unlimited; Photo courtesy of MU Extension and Agricultural Information.

INTRODUCTION NASA Space Flight Center

CHAPTER 1 **1.1** Courtesy of Conservation International. **Page 3** From second from top, Jack de Coningh; © Lewis Trusty/ Animals Animals; © Nick Brent; © Raymond Gehman/ Corbis. **1.2** (a) Rendered with Atom In A Box, copyright Dauger Research, Inc.; (d) © Science Photo Library/ Photo Researchers, Inc.; (e) © Bill Varie/ Corbis; (f–h) © Jeffrey L. Rotman/Corbis; (i) © Peter Scoones; (j–k) NASA. **1.4** © Y. Arthus-Bertrand/ Peter Arnold, Inc. **1.5** © Jack de Coningh. **1.7** (a) clockwise from top left, © Dr. Richard Frankel; © David Scharf, 1999. All rights reserved; © Susan Barnes; © SciMAT/ Photo Researchers, Inc.; (b) left, © R. Robinson/ Visuals Unlimited, Inc.; right, © Dr. Harald Huber, Dr. Michael Hohn, Prof. Dr. K. O. Stetter, University of Regensburg, Germany; (c) above, left, clockwise from top, © Lewis Trusty/ Animals Animals; © Emiliania Huxleyi photograph, Vita Pariente, scanning electron micrograph taken on a Jeol T330A instrument at Texas A&M University Electron Microscopy center; © Carolina Biological Supply Company; © Oliver Meckes/ Photo Researchers, Inc.; Courtesy of James Evarts; right, © John Lotter Gurling/ Tom Stack & Associates; inset, © Edward S. Ross; below, left, from left, © Robert C. Simpson/ Nature Stock; © Edward S. Ross; right, © Stephen Dalton/ Photo Researchers, Inc. **1.8** (a) Photographs courtesy Derrell Fowler, Tecumseh, Oklahoma; (b) © Nick Brent. **1.9** (a) © Lester Lefkowitz/ Corbis; (b) Centers for Disease Control and Prevention; (c) © Raymond Gehman/ Corbis. **1.10** top, © Superstock. **1.11** (a) © Matt Rowlings, www .eurobutterflies.com; (b) © Adrian Vallin; (c) © Antje Schulte. **Page 17** © Gary Head. **Page 18** Scientific Paper; Adrian Vallin, Sven Jakobsson, Johan Lind, and Christer Wiklund, Proc. R. Soc. B (2005 272, 1203, 1207). Used with permission of The Royal Society and the author.

Page 19 UNIT I © Wim van Egmond, Micropolitan Museum

CHAPTER 2 **2.1** © Owaki-Kulla/ CORBIS. **Page 21** From second from top, © Michael S. Yamashita/ Corbis; © Bill Beatty/ Visuals Unlimited; © R. B. Suter, Vasar College; © W. K. Fletcher/ Photo Researchers, Inc. **2.4** Courtesy © GE Healthcare. **Page 24** Left, © Michael S. Yamashita/ Corbis. **Page 25** © Hubert Stadler/ Corbis. **2.7** (a) Left, upper, Gary Head; lower, © Bill Beatty/ Visuals Unlimited. **2.10** (b) Right, © Steve Lissau/ Rainbow; (c) Right, © Dan Guravich/ Corbis. **2.12** (a) © Lester Lefkowitz/ Corbis; (b) © R. B. Suter, Vasar College. **2.13** Photos from © JupiterImages Corporation. **2.14** Left, Michael Grecco/ Picture Group; right, © W. K. Fletcher/ Photo Researchers, Inc.

CHAPTER 3 **3.1** © ThinkStock/ SuperStock. **Page 35** From top, © Tim Davis/ Photo Researchers, Inc.; © JupiterImages Corporation; Kenneth Lorenzen. **3.2** (b) © JupiterImages Corporation. **3.3** (a), © National Cancer Institute/ Photo Researchers, Inc.; (b–d) Hemoglobin models: PDB ID: 1GZX; Paoli, M., Liddington, R., Tame, J., Wilkinson, A., Dodson, G., Crystal structure of T state hemoglobin with oxygen bound at all four haems. *J.Mol.Biol.*, v256, pp. 775–792, 1996. **3.5** © Tim Davis/ Photo Researchers, Inc. **3.7** Left, © JupiterImages Corporation. **3.8** © JupiterImages Corporation. **3.11** © Kevin Schafer/ Corbis. **Page 43** Bottom left, Kenneth Lorenzen. **3.16** (a–d, bottom) PDB files from NYU Scientific Visualization Lab. **3.17** (b, right) After: *Introduction to Protein Structure*, 2nd ed., Branden & Tooze, Garland Publishing, Inc.; (c, left) PDB ID: 1BBB; Silva, M. M., Rogers, P. H., Arnone, A.; A third quaternary structure of human hemoglobin A at 1.7-Å resolution; *J Biol Chem* 267 pp. 17248 (1992); (c, right) After: *Introduction to Protein Structure*, 2nd ed., Branden & Tooze, Garland Publishing, Inc. **3.18** PDB ID: 1BBB; Silva, M. M., Rogers, P. H., Arnone, A.; A third quaternary structure of human hemoglobin A at 1.7-Å resolution; *J Biol Chem* 267 pp. 17248 (1992). **3.19** (c) © Dr. Gopal Murti/ SPL/ Photo Researchers, Inc.; (d) Courtesy of Melba Moore. **3.20** PDB files from Klotho Biochemical Compounds Declarative Database. **3.22** PDB ID:1BNA; H. R. Drew, R. M. Wing, T. Takano, C. Broka, S. Tanaka, K. Itakura, R. E. Dickerson; Structure of a B-DNA Dodecamer. Conformation and Dynamics; PNAS V. 78 2179, 1981. **Page 50** © JupiterImages Corporation.

CHAPTER 4 **4.1** Left, © JupiterImages Corporation; right, © Stephanie Schuller/ Photo Researchers, Inc. **Page 53** From top, Tony Brian and David Parker/ SPL/ Photo Researchers, Inc.; © JupiterImages Corporation; R. Calentine/ Visuals Unlimited; © ADVANCELL (Advanced In Vitro Cell Technologies; S.L.) www .advancell.com; © Dylan T. Burnette and Paul Forscher. **4.2** © Tony Brian and David Parker/ SPL/ Photo Researchers, Inc. **4.3** (a) Parke-Davis; Above, Linda Hall Library, Kansas City, MO; (b) © Michael W. Davidson, Molecular Expressions; above, © The Royal Society. **4.7** (a) Above, © JupiterImages Corporation; (b) © Geoff Tompkinson/ Science Photo Library / Photo Researchers, Inc. **4.8** (a–b, d–e) Jeremy Pickett-Heaps, School of Botany, University of Melbourne; (c) © Prof. Franco Baldi. **4.9** (0.1m) Robert A. Tyrrell; (1m) © Pete Saloutos/ Corbis; (100m) Courtesy of © Billie Chandler. **4.11** (a) ArchiMeDes; (b,c) © K.O. Stetter & R. Rachel, Univ. Regensburg. **4.12** (a) Rocky Mountain Laboratories, NIAID, NIH; (b) R. Calentine/ Visuals Unlimited. **4.13** Courtesy of © Roberto Kolter Lab, Harvard Medical School. **4.14** (a) Dr. Gopal Murti/ Photo Researchers, Inc.; (b) M.C. Ledbetter, Brookhaven National Laboratory. **4.16** Right, © Kenneth Bart. **4.17** (a) Don W. Fawcett/ Visuals Unlimited; (b) © Martin W. Goldberg, Durham University, UK. **4.18** (a) © Kenneth Bart; (b,d) Don W. Fawcett/ Visuals Unlimited; (e) Micrograph, Gary Grimes. **4.19** © Conner's Way Foundation, www.connersway.com. **4.20** Micrograph, Keith R. Porter. **4.21** © Dr. Jeremy Burgess/ SPL/ Photo Researchers, Inc. **4.22** (c) © Russell Kightley/ Photo Researchers, Inc. **4.23** George S. Ellmore. **4.24** Left, © Science Photo Library/ Photo Researchers, Inc.; Right, Bone Clones®, www.boneclones .com. **4.25** © ADVANCELL (Advanced In Vitro Cell Technologies; S.L.) www.advancell .com. **4.26** Below, © Dylan T. Burnette and Paul Forscher. **4.28** (a) © Dow W. Fawcett/ Photo Researchers, Inc.; (b) Mike Abbey/ Visuals Unlimited. **4.29** © Don W. Fawcett/ Photo Researchers, Inc. **4.30** From "Tissue & Cell," Vol. 27, pp. 421–427, Courtesy of Bjorn Afzelius, Stockholm University. **Page 75** Right, P. L. Walne and J. H. Arnott, *Planta*, 77:325–354, 1967.

CHAPTER 5 **5.1** Clockwise from top left, Courtesy of © The Cody Dieruf Benefit Foundation, www.breathinisbelievin.org; Courtesy of © Bobby Brooks and The Family of Jeff Baird; Courtesy of © Steve & Ellison Widener and Breathe Hope, http://breathehope.tamu.edu; Courtesy of © the family of Brandon Herriott; Courtesy of © The Family of Savannah Brooke Snider; Courtesy of The family of Benjamin Hill, reprinted with permission of © Chappell/ Marathonfoto. **Page 77** From second from top, © Andrew Lambert/ Science Photo Library/ Photo Researchers, Inc.; © R.G.W. Anderson, M.S. Brown, and J.L. Goldstein. *Cell* 10:351 (1977); © Claude Nuridsany & Marie Perennou/ Science Photo Library/ Photo Researchers, Inc. **5.7** © Andrew Lambert/ Science Photo Library/ Photo Researchers, Inc. **5.9** PDB files from NYU Scientific Visualization Lab. **5.10** After: David H. MacLennan, William J. Rice, and N. Michael Green, "The Mechanism of Ca2+ Transport by Sarco (Endo) plasmic Reticulum Ca2+-ATPases." *JBC* Volume 272, Number 46, Issue of November 14, 1997, pp. 28815–28818. **5.13** © R.G.W. Anderson, M.S. Brown, and J.L. Goldstein. *Cell* 10:351 (1977). **5.14** (a) © Biology Media/ Photo Researchers, Inc. **5.17** (a) Art, Raychel Ciemma; (b-d) M. Sheetz, R. Painter, and S. Singer, *Journal of Cell Biology*, 70:193 (1976) by permission of The Rockefeller University Press. **5.18** (a) Gary Head; (b,c) © Claude Nuridsany & Marie Perennou/ Science Photo Library/ Photo Researchers, Inc. **Page 90** © Children's Hospital & Medical Center/ Corbis. **Page 91** Frieder Sauer/ Bruce Coleman Ltd.

CHAPTER 6 **6.1** Left, © BananaStock/ SuperStock. **Page 93** From top, © Martin Barraud/ Stone/ Getty Images; © Scott McKiernan/ ZUMA Press; © JupiterImages

Tübingen; (b) © Visuals Unlimited; (c) © Eye of Science/ Photo Researchers, Inc.; (d) right, Courtesy of Edward B. Lewis, California Institute of Technology; others, © Carolina Biological / Visuals Unlimited. **15.9** (a–e) © Maria Samsonova and John Reinitz; (f) © Jim Langeland, Jim Williams, Julie Gates, Kathy Vorwerk, Steve Paddock, and Sean Carroll, HHMI, University of Wisconsin-Madison. **15.10** PDB ID: 1CJG; Spronk, A.A.E.M., Bovin, A.M.J.J., Radha, P.K., Melacini, G., Boelens, R., Kaptien, R.: The Solution Structure of Lac Repressor Headpiece 62 Complexed to a Symmetrical Lac Operator Structure (London) 7 pp. 1483, (1999). Also PDB ID: 1LBI; Lewis, M., Chang, G., Horton, N.C., Kercher, M.A., Pace, H.C., Schumacher, M.A., Brenan, R.G., Lu, P.: Crystal structure of the lactose operon repressor and its complexes with DNA and inducer. *Science* 271 pp. 1247 (1966); lactose pdb files from the Hetero-Compound Information Centre-Uppsala (HIC-Up). **Page 237** © Lowe Worldwide, Inc. as Agent for National Fluid Milk Processor Promotion Board.

CHAPTER 16 **16.1** © Courtesy of Golden Rice Humanitarian Board. **Page 241** Top, © Professor Stanley Cohen/ SPL/ Photo Researchers, Inc.; from third from top, Courtesy of © Genelex Corp.; Argonne National Laboratory, U.S. Department of Energy; Photo courtesy of MU Extension and Agricultural Information. **16.3** (a) © Professor Stanley Cohen/ SPL/ Photo Researchers, Inc.; (b) with permission of © QIAGEN, Inc. **16.9** Courtesy of © Genelex Corp. **16.10** Right, © Volker Steger/ SPL/ Photo Researchers, Inc. **16.11** (a) Argonne National Laboratory, U.S. Department of Energy; (b) Courtesy of

Joseph DeRisa. From *Science*, 1997 Oct. 24; 278 (5338) 680–686. **Page 250** Photo Courtesy of Systems Biodynamics Lab, P.I. Jeff Hasty, UCSD Department of Bioengineering, and Scott Cookson. **16.12** (d) © Lowell Georgis/ Corbis; (e) Keith V. Wood. **16.13** (a) The Bt and Non-Bt corn photos were taken as part of field trial conducted on the main campus of Tennessee State University at the Institute of Agricultural and Environmental Research. The work was supported by a competitive grant from the CSREES, USDA titled "Southern Agricultural Biotechnology Consortium for Underserved Communities," (2000–2005). Dr. Fisseha Tegegne and Dr. Ahmad Aziz served as Principal and Co-principal Investigators respectively to conduct the portion of the study in the State of Tennessee; (b) Dr. Vincent Chiang, School of Forestry and Wood Products, Michigan Technology University. **16.14** (a) © Adi Nes, Dvir Gallery Ltd.; (b) Transgenic goat produced using nuclear transfer at GTC Biotherapeutics. Photo used with permission; (c) Photo courtesy of MU Extension and Agricultural Information. **16.15** R. Brinster, R. E. Hammer, School of Veterinary Medicine, University of Pennsylvania. **Page 254** © Jeans for Gene Appeal. **Page 255** © Courtesy of Golden Rice Humanitarian Board. **16.16** (a) Laboratory of Matthew Shapiro while at McGill University. Courtesy of Eric Hargreaves, www.pageoneuro plasticity.com.

Appendix V Hemoglobin models: PDB ID: 1GZX; Paoli, M., Liddington, R., Tame, J., Wilkinson, A., Dodson, G., Crystal structure of T state hemoglobin with oxygen bound at all four haems. *J.Mol.Biol.*, v256, pp. 775–792, 1996.

Appendix VI Electron transfer chains: PDB ID: 1A70; Binda, C., Coda, A., Aliverti, A., Zanetti, G., Mattevi, A., Structure of the mutant E92K of [2Fe-2S] ferredoxin I from Spinacia oleracea at 1.7 Å resolution. *Acta Crystallogr.*, Sect.D, v54, pp. 1353–1358, 1998. PDB ID: 1AG6; Xue, Y., Okvist, M., Hansson, O., Young, S., Crystal structure of spinach plastocyanin at 1.7 Å resolution. *Protein Sci.*, v7, pp. 2099–2105, 1998. PDB ID: 1ILX; Vasil`ev, S., Orth, P., Zouni, A., Owens, T.G., Bruce, D., Excited-state dynamics in photosystem II: insights from the x-ray crystal structure. *Proc.Natl.Acad.Sci.*, USA, v98, pp. 8602–8607, 2001. PDB ID: 1Q90; Stroebel, D., Choquet, Y., Popot, J.-L., Picot, D., An Atypical Haem in the Cytochrome B6F Complex, *Nature*, v426, pp. 413–418, 2003. PDB ID: 1QZV; Ben-Shem, A., Frolow, F., Nelson, N., Crystal structure of plant photosystem I, *Nature*, v426, pp. 630–635, 2003. PDB ID: 1IZL; Kamiya, N., Shen, J.-R., Crystal structure of oxygen-evolving photosystem II from Thermosynechococcus vulcanus at 3.7-Å resolution, *Proc.Natl.Acad.Sci.*, USA, v100, pp. 98–103, 2003. PDB ID: 1GJR; Hermoso, J.A., Mayoral, T., Faro, M., Gomez-Moreno, C., Sanz-Aparicio, J., Medina, M., Mechanism of coenzyme recognition and binding revealed by crystal structure analysis of ferredoxin-NADP+ reductase complexed with NADP+., *J.Mol.Biol.*, v319, pp. 1133–1142, 2002. pdb ID: 1C17; Rastogi, V.K., Girvin, M.E., Structural changes linked to proton translocation by subunit c of the ATP synthase., *Nature*, v402, pp. 263–268, 1999. PDB ID: 1E79; Gibbons, C., Montgomery, M.G., Leslie, A.G., Walker, J.E., The structure of the central stalk in bovine F(1)-ATPase at 2.4 Å resolution., *Nat. Struct.Biol.*, v7, pp. 1055–1061, 2000.

Index
Page numbers followed by an *f* or *t* indicate figures and tables. ▪ indicate applications. Bold terms indicate major topics.